AF577262

# THE PHYSIOLOGY OF DISEASE

# THE PHYSIOLOGY OF DISEASE


IAN C. RODDIE

D.Sc., M.D., F.R.C.P.I.

*Dunville Professor of Physiology, The Queen's University of Belfast, and Consultant in Physiology to the Northern Ireland Eastern Health and Social Services Board*

WILLIAM F. M. WALLACE

B.Sc., M.D., M.R.C.P.

*Senior Lecturer in Physiology, The Queen's University of Belfast, and Consultant in Physiology to the Northern Ireland Eastern Health and Social Services Board*

LLOYD-LUKE (MEDICAL BOOKS) LTD

49 NEWMAN STREET

LONDON

1975

PRINTED AND BOUND IN ENGLAND BY
THE LAVENHAM PRESS LIMITED

ISBN 0 85324 109 0

# PREFACE

THERE IS a growing tendency to introduce medical students to clinical work at an early stage in their course. This policy encourages the student to harness his vocational interest in medical matters to the acquisition of the necessary knowledge of the basic medical sciences. It also tends to blur the artificial distinction between preclinical and clinical training. The purpose of this book is to assist the student to integrate the preclinical and clinical subjects. Our aim has been to explain the physiology of disease in language and concepts that a junior student can redily understand, and thereby to ease the transition from pre-clinical to clinical work. We hope the book will be useful throughout the clinical years and help those people who are interested in improving their knowledge of physiological mechanisms in disease.

The book deals more with the events which occur when normal physiology breaks down than with the disease processes that cause the breakdown. Thus the student who is familiar with normal physiology should find it relatively easy to understand the material. The newcomer to clinical studies is sometimes overwhelmed and depressed by the great number and variety of diseases that can attack man. The transition is made easier if he realizes that damage to a body system usually leads to a logical and recognizable collection of effects which is relatively independent of the agent causing the damage. There are only a few *basic* clinical pictures associated with any system and this book attempts to describe and explain them.

The concepts upon which this book is based have developed as a result of our experience of teaching applied physiology to medical students in Belfast. We thank our student and professional colleagues for their helpful comments during the production of this book. We are also very grateful to the following clinical colleagues who looked at drafts of sections dealing with their own speciality and whose suggestions, which were most helpful, have been incorporated in the text: Drs. B. N. Barwin, J. S. Geddes, D. R. Hadden, Messrs. S. S. Johnston, A. G. Kerr, Drs. Jean H. M. Langlands, J. A. Lyttle, Professor A. H. G. Love, Drs. Elizabeth E. Mayne, J. McEvoy and T. A. McNeill.

We would still be struggling with our untidy preliminary drafts were it not for the skilful typing assistance of Miss Margaret Wilkinson and the draughtmanship of Messrs. C. Ferris, S. McGrann and H. Turtle. We would also like to thank Mr. Douglas Luke of Lloyd-Luke (Medical Books) Ltd. for his courtesy and helpfulness extended to us continuously throughout the preparation of this book.

Finally, we would appeal to our readers, especially students, to write to us about any parts of the book which seem inaccurate, incomplete, or difficult to follow.

IAN C. RODDIE
WILLIAM F. M. WALLACE

Belfast, *August 1974*

# CONTENTS

Preface v

## *Section I*—INTRODUCTION

1 The Missing Link 3
2 Normal Values 4
3 Reserve, Compensation and Failure 6
4 The Greeks (or Romans) Had a Word For It 11
5 Further Reading 12

## *Section II*—THE EFFECT OF AGE ON DISEASE PATTERNS

6 Introduction 15
7 Disease Before Birth 17
8 Disease in the Newborn 21
9 Disease in Childhood 24
10 Ageing 25
11 Disease in the Elderly 30
12 Death 34
Recent Leading Articles 36

## *Section III*—DISORDERS CAUSED BY PHYSICAL AND EMOTIONAL STRESS

13 Temperature 41
Disorders caused by heat
Disorders caused by cold
14 Barometric Pressure 58
Disorders caused by breathing at high pressure
Disorders caused by breathing at low pressure
15 Gravity 68
Disorders caused by normal gravity
Disorders caused by increased gravity
Disorders caused by decreased gravity
16 Disorders Caused by Lack of Food 73
17 Disorders Caused by Lack of Water 75
18 Disorders Caused by Lack of Sleep 78
19 Disorders Caused by Trauma 80
20 Disorders Caused by Ionizing Radiation 85
21 Disorders Caused by Emotional Stress 90
Recent Leading Articles 94

## *Section IV*—DISORDERS OF THE CARDIOVASCULAR SYSTEM


22 High Blood Pressure (Hypertension) 97
23 Peripheral Circulatory Failure 110
24 Central Circulatory Failure 121
25 Cardiac Arrest 131
26 Local Circulatory Failure 133
27 Electrocardiography 141
28 Murmurs 151
Recent Leading Articles 153


## *Section V*—DISORDERS OF THE RESPIRATORY SYSTEM


29 Disturbance of Upper Respiratory Tract Function 157
30 Disturbance of Lower Respiratory Tract Function 161
Recent Leading Articles 193


## *Section VI*—DISORDERS OF THE ALIMENTARY SYSTEM


31 Deficient Nutrition 197
32 Excessive Nutrition 204
Obesity
Vitamin toxicity
33 Vomiting 211
34 Diarrhoea 218
35 Constipation 221
36 Alimentary Tract Obstruction 224
37 Diseases Causing Destruction of the Wall of the Gut 228
38 Impaired Liver Function 232
39 Pancreatic Disease 238
40 Increased Circulatory Bilirubin (Jaundice) 241
Recent Leading Articles 243


## *Section VII*—DISORDERS OF THE NERVOUS SYSTEM


41 Disturbance of Lower Motor Neurone Function 247
42 Disturbance of Pyramidal System Function 253
43 Disturbance of Extrapyramidal System Function 262
44 Disturbance of Cerebellar System Function 267
45 Disturbance of Sensory System Function 271
46 Pain 275
47 Spinal Cord Transection 286
48 Depression of the Nervous System 288
49 Hyperactivity in the Nervous System 295

50 DISORDERS AFFECTING CEREBROSPINAL FLUID 300
Reduction in CSF volume
Increase in CSF volume
Contamination of CSF
RECENT LEADING ARTICLES 307

## *Section VIII*—DISORDERS OF THE SPECIAL SENSES

51 DISTURBANCES OF HEARING 311
Deafness
Distortion of hearing
52 DISTURBANCES OF VESTIBULAR FUNCTION 317
Loss of vestibular function
Excessive or unbalanced vestibular function
53 DISTURBANCES OF VISION 321
Loss of vision
Refractive errors
Astigmatism
Squinting
Colour blindness
False visual impressions
54 DISTURBANCES OF SMELL 337
Loss of the sense of smell
Distortion of the sense of smell
55 DISTURBANCES OF TASTE 338
RECENT LEADING ARTICLES 338

## *Section IX*—DISORDERS OF THE LOCOMOTOR SYSTEM

56 DISTURBANCE OF JOINT FUNCTION 341
57 DISTURBANCE OF BONE FUNCTION 347
58 NEUROMUSCULAR DISORDERS 351
RECENT LEADING ARTICLES 353

## *Section X*—DISORDERS OF THE BLOOD

59 DEFICIENCY OF HAEMOGLOBIN (ANAEMIA) 357
Microcytic anaemia
Macrocytic anaemia
Normocytic anaemia
60 INACTIVATION OF HAEMOGLOBIN 370
61 INCREASE IN HAEMOGLOBIN (POLYCYTHAEMIA) 372
62 DEFICIENCY OF GRANULOCYTES (AGRANULOCYTOSIS) 375
63 ABNORMAL PRODUCTION OF WHITE CELLS (LEUKAEMIA) 377
64 ABNORMAL BLEEDING (HAEMORRHAGIC DISORDERS) 379

65 INTRAVASCULAR CLOT FORMATION (THROMBOSIS) 384
RECENT LEADING ARTICLES 388

## *Section XI*—DISORDERS OF IMMUNITY

66 THE NORMAL IMMUNE RESPONSE 393
67 DEFICIENT IMMUNITY 397
68 EXCESSIVE RESPONSE TO FOREIGN ANTIGENS (HYPERSENSITIVITY) 400
The anaphylactic reaction
The cytolytic reaction
Serum sickness reaction
Sensitized lymphocyte reaction
69 EXCESSIVE IMMUNE RESPONSE TO BODY TISSUES (AUTOIMMUNITY) 407
70 ABNORMAL PRODUCTION OF IMMUNOGLOBULINS 409
71 TISSUE TRANSPLANTATION 411
72 BLOOD TRANSFUSION 414
Blood donation
Blood transfusion
RECENT LEADING ARTICLES 416

## *Section XII*—DISORDERS OF THE BODY FLUIDS

73 INTRODUCTION TO WATER AND ELECTROLYTE DISTURBANCES 419
74 DISTURBANCE OF SODIUM BALANCE 423
Sodium ion depletion
Sodium ion retention
75 DISTURBANCE OF WATER BALANCE 428
Water depletion
Water retention
76 DISTURBANCE OF POTASSIUM BALANCE 433
Potassium ion depletion
Potassium ion retention
77 DISTURBANCE OF MAGNESIUM BALANCE 441
Magnesium ion depletion
Magnesium ion retention
78 DISTURBANCE OF HYDROGEN ION BALANCE 442
Hydrogen ion depletion (alkalosis)
Hydrogen ion retention (acidosis)
79 DISTURBANCE OF CALCIUM BALANCE 455
Calcium ion depletion
Calcium ion retention
80 INCREASED INTERSTITIAL FLUID VOLUME (OEDEMA) 460
RECENT LEADING ARTICLES 464

## *Section XIII*—DISORDERS OF THE URINARY SYSTEM

81 URINARY TRACT INFECTION 467
82 URINARY TRACT OBSTRUCTION 469

83 Neurological Disturbances of Micturition 473
84 Gradual Nephron Failure (Chronic Renal Failure) 480
85 Sudden Nephron Failure (Acute Renal Failure) 492
86 Glomerular Failure (The Nephrotic Syndrome) 495
87 Tubular Failure 498
Recent Leading Articles 499

## *Section XIV*—DISORDERS OF THE ENDOCRINE SYSTEM

88 Thyroid Gland 503
Deficient secretion
Excessive secretion
Calcitonin secretion
89 Parathyroid Gland 509
Deficient secretion
Excessive secretion
90 Adrenal Cortex 512
Deficient secretion
Excessive secretion
91 Adrenal Medulla 518
Excessive secretion
92 Pituitary Gland 520
Deficient anterior lobe secretion
Deficient posterior lobe secretion
Excessive growth hormone secretion
93 Pancreatic Islet Cells 528
Deficient beta cell secretion
Excessive beta cell secretion
Alpha cell secretion
94 Gonads 536
Deficient testicular secretion
Deficient ovarian secretion
Excessive testicular secretion
Excessive ovarian secretion
95 Miscellaneous 541
Hypothalamus
Kidney
Endocrine secretion by tumours
Recent Leading Articles 542

## *Section XV*—DISORDERS OF THE REPRODUCTIVE SYSTEM

96 Problems of Fertility 545
Infertility
Hyperfertility (over-population)

97 Problems in Maternal Adaptation to Pregnancy 554
Unmasking of maternal disease
Hypertension, oedema and proteinuria (pre-eclampsia)
Accumulation of amniotic fluid (polyhydramnios)
Deficiency of haemoglobin (anaemia)
Infection of the urinary tract
Vomiting
Placental haemorrhage
Intravascular blood clotting (venous thrombosis)
Abortion
98 Problems in Fetal Growth and Development 559
99 Problems in Labour 561
Early labour (prematurity)
Late labour (postmaturity)
Prolonged labour
100 Problems in Adaptation to Extra-uterine Life 568
Failure to breathe
Congenital malformation
Birth injuries
Infection
Haemolytic disease
Haemorrhage
101 Problems in Breast Feeding 571
Failure of breast feeding
Suppression of lactation
Recent Leading Articles 572

Index 575

*Section I*

# INTRODUCTION

# 1. THE MISSING LINK

Too often Physiology and the clinical subjects appear separate and clearly demarcated from one another so that the student leaves behind and forgets much of his physiological knowledge before he has the opportunity to relate it to his developing knowledge of Clinical Medicine. As the mechanisms of disease become more fully understood, applied Physiology is acquiring an increasing importance in clinical subjects. The aim of this book is to help the student to develop and maintain for himself the link between Physiology and Clinical Medicine. The book has been written primarily for the student who has completed a course in basic Physiology and is beginning his clinical studies. It is assumed that the reader has a general knowledge of normal human Physiology and its vocabulary, and an attempt is made to describe the effects and treatment of disease in the words and concepts of Physiology.

In this book we have concentrated on this relationship between Physiology and Medicine. Considerable space has been given to attempts to explain important concepts which students tend to find difficult to grasp. On the other hand, long and complete lists of the causes of disease have been largely omitted, as have descriptions of diseases and treatments whose explanation is as yet poorly understood. At the same time we have tried to maintain a sense of proportion by emphasizing items to some extent according to their clinical importance. We hope that a clinician reading this book will not find himself in a strange and rarified atmosphere of uncommon conditions and impracticable investigations.

Our plan of approach is to consider various disordered functions, for example, anaemia, heart failure, vomiting, renal failure, under the following headings:

1. **Definition:** Definition of condition in terms of physiological abnormality.
2. **Effects:** Description of effects in terms of disturbance of normal function.
3. **Causes:** Underlying causes of condition considered in physiological terms.
4. **Investigations:** Analysis of body fluids and other procedures which help to increase knowledge of abnormalities of function.
5. **Treatment:** Brief outline of treatment seen as an attempt to overcome or compensate for physiological abnormality.

When a student is interested in a specific clinical symptom or sign, such as cyanosis or dyspnoea, he should consult the index.

Many diseases, like many normal functions, cannot yet be explained satisfactorily in physiological terms. Students often waste time in a fruitless attempt to penetrate the screen of words that teachers use to mask their ignorance. We have tried as far as possible to distinguish clearly between what is known and what is not, between what is fact and what is speculation. This has been done in the hope that a student who has read a section will not feel inadequate because he has failed to comprehend what is in fact incomprehensible.

# 2. NORMAL VALUES

WE DO NOT generally refer to normal values for the various tests mentioned in this book because we are concerned mainly with the mechanisms of disease rather than precise criteria for diagnosis. Normal values are given in most textbooks of Medicine. In addition, the usual ranges of normal results are being increasingly often quoted with the patient's results on laboratory report forms. This is useful, because results for some tests vary from laboratory to laboratory and also it avoids the unnecessary task of trying to remember the ranges for all possible tests. There is by no means general agreement on the normal ranges for some tests as Table I shows with respect to the platelet count. In such cases, if one wishes to remember the normal range, it is reasonable to select one which is easy to remember, bearing in mind that all normal ranges are only approximate guides.

When interpreting the results of tests, it must be remembered that, due to normal variation, a few healthy people will have results outside the expected range. The upper and lower "limits of normality" are diffuse zones rather than linear boundaries. If the mean of a group of "normal" people $\pm$ 2 standard deviations is regarded as the "normal range" then 5 per cent of normal people must lie outside the normal range.

The more biochemical tests are carried out on an individual, the more chances the individual has of being classified as abnormal on at least one occasion. It is probably better to think of average values and standard deviations, rather than normality and abnormality.

A recent review article (*British Medical Journal* 1972, **1**, 328) discusses the problem with respect to the white cell count. It quotes the case of an apparently healthy man whose white cell count over a period of seven years was usually above 25,000/mm$^3$—more than twice the generally accepted upper limit of

TABLE I

Quoted ranges for the normal platelet count in some textbooks of Physiology (P) and Medicine (M) as an example of the absence of general agreement on normal ranges. To be "normal" by all these standards, the platelet count would have to lie between 200,000 and 300,000. On the other hand, counts between 100,000 and 500,000 would fall within at least one of the normal ranges.

| *Textbook* | *Platelet range/mm$^3$* |
|---|---|
| Bell, Davidson & Emslie-Smith (P) 1972 | 100,000-500,000 |
| Best & Taylor (P) 1966 | 200,000-400,000 |
| Cecil & Loeb (M) 1967 | 150,000-450,000 |
| Davidson (M) 1971 | 150,000-400,000 |
| Guyton (P) 1971 | 150,000-300,000 |
| Harrison (M) 1970 | 140,000-440,000 |
| Horrobin (P) 1968 | 150,000-500,000 |
| Price (M) 1973 | 150,000-400,000 |
| Samson Wright (P) 1971 | 150,000-350,000 |

normality. It is of significance that he was anxious about this abnormality. If it is necessary to carry out further investigations in such cases, considerable experience is required to know when investigations should cease; the patient must be reassured as far as possible. In some cases, it may be best to ignore a single abnormal result out of keeping with the patient's general condition.

For some tests, such as lung function tests, results depend so markedly on body size, age and sex, that expected results must be worked out for each individual. Blood indices, such as haemoglobin concentration, are higher in men than in women. Racial differences also occur. In addition to variations in the basic adult normal values, some tests have quite different results in children, and particularly in the newborn, as compared with adults. Some blood constituents, e.g. cortisol, fluctuate markedly throughout the day, and this regular diurnal variation means that sampling time must be known when the significance of the result is being considered. If there is doubt about the significance to be attached to the result of a particular test, it is often very helpful to consult the laboratory staff where the test was carried out.

## 3. RESERVE, COMPENSATION AND FAILURE

IN STUDYING the physiological consequences of breakdown of normal function in an organ, one may oversimplify by equating normal function with health and loss of function with illness. The body is a very durable machine with a formidable array of overlapping mechanisms for maintaining a state of normality. Thus when one mechanism breaks down, another usually takes over its function. This provides the body with defence in depth against injury. If this were not the case, man would be very vulnerable as there would not be any way of compensating for bodily damage.

This fact has considerable implications in understanding the effects of disease. Because the body can readily adapt to damage, damage or disease may be present without any obvious decrease in the ability of the body to carry out its normal activity. Damage usually has to be severe and widespread before signs of organ failure become manifest. When looking at the effect of disease processes on physiological functions, it is just as important to understand how the body compensates for loss of function as to understand how failure of that function affects the body as a whole.

Bearing this in mind, it is useful to consider the gradual erosion of function by disease as a process which occurs in stages. First there is the stage of falling reserve, then the stage of compensation and finally the stage of manifest failure. These stages will now be considered in some examples of disturbed function.

### 1. Loss of Renal Function

#### STAGE OF FALLING RESERVE

One of the functions of the kidneys is to excrete the urea produced by the deamination of amino-acids. Although urea is relatively harmless, its handling by the kidney provides a useful indication of the ability of the kidney to excrete more harmful but less easily measurable substances. On an average diet, a normal person would excrete about 20 g/day. If, as the result of disease or injury he lost one of his kidneys, he still would have no difficulty in excreting the normal daily output of urea. There would be no evidence from this function that he had lost half his renal tissue.

The reason for this is the enormous reserve of renal function; there is about 75 per cent more renal tissue than is needed to deal with the average day-to-day requirements for excreting urea. Because of this, widespread disease and destruction may be present in the kidneys without affecting urea excretion.

However the person with one kidney is a much less versatile and durable animal than the person with two. Firstly, with loss of reserve function, he has not the same ability to cope with extreme conditions, e.g. water, protein, or salt excess. Secondly, he is less able to cope with further renal damage than a normal person. By losing one kidney, he is made more vulnerable and his chances of survival are correspondingly decreased.

### Stage of Compensation

When disease has eroded the reserves of function in an organ, function can still be maintained at an adequate level to support life by compensatory mechanisms. In the case of the kidneys and urea excretion, the normal output of about 20 g/day can be maintained by the employment of a variety of stratagems which increase the output of urine.

A rise in the blood urea level allows a higher concentration of urea to be presented to the nephrons which still function. The osmotic effect of this urea will result in an increase in urine output. A high level of urea in the plasma may compensate for a low urea clearance value; urea excretion = (plasma concentration) (clearance volume). In severe renal disease, limitation of renal blood flow may increase the release of renin and thus the arterial pressure. This tends to increase filtration at functional glomeruli and thus increase urinary output. Loss of renal function includes loss of the ability to concentrate urine. In these circumstances the reabsorption of urea by the renal tubules is decreased and the excretion in the urine correspondingly increased.

### Stage of Manifest Failure

This can be considered a stage of ***decompensation,*** i.e. failure of function supervening when the compensatory mechanisms have been exhausted or overwhelmed. In the case of the kidneys and urea excretion, decompensation means that the normal daily production of urea cannot be excreted. The output falls below 20 g/day and urea steadily accumulates in the blood until death occurs.

It is at this stage that outside help is required if the patient is to survive, because the reserve and compensation mechanisms are exhausted. For example, he can be given a diet which is low in protein but sufficiently high in carbohydrate that he does not need to use his body protein to provide energy. These measures will reduce the rate of urea production in the body. His blood vessels may be connected to an "artificial kidney" so that his blood is dialysed outside the body. If a suitable donor is found, transplantation of a healthy kidney into the patient may permit renal function to rise again to a level which is compatible with life.

The stages described above are not always discrete and vary considerably from organ to organ. However, even if the stages overlap to some extent, it is useful to consider loss of function in terms of the stages. The concept can be applied to loss of function in nearly any system.

## 2. Loss of Muscle Joint Sense

Sensory information from receptors in muscles and joints about the position of the body are carried to the brain in the posterior columns of the spinal cord. This information is necessary for the execution of appropriate muscular movements in voluntary activity such as walking. Various diseases such as multiple sclerosis or sub-acute combined degeneration of the spinal cord cause progressive damage to these tracts. In the ***stage of falling reserve,*** the patient is able to walk normally without showing any sign of the spinal cord damage. However, his ability to carry out more complex motor acts is reduced. Thus, if he is asked repeatedly to turn about rapidly as he walks, his movements become visibly awkward.

In the *stage of compensation*, simple walking can take place fairly normally if the patient uses his eyes to compensate for loss of muscle joint sense in his legs. At this time as he walks he may strike the ground with his feet more vigorously than a normal person. This allows him to obtain more sensory information about the position of his legs during walking. In the dark, or if he closes his eyes, his ability to compensate is reduced and his clumsiness becomes evident.

If the disease eventually interferes with vision, as may happen in multiple sclerosis, the ability to walk will be lost completely. However, in most cases vision permits enough compensation for some walking ability to be maintained and the *stage of manifest failure* is not reached.

### 3. Loss of Myocardial Function

The heart normally pumps about five litres of blood per minute into the systemic circuit. This is adequate for the resting requirements of the body. In the early stages of myocardial damage, the heart continues to do this. However, it cannot increase output to deal with extreme conditions so well as before. The patient's ability to carry out severe exercise is reduced (*stage of falling reserve*).

Eventually the normal resting cardiac output is maintained by *compensatory mechanisms*. Dilatation of the heart and hypertrophy of its muscle and fluid retention in the body may occur to maintain stroke volume. A reflex increase in heart rate and force of contraction due to increased activity in the cardiac sympathetic nerves also helps to prevent a fall in cardiac output.

Eventually these compensatory mechanisms cannot even support the normal resting cardiac output. Underperfusion of the tissues with blood then occurs and leads to death.

### 4. Loss of Mental Power

It has been found that the number of neurones in the brain keeps falling as one grows older. As these cells cannot undergo mitosis there is no way of replacing the neurones that disappear. So people have to carry out their lives against a background of diminishing brain function. First there is the *stage of falling reserve*. During this stage there is a decline in the ability to learn new skills. This stage begins quite early and by the third decade, a person's ability to absorb new information is greatly reduced as compared with that of a child. Then there is the *stage of compensation* in which he uses experience, cunning and acquired skill to make good his reduction in mental power. At this stage his mental ability appears quite adequate in dealing with routine matters. However, his inadequacy becomes evident if he is made to carry out more complex mental gymnastics. Finally the stage comes when even maximum use of cunning and acquired knowledge cannot hide the inadequacy from the outside world.

### 5. Loss of Limb Function

The ability of man to cope with and manipulate his environment lies mainly in his limbs and the brain systems which control them. Impairment or loss of limb function might therefore be expected to make the individual unable to cope with his environment. However, such is man's resilience as a machine, that he can suffer gross impairment of limb function without having to succumb to the environment.

In the *stage of falling reserve,* when the limb muscles first start to weaken, the individual may show no outward sign of the deficiency; he limits his activity to suit the power of his muscle. Though he can accomplish ordinary everyday tasks, his reserve for strenuous or highly skilled work is limited; he is no longer capable of athletic or dexterous excellence.

It is in the *stage of compensation* for loss of limb function that man shows perhaps his greatest versatility. It seems that if the brain wishes a movement to be made, it will employ any combination of muscles that work to effect that movement. Thus if muscles in the arm become weak, muscles inserted on the shoulder girdle can be recruited to move the arm. If the leg muscles normally used for walking become weak, trunk muscles acting on the pelvis can succeed in swinging the leg. If muscular weakness makes the legs a less stable support for the body by reducing the ability to lock the knee joints, the body may compensate for this by having the feet slightly separated and the knees held together so that one leg may support the other. Essentially these compensatory adaptations are functions of the brain. The brain deploys the muscles in new ways to make the movements it requires. Though these adaptations are in one sense "learned", the individual is normally quite unconscious of learning them or using them.

As with most aspects of learning, the young are more adept than the old. The saying that an "old dog cannot learn new tricks" has some relevance here. The earlier in life limb power is diminished, the better can the individual learn to harness his remaining muscle power to meet his needs. It is quite impressive to see a child with severe muscular weakness in the legs achieve the standing position by virtually climbing up his own body by means of his arms.

If limbs are lost, compensation can be achieved by having remaining limbs take over the function of the missing limbs. At a simple level this may involve a right-handed person having to learn to write with his left hand. In more extreme cases, it may involve learning to use feet to do the jobs normally carried out by the hands. The feet have a considerable potential for skilled "manual" tasks and typing and painting by foot are not unknown. Again, the younger the individual the better can he adapt in this way. Children with congenital abnormalities that result in rudimentary or absent limbs are remarkably adept at using whatever facilities they have to the best advantage. In the absence of ordinary limbs they can move by rolling their bodies and use strategically placed hooks on the wall to assist with dressing and undressing.

In the *stage of manifest failure,* the disability is such that the individual is unable to cater for his everyday needs and loses his independence. At this time external aids become essential to support his ordinary activities. He may require constant nursing attention to deal with his feeding, toilet and excretory functions.

However, his ability to affect his environment may be greatly extended by the use of servo mechanisms. These are mechanisms where the input of a small quantity of energy leads to the activation of a powerful motor and the output of a large quantity of energy. A motor car is a well-known example of such a mechanism. A light press on the accelerator leads to a controlled and powerful surge of energy expenditure by the car. In a variety of ways, motors can be controlled by signals made by the weakened muscles. A motorized wheel-chair is an obvious example of such a device. However, a person with no limbs may activate a motor to turn over the pages of his book or operate a typewriter by blowing into a tube connected to a pressure-sensitive device. Great ingenuity has been used in recent years in inventing servo devices to restore some independence to handicapped people.

### Inappropriate Compensation

In some cases the body's compensatory efforts may do more harm than good. Thus in the case of unilateral renal artery narrowing, systemic arterial pressure tends to rise and this tends to counteract the reduction in blood flow to the affected kidney. However, the slight improvement in renal blood flow is bought at the risk of widespread serious damage due to a severe rise in arterial pressure, and early removal of the affected kidney may be necessary to avoid the unwanted compensatory hypertension.

Again, in some cases of heart failure, the impaired cardiac output leads to a compensatory increase in extracellular fluid and hence plasma volume. This extra fluid may cause troublesome effects and the patient may feel better when drugs are given to eliminate the excess fluid.

# 4. THE GREEKS (OR ROMANS) HAD A WORD FOR IT

THE ABILITY to think logically and communicate with other people depends largely on understanding the exact meaning of words. Much of the confusion and misunderstanding in Medicine, as in other branches of life, is due to words being used imprecisely. Over many years, doctors have built up a jargon and vocabulary of their own. This has advantages in that it allows them to think and communicate with an economical use of words and symbols; thoughts do not become encumbered by verbiage. It also has disadvantages in that the jargon may mean different things to different people and even to the same person at different times. It is very important, therefore, that the student knows the meaning of any jargon he uses; if there is any doubt about meaning, he should express his observations and thoughts in simpler words. This may (or may not!) take longer, but it is infinitely preferable.

To help the reader understand the meaning of medical terms, we frequently refer throughout this book to the Greek (Gr.) or Latin (L.) word from which the terms are derived. When medical terms were first coined, an effort was made to express as much meaning in as little space as possible. Greek and Latin words were used as these languages already contained a basic anatomical and medical vocabulary and were the media used for teaching medicine internationally until relatively recent times. When the *meaning* of the medical term is understood, it is seen, not so much as an obstacle to be overcome but as a shorthand way of summing up a clinical finding concisely. For example, 'diabetes' is the Greek for a siphon, or, in the medical context, profuse urine. Since 'mellitus' means sweet and 'insipidus', not surprisingly, means tasteless, there should be no doubt that glucose is found in the urine in diabetes mellitus but not in diabetes insipidus.

Again, 'cyanosis' means a blue discoloration of the skin (Gr. *kyanos* = blue). Hence the word cyanosis is not strictly limited to the appearance of the skin when there is much reduced haemoglobin in the superficial vessels, but, less commonly, can be used to describe the bluish appearance due to an abnormal pigment in the blood, e.g. methaemoglobin or sulphaemoglobin.

The origin of medical terms can be obtained from a medical dictionary, e.g. *Dorland's Illustrated Medical Dictionary* published by Saunders. A good medical dictionary is one of the most useful books a medical student can buy.

## 5. FURTHER READING

As well as acquiring a basic knowledge of medical practice, the student of medicine requires the ability to modify his structure of knowledge as the subject itself advances. An interesting and efficient way to keep abreast of the main features of advancing medical knowledge is to read the leading articles in periodicals such as the *British Medical Journal* and *Lancet*. A few minutes spent in this way each week is perhaps the best investment of time that the medical student or doctor can make in his efforts to keep his knowledge up-to-date. These brief articles provide an authoritative review of important developments in medical science and are often written from the viewpoint of applied Physiology. The articles are generally followed by a number of references to more detailed descriptions of the research work on which the leader was based.

We give a short selection of recent leading articles at the end of the main sections in this book. These are not comprehensive but will enable the student to sample this method of learning.

By reading these leaders and keeping abreast of new material in periodicals as they appear, the student soon learns that medical knowledge and ideas are in a state of perpetual evolution. Though many apparent advances are merely expressions of current medical fashion, new knowledge often helps to make sense of phenomena which were formerly puzzling and did not fit together.

*Section II*

# THE EFFECT OF AGE ON DISEASE PATTERNS

# 6. INTRODUCTION

HOSPITALS generally segregate their patients into four main age groups, although some overlap between groups tends to occur. The age groups are neonatal (strictly speaking this refers to the first four weeks after birth), childhood, adulthood and the elderly. One might add a fifth group—fetal, these patients being looked after in the antenatal wards of maternity hospitals. Segregation according to age is based partly on social grounds (though these could be questioned) and partly on the differing disease patterns found in the different age groups.

Most diseases can occur at any age and cause fundamentally the same problems at all ages. However, three main factors determine that both the frequency of a disease and its effects on the individual vary markedly throughout life, namely the age distribution of *congenital abnormalities*, *degenerative diseases* and *homeostatic efficiency* (Fig. 1).

### Congenital Abnormalities

These are biochemical or structural abnormalities with which an individual is born (L. *con* = with, *genitus* = born). Some of these abnormalities are compatible with fetal life but not with the greater demands of life after birth. Infants with such abnormalities tend to die in the first few days after birth. Other severe abnormalities almost invariably lead to death in the first few years of life. Some abnormalities are of little or no consequence and do not affect the life span of the patient. The trumpet-shaped distribution shown in Fig. 1(1) is being modified gradually by increasing skill in treating patients with congenital abnormalities. Some abnormalities can be virtually cured by appropriate medical or surgical treatment. The effect of some abnormalities can be greatly lessened and life correspondingly prolonged. In some cases it is difficult to decide how energetic should be the treatment of a child with a severe mental or physical handicap.

### Degenerative Diseases

These diseases are due to changes in body tissues so that they become less able to carry out their normal functions. Particularly important are changes in the walls of arteries such as the coronary and cerebral arteries whereby they become narrow and rigid, and eventually fail to supply sufficient blood to maintain function in the organ they supply. In the older age groups, the majority of people die from such conditions.

### Homeostatic Efficiency

Most body functions, especially those of the brain, can be carried out normally only if the internal environment of the body is carefully maintained within fairly narrow limits. The body has a formidable array of homeostatic mechanisms which restore the internal environment to normal when it is disturbed by external and internal stresses. Such mechanisms are of great importance in determining the outcome when the body is stressed by disease. The efficiency of homeostatic mechanisms varies with age; in general it tends to increase in early life and decline after early adulthood, at first gradually, then more rapidly. It is well known that newborn babies and the elderly may be seriously threatened by conditions which have little or no effect on young adults. A number of body functions follow approximately the inverted-U line shown at the bottom of Fig. 1. This line could represent approximately the changes throughout life of body functions as diverse as ability of the lung to transfer oxygen to the blood, ability to vary urinary excretion of salt and water, ability to withstand disturbances of body fluids, ability to withstand extremes of temperature, and skeletal muscle strength.

Thus both infants and the elderly are particularly prone to suffer severely from such conditions as dehydration due to vomiting and hypothermia due to a cold environment.

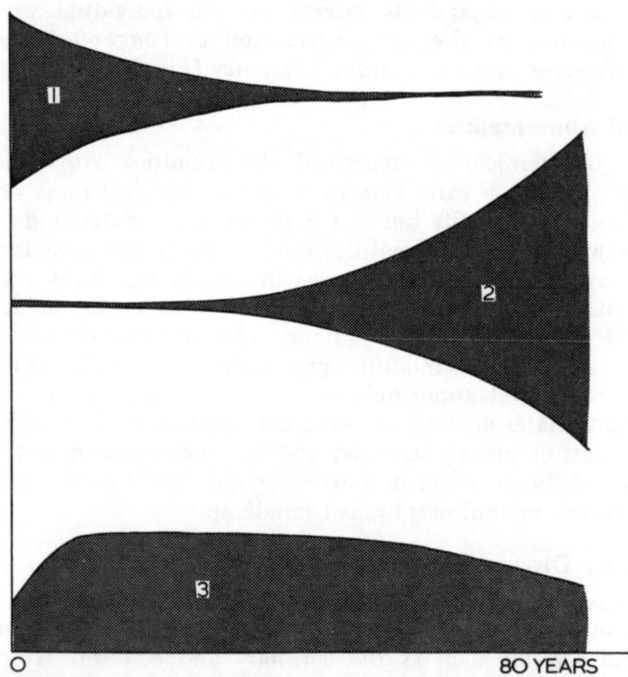


FIG. 1.—Three factors which cause disease patterns to vary with age. 1. *Congenital abnormalities* are common in children's wards but much rarer in adult wards. 2. *Degenerative diseases* follow the reverse distribution. 3. *Homeostatic efficiency* increases in early life and subsequently declines gradually.

# 7. DISEASE BEFORE BIRTH

## Definition

This chapter includes disease of the developing individual prior to birth. Technically the developing organism is referred to as an embryo for the first 8 weeks of its existence and a fetus thereafter. This arbitrary distinction has, of course, no biological basis—the development of the human embryo-fetus is a steady progression. However, around the eighth week the individual has a distinctly human appearance, and all major organs are recognizable—the rest of the fetal stage is concerned with their growth and with the development of the finer details of structure.

## Effects

The effects of adverse conditions on the developing organism are related to the stage of development at which they occur. Damage in the first 6-8 weeks or so tends to cause major structural and functional abnormalities. The reason is that this period is a crucial stage of development, when all the major organs appear and acquire their major characteristics. Around the 4th week the heart starts beating and there is a rudimentary circulation and nervous system. These systems develop rapidly in the next four weeks and in this same period the eye, ear and limbs acquire in rudimentary form their main adult features.

The approximate appearance of the developing human organism between the 4th and 12th weeks is shown in Fig. 2. One can envisage that arrest of development of the limb buds at around 4 weeks could result in the birth of a baby without any limbs (*amelia*—Gr. *melos* = limb). Arrest of limb development around the sixth week could lead to the condition commonly seen in "thalidomide babies"—hands and feet attached to the body with little or no intervening limb. The condition is referred to as *phocomelia* (Gr. *phoke* = seal), i.e. limbs like seal flippers. On the other hand, disease after the 8th-12th week could hardly alter the basic structure of the limb, although the limb could be damaged or could fail to grow adequately.

The same considerations apply to the other major parts of the body. Gross developmental abnormality originates before the 8th week. Developmental abnormalities are often multiple, since several major bodily features may be at a critical stage of development when development is interfered with.

Infectious agents do not often penetrate the barrier provided by the placenta and the membranes which surround the fetus. Viruses cross more readily than bacteria and, in later pregnancy, produce in the fetus similar effects to those seen in postnatal life. A child may be born with a measles rash or with muscle paralysis due to infection with the poliomyelitis virus.

Finally, when disease is severe enough to cause death in embryonic or early fetal life, death of the fetus tends to be followed by premature expulsion of the products of conception from the uterus (***abortion***).

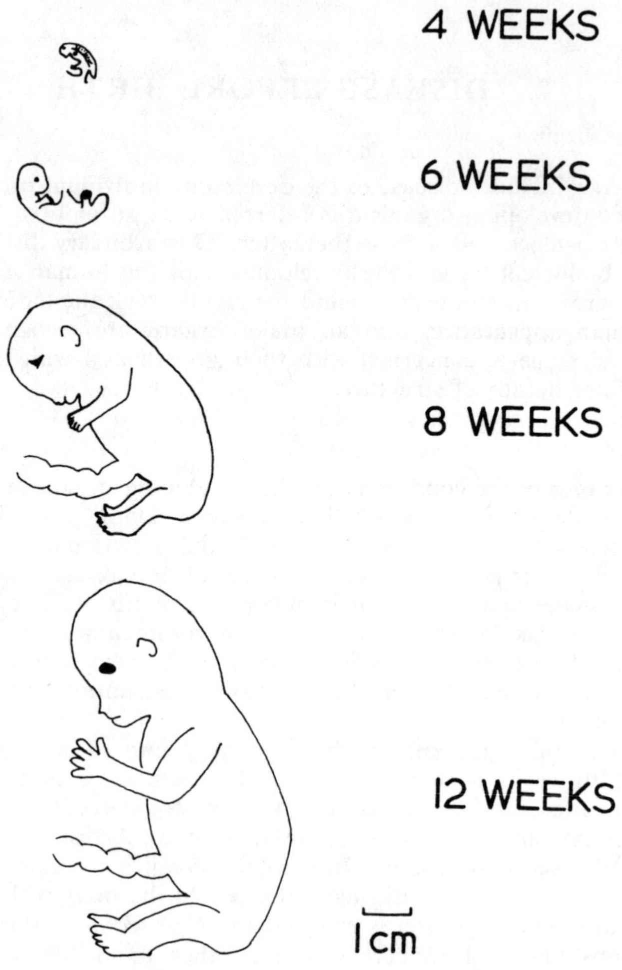


FIG. 2.—Development of the human organism in the first 12 weeks of life after conception. The major internal and external features are recognizable at 8 weeks.

### CAUSES

Developmental abnormalities have three main causes. Firstly, they may result from the inheritance of abnormal genes carried by the mother, or father, or both. Secondly, they may be due to damage to the particular germ cell from which the fetus developed, e.g. to the effect of radiation on a parental gonad. Thirdly, the abnormalities may result from disease affecting the mother or the child during its development.

Drugs, such as thalidomide, given to the mother, and viral infections such as German measles (rubella) can affect the development of the child. The tendency of an agent to cause developmental abnormalities is referred to as teratogenicity (Gr. *teras* = monster). Drugs and hormones given to the mother can also damage, stimulate or depress specific fetal organs, just as they would if given directly to the fetus. Thus a drug given to suppress excessive thyroid activity in the mother will tend to suppress activity in the fetal thyroid, causing fetal deficiency of thyroid hormones.

Radiation damage to the fetus may cause death, developmental abnormality or an increased tendency to develop cancer in postnatal life. Haemolytic anaemia due to maternal antibodies against fetal red cells is another example of fetal disease, but is dealt with elsewhere.

Maternal malnutrition or general disease such as diabetes mellitus tend to impair fetal growth. The babies of mothers who smoke cigarettes are smaller than average. This may be at least partly due to the carbon monoxide content of cigarette smoke which impairs the ability of the blood to carry oxygen and deliver it to the tissues. Heart, respiratory or blood disease in the mother may also interfere with the delivery of oxygen to the placenta and hence depress fetal growth.

When general disease of the mother is severe and life-threatening, the fetus tends to be expelled from the uterus, regardless of its stage of development. Though the mechanism is not known, this could be regarded as a compensatory effect aimed at preserving the mother's life, regardless of the consequences for the fetus.

The most hazardous period for the fetus is during labour when contractions of the uterus interfere with placental circulation. As a result, $Po_2$ in the fetal blood falls and $Pco_2$ rises. The brain is disturbed most by these abnormal gas tensions; if prolonged, permanent brain damage may result.

## INVESTIGATIONS

The fetus is relatively inaccessible to investigations, but increasing information is being obtained by a variety of methods, some of which are available at certain centres only.

*X-ray examination* of the fetus has long been used to show the presence of severe disease, such as deformities of the head and signs of death of the fetus, as well as confirming the diagnosis of twins, etc. Since X-rays may themselves damage the fetus, such studies are reserved for cases where abnormality is definitely suspected, and where the information obtained may justify the slight risk of fetal injury.

*Ultrasonic scanning* can also give information about soft tissues, e.g. the placenta. It appears to be less hazardous to the fetus than X-ray examination.

*Biochemical tests on the blood or urine of the mother,* may suggest impairment of placental function, a condition likely to lead to fetal damage. For example, normal levels of circulating oestrogen during pregnancy depend on normal placental function. Hence a reduced maternal oestrogen level suggests impaired placental function.

*Examination of a sample of amniotic fluid* may reveal changes due to fetal diseases, as in haemolytic disease. Fetal cells may be obtained from the amniotic fluid and, if desired, cultured. Examination of such cells may demonstrate chromosomal abnormalities and enzymatic abnormalities associated with serious congenital disease. The child's sex can be ascertained from the sex chromosomes in the cells.

*During labour,* it is desirable to detect signs of fetal distress as soon as possible, so that appropriate action may be taken to relieve the distress. The fetal heart rate may be monitored and samples of blood obtained from the fetus, e.g. from a scalp vein if the head is presenting in the cervix. Abnormally slow or fast heart rates suggest fetal distress, as do abnormal levels of oxygen and carbon dioxide in fetal blood.

## Treatment

"Treatment" of prenatal disease is almost entirely confined to prevention of such disease and to inducing early birth of the baby when this may improve its chances of survival as a healthy infant.

Prevention of disease includes ensuring adequate nutrition of the mother, avoiding viral infections if possible and keeping drug therapy of the mother to a minimum.

When there is evidence of haemolytic disease, induction of labour before the expected date of delivery will terminate the entry of maternal antibody into the infant's circulation. Unduly early delivery must be avoided because of the risks associated with immaturity of the infant.

When there are signs of fetal damage during labour—the risk tends to increase with the length of labour—prompt delivery by an abdominal (Caesarean) operation in which the fetus is delivered through an opening made into the uterus may save the life of the fetus.

When there are signs that the fetus has died during pregnancy, delivery is usually induced as early as possible. When evidence of severe congenital disease has been obtained at an early stage of the pregnancy, the question of producing abortion arises.

# 8. DISEASE IN THE NEWBORN

## Definition

THIS SECTION concerns disease in the neonatal period, which is the first four weeks after birth.

## Effects

During fetal life, the placenta functions as alimentary tract, lungs, kidneys and liver for the fetus. Harmful chemicals diffuse into the maternal blood stream to be excreted by the maternal kidneys, liver and lungs. Oxygen and predigested nutrients are available as required. Temperature regulation and physical protection are provided.

At birth, the baby must suddenly depend on its own resources for the vital functions of nutrition, excretion, oxygenation and temperature regulation. If one or more of the necessary organs has failed to develop adequately, the effects of failure of that organ suddenly become apparent. The situation is rather like that of someone who, having lived by sponging on his parents for many years, is forced to leave home. The sudden need to activate his latent potential for work is an enormous challenge.

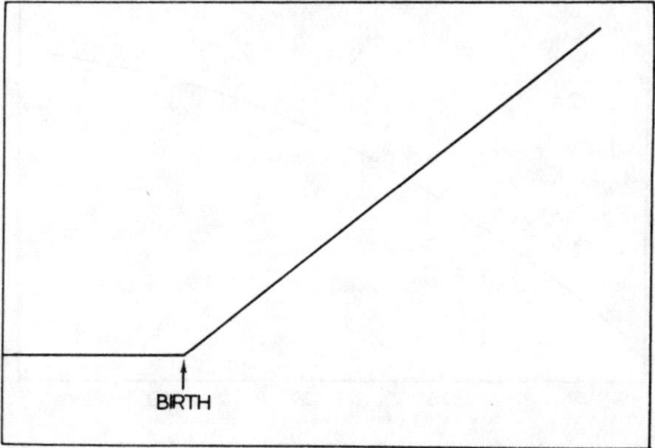


FIG. 3.—Diagram illustrating the "unmasking" effect of birth on a fetal inadequacy previously compensated for by the placenta. The graph could represent the bilirubin level associated with hepatic disease, the urea level associated with renal disease, etc.

Thus the effects of much neonatal disease are due to the unmasking of organ failure which was previously compensated for by the placenta (Fig. 3). Much of this type of disease cannot be treated adequately and accounts for the relatively high death rate of infants in the neonatal period. The death rate is maximal in the first day and falls rapidly in the next few weeks as the less serious conditions are compensated for.

## Causes

The main causes of neonatal disease are congenital abnormalities and immaturity (or prematurity as it is conventionally referred to).

Immaturity at the time of birth is due mainly to the fetus being born before the normal period of gestation has been completed. Because it is not usually possible to be certain of the precise date of conception, prematurity in the infant is conventionally diagnosed by weight. Thus an infant weighing 2.5 kg or less is by definition premature, although obviously this is an arbitrary definition and treatment must be related to the general condition. Nevertheless, most infants who weigh appreciably less than 2.5 kg at birth show signs of immaturity.

The reason why immaturity presents serious problems is that most vital bodily functions develop rapidly in the last few weeks of normal gestation, so that several weeks of immaturity correspond to marked inefficiency of such organs as the kidney, the liver, the intestine and the mechanisms which maintain constant central body temperature (Fig. 4).

Thus the premature infant is markedly limited in his ability to excrete concentrated or dilute urine. He may become jaundiced due to limited excretion of bilirubin. He cannot digest milk adequately and his capital body temperature readily rises or falls if he is not kept in an incubator at an appropriate temperature.

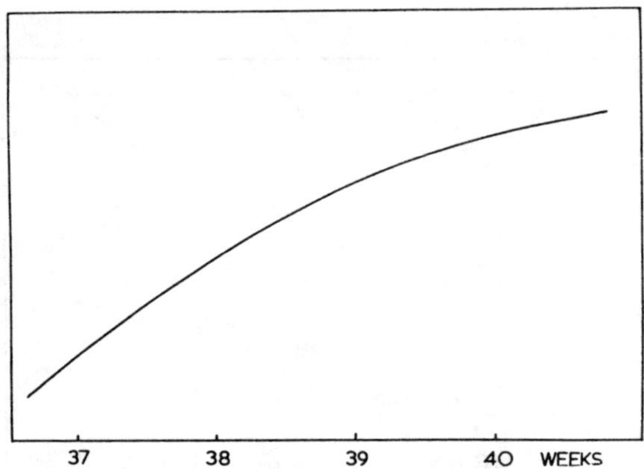


**FIG. 4.—Graph indicating the development of a typical body function around the time of birth. The function represented could be expansibility of the lungs, maximal urinary concentration, environmental thermal gradient at which body temperature can be maintained, etc. Note that the curve rises steeply in the weeks before the time of normal delivery (40th week).**

It should be noted (Fig. 4) that even infants born at the full 40 weeks have by no means reached full maturity of the various body functions and are relatively vulnerable to adverse environmental conditions (temperature, nutrition, etc.) especially in the first few days of life.

The commonest cause of death in premature infants is failure of gas exchange. This is related to a deficiency of the surfactant material which lowers surface tension in the alveoli and greatly reduces the work of breathing. Production of surfactant in the lungs follows the pattern shown in Fig. 4, so that the more premature infant may have a serious lack of it. In such cases death often occurs within a few days because the lungs fail to function due to the enormous effort required to expand them.

Less commonly, failure of blood oxygenation is due to an abnormality of the central blood vessels, whereby blood continues to bypass the lungs as in the fetal state.

### Investigations

A wide range of investigations is used, comparable to that used in adults. When the infant's life is threatened, investigation and treatment (if possible) are of the utmost urgency.

### Treatment

Immaturity requires careful treatment, with conditions initially approximating as closely as possible to those in the uterus. The infant is kept at an environmental temperature a little below body temperature; the more premature the infant, the nearer to body temperature must it be kept. Thus a 1 kg infant requires an environmental temperature around 30°C initially, whereas a mature infant can tolerate a temperature around 25°C. Fluids should be given in adequate amounts—neither too much nor too little is tolerated. Glucose, rather than milk, is given initially. Occasionally jaundice due to hepatic immaturity requires treatment by exchange transfusion.

A number of congenital defects, such as mechanical obstruction of the gut, connections between the alimentary and respiratory tracts and obstructtion of the urinary tract, require urgent surgical correction.

Difficulties in ventilation due to lack of surfactant may, if not unduly severe, be overcome by artificial ventilation. If the child survives the initial difficulty, maturation of the lungs leads to increased surfactant formation and the artificial ventilation can be discontinued. As with artificial ventilation in general, careful monitoring of the blood gases and pH is essential.

# 9. DISEASE IN CHILDHOOD

DISEASE patterns in childhood are intermediate between those in the neonatal period and those in adulthood. They will not therefore be discussed in detail, but several points will be made.

One of the main differences between treating an adult and a young child is that the child may be unable to give any account of his symptoms, or his account may be inaccurate and misleading. Diagnosis then depends largely on observations made by those looking after the child, e.g. parents and nurses. Thus diagnosis of conditions such as appendicitis may be very difficult in young children.

Difficulty in giving a history may be one factor in the higher death rate from disease which has been observed in mentally retarded children. The lower the intelligence, the higher the mortality. Another factor may be that such children are less able to avoid hazards against which other children can be warned.

*Congenital disease,* while relatively rare in the general population, makes up a considerable proportion of medical practice amongst children. This is because many of these children require regular supervision and frequent admissions to hospital. A number of surgical operations to correct congenital defects are deferred until childhood because the condition is not immediately threatening life and because such operations are often very much easier technically when the individual has grown to a larger size.

*Inadequate nutrition* is particularly harmful in childhood. Just as in early fetal life the body organs acquire their main characteristics, so in childhood the main physical and mental features of the adult state are determined. Malnutrition in childhood tends to cause both physical and mental stunting. The increasing height of school-children and adults in many countries over the course of the last century has been attributed to improved nutrition.

Chronic diseases, such as hormone deficiencies (e.g. adrenal, thyroid) and severe lung, heart and kidney disease also lead to stunting of growth. Here also, optimal conditions for cellular multiplication and maturation are not achieved.

*Infectious disease* is a particular problem of childhood. The individual as a fetus and an infant has been fairly well protected against infection, initially by the placenta and subsequently by the passive immunity conferred by maternal antibodies received in the uterus. This immunity wears off during the first year, as maternal antibody is broken down and infections tend to be frequent until the child acquires his own active immunity to the common infections. Production of active immunity by inoculation of altered infectious agents or their products is gradually changing the pattern of infectious illnesses in childhood.

# 10. AGEING

## Definition

Ageing is the process whereby body tissues and hence body functions gradually deteriorate with the passage of time in apparently healthy people.

## Effects

Ageing leads to a reduced ability of the individual to vary body function. This stems from a reduced reserve capacity of all body functions. The various functions deteriorate at different rates, the pattern being different in different individuals.

The deterioration of lung function with age has been well documented between the ages of 20 and 60 and is an example of the effects of ageing in reducing body reserves. Normal values from individuals with no obvious chest disease show a steady fall with age in their vital capacity and their forced expiratory volume in one second; residual volume increases. Thus less and less of the total lung capacity can be used in breathing; the ability of the lungs to transfer gas to the pulmonary capillaries falls in parallel (Fig. 5). *If* the changes in lung function were to follow the same trends after the age of 60, lung function in the average person would fall below the level needed to maintain life between the ages of 150 and 200!

Other vital functions have been less precisely related to age, but there is no doubt that such diverse features as cardiac function, renal function, hepatic function, visual and auditory acuity, and bone strength deteriorate with age. One measurement in the circulatory system which changes with age is the blood pressure. Both systolic and diastolic values increase with age throughout life, but systolic pressure rises more rapidly, so that pulse pressure increases with age. At least one reason for this increase in pulse pressure is a fall in aortic elasticity—stretching of the aorta in systole reduces the systolic pressure rise; the elastic recoil in diastole increases the diastolic pressure.

Slow but steady intellectual deterioration is an obvious feature of ageing. Increasing experience tends to mask this effect in middle life but, especially in later life, the ability to react to new situations diminishes progressively. This decreasing intellectual capacity, well-known as "second childhood" tends to produce, as in the child, a decreasing ability to give an accurate account of symptoms and an increasing liability to accidental injury. The tendency to falls is accentuated by impairment of co-ordination and vision; the tendency to poisoning by coal gas or exhaust fumes is increased by impairment of smell.

Combination of these effects of ageing leads to increasing sensitivity to the effects of diseases which further impair reserves. Because many body systems are deteriorating in parallel, multiple co-existing disease processes are increasingly common with advancing age.

Response to physical stress, such as surgery, or injury, is impaired in the elderly. A prolonged period of unconsciousness due to head injury may be followed by a return to more or less normal brain function in a child or young adult; with increasing age, the degree of recovery diminishes. Ability to adjust psychologically and physically to loss of a limb or limbs is remarkably great in the young; much less so in the elderly. One of the characteristic features of old age is an accentuation of this inability to deal with new situations. Just as severely mentally retarded children have a reduced expectation of life, so a severe loss of mental capacity (dementia) is generally followed by death within a few years.

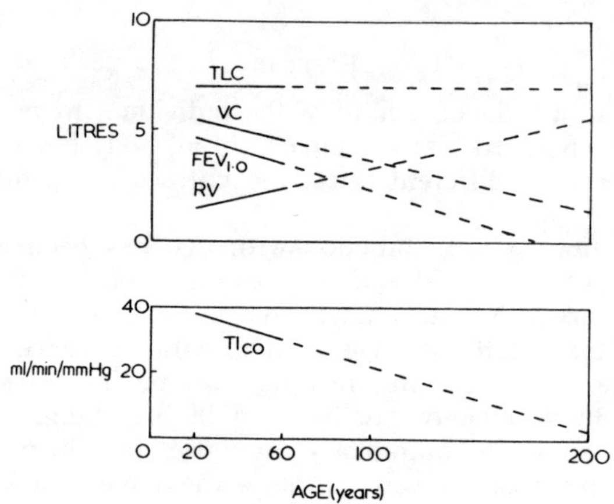


**FIG. 5.—Changes in lung function with age. The solid lines show observed changes for a 6′ (1.83 m) man (Cotes, J. E. in *Lung Function* (1968). Oxford: Blackwell Scientific Publications). The interrupted lines show the changes which would occur if the trends between 20 and 60 continued unchanged.**

Most types of cancer increase in commonness with age. This suggests the possibility that mechanisms—possibly immunological—which help to control cancer become less effective with ageing.

Elderly patients are often unduly sensitive to drugs. This is related to the fact that metabolism and excretion of the drug is slowed down so that the drug persists longer in the body.

The stage in life at which these effects of ageing will interfere with health is influenced by the state of affairs in early adult life, which, in turn, is related to health and nutrition in childhood. Organ reserves in old age are related to the state of the organ at maturity. If the lungs are large and efficient in early adult life, they are more likely to function adequately later in life. Bone strength at maturity is one of the factors which determine whether or not the individual will have weak, easily-broken bones in old age.

## Causes

The effect of ageing can be related largely to a gradual loss of cells which cannot be replaced and to a gradual change in extracellular materials, including a loss of elasticity and changes in calcium distribution.

**Loss of Cells**

The central nervous system provides an example of cells which cannot be replaced. Throughout life, the population of nerve cells decreases, as cells die and are not replaced. This adverse effect on brain function is counteracted in early life by the maturation and growth of surviving cells and throughout most of life by the organization and reorganization of the available neurones (Fig. 6). Thus the graph of mental ability has the form of an inverted U; the greatly increased efficiency developed mainly in the first two decades accounts for the steep initial rise; the diminution in cell population accounts for the terminal fall.

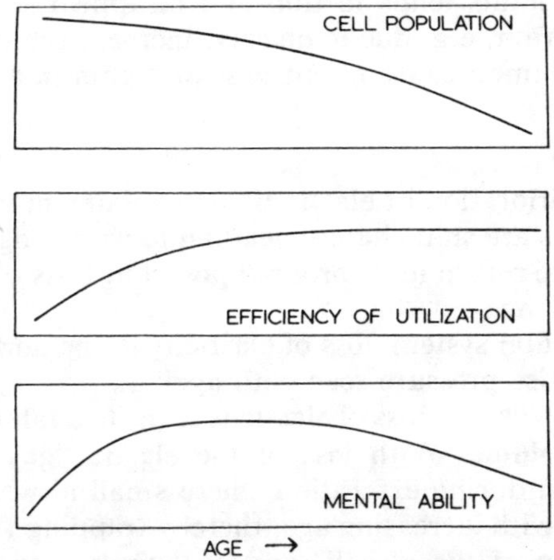


FIG. 6.—Arbitrary graphs, illustrating how changes with age in mental ability can be related to nerve cell population and the efficiency of utilization of these cells.

The shape of the "mental ability" curve will depend on the ability being tested. Conventional intelligence tests usually produce a maximum score somewhere around the age of 20. Tests which are related largely to acquired experience, e.g. competence in a complex system of knowledge such as advanced law or medicine, produce a maximum perhaps a decade later. Ability to solve problems which can be experienced only gradually over the course of years, e.g. the ability to find the best solution to complex administrative problems, may reach a peak somewhat later. Despite this, the ability to react to largely new situations declines steadily after the second or third decade. After this age people have increasing difficulty in adjusting to radical changes in their environment. This raises increasing social problems at the present time when developing knowledge is making certain skills redundant at an increasing rate; those with a relatively low mental ability may be incapable of learning the new skills associated with a new job.

Anything which damages brain cells adds to the incapacity caused by the ageing process. Thus a prolonged lack of glucose (hypoglycaemia) or oxygen (hypoxia) hastens the deterioration of brain function, the effects being more noticeable in those with least reserve, namely the elderly. The hypoxia sometimes associated with a general anaesthetic may severely reduce mental capacity in the elderly.

Other central nervous system functions share in this reduction in efficiency with age. Cerebellar deterioration and impairment of postural reflexes contribute to increased clumsiness and an increased tendency to stumble and fall.

Another example of a falling cell population is provided by the heart. Cardiac muscle cells cannot divide, so that, when small areas of cardiac muscle die due to loss of blood supply, the tissue is replaced by non-contractile scar tissue and the functional capacity of the heart is thereby reduced. The hearts of elderly people contain more fibrous tissue than those of the young.

The cartilage of joints tends to thin and disappear with ageing. Previous injury or excessive strain, e.g. due to obesity, increase the speed of this change. This process is a common cause of stiffness and pain in aged joints.

### Loss of Elasticity

Widespread deterioration of elastic tissue explains many of the features of ageing. Most obvious are skin changes leading to wrinkling. A pinched-up fold of skin takes longer to return to its original position. Loss of elastic tissue in the breasts leads to sagging.

In the cardiovascular system, loss of elasticity in the aorta and large arteries causes the rise in pulse pressure seen with ageing.

In the respiratory system, loss of elasticity leads to a fall in vital capacity and a rise in residual volume. With loss of the elastic "guy-ropes" which hold smaller airways open during expiration, these small airways close earlier and earlier in expiration with increasing age, thereby trapping the air beyond them, impairing ventilation of the alveoli, and contributing to the gradual fall in arterial $Po_2$ and rise in $Pco_2$ which accompanies ageing.

Another site of loss of elasticity with ageing is the lens of the eye. The ability to focus on nearby objects depends on the lens springing back into a more convex shape when the pull of its suspensory ligaments is relaxed. This occurs less readily with advancing years, leading to a recession of the near point.

### Changes in Calcium Distribution

Ageing is associated with progressive loss of calcium from bones and a tendency to deposit calcium in tissues from which it is absent in early life.

Loss of calcium contributes to the fragility of elderly bones so that fractures result from minor falls. Healing is slow, in contrast to the situation in children where healing is rapid and initial deformities are corrected by remodelling.

The arteries are a major site of new calcium deposition. Areas of calcification may be felt in the limb arteries or seen on X-ray in the aorta, coronary and cerebral arteries. This calcification of itself may not seriously reduce organ blood flow under resting conditions but would prevent any marked change in the calibre of the vessels affected.

As with loss of elasticity, the lens may share in the tendency for calcium to be deposited in the elderly. Calcification of the lens contributes to opacity in some cases of cataract.

The question arises as to the underlying cause of these phenomena which account for the various features of ageing. Various endocrine changes, particularly in the gonads, thyroid and adrenals, have been blamed, but administration of the corresponding hormones does not halt or reverse the effects of ageing and may produce the adverse effects associated with an excess of the hormone or hormones administered. It seems more likely that a fall-off in hormonal levels is part of the process of ageing rather than its cause.

More likely is the possibility that ageing is due to a gradual mutation of the genetic material of cells to less favourable forms so that cell function degenerates. This hypothesis is supported by the fact that agents which cause genetic mutations—e.g. various radiations, including X-rays—also tend to hasten the ageing of tissues. Intense solar radiation accelerates ageing of the skin and lens in countries where these organs are exposed to high doses of such radiation. Again, mongolism, which is due to a genetic abnormality, is increasingly common with increasing maternal age, suggesting a genetic deterioration with age in the immature ova.

On this view the human organism, like all known forms of life, has a "built-in obsolescence" whereby ageing and eventual death of the individual is inevitable.

## Investigations

As increasing numbers of investigations are carried out on elderly patients it is important to recognize changes which are to be expected with ageing so that these may be distinguished from changes due to specific disease processes.

Since different organs age at different rates in the same individual, no single test to diagnose "ageing" exists or is to be anticipated. Nevertheless, when considering surgery in an elderly patient it is desirable to make some assessment of cardiac and pulmonary function to try to avoid surgery where the risks of surgery exceed its benefits.

## Treatment

Searches for a "cure" for ageing have rivalled searches for the "philosopher's stone" to turn base metals into gold. The desirability of greatly prolonging life even if this were possible must be questioned in view of the problem of over-population and the social problems of excessive numbers of elderly persons. If ageing and death are regarded as normal physiological processes, it is a mistake to approach their medical management in the same way as specific diseases are approached. Rather, it is a matter of taking into account the effects of ageing when dealing with specific disease processes. Modification of drug treatment and of surgical treatment may be required.

# 11. DISEASE IN THE ELDERLY

## DEFINITION

THIS CHAPTER is concerned with the pattern of disease seen in patients from around the age of 70 onwards. It is common knowledge that people age at different rates and that rigid adherence to artificial dividing lines beyond which, e.g., a certain operation would not be carried out, cannot be justified in every case. Nevertheless, a patient's age has an important influence on diagnosis and treatment. In fact, asking a patient's age is one of the first priorities in taking an adequate case history. Most people age at a fairly constant rate. This can be confirmed by the student going round a ward of adult patients and estimating their ages. He should almost always be correct to within a decade.

There is no doubt that a ward unit of patients in their seventies and eighties will present a very different pattern of diseases and appropriate treatments from a ward unit of patients in their twenties and thirties and it is the former pattern of diseases which is considered in this chapter.

## EFFECTS

### 1. Multiple Disorders

Disease in elderly patients is in some ways comparable to a breakdown in an elderly car in which one fault may throw strain on other worn parts, which in turn give way. A visit to the garage for repairs may well bring to light other defective parts which require replacement. Disease in the elderly is characterized by failure of multiple body systems to varying degrees. Anaemia or excessive thyroid activity in the elderly is quite likely to "unmask" cardiac disease, even though cardiac function had been adequate for the patient's moderate activity before the anaemia developed. A slight stroke or chest infection in an elderly person living alone may cause him or her to remain unduly long in an unheated bedroom in winter, possibly without adequate clothing. A serious fall in central body temperature (hypothermia) is then likely to occur.

Putting an elderly patient to bed for any reason may make mild joint stiffness very much worse and also carries the risk of allowing blood to clot in obstructed leg veins. Admitting the elderly patient to hospital may place an excessive strain on failing mental powers. In the unfamiliar surroundings the patient may become confused and unco-operative. Mental deterioration for any reason compromises the chances of recovery from physical illness. Symptoms may be inadequately described. The patient may damage himself by falling out of bed or placing undue strain on a recent operation scar or recently set fracture. Medication may be inadequate if the patient pulls down an intravenous infusion set or refuses to take medicine by mouth.

### 2. Predominance of Degenerative Diseases

As mentioned in the introductory chapter, the frequency of degenerative diseases increases sharply with age. One tends to think of the typical patient with heart disease due to coronary artery degeneration and blockage as a man in late middle life. Heart disease due to inadequate perfusion of the myocardium (ischaemic heart disease) is indeed common in this age group. It is, however, even commoner, on a percentage basis, in the older age groups. There is a steady increase with age in the percentage of the population suffering from ischaemic heart disease. The reason why such disease in 90-year-old patients is not often encountered is that there are relatively few 90-year-olds.

Other degenerative diseases which are very common in the elderly include ischaemic disease of the brain and wear-and-tear changes in joints. The degenerative diseases tend to merge gradually into the changes regarded as normal effects of ageing. Thus the gradual deterioration of brain function which inevitably accompanies ageing may be accompanied by ischaemic damage and it may be impossible to say where one process ends and the other begins.

Cancer may be regarded in some senses as a type of degenerative disease and it is certainly much commoner with advancing age. Its cause in some cases could be degeneration of the genetic material in cells and a deterioration in immune mechanisms which would tend to destroy such abnormal cells and prevent them from multiplying.

### 3. Poor Compensation for Disease

The parallel falls of reserves in the various body systems with age result in a poor response to conditions which would be adequately compensated for earlier in life. The organ directly involved in the disease process may have little reserve function and other systems which might otherwise compensate may be unable to do so. Reference has already been made to the fact that in elderly persons the heart may be unable to compensate for anaemia by the usual increase in cardiac output. The heart may fail during the compensatory attempt.

Thus disease in the elderly may set off a chain reaction of decompensation in other systems. It is not uncommon for an elderly patient to be admitted to hospital for one condition and die a week or two later from disease in a different system. Clotting of blood in coronary or cerebral arteries or in leg veins may complicate a variety of other conditions. The fact that the patient may have to be kept in bed promotes these complications.

An impaired ability to compensate for disturbances of body fluids exacerbates conditions involving vomiting and diarrhoea.

### 4. Unusual Presentation of Disease

Diagnosis is sometimes delayed in elderly patients because they do not have the typical symptoms and signs. Appendicitis is such a condition. As with young children, it tends to present vague features initially in the elderly and diagnosis may be delayed until serious complications have developed. Many other diseases may similarly be slow to make their presence obvious. One reason is probably that the body's reaction to the disease is less vigorous and hence painful conditions such as severe inflammation are less marked—another example of impaired compensation. In some cases an absent or poor account by the patient of his symptoms may contribute to the delay in diagnosis.

**5. Adverse Reactions to Drugs**

Complications of drug therapy are particularly common in the elderly for several reasons. Firstly, elderly patients are often treated with a wide variety of different drugs for their multiple disorders. Secondly, their reduced capacity to metabolize and excrete drugs may readily lead to unexpectedly high circulating levels of drugs. Thirdly, compensation for abnormal states produced by drugs is relatively poor.

## CAUSES

The disease patterns seen in the elderly are related to (a) the effects of ageing which impair organ reserves as already discussed and (b) the many degenerative diseases which augment the effects of the ageing process. The underlying cause of many of these conditions remains obscure. Cigarette smoking is one identified cause of degenerative disease. It hastens degeneration of the bronchial mucosa, interfering with ciliary action and drainage of secretions. Degeneration of the coronary arteries and myocardium are also associated with cigarette smoking. If the various toxic agents or possible deficiency states which account for degenerative disease could be identified and removed, it is likely that the average life-span would increase, with breakdown of function more often deferred to extreme old age. However, unless the underlying process of ageing were halted, it is unlikely that life would be prolonged far into the second century.

## INVESTIGATIONS

Two points may be made about investigations in the elderly patient. Firstly, assessment of functional reserve where possible, e.g. by respiratory or renal function tests, provides a better basis for deciding on such matters as whether to advise surgery than does reliance on age alone. Secondly, elaborate and potentially dangerous investigations, which might be justified at an earlier age should be avoided, unless the benefit to the patient clearly outweighs the discomfort and the risks, which may be greater with advancing age.

## TREATMENT

**Palliation rather than cure.**— In many cases, cure of the condition is not possible and treatment is aimed at relieving discomfort and, where possible, compensating for impaired function.

**Minimal bed rest.**— One of the most dangerous forms of treatment for an elderly patient is complete bed rest. Serious, and likely complications include clotting of blood in leg veins with subsequent blockage of pulmonary arteries by detached fragments of clot, increased joint stiffness, loss of calcium salts from bones due to removal of gravitational stresses, breakdown of skin at pressure points and mental deterioration. If at all possible, the patient should at least sit out of bed for part of the day.

**Minimal drug therapy.**—Caution in the prescription of drugs is even more necessary in the elderly than in the young adult. Because of reduced rates of excretion and metabolism, lower dosage is in some cases appropriate. Adverse effects may be more severe because the reserves of the affected organ are reduced. In addition, it should be borne in mind that elderly patients are often incapable of adhering to complex dosage schedules. Further, there is no point in giving long-term drug therapy which may cause danger and discomfort if the only benefit to be expected lies in preventing complications in a remote future to which the elderly patient is unlikely to survive.

**Surgery.**—The hazards of surgery should not exceed its possible benefits for the elderly patient. However, more adventurous surgery may be justified in the elderly than in the young in situations where there is much to gain if the operation is successful. The potential gain has to be considered alongside the limited life expectancy.

Wounds heal well in the elderly provided that the general state of nutrition is adequate. However, greater care is needed to maintain the blood volume, circulation and respiration because of the reduced efficiency of their homeostatic mechanisms. Special care is needed after the operation to keep the patient active so that he does not develop infection in the lungs and clot formation in the veins.

**A time to suspend treatment.**—If an elderly patient suffering from advanced disease collapses and dies, it is inappropriate to apply vigorous efforts to maintain ventilation and restart the heart. Depletion of essential reserves of function makes such efforts unprofitable, and they may cause unnecessary distress to the patient and his relatives. There is considerable debate about when to suspend treatment in elderly patients where the issues are less clear-cut.

**Relatives.**—The relatives of elderly patients requiring active home care run the risk of mental or physical break-down due to the stress of such situations. Such illness in the relatives may be avoided by admitting the patient to public care for periods during which the reserves of the relatives may be restored.

# 12. DEATH

## Definition

THE DEFINITION of death is often given in terms such as "the cessation of life". Thus one indefinable term is "defined" in terms of another indefinable term. Concepts of death vary with the point of view from which the subject is approached. The lawyer demands an event which can be precisely located in time. The doctor regards death as the point in time beyond which further treatment is useless. However, from the physiological point of view, death of the human body is undoubtedly a gradual process.

### Physiological Death

From the physiological point of view, a cell, tissue or organ is regarded as dead when no evidence of function can be detected, e.g. it does not produce any detectable response to stimuli which normally produce a response. Thus, when an isolated perfused artery which normally constricts in response to noradrenaline ceases to respond to noradrenaline at any dosage, it is regarded as dead. Using the same preparation, it is possible to study an artery removed some hours after an individual has been medically certified as dead. Frequently such an artery will respond by constriction to noradrenaline and other stimuli, demonstrating that this part of the body is still alive.

Similarly, when a person has been pronounced dead from some condition such as severe head injury, it may be possible to remove the heart, lungs, liver, kidney etc. before these undergo any serious changes. The organs may show themselves to be very much alive by functioning normally when transplanted into another person to replace a diseased organ which has been removed.

Survival of the complex internal organs is normally much briefer than that of peripheral tissues such as a superficial artery or vein. There are at least two reasons for this. Firstly, such organs as the heart and kidneys have a higher metabolic rate than many tissues. When deprived of their circulation, toxic materials rapidly accumulate. Secondly, these organs usually remain at around core temperature for some time after death; peripheral tissues are generally at a lower temperature initially and cool more rapidly. The more rapidly a tissue is cooled, the better its chances of survival because of the reduction in metabolic rate at lower temperatures. The optimal temperature for preservation is just above 0°C; freezing produces severe damage unless special precautions are taken.

The brain, being the most sensitive of all body tissues to failure of the circulation, undergoes organ death at the most rapid rate. If the circulation to a healthy brain is suddenly arrested, unconsciousness results in a few seconds. If the circulation remains arrested for more than a few minutes and is then restored, normal brain function is unlikely to return, indicating that severe damage has occurred.

Just as parts of the body can survive the rest, so the reverse may happen. When a limb or part of a limb suffers a serious loss of blood supply, the tissues die. The condition is termed gangrene and is the usual reason for amputation in the elderly.

The concept of gradual death of the body is illustrated in Fig. 7.

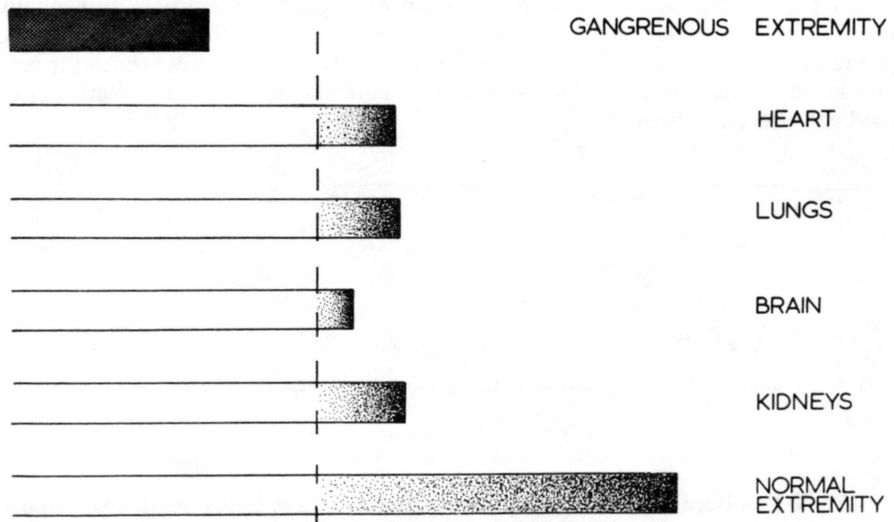


FIG. 7.—Illustrating the concept of gradual death of body organs. Cessation of heart beat and respiration are indicated by the interrupted line. Degeneration and death of tissues is indicated by darkening shading. The time scale will be influenced by the temperature of the tissues.

### Medical Death

From a practical point of view, it is necessary to determine a point at which the individual patient is regarded as dead, and further treatment is inappropriate.

In many cases, the situation is relatively simple. The patient has been suffering from a potentially fatal disease. Heart beat and respiration abruptly cease, within a few minutes of each other. Listening with a stethoscope confirms the complete absence of heart and respiratory sounds, and the patient is pronounced dead. In such a case, medical (and legal) death would correspond with the interrupted line in Fig. 7.

In three types of situation, determination of death is more difficult:

**(i) Very gradual death.**—In certain cases, particularly when body temperature falls markedly before death (e.g. hypothermia from exposure to a cold environment), the transition from life to death may be almost imperceptible and recovery from a state of deep coma with a minimal cardiac output is possible. This is the kind of patient who occasionally "wakes up" in the mortuary! In doubtful cases repeated examinations may be necessary before it is possible to say for sure that recovery is impossible.

**(ii) When the question of cardiac resuscitation arises.**—An attempt to restart the heart is usually made when a patient who has been otherwise fit suddenly collapses due to cessation of the cardiac output due, e.g., to ventricular fibrillation. Such a case is illustrated in Fig. 8, where a patient's cardiac output is abolished by two successive cardiac arrests $A_1$ and $A_2$. The first effort at resuscitation, $R_1$, is successful; the second, $R_2$, unsuccessful. Prior to, or in the absence of resuscitation, the patient would be regarded medically and legally dead at the point $A_1$. In the presence of resuscitative efforts, he would be pronounced medically dead at the end of period $R_2$, but legal death might well be placed at the point $A_2$. This shows the arbitrary nature of the concepts of medical and legal death.

FIG. 8.—This illustrates falls in cardiac output produced by successive cardiac arrests, $A_1$ and $A_2$. Two efforts to restart the heart are made; $R_1$ is successful, $R_2$ is not. Difficulties in pin-pointing a precise time of death in such a case are discussed in the text.

**(iii) When cardiac and/or respiratory function are being artificially maintained.**— Such cases may arise when a person has suffered severe brain damage which would arrest respiration due to depression of the control areas in the brain stem. In some cases the artificial maintenance of one or both of these functions enables useful recovery of brain function to occur. In other cases it gradually becomes evident that this will not occur. It is then difficult to know when to switch the machine off. The concept of *cerebral death* is then applied. The machine is switched off on the basis that death of the brain means death of the individual. Criteria for diagnosing cerebral death include prolonged absence of any response, reflex or otherwise, by the patient, and prolonged absence of any electrical activity in the brain as recorded by the electroencephalogram. Naturally such criteria must be somewhat imperfect and arbitrary. A further complication is that organs from such a patient may be used in transplant operations. The desire to remove such organs in as healthy a state as possible may conflict with the desire to ensure that the patient is given every possible chance of survival.

## RECENT LEADING ARTICLES

Biological age of man. *Lancet,* 1970, **2,** 558.
Brain damage and brain death. *Lancet,* 1974, **1,** 341.
Care of the dying. *Lancet,* 1971, **2,** 753.
Death of a human being. *Lancet,* 1971, **2,** 590.
Determination of death. *Lancet,* 1970, **1,** 1092.
Donors for organ grafting. *Brit. med. J.,* 1972, **1,** 582.
Drugs and the elderly mind. *Lancet,* 1972, **2,** 126.
Fashions in infant feeding. *Brit. med. J.,* 1973, **2,** 727.
Food for thinking. *Brit. med. J.,* 1973, **1,** 126.
Geriatrics is medicine. *Lancet,* 1974, **1,** 663.

Limitations of resuscitation. *Lancet,* 1972, **1,** 1169.
Nutrition and the developing brain. *Lancet,* 1972, **2,** 1349.
Old bones. *Lancet,* 1972, **1,** 1059.
Operating on the elderly. *Brit. med. J.,* 1972, **2,** 480.
Organ preservation. *Lancet,* 1973, **2,** 715.
Respiratory distress syndrome and continuous positive airway pressure. *Lancet,* 1971, **2,** 477.
Reversible death. *Lancet,* 1970, **2,** 1172.
Smoking hazard to the fetus. *Brit. med. J.,* 1973, **1,** 369.
Sudden whitening of hair. *Brit. med. J.,* 1973, **1,** 504.
Talking about death. *Brit. med. J.,* 1974, **1,** 131.
The dying patient. *Lancet,* 1972, **2,** 1238.
The elderly, at home or in hospital. *Lancet,* 1972, **1,** 479.
What do the elderly need? *Lancet,* 1973, **2,** 833.

## *Section III*

# DISORDERS CAUSED BY PHYSICAL AND EMOTIONAL STRESS

MAN IS WELL adapted as an animal to the conditions found on this planet. His physiology is neatly tailored to the temperatures, pressures, radiations, gravitational forces, etc. that are found here; he needs very complex support systems when he leaves, even to go as far as the moon. However, conditions can vary from place to place and from time to time on Earth and man's survival has resulted largely from his homeostatic ability, i.e. his ability to maintain a relatively stable internal environment in the face of external stresses. It has permitted him to range freely over the planet and work effectively in a variety of environmental conditions. A stable internal environment is essential for the normal functioning of the human brain.

Homeostatic mechanisms tend to minimise changes in the internal environment when the external environment deviates from the normal. However, there are limits to the ability of any homeostatic mechanism to deal with change. When these limits are exceeded, homeostasis is overwhelmed (Fig. 9) and disorders due to changes in the internal environment occur. This section deals with some of these disorders.

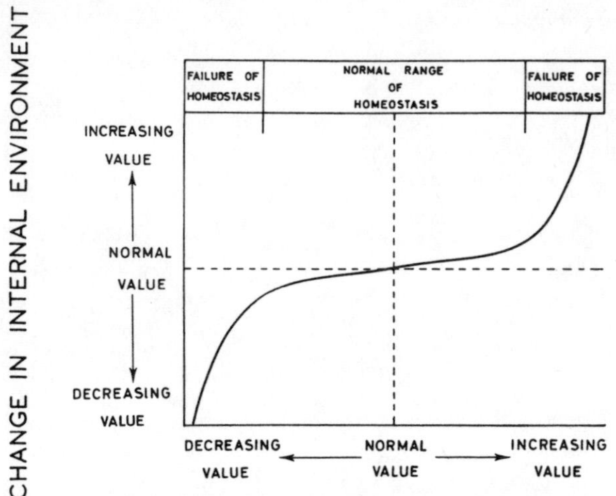


**FIG. 9.—The effect of external stress on the internal environment of the body. Over the normal range of homeostasis, relatively large changes in the external environment produce relatively small changes in the internal environment. The internal environment is "buffered" by homeostatic mechanisms. However, when the limits of the homeostatic mechanisms are reached, further changes in the external environment produce disorders due to dramatic changes in the internal environment.**

*Section III*

# DISORDERS CAUSED BY PHYSICAL AND EMOTIONAL STRESS

# 13. TEMPERATURE

## DISORDERS CAUSED BY HEAT

### Definition

MAN'S ACTIVITY presents him with a heat problem. Due to the inefficiency of most of his energy transformations, 80 per cent or so of all the energy he uses is dissipated as heat. If heat loss were prevented, body temperature would rise by several degrees centigrade per hour. Normally heat gain is balanced by heat loss from the skin to the environment. If, for any reason, heat gain should tend to exceed heat loss, the body increases heat loss by increasing the blood flow to the skin and increasing the rate of sweating. However, in certain conditions the heat problem is greater than the normal physiological processes of the body can manage and this gives rise to heat stress.

### Effects

**Heat Adaptation**

When a person moves suddenly from a temperate to a hot climate he is unable for some time to carry out sustained work. Before adaptation occurs, he tends to be apathetic, depressed and easily fatigued. His appetite is poor and he faints readily. On examination, his heart rate is fast, his face is flushed and his body temperature is raised. The hyperventilation that occurs in hot humid conditions may lead to washing out of carbon dioxide, respiratory alkalosis and tetany.

However, after four or five days of graded work in the new conditions, the person starts to adapt to them. He becomes able to work without excessive fatigue and his attitude is much more cheerful and alert. When he works, the rise in heart rate is about normal. The skin is no longer flushed and he can stand without fainting. The body temperature returns to near normal. His maximum sweat rate rises so that he may lose as much as 12 litres of sweat per day. He experiences considerable thirst, though this does not provide an adequate drive to replace his water deficit. Because of this he still faces the danger of dehydration.

The mechanism of heat adaptation is not understood, but many factors have been considered. There appears to be an increased sensitivity and efficiency of the thermoregulatory mechanisms. It is possible that there is a reduction in the basal metabolic rate secondary to a reduction in the output of thyroid hormone. It has been postulated also that the facility for heat loss increases with adaptation. Certainly there is an increase in the amount of sweat secreted per day, but it does not seem enough to explain the improvement in performance. An expansion of the blood volume is thought to facilitate the increase in cardiac output. There may be an increased tolerance of the effects of the physiological changes caused by heat. The evidence for and against these possibilities is often contradictory and inadequate.

Age affects the degree of adaptation that can be obtained. Poor adaptation is seen in the very young and the very old. A highly pigmented skin does not ensure ready heat adaptation; black soldiers from a temperate climate do not adapt to hot environments more easily than white soldiers.

Adaptation to heat is short-lived on return to a temperate climate. There is rapid loss over the first few days and virtual disappearance in two to three weeks.

### Heat Syncope (Gr. *synkope* = cutting short)

This refers to the fainting which is sometimes a feature of heat stress before adaptation occurs. It is due to the fall in peripheral resistance associated with the vasodilatation in skin. It may be aggravated by a decrease in blood volume due to dehydration. It is usually precipitated by standing for some time, as this allows blood to pool in the vessels of the lower extremities.

### Heat Cramps

These are muscular cramps which may occur in people who sweat profusely doing heavy work in very hot conditions. The condition is sometimes called "stokers' cramp" because it was seen in the stokers who tended boilers in steamships. The cramps, which may occur in limb muscles or in the abdomen, are not related to muscular exercise. The abdominal cramps may simulate a surgical emergency. The condition is thought to be due to salt depletion; excessive sweating leads to loss of salt and water.

The higher the rate of sweat loss, the higher is the *concentration* of salt in the sweat. Sweat glands are in some respects analogous to nephrons. Sweat is formed mainly in the deeper parts of the gland and salts are then reabsorbed in the ducts. A high flow rate means less complete reabsorption of salts. Hence doubling the rate of sweating will more than double the loss of salts, mainly sodium chloride.

If the water deficit alone is made good by drinking, a salt deficit will occur. Heat cramps may be prevented by giving sodium chloride tablets to people required to work in hot conditions.

### Heat Exhaustion

This describes the condition which occurs when people lose their heat adaptation in a hot environment. In the typical case a person who has adapted to the hot conditions develops the symptoms and signs of the non-adapted state. Weakness and hypotension with a moderate rise in body temperature are usual. An intercurrent infection, alcoholic excess, dehydration or loss of sleep may trigger off heat exhaustion. It is probably always associated with some degree of water and electrolyte disturbance.

### Pyrexia

When the body loses heat at a slower rate that it gains heat, body temperature rises (pyrexia, Gr. *pyressein* = to be feverish). Many of the effects of such a rise are due to the cause of the imbalance but the rise in body temperature causes a number of effects *per se.*

(a) There is a rise in the basal metabolic rate. All chemical reactions proceed faster at higher temperatures and the body's metabolism is no exception. The basal metabolic rate rises by about 15 per cent for each °C rise in body temperature. Perhaps because of the fall in food intake with loss of appetite, the carbohydrate stores fall and protein and fat are metabolized for energy. Muscle and fat in the body tend to be broken down and the urinary excretion of nitrogen rises. The resulting incomplete oxidation of fat may lead to ketosis.

(b) A rise in body temperature interferes with the normal functioning of the central nervous system. The higher cells in this system require strict homeothermy. As the temperature rises, the normal powers of concentration and logical thought are lost. Later the patient may become confused and delirious. The speech may become incoherent. Eventually stupor and coma may supervene.

(c) There is still no clear evidence that an increase in body temperature increases the body's ability to deal with infection. On teleological grounds this idea is attractive because many infections are accompanied by fever. Fevers therapeutically induced by malarial parasites have been used in the treatment of syphilis.

### Heat Stroke

This is a rare condition in people exposed to severe prolonged heat stress where the most striking feature is a massive rise in body temperature. It may rise to over 43°C. This is very bad for the central nervous system, and irritability, visual disturbances, mental confusion, collapse, coma and death may be the clinical progression. The skin is hot and dry and there is a full bounding pulse. The critical temperature for damage to the central nervous system is in the 42-45°C range. The damage probably results from denaturation of protein, degradation of enzymes and alteration in cell membrane characteristics.

The cause of the hyperpyrexia is thought to be a cessation of sweating. The reason for this is obscure. Some suggest that it is due to exhaustion of the sweat glands but it may be due to a failure of the temperature-regulating centre. It is often associated with a fall in the volume of extracellular fluid. Exposure to prolonged heat may result in over 12 litres of sweat being lost per day. The thirst engendered by such loss is often not sufficient to make good the deficit. When the water loss exceeds about 15 per cent of the body weight, dramatic increases in body temperature usually occur. Once the body temperature rises considerably the increase in metabolism at the higher temperatures results in greater heat production. This positive feedback may account for the rapid rise in body temperature that is often seen in heat stroke.

## Causes

**High environmental temperature.**— The body normally loses heat by conduction, convection and radiation from the skin. The rate of heat loss by these means depends on the temperature gradient between the skin and the environment. As the environmental temperature rises to body temperature (37°C), heat loss by these means falls to zero. The body will gain heat from the environment when the environmental temperature rises above body temperature.

**High relative humidity.**—Even when the environmental temperature is higher than body temperature, the body can still lose heat through the evaporation of sweat on the skin. For water to evaporate, energy is required and this energy, the latent heat of evaporation, is abstracted from the skin. If the sweat runs off the body and does not evaporate it is quite useless for heat loss purposes. The ease with which sweat evaporates depends on the relative humidity of the air. If the air is dry, water evaporates easily but if the air is saturated with water vapour, evaporation is impossible. For these reasons temperature and humidity have to be taken into consideration when assessing a climate (Fig. 10). Thus, 35°C with 95 per cent humidity is just as bad as 45°C with 40 per cent humidity. It is equally difficult to carry out sustained work in each of these sets of environmental conditions.

**Reduced air movement.**—The removal of heat from the skin may be helped by disturbance of the warm layer of air that normally envelops the body. Disturbance of this layer effectively increases the temperature gradient between the skin and its immediate environment. It also favours evaporation of sweat. Absence of breeze or wind thus reduces the rate of heat loss from the skin and would tend to shift the curves shown in Fig. 10 to the left. Conversely, strong winds would tend to shift the curves to the right.

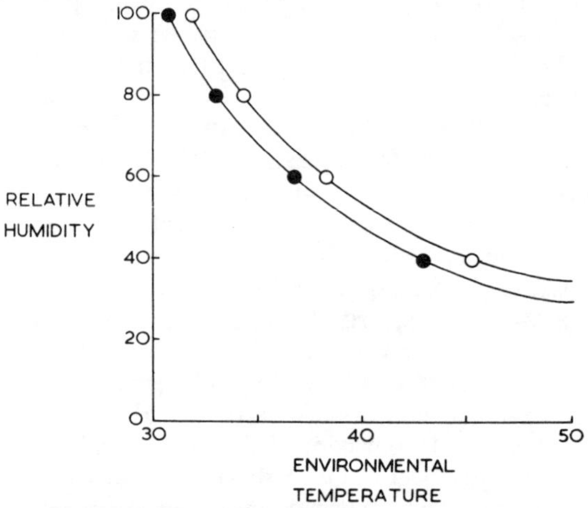


FIG. 10.—The upper limits of environmental temperature and humidity for ability to work in unclothed (O) and clothed (●) warm-adapted men.

Clothes act by trapping air in the immediate environment of the body and preventing it from moving away. Increasing the quantity and quality of the clothes worn is thus equivalent to reducing the amount of air movement. In addition, however, clothing tends to prevent heat gain from solar radiation. White clothing, traditionally worn in hot climates, is particularly effective in reflecting solar radiation. In general, suitable clothing may help to maintain thermal balance in bright sunlight but has the reverse effect in the shade. The trapped air acts as an insulator and thus reduces heat loss to the environment.

**High work loads.**—The amount of heat produced by the body depends on the amount of work carried out. With hard work the daily heat production of the body may be more than double that of someone leading a sedentary life. Clearly, increasing severity of work would shift the curves shown in Fig. 10 to the left. In other words, man's ability to withstand high temperature and humidity is progressively reduced with increasing work loads.

**Pyrogens.**—Certain bacteria produce substances which result in a rise in body temperature. These substances are called pyrogens and are thought to be lipopolysaccharides. They can affect the set point of the temperature-regulating centre in the hypothalamus. It is not clear whether they do this directly or indirectly by inducing leucocytes to release a pyrogen. When bacterial pyrogens are injected, the hypothalamic centre varies heat gain and heat loss so that the body temperature is regulated about some value higher than normal. The rise in body temperature is induced by shivering which increases heat gain and peripheral vasoconstriction which decreases heat loss. The onset of these effects (sometimes referred to as a rigor) is often abrupt and the subject feels chilled and cold despite the fact that his body temperature rises above normal.

Cellular damage in the body is also thought to release pyrogen. Thus fevers occur, for example, when heart muscle is damaged due to coronary thrombosis.

## INVESTIGATIONS

### Measurement of Body Temperature

The temperature of the peripheral parts (shell) of the body is constantly changing in the processes regulating heat loss from the body. It is only the central parts of the body (core) where tissue temperature is homeostatically controlled. Measurement of core temperature presents many technical difficulties because the core is relatively inaccessable. Use is usually made of one of the body orifices to introduce a temperature measuring device towards the core. These orifices vary in their reliability, convenience and social acceptability. Ideally, *oesophageal temperature* recorded by a thermocouple swallowed so that it lies in the oesophagus just behind the heart should be measured. Because of the difficulties in swallowing such a device, other sites are usually chosen. *Mouth temperature* recorded with a clinical thermometer held under the tongue approximates closely to core temperature if the mouth has been closed for the previous four minutes and warm or cold fluids have not been recently drunk. However, it tends to underestimate core temperature by about half a degree centigrade. *Rectal temperature* is also used to estimate core temperature but it often exceeds that of the core because of local heat produced by bacterial metabolism. *Vaginal temperature* can be measured conveniently in women and may be used to map out the menstrual cycle in the rhythm method of birth control. For the longer term estimations of core temperature that may be required on scientific expeditions to hot climates, *tympanic membrane temperature* measured by a thermocouple on a device inserted into the external auditory meatus is sometimes used. A less direct method which has been used is to hold the bulb of a clinical thermometer in the stream of urine during micturition.

When interpreting the meaning of a core temperature it should be borne in mind that the range of normal is quite wide. There is a daily (circadian) rhythm of body temperature. It fluctuates by about 1°C, having a trough at about 4 a.m. and a peak about 6 p.m. Core temperature also varies with the stage of the menstrual cycle, being a degree or so higher in the progestogen-dominated latter half of the cycle. Body temperature can also be raised by severe exercise. Individual differences are superimposed on these fluctuations.

**Assessment of Environmental Conditions**

The comfort of an individual depends on the ability of the environment to cool the body. If this is excessive, the subject is uncomfortably cold and vice versa. A variety of measurements can be made to describe this aspect of the environment. They include:

(*a*) *Ambient temperature.*—The cooling power obviously depends on the ambient temperature. However, it also depends on the relative humidity and the amount of air movement. Thus measurements of environmental temperature alone do not provide an adequate description of cooling power.

(*b*) *Relative humidity.*—Sweat may evaporate only if the air is not saturated with water vapour. As the cooling power of sweating depends entirely on the latent heat of evaporation, the degree of saturation of the air with water vapour is an index of the cooling power of the environment. The degree of saturation is usually expressed as relative humidity. For this, the amount of water vapour present in air is expressed as a percentage of the amount of water vapour present in the same volume when saturated at the same temperature and pressure.

To measure relative humidity, an instrument called a *whirling hygrometer* is used. This consists of two thermometers held in a wooden frame which can be rotated in a similar fashion to the noisy rattles favoured by certain spectators at football matches (Fig. 11). The bulb of one thermometer is covered by a moist stocking and the thermometers are then whirled until their readings become constant. If the atmosphere is saturated with water vapour there will be no difference between the readings from the wet and dry bulbs. The dryer the air the greater will be the difference. From the temperatures of the bulbs after whirling the relative humidity can be calculated.

(*c*) *Cooling power.*—A device called the *katathermometer* can be used to describe the cooling power of the environment on the body in terms of heat loss per $cm^2$ of surface per second (Fig. 12). It is an alcohol-filled thermometer with a large bulb. The thermometer is first heated until it is a little above body temperature and alcohol enters the upper bulb. It is then suspended in the atmosphere and a measurement is made of time taken for the temperature to fall between the 38° and the 36°C marks on the thermometer stem.

With the bulb dry (A), the time taken for the alcohol to fall will depend on the environmental temperature and air movement. It indicates the amount of heat which would be lost by a surface at body temperature to the environment by convection and radiation.

With the bulb wet (B), the time taken for the alcohol to fall depends on the environmental temperature, air movement and dryness. It indicates the amount of heat that would be lost from a wet surface at body temperature by convection, radiation and evaporation.

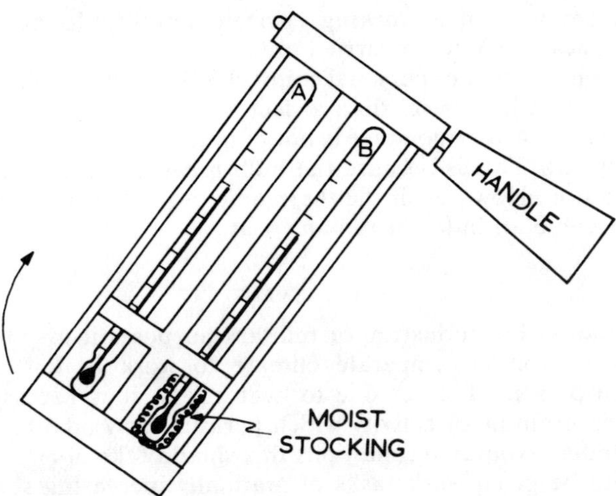


FIG. 11.—Whirling hygrometer. Two thermometers are rotated (whirled) in their wooden frame, the bulb of one being covered by a moist stocking. The difference in the temperature recorded by the two thermometers is a function of the saturation of the atmosphere with water vapour.

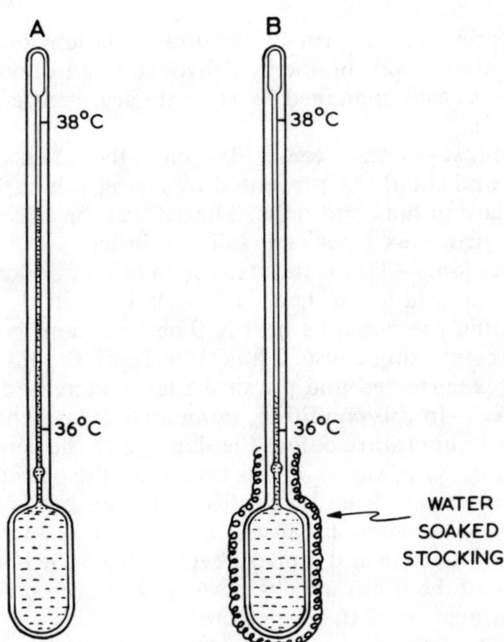


FIG. 12.—Dry bulb (A) and wet bulb (B) katathermometers. Dry bulb readings are an index of cooling due to convection and radiation. They are a function of the environmental temperature and air movement. Wet bulb readings are an index of cooling due to convection, radiation and evaporation. They are a function of environmental temperature, air movement and humidity.

The cooling power of a working environment should be related to the amount of physical work to be carried out.

From the point of view of personal comfort it is important that cooling power should be greater at head level than at feet level.

(*d*) *Radiation.*—A thermometer is not of much use to measure radiant heat in the environment. This is because the bulb reflects a large proportion of the incoming radiation. However, if the thermometer is placed in a blackened box its readings become an index of radiant heat.

## Treatment

**1. Prevention.**—For industrial or military purposes it may be necessary to take individuals from a temperate climate to work in hot and/or humid conditions. To prevent diseases due to heat stress, it is necessary to provide them with a programme of activity which permits early adaptation to the new conditions. Under favourable conditions this should take about two weeks. The subjects should be given work tasks of gradually increasing severity over this time. They should be encouraged to drink to avoid dehydration and take salt tablets to avoid sodium chloride depletion. Factors which increase physical fitness such as sleep, prevention of infection, avoidance of alcohol, seem to help. There is no reason why the process of adaptation cannot be undergone during relatively brief daily periods in hot climatic chambers in the temperate climates before the subjects actually travel.

**2. Heat syncope.**—This form of postural hypotension due to peripheral vasodilatation, fall in blood volume by dehydration and pooling of blood in the lower extremities is easily managed by recumbency, gentle cooling and correction of dehydration.

**3. Heat cramps.**—These cease dramatically when additional sodium chloride is given and should be prevented by giving salt supplements to people who must work hard in hot conditions. The salt can be added to the diet, taken as salt tablets or drunk as 1 per cent saline solution.

**4. Heat exhaustion.**—This is the state of collapse and loss of ability to work that occurs when adaptation to heat is lost. It is often due to some factor or factors undermining the patient's health. The treatment is usually directed at correcting the precipitating cause. Thus, infections should be cured, salt and water disturbances corrected and physical fitness increased.

**5. Heat stroke.**—In this condition, immediate treatment must be instituted to lower the body temperature before the damage to the central nervous system becomes irreversible. One way to do this is to wrap the patient in wet sheets and expose him to the draught from an electric fan. Another is to pack him in ice. The peripheral vasoconstriction caused by these procedures may increase the insulating power of the skin and hinder heat loss from the body core. Drugs may be administered and the limbs may be massaged to overcome this and facilitate the dissipation of heat from the body core.

## DISORDERS CAUSED BY COLD

The body can be harmed by cold in two main ways. Firstly, excessive cooling of the peripheral parts may cause local tissue damage in these areas. *Immersion foot* and *frost-bite* are examples of this type of damage. Secondly, cooling may affect the whole body to such an extent that the normal regulatory mechanisms which maintain body temperature are overwhelmed and body temperature falls. *Hypothermia* is the state where the body core temperature falls below 35°C.

## *Immersion Foot*

### Definition

This is the condition where the lower extremities are damaged by prolonged immersion of the lower limbs in water below 15°C.

### Effects

When a limb is exposed to cold water in the range of 10-15°C, intense vasoconstriction occurs in the limb so that blood flow virtually ceases. This helps to prevent a fall in central body temperature because cool blood is not returned from the cold extremity. However, if the exposure persists for some time, the absence of blood flow is associated with local tissue damage.

The affected parts become numb and swollen. Later, the skin becomes discoloured blue (cyanosed) and may blister. Eventually the tissues of the extremities may die (become gangrenous). The numbness may be due to lack of blood supply to the nerves but cold by itself may interfere with impulse transmission in nerves. The swelling has been attributed to a number of mechanisms. Absence of blood flow may result in capillary damage so that their walls become more permeable to plasma protein. Release of histamine from damaged tissue in the area around capillaries would have the same effect because it also increases capillary permeability. With loss of protein into the interstitial fluid, the colloid osmotic gradient drawing fluid into the capillaries is lost and oedema ensues. Local cold also causes venoconstriction. This would tend to raise hydrostatic pressure in the capillaries and increase fluid loss to the tissues.

Even though blood flow to the extremities has virtually ceased, the skin may be bright red or blue. This is due to engorgement with blood of the small vessels, mainly capillaries and venules in the skin. However, this blood is fairly static and does not contribute much to the metabolic needs of the part. The fact that the blood appears bright red does not necessarily mean that the tissues are not endangered by oxygen lack. Cooling the blood shifts its oxygen dissociation curve to the left so that it will not release its oxygen readily (Fig. 13). Thus in the cold limb, the blood can appear bright red though the oxygen partial pressure in the tissues is very low.

Despite the reduced metabolism and oxygen requirements in the cool tissues, the absence of blood flow results eventually in very low oxygen pressures in the tissues. Then, the haemoglobin becomes reduced and the parts become cyanosed. Tissue damage and death from hypoxia occur eventually if exposure persists.

If exposure does not persist and the extremities are warmed again, the skin shows many of the features of the "triple response" seen when histamine is injected under the skin. The swelling and oedema become considerably worse.

This is thought to be due to a number of factors. Restoration of blood flow allows much more protein and fluid to escape from the damaged capillaries. Further, it is thought that constriction of the veins persists for longer than constriction of the arterioles. Thus there is an increased capillary hydrostatic pressure which facilitates fluid loss to the tissues. Severe pain is also associated with rewarming. This is probably due to the accumulation of pain-producing products of tissue damage in the affected parts. During exposure to cold the subject is unaware of them because of the numbness induced by cold. However, when function is restored to the pain nerve fibres by the returning blood supply, severe pain is experienced until the products are removed or inactivated.

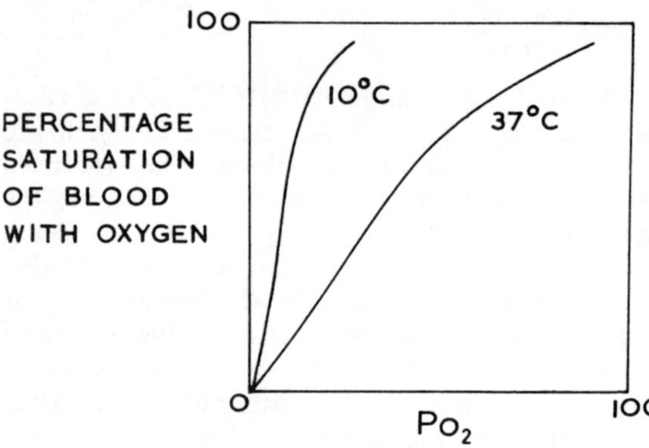


FIG. 13.—Effect of cooling blood on its oxygen dissociation curve. At 10°C the blood will not release its oxygen unless the partial pressure of oxygen is reduced to very low values.

## Causes

"Immersion foot" was the name given to the condition of the lower extremities seen when people had been shipwrecked and forced to spend several hours with their feet immersed in water below 15°C. A similar situation is seen in soldiers or hill climbers wearing wet or leaking footwear for long periods in near freezing conditions. This syndrome is called "trench foot".

Cold and wetness seem essential ingredients of the cause. Water is a better heat conductor than air. Thus a hand feels colder when moved from air at 10°C to water at 10°C. Though the temperature is the same, more heat passes from the hand to water than to air per unit time. If the water surrounding the limbs is held still by clothes or other means, a warm microclimate of water builds up which reduces heat loss. However, the more the limbs are moved or the more the water in contact with the skin is disturbed, the greater will be the heat loss. The evaporation of water from wet skin surfaces results in heat loss due to the latent heat required for evaporation.

## Treatment

When conditions predispose to immersion foot, attempts should be made to prevent it. The clothing and boots covering the feet should be retained so that a still layer of warm water may accumulate around the lower limbs. To serve the same end, vigorous movements of the limbs should also be avoided. For people exposed to near freezing conditions for long periods on land, it is essential to see that footwear does not leak and that socks are washed and changed frequently. By entrapping a warm layer of air around the feet, clean, dry socks form a good protection against cold.

If damage has occurred, gentle warming of the damaged area is required. This may well be very painful and in severe cases may require a strong pain suppressant such as morphine. The swelling which occurs during recovery may be diminished by elevating the affected parts. This reduces the hydrostatic pressure in the capillaries and thus reduces fluid loss into the tissues. If tissue death has occurred, the part should be kept dry and clean with surgical spirit until a clear line of demarcation appears between the living and dead tissue, after which amputation may be required.

## *Frost-bite*

### Definition

Frost-bite occurs when tissues are exposed to such severe cold that they freeze.

### Effects

Exposed parts with large surface area to volume ratios such as the nose, ears, lips, fingers and toes are usually affected first. The affected parts go white and hard. There may be a feeling of numbness but pain is uncommon because the cold inactivates the pain nerve fibres. Thus it is often a companion who first notices the condition by its appearance. The freezing may affect the superficial layers of skin only or the skin and subcutaneous tissue.

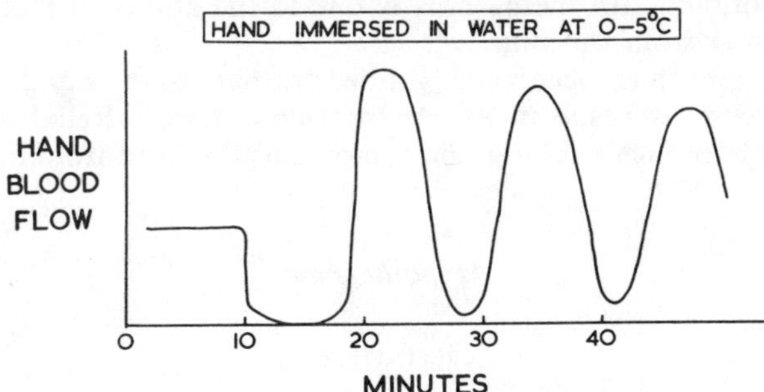


FIG. 14.—The effect of immersing the hand in water at 0—5°C on hand blood flow. The periodic rise and fall in blood flow is known as the hunting reaction.

Frost-bite occurs only when the normal protective mechanisms of the extremities against freezing are overcome. The peripheral parts of the body are normally protected against freezing conditions by the periods of vasodilatation induced by these temperatures (Fig. 14). When an extremity is placed in ice water, the blood flow through that extremity initially falls because of the direct effect of cold on the blood vessels. After some minutes it becomes painful. Then a strong vasodilatation occurs locally so that warm blood surges through the cold extremity and the pain disappears until the blood flow falls again for a further period. This cyclic rise and fall in the blood flow through areas exposed to near freezing conditions is known as the ***hunting reaction***; "hunting" is a word used by engineers to describe the oscillation of a controlled quantity around a set value in a control or servo system. The mechanism of this response to cold reaction is not known. It may be that the reduction in blood flow permits tissue damage and that the products of the damage induce strong vasodilatation. When the products of damage are washed away the blood vessels constrict again and the cycle continues. Its effect is to keep the temperature of the extremities above the point where freezing takes place. However, it does allow blood which has been cooled by circulating in the cold extremities to return to the body core and lower the central body temperature.

As with frozen pipes in a plumbing system, the greatest damage occurs during the freezing stage as water expands, but the effects become apparent only during thawing. The disruption of the cells by the ice crystals causes tissue damage. The products of cell destruction give rise to severe pain, vasodilatation, increased capillary permeability and oedema of the affected parts. Blisters appear on the skin and the dead (necrotic) tissues may separate from those that are still alive.

#### Treatment

This is almost the same as the treatment of immersion foot. Again prevention is more important than cure. It involves the wearing of suitable loose windproof clothing which permits a layer of still warm air to accumulate around the body. Minimal exposure of the body is desirable. It is important to avoid contact with objects which permit heat to be conducted away from the skin rapidly. Metallic objects are notorious for their ability to cause frost-bite in skin which makes contact with them. This is due to the ability of metal to conduct heat quickly away from the skin.

Once frost-bite has occurred, gentle, rather than rapid, warming is desirable and seems to result in less destruction of tissue. Relief of pain, elevation of the part to reduce swelling and removal of the fluid from blisters may be necessary.

### *Hypothermia*

#### Definition

Normally the central body temperature fluctuates within a degree or so of the mean value 37°C. When it falls below 35°C the condition is arbitrarily called *hypothermia.*

#### Effects

The effects of body cooling vary considerably from species to species and with the rate at which the cooling is effected. This accounts for much of the confusion and conflict in the description given of hypothermia. However, the following sequence of events might be seen in a person whose central body temperature gradually fell. At 35°C, the subject would display apathy, some disorientation and tiredness. At about 33°C, he would lose consciousness. At about 31°C, the temperature-regulating centre would cease functioning and the subject would take up the temperature of his environment. At 29°C, disturbances of heart rhythm would begin and at about 27°C, the heart would cease to function effectively because of ventricular fibrillation.

**General appearance.**—Hypothermic people look ill and their body feels like that of a cadaver. There is usually no shivering and the muscles feel stiff. The skin is pale and the subcutaneous tissue feels doughy, perhaps because the fat it contains is less fluid. The muscular and pupillary reflexes are slow and sluggish.

**Metabolism.**—The rate of all chemical processes depends on temperature and all metabolic processes in the body therefore slow down as the body temperature falls. The oxygen consumption of the body is therefore reduced and the blood flow requirements of the various organs becomes smaller. Normally the brain can tolerate an interruption of its blood for only about 4 minutes without sustaining irreversible damage. With a body temperature of 30°C it can tolerate interruption of blood flow for about 8 minutes.

**Respiratory system.**—The depression of the central nervous system leads to a depression of the neural ventilatory drive. Breathing is slow and weak with a prolonged expiratory phase. Because of this, carbon dioxide tends to accumulate in the body, despite the depression of cellular metabolism, and results in acidosis.

**Cardiovascular system.**—As the temperature falls the activity of the cardiovascular system is gradually reduced. The heart is slowed, partly because of the direct action of the blood temperature on the sino-atrial node. Cardiac output is reduced; perhaps because the decrease in peripheral blood flow and the muscular inactivity reduce the venous return to the heart. For an unknown reason there is a shift of fluid from the blood to the tissues; oedema of the peripheral tissues is common in patients with hypothermia. The resulting haemoconcentration and increase in blood viscosity may account for the stasis and sludging of blood in the capillary vessels.

One of the most sinister aspects of hypothermia is the disorder of cardiac rhythm which it induces. The mechanism of this is obscure. Speculations have been made about the effect of cold on the distribution of ions such as potassium and calcium across the membranes of cardiac pacemaker cells. Atrial fibrillation may occur quite early and ventricular fibrillation may lead to death once the body temperature falls to about 27°C.

**Urinary system.**—The output of urine is low in people with profound hypothermia. It probably results from the reduction in renal blood flow.

**Nervous system.**—Cooling depresses the nervous system. The higher cerebral functions are the first to be affected. Thus small falls in body temperature lead to loss of judgement and critical ability. This is very dangerous for people lost in arctic conditions as the loss of rational behaviour with a fall in body temperature may jeopardize their chances of survival.

As the central body temperature falls, there is gradual depression of all cerebral and spinal activity. Consciousness is lost and the reflexes become sluggish and difficult to elicit. Temperature regulation and ventilation eventually cease. The effects of hypothermia are summarized in Fig. 15.

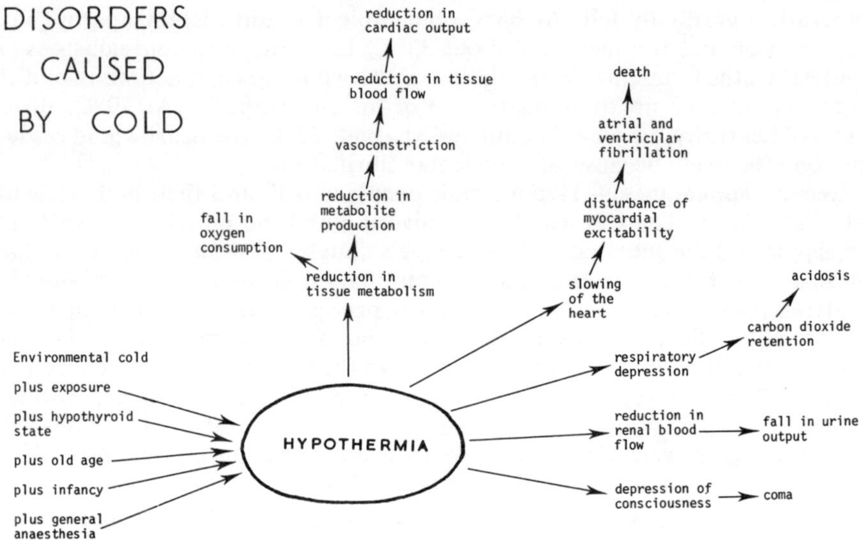


FIG. 15.—The effects of hypothermia.

## CAUSES

**Cold environmental conditions.**— Man's ability to withstand cold depends to some extent on his physiological thermoregulation but also on his ability to use clothes and shelter to diminish heat loss. Naked, unprotected man cannot survive the rigors of an arctic climate. Factors which facilitate heat loss include a low environmental temperature, rapid air movement (wind), a water rather than an air environment, body movement which disturbs the warm micro-climate of the body, clothes which provide poor insulation (especially wet clothes), and inadequate shelter. Many of these factors may affect climbers marooned in mountains or sailors shipwrecked at sea.

**Hypothyroidism.**— Patients with an inadequate level of circulating thyroid hormones experience a fall in their basal metabolic rate. The reduction in heat production which is often associated with lack of voluntary activity usually results in a fall in body temperature.

**Hypothermia in old age.**— Old people living alone in poorly heated rooms on inadequate diets are likely to become hypothermic. This may be because temperature regulation is less efficient in the old, but the combined effects of inadequate clothing, inactivity, a cold environment, the lack of the specific dynamic action of food and a degree of hypothyroidism are probably more important.

**Hypothermia in infancy.**— Temperature regulation in infancy is poor. This is because infants have a relatively large surface to volume ratio which facilitates heat loss, an inability to shiver and very little subcutaneous fat. In addition, the temperature regulating mechanism seems less sensitive in the infant than in the adult. Most of its heat production in response to cold stress seems to come from brown fat. This fat, rich in mitochondria, is found in the neck region and between the scapulae of infants. In response to cooling, sympathetic adrenergic nerves to brown fat are reflexly excited. This leads to increased metabolism of the fat and the generation of heat.

Despite the brown fat mechanism, infants are not well adapted to withstand cold. This is especially true of premature infants who are practically poikilothermic. Thus infants left in cold bedrooms or outside the house in perambulators in the mistaken belief that they benefit from fresh air are likely to become hypothermic.

**General anaesthesia.**—During general anaesthesia, the temperature regulating centre does not work adequately and hypothermia results unless the environment is warm.

**Induced hypothermia.**— In the early days of heart surgery, hypothermia was sometimes induced to lower the metabolic needs of the body. This allowed the surgeon to increase the time he could stop the heart and interrupt the circulation to the brain while he carried out his surgery on the heart. It is not needed much nowadays because very reliable blood pumping and oxygenation systems have been devised for bypassing the heart and lungs. Cooling was achieved by circulating cold brine through tubes in a blanket enveloping the patient or by packing him in ice. It was necessary to paralyse the skeletal muscles with curare to obtain a satisfactory fall in body temperature; otherwise the shivering induced by the cold made a fall difficult to achieve.

Science fiction writers and morticians in the United States of America have wondered about and played with the idea of reducing body temperature to a level where animation is suspended. The cooling down aspect of the problem is relatively easy. Animals and people are readily frozen and stored at sub-zero temperatures where they keep indefinitely. The main difficulties are experienced in the re-warming phase of the operation. This has not yet been successful for large animals. However, small animals such as hamsters and rats have been cooled to sub-zero temperatures and rewarmed again to a viable state. Before cooling they are given an anti-freeze solution, and then the cooling procedure is executed rapidly. Ice crystals do not form under these conditions and warming back to life can be accomplished by microwave diathermy.

Human tissues such as cartilage cells which may be required for grafting purposes can also be kept indefinitely at sub-zero temperatures. Again it is necessary to substitute anti-freeze for some of the intracellular fluid before cooling, to prevent the formation of ice crystals. At present many tissues and organs, e.g. kidneys for grafting, cannot be preserved in this way and must be prevented from freezing.

## INVESTIGATIONS

### Assessment of the Environment

Methods similar to those described in the investigation of disorders caused by heat are used. Again air temperature alone is a poor index of the cooling power of the environment. Wind speed is a very important factor. In terms of comfort an air temperature of —60°C with no wind feels no colder than an air temperature of 0°C with a wind velocity of 7 metres/second (about 16 miles per hour). In a cold environment, the atmospheric humidity is quite unimportant in terms of thermal comfort because sweating plays little part in thermoregulation at these temperatures. Direct and reflected solar radiation are important measurements; heating by direct solar radiation is an important source of heat gain in arctic conditions because of the clear air and because reflected radiation from the snow can be considerable.

**Assessment of clothing**

The thermal insulation afforded by clothes can be measured and expressed in terms of the unit "*clo*". The clo is the amount of thermal insulation necessary to maintain comfort of a resting subject in a room at 21°C with an air speed of 1 metre/second and a humidity of less than 50 per cent. It is a rough working unit and there are a number of difficulties in using it in a meaningful way.

## Treatment

### (a) Prevention of Hypothermia

For people exposed to severe heat-losing conditions protection has to be sought in behavioural rather than in physiological adaptation. There are some racial exceptions to this. The Australian aborigine sleeps near nude in the intense cold of the Australian desert night. He adapts to this by letting his body temperature fall so that the skin-to-air temperature gradient and therefore heat loss is reduced. The heat deficit is made good by morning exercise. Another interesting adaptation is that seen in the Korean diving women. They swim in extremely cold water for long periods. They seem to be able to increase their heat production by non-shivering thermogenesis and, on exposure to cold, the insulating properties of their skin and subcutaneous tissues increases more than in ordinary people.

For ordinary people there are fairly simple rules for survival in heat-losing situations.

**Good protective clothing should be worn.**—This should trap warm air in still layers round the body but still permit water vapour to escape out. In cold water all clothes should be kept on. Even if they fill with water, they will trap a layer of warm water around the body and reduce the temperature gradient between the skin and its surroundings.

**Good shelter should be sought.**—The importance of wind velocity in cooling has been stressed previously. An attempt to get out of the wind is therefore important. A hole dug in the snow affords good protection. Not only will it keep away wind but it will form a very good thermal insulator. The tiny air bubbles in compact snow make it a very poor heat conductor.

**Effective surface area should be reduced.**—By curling up into a ball, a subject reduces the surface area from which heat may be lost. Further, if several people are involved they should huddle together to reduce their collective surface area like puppies in a litter.

**Bodily movements should be reduced.**—Though exercise, by increasing metabolism, increases heat production, it should be avoided in severe heat-losing conditions. This is because exercise increases heat loss by disturbing the microclimate around the body as well as squandering the energy reserves of the body. When a party gets lost in the mountains the survivors are usually those who stay in a sheltered place quietly until they are found and the casualties are those who attempt to find their own way down. Similarly, in very cold water it is advisable to stay still rather than swim around. The disturbance of the warmed layer of water surrounding the body by swimming movements greatly increases heat loss and leads to earlier hypothermia.

**Food intake should be high.**— The energy cost of living in severe heat-losing conditions is high, e.g. double that in a thermoneutral environment. Fat in the form of butter with its high energy content should be a staple ingredient of the diet.

**Alcohol should be avoided.**— By causing cutaneous vasodilatation, alcohol may give the sensation of warmth but it facilitates heat loss through the skin.

### (b) Rewarming in Hypothermia

In someone who has become hypothermic relatively quickly, the most effective treatment is to raise his body temperature quickly by the application of external warmth. Defibrillation with electric shocks may be necessary if ventricular fibrillation occurs.

In elderly patients with spontaneous hypothermia, more gradual rewarming is recommended. In these patients it is considered that increasing the peripheral circulation too rapidly would throw an intolerable strain on the heart. In this way the cerebral and coronary circulations might be jeopardized. Recovery from spontaneous hypothermia is still rare despite treatment, so further investigation is needed.

# 14. BAROMETRIC PRESSURE

## DISORDERS CAUSED BY BREATHING AT HIGH PRESSURE

### DEFINITION

MAN IS WELL adapted to breathe the atmospheric gas pressures found at sea level. If, however, he has to work under water where gas pressure rises by one atmosphere every 10 metres, certain problems are experienced during descent and ascent. These are seen in men employed in the various forms of diving and in underwater tunnels and caissons from which water is excluded by compressed air. Breathing at high pressure is not a problem in submarines where the air is pressurized to sea level values. However, it is a problem for people who have to escape from a disabled submarine lying on the bottom of the sea.

### EFFECTS

To understand the effect of submersion on man it is useful to consider the information given in Table II. Below sea level the pressure rises by one atmosphere for every 10 metres of depth so that at a depth of 90 metres air has to be breathed at a pressure of 10 atmospheres. Because the density of air is much lower than that of water, distances above sea level have relatively little effect on pressure. Whereas pressure is doubled 10 metres below sea level, it is necessary to ascend 6000 metres to halve the pressure.

The problems associated with breathing at high pressure can be divided into those experienced during descent and those experienced during ascent.

TABLE II

Effect of height above (+) and depth below (—) sea level on gas pressures and volumes. The functional residual capacity (F.R.C.) is the volume of gas in the respiratory tract at the end of a normal expiration at rest.

| Depth | Pressure (Atmospheres) | Volume Occupied by Sea Level F.R.C. (litres) | $P_{N_2}$ mm Hg | $N_2$ in Solution in Body (litres) | $P_{O_2}$ mm Hg | $P_{CO_2}$ mm Hg |
|---|---|---|---|---|---|---|
| + 12000 m | 0.25 | 12 | 150 | 0.25 | 37 | 0.05 |
| + 6000 m | 0.5 | 6 | 300 | 0.5 | 75 | 0.1 |
| Sea Level | 1 | 3 | 600 | 1 | 150 | 0.2 |
| — 30 m | 4 | 0.75 | 2400 | 4 | 600 | 0.8 |
| — 60 m | 7 | 0.43 | 4200 | 7 | 1050 | 1.4 |
| — 90 m | 10 | 0.30 | 6000 | 10 | 1500 | 2.0 |
| — 120 m | 13 | 0.23 | 7800 | 13 | 1950 | 2.6 |

**During Descent**

**(a) Due to the decrease in gas volumes.**—When a diver descends 10 metres the pressure of the air in his chest doubles. From Boyle's law ($P_1V_1 = P_2V_2$) it can be seen that this reduces the gas volume in his lungs to half. Divers thus experience what is called "*the squeeze*", a feeling of acute compression of the chest. The diver attempts to combat this by inhaling air rapidly as he descends. If he descends with the chest in a position of full inspiration and the glottis closed, the chest will cave in at a depth of about 30 metres. If the pharyngotympanic tubes are blocked by inflammatory or other causes, the reduction in the gas volume in the middle ear leads to rupture of the tympanic membranes. Pain may also be experienced in cranial sinuses that are not in free communication with the nasal passages.

To eliminate $CO_2$ satisfactorily from the lungs, the alveoli must be ventilated with an adequate volume of gas. At sea level, this is accomplished at rest with a total ventilation of about 6 litres/min. At a depth of 30 metres the increase in pressure reduces volume to one-quarter of its sea level value; it is therefore necessary to pump 24 litres of air/min at sea level to provide a diver at 30 metres depth with 6 litres/min. The fact that the air is made more dense by compression does not improve its ability to wash out $CO_2$.

**(b) Due to the increase in gas viscosity.**—As the air is compressed it becomes more viscous. This makes the work of breathing at depth considerably greater. The maximum breathing capacity is reduced by about 50 per cent at a depth of 30 metres and this reduces the exercise tolerance of the individual considerably.

**(c) Due to the rise in $P_{N_2}$ in inspired air.**—At sea level with a $P_{N_2}$ of about 600 mm Hg in the alveoli, the amount of $N_2$ in solution in the body is about 1 litre. If the $P_{N_2}$ is doubled, the amount of $N_2$ going into solution in the body is also doubled. $N_2$ is not very soluble in water but is more soluble in fat, so much of the additional $N_2$ goes into solution in fatty tissues including the central nervous system.

$N_2$ is a relatively inert gas and at a pressure of one atmosphere exerts no obvious biological effect. However, at high pressure, when more goes into solution it produces depressant effects on the brain like a general anaesthetic such as alcohol. This is known as *nitrogen narcosis.* At a depth of about 30 metres where about 4 litres of nitrogen are dissolved in the body this might manifest itself as undue joviality. At 60 metres the ability to control muscles is greatly reduced and at 90 metres the diver would become useless as a worker. Unconsciousness would occur at about 120 metres.

On exposure to a high nitrogen pressure the additional nitrogen does not go into solution immediately. It takes up to about one hour for equilibrium to be reached. For this reason, nitrogen narcosis is not seen if the exposure to high pressures is suitably brief.

**(d) Due to the rise in $P_{O_2}$ in inspired air.**—Breathing oxygen at high (e.g. several atmospheres) pressure can cause two types of disturbance. It may affect the central nervous system to cause convulsions or it may cause inflammatory changes in the lung giving a pneumonia-like condition. For divers, the central nervous system disturbances are the only real worry. As prolonged exposure to high oxygen pressures are required to cause pulmonary damage, the onset of convulsions does not permit sufficient time for pulmonary damage to develop.

Oxygen toxicity is not usually seen in divers breathing air. As will be seen from Table II, a diver needs to work at a depth of about 40 metres to experience an oxygen pressure of about 750 mm Hg, the $Po_2$ of 100 per cent oxygen at sea level. Prolonged work at this depth is not usually carried out by divers. As convulsions are not seen in man unless oxygen is breathed at pressures greater than one atmosphere, oxygen toxicity is usually not seen in divers using air.

However, it may be seen when divers use oxygen-rich mixtures or when oxygen is used during decompression to aid the elimination of inert gas. It may also be seen in patients exposed to hyperbaric oxygen in the treatment of conditions where oxygen transport is impaired.

*Central nervous system.*—Oxygen toxicity may cause nausea, irritability and feelings of pins and needles (paraesthesiae). Muscular twitching culminating in convulsions eventually supervene. Such convulsions are, of course, extremely dangerous to people enclosed in diving apparatus. The probability of convulsions appearing depends on both the oxygen pressure breathed and the duration of the exposure (Fig. 16). It is also increased by exercise during the exposure and by adding carbon dioxide to the inspired gas.

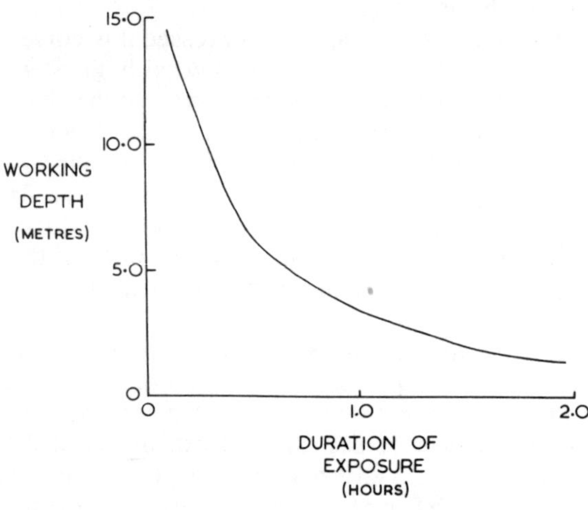


FIG. 16.—Safety curve for divers breathing pure oxygen under water. Convulsions are liable to develop in conditions to the right of the curve. The conditions to the left of the curve are safe.

The mechanism of the convulsions is not understood. They have been attributed at various times to inactivation of oxidative enzymes, reduction in cerebral blood flow and oxidation of the tissues by oxidizing radicals formed at high oxygen pressures.

*The lungs.*—Animals made to breathe 100 per cent oxygen at sea level for periods greater than 24 hours, develop a pulmonary inflammation and oedema in the alveoli and bronchial mucosa. It is thought that the high oxygen pressure results in the formation of free oxidizing radicals which oxidize or burn the pulmonary tissues.

It has not been described in man, probably because man has not been made to breathe 100 per cent oxygen for very long periods. It is known that breathing 45 per cent oxygen for prolonged periods has no adverse effects on the lungs.

*The blood.*—Just as the breathing of low oxygen pressures leads to the release of erythropoietin and the stimulation of bone marrow erythropoiesis, breathing at high $Po_2$ depresses erythropoiesis and the red cell count.

**(e) Due to the rise in $Pco_2$ in inspired air.**—It can be seen from Table II that $CO_2$ forms such a small percentage of air that even at extreme depths the $Pco_2$ is of negligible physiological significance. However, the increasing difficulty of providing adequate ventilation of the alveoli at depth may result in $CO_2$ retention during the dive.

**During Ascent**

**(a) Due to the increase in gas volumes.**—When a diver ascends from a depth of ten metres to the surface, the volume occupied by the gas in his lungs doubles. It is therefore essential that he keeps his glottis open as he ascends to allow the increased volume to escape. If he does not, the distension of the lungs may cause them to rupture, allowing air to escape into the pleural cavity (*pneumothorax*). Gas may also be forced as bubbles into the pulmonary blood vessels where they will obstruct the flow of blood (*air embolism*). If the air enters the heart, the heart's action may form in the ventricles a froth whose viscosity is such that the heart is unable to evacuate its chambers.

The expansion of gas in the middle ear will rupture the tympanic membranes if the pharyngo-tympanic tubes are not patent. Severe sinus pain will also be experienced if the openings of these chambers into the nose are not patent. Expansion of gas in the intestines may cause abdominal pain.

**(b) Due to the fall in $P_{N_2}$.**—At sea level about 1 litre of nitrogen is dissolved in the body. About half is dissolved in water and the other half is dissolved in fat which comprises about 15 per cent of the total body weight. As divers descend, approximately one additional litre of $N_2$ goes into solution for each 10 metres of depth. When the diver ascends from 10 metres depth to the surface, therefore, about one litre of nitrogen may come out of solution rather like $CO_2$ bubbles form in a bottle of carbonated drink when the pressure is reduced by removing the cap. However, the formation of actual bubbles in man happens only when decompression is very rapid, i.e. when the $P_{N_2}$ in the lungs is reduced to less than one-third the $P_{N_2}$ in the body. This causes *decompression sickness* and the symptoms are due to the formation of nitrogen bubbles in the tissues, especially the fatty ones. These bubbles may disrupt the tissues involved and cause permanent damage.

The likelihood of decompression sickness occurring depends on the amount of additional nitrogen that goes into solution in the body which in turn depends on the depth and duration of the dive (Fig. 17). At depths less than about 15 metres, a diver may stay submerged more or less indefinitely without risk. Similarly, enormous depths can be tolerated if the exposure lasts for only one or two minutes.

The most common complaint in decompression sickness is of pains in the arms and legs (*the bends*) especially near joints. This may be due to local bubble formation in periarticular fat. Much less commonly, pain behind the sternum and breathlessness (*the chokes*) is experienced. This has been attributed to the formation of gas bubbles in the pulmonary artery blood but submucosal bubbles in the bronchial tree may be the cause. In rare cases damage to the central nervous system may cause paralysis or loss of consciousness and the paralysis may be permanent.

The fall in $Po_2$ and $Pco_2$ during ascent normally gives rise to no difficulty. Oxygen coming out of solution is so small in amount compared with nitrogen as to be of no significance.

The problems that may arise during descent and ascent during diving are summarized in Figs. 18 and 19.

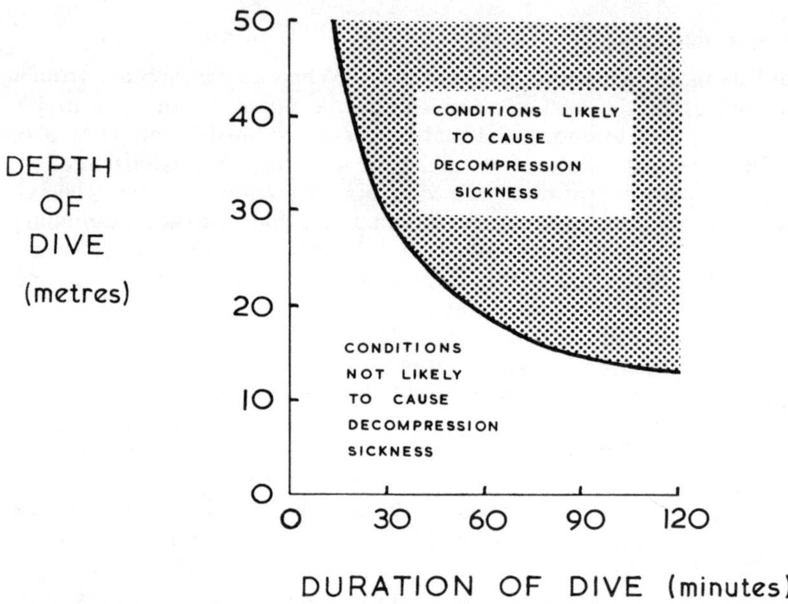


FIG. 17.—The effect of dive duration and dive depth on the likelihood of developing decompression sickness during ascent. Decompression sickness is likely to occur with dives lying to the right of the curve.

## Treatment

To ensure safety for under-water workers several general points should be remembered.

**1. Air passages should be patent.**—During ascent and descent the glottis should be kept open and the middle ears and sinuses should have free communication with the pharynx. In this way the problems of changing gas volumes in closed spaces can be avoided.

**2. Avoid nitrogen narcosis.**—Divers breathing air should not have to work at such depths that their mental ability is impaired by nitrogen narcosis. If very deep work is required, helium should be substituted for nitrogen because it is both less soluble in the body and less narcotic than nitrogen.

**3. Avoid oxygen toxicity.**—Divers breathing oxygen should not work at depths or for durations that might induce the convulsion of oxygen toxicity.

**4. Avoid decompression sickness.**—Divers breathing air who have worked at depths for sufficient time to allow large amounts of nitrogen to go into solution

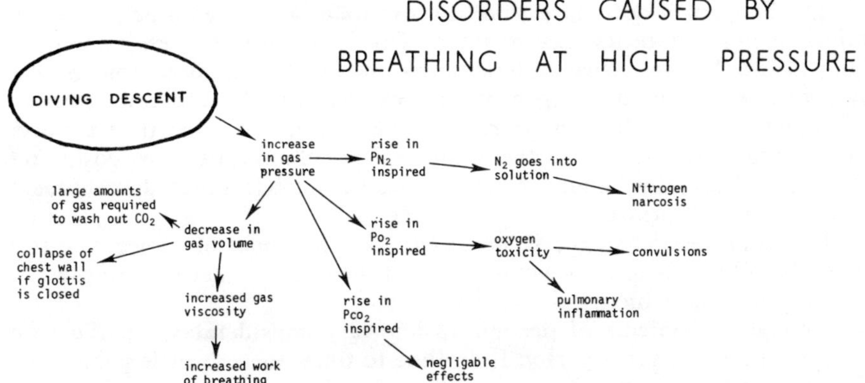


FIG. 18.—Problems that may arise during the descent phase of diving.

should be decompressed slowly. Tables have been constructed to show how to do this safely. The guiding principle is that the $P_{N_2}$ in the lungs must never be less than one-third of the $P_{N_2}$ in the rest of the body. It is found that submersion at 100 metres for 20 minutes requires several hours of gradual decompression, whereas submersion at 50 metres for 20 minutes requires only several minutes of gradual decompression.

Gradual decompression can be carried out by raising the diver by stages to the surface or by bringing him immediately to a decompression chamber at the surface and then decompressing him in relative comfort by stages.

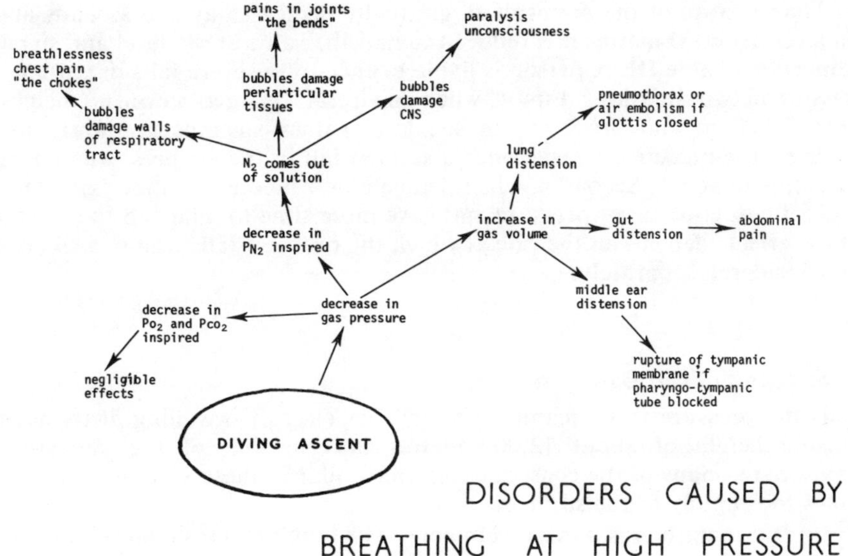


FIG. 19.—Problems that may arise during the ascent phase of diving.

The dangers of decompression sickness are also reduced if helium is substituted for nitrogen in the gas breathed. This is because it is less soluble in the body fluids, therefore fewer bubbles are formed with acute decompression. As a smaller molecule than nitrogen, it diffuses more rapidly and can therefore be eliminated from the body more rapidly. This means that the time for decompression can be shorter. A further advantage of helium is its low viscosity so that the work of breathing helium at high pressure is much less than that of breathing nitrogen at high pressure.

If decompression sickness does occur, the treatment of choice is to recompress the subject as quickly as possible and then recommence decompression in a more gradual fashion.

Though the effects of decompression vary considerably from person to person and in the same person from time to time, there is little good evidence that man adapts to the stresses associated with hyperbaric conditions. Some animals have adapted well to these conditions. Whales may stay submerged for about two hours. They avoid the possibility of decompression sickness and nitrogen narcosis by emptying their lungs before the dive. Though they do not carry an obvious store of oxygen, diving animals can shut off the circulation to their muscles, skin and kidneys during the dive. In this way the oxygen they carry can be reserved for the heart and brain. There is some evidence that this diving reflex occurs in man. However, it has little significance for the professional diver.

## DISORDERS CAUSED BY BREATHING AT LOW PRESSURE

### Definition

The pressure of the atmosphere gradually decreases as one ascends above sea level. At 6000 metres it is reduced to half the value at sea level and so on in proportion (Table II). A person is liable to encounter severe falls in barometric pressure in two situations. Firstly, when an aircraft flying at a cruising height of 6000-12,000 m and pressured to sea level values sustains an injury to its fuselage, the passengers experience a sudden fall in the air pressure and have little time to adapt. Secondly, when people climb mountains they experience a gradual reduction in air pressure and have more time to adapt to the changes. As the effects depend on the rate at which the pressure falls, the two situations are considered separately.

### Effects

#### Sudden Falls in Barometric Pressure

If the pressure in the cabin of a large jet aircraft travelling at its normal cruising height of about 12,000 metres falls to that of the surrounding atmosphere, many of the consequences are similar to those seen in a diver who comes too rapidly to the surface.

**(a) Due to the increase in gas volume.**—The functional residual volume in the lungs will suddenly increase from 3 to 12 litres (Table II). If the glottis is closed this will burst the lungs or cause air embolism. The expansion of gas in the middle ears, sinuses and intestines may well rupture the ear drums and the walls of the alimentary tract. An additional hazard is that the expansion of the air in the cabin creates a strong outward current through the hole in the fuselage which may carry passengers in the vicinity out into space.

**(b) Due to the fall in $P_{N_2}$.**—Three-quarters of the nitrogen in solution in the body comes out of solution and if this is sufficiently rapid, nitrogen bubbles form, causing decompression sickness indistinguishable from that seen in divers.

**(c) Due to the fall in $P_{O_2}$.**—The situation here is somewhat different from that of the diver. At 12,000 metres the barometric pressure is about 150 and the $P_{O_2}$ is about 37 mm Hg. As the alveolar water vapour pressure stays constant at about 47 mm Hg at 37°C and the alveolar $P_{CO_2}$ remains about 40 mm Hg, more than half the alveolar gas pressure would be exerted by these two gases. The highest alveolar $P_{O_2}$ that could be achieved breathing air would be about 1/5 of the remaining 60 mm Hg or so of gas tension, i.e. about 12 mm Hg. This acute hypoxia would desaturate the haemoglobin and result in loss of consciousness in a few seconds. However, breathing 100 per cent oxygen, the remaining 60 mm Hg or so of gas tension could be exerted by $O_2$ and this alveolar $P_{O_2}$ would be sufficient to sustain life until the aircraft had descended to more favourable conditions.

### Gradual Falls in Barometric Pressure

When people climb mountains, the reduction in barometric pressure is so gradual that no ill effects result from decompression. However, problems are met because of the reduction in $P_{O_2}$. No effects are noticed below about 3000 metres. At 4000 metres definite signs of oxygen lack appear. The highest level at which permanent human habitation is found is about 6000 metres. Above 7000 metres, human life cannot be supported unless oxygen is added to the inspired air. Mount Everest, the highest mountain on earth, has a height of just under 10,000 metres.

#### (a) *Mountain Sickness*

This is a combination of symptoms that appear within a day after ascending to a height greater than about 4000 metres. Appetite disappears and there may be nausea and vomiting. Headache and irritability are common and the subjects are troubled with sleeplessness. They complain of breathlessness especially on exertion. This condition usually lasts about a week. Gradually the subjects appear to acclimatize to the new conditions. The cause of mountain sickness is probably the hypoxia and the compensatory mechanisms that it elicits.

The fall in the arterial $P_{O_2}$ stimulates the peripheral chemoreceptors to cause hyperventilation. Though this brings more $O_2$ to the alveoli, it washes out $CO_2$ and causes a respiratory alkalosis. This alkalosis probably contributes to the symptoms of mountain sickness; these symptoms can be alleviated by drugs which correct the alkalosis by blocking the effect of carbonic anhydrase and reducing the blood bicarbonate level.

The adaptation to altitude which coincides with the disappearance of the symptoms takes a week or so to develop. Firstly, there is *an increase in ventilation* over and above that which occurred on moving to altitude. This is thought to be due to an increase in the sensitivity of the respiratory centre to $CO_2$. This is the reverse of the decrease in sensitivity to $CO_2$ seen in patients with chronic respiratory disease. The increase in ventilation persists for some days after the mountaineer has returned to sea level. Secondly, the kidneys help to overcome the effects of the respiratory alkalosis due to the hyperventilation. Bicarbonate is excreted to give *an alkaline urine.* As the plasma bicarbonate falls, the ratio of carbonic acid to bicarbonate in the blood is returned towards its normal value. The hydrogen ion concentration in the blood is thus corrected.

Thirdly, there is *an increase in the red cell* count which increases the oxygen-carrying power of the blood. This is due to the release of erythropoietin from the kidneys in response to the low $Po_2$ in renal tissues. Though the erythropoietin is released almost immediately, some weeks are required for the red cell count to rise.

With these adaptive changes, humans can live quite effectively at heights of 6000 metres or more. With their low arterial $Po_2$ and polycythaemia they can do quite hard work in mines, reproduce and generally enjoy life. They tend to develop pulmonary hypertension and suffer from right heart failure. This may result from the constriction of the pulmonary blood vessels by the low $Po_2$ in alveolar air.

### (b) *Pulmonary Oedema*

With the increase in the popularity of mountain climbing, death from pulmonary oedema in climbers is becoming more common. It is usually seen in healthy young men within two or three days of moving to a height of more than 3,600 metres. It is often precipitated by swimming or exposure to cold.

It starts rather like ordinary mountain sickness with loss of appetite, nausea, gastro-intestinal upsets, insomnia, muscular weakness, headache and irritability. However, a condition rather like acute pneumonia then supervenes, but unlike pneumonia, does not respond to antibiotics.

A cough develops and breathing becomes periodic. Mental confusion, hallucinations, convulsions and coma may follow. It all happens very rapidly and is often fatal. At postmortem, cerebral and pulmonary oedema are found.

The mechanism of this pulmonary oedema at altitude is not understood.

The effects of breathing at low pressures are summarized in Fig. 20.

## TREATMENT

### Sudden Reduction in Barometric Pressure

When a large hole appears in the fuselage of a high-flying aircraft, the pilot should try to reduce altitude as quickly as possible to one where the $Po_2$ is sufficient to support life. This has the added advantage of recompressing the passengers and thus preventing decompression sickness. The passengers should also be given oxygen to breathe through oxygen masks. At 10,000 metres consciousness would be lost in about 40 seconds breathing air, whereas consciousness is maintained indefinitely breathing pure oxygen at heights under 15,000 metres. The risk of death is diminished if the passengers are wearing seat belts at the time of fuselage damage; there is less danger of being blown out through the hole by the expanding air.

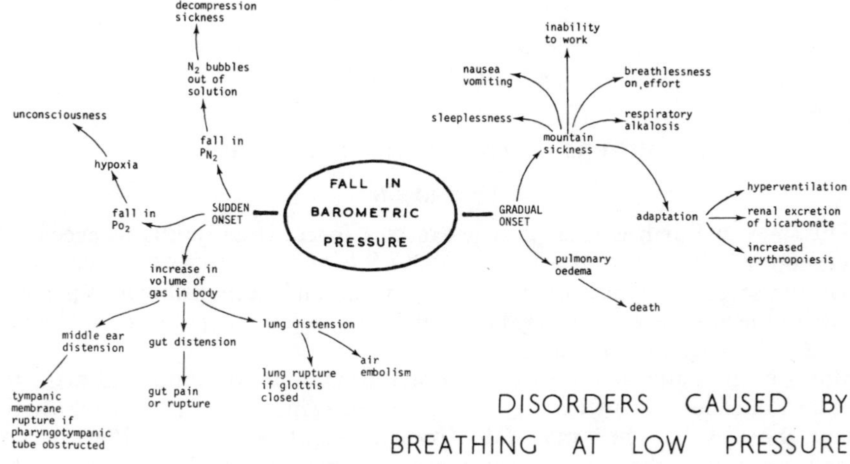


FIG. 20.—The effects of sudden and gradual reductions in barometric pressure.

## Gradual Reduction in Barometric Pressure

### (a) *Mountain Sickness*

Firstly, it should be realized that normal people will be incapacitated by mountain sickness for about a week after they move to a high altitude and hard work should not be expected of them during this time. Attractive food should be offered because this makes the nausea and loss of appetite a smaller problem. Early mountain climbers added to their own problems by having a diet of pemmican, dried fat and meat in a compressed form. Hypnotic drugs are valuable to obtain sleep. There is evidence that correction of the respiratory alkalosis can help alleviate symptoms. Drugs antagonizing carbonic anhydrase have been used for this, as explained above.

### (b) *Pulmonary Oedema*

Mountain climbers should be aware that pulmonary oedema can affect healthy and fit young men and should be taught to recognize the first symptoms and precipitating causes.

If it has developed, the climber should be brought down again to a lower altitude if conditions permit. Oxygen to breathe is a help, but the mechanism of this help is obscure. Artificial ventilation with positive pressure helps reduce the formation of oedema fluid by decreasing the hydrostatic gradient across the pulmonary capillaries. Drugs which stimulate cardiac contraction do not appear to help the condition which is indirect evidence that the pulmonary oedema is not due to left heart failure. Diuretics may help by lowering the circulating blood volume.

# 15. GRAVITY

## DISORDERS CAUSED BY NORMAL GRAVITY

### Definition

THE MASS OF Earth is such that it exerts a force which tends to accelerate objects towards its centre at a rate of about 9.8 metres per second per second (32 feet per second$^2$). This is the force of gravity and it is responsible for a person's weight; in the abscence of gravity he is weightless and if the gravitational force is doubled, his weight is doubled.

Man's body is adapted to meet the gravitational stress found on Earth; the bones and muscles are sufficiently strong to withstand them. If we exclude the damage that results when man falls from a height, accelerates towards and makes impact with the ground, the main problem imposed by gravity on the body is connected with the circulation. On standing up, the blood tends to gravitate to the vessels in the lower part of the body and the fall in venous return to the heart reduces cardiac output and arterial pressure. Normally, reflex and other measures compensate for this to ensure an adequate arterial pressure to perfuse the brain. Failure of compensation causes *postural (orthostatic) hypotension.*

### Effects

The effects of posture on the circulation are shown in Fig. 21. When the subject is horizontal, most of his blood is at about the same level as the heart, and the pressures in vessels in the head end of the body are similar to those in corresponding vessels in the foot end. The perfusion pressure (arteriovenous pressure difference) for the feet is the same as for the top of the head, i.e. about 90 mm Hg. The capillary hydrostatic pressure is about equal to the colloid osmotic pressure of blood so that there is little net movement of fluid across the capillary wall.

When the person stands up, all this is changed. Because of gravity (or the weight of blood), the hydrostatic pressure of blood tends to rise in the feet and fall in the head. In the feet the pressure in the arteries, capillaries and veins rises by the hydrostatic equivalent of the column of blood below the heart, i.e. about 100 mm Hg. It should be noted that this does not increase the perfusion pressure for the feet; the artery and vein pressures rise by the same amount. The massive venous pressure (around 100 mm Hg) aggravates *varicose veins* in a person suffering from the condition. The increase in pressure tends to distend the vessels of the lower extremities so that a larger fraction of total blood volume is accommodated there. In addition, the capillary hydrostatic pressure (125 mm Hg) now greatly exceeds the colloid osmotic pressure (25 mm Hg) so that there is net movement of fluid out of the blood and the blood volume shrinks. Both these factors tend to reduce venous return to the heart; the diastolic filling pressure, cardiac output and arterial pressure fall and, if these changes are sufficiently severe, circulatory failure results.

The fall in pressure is sensed by the arterial baroreceptors and this results in a reflex increase in heart rate and force, arteriolar tone and venous tone. This tends to increase arterial pressure and limit the pooling of blood in the lower extremities. The effects are mediated through the sympathetic nervous system and, if the system does not function well, circulatory failure may result. Another factor tending to limit pooling of blood is the muscle pump; contractions of the leg muscles compress the leg veins. This drives blood towards the heart because venous valves prevent retrograde flow.

The effects of the erect posture on the circulation *above* the heart is somewhat different (Fig. 21). At the top of the head, the arterial pressure falls by the hydrostatic equivalent of the column of blood above the heart, i.e. by about 40 mm Hg. There is no corresponding fall in the venous pressure because the veins collapse and a negative pressure cannot be transmitted along a collapsed

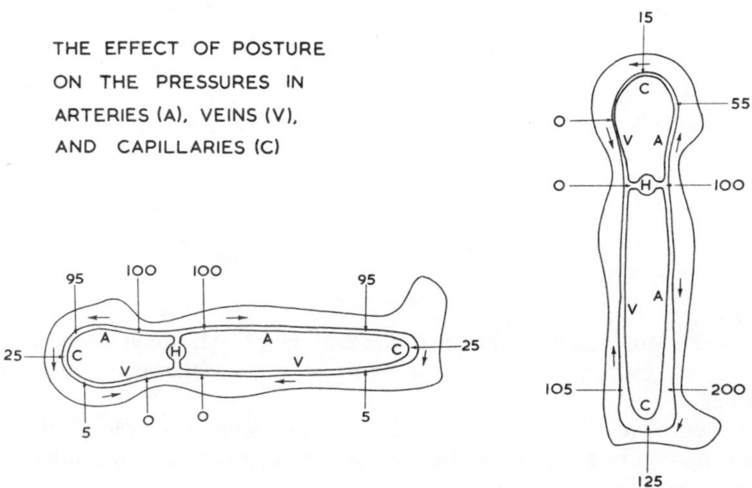


FIG. 21.—The effect of changing from the recumbent to the standing position on the pressures in arteries (A), veins (V) and capillaries (C). In the lower extremities, the perfusion pressures are not altered but the distending (transmural) pressure is increased in all vessels. Above the heart, the perfusion pressure falls.

tube; the venous pressure is therefore about zero. Thus the arteriovenous pressure gradient perfusing the top of the head is about halved; this tends to reduce blood flow. The hydrostatic pressure within the small blood vessels falls; this tends to cause collapse of the vessels and a net movement of tissue fluid into the capillaries. When the hand is elevated in the erect posture, the blood flow reduction is even more marked and prolonged intense physical work by the hand in this position is impossible.

Brain blood flow is maintained by autoregulation in which the tendency for the carbon dioxide tension to rise causes local vasodilatation.

The brain vessels are protected to some extent against mechanical damage due to changing effects of gravity by the cerebrospinal fluid; the fall in intravascular pressure is matched by a corresponding fall in CSF pressure so that the pressure gradients across the vessel walls (transmural pressures) are little changed.

In summary, when changing from the lying to the standing position, gravity causes pooling of blood in the lower extremities, an increase in capillary filtration in the lower extremities and a fall in the tissue perfusion pressure above the heart. All these factors tend to limit blood flow to the brain and, if compensation is less than adequate, cause dizziness, light-headedness and in extreme cases *fainting*.

Fainting can occur in healthy people who have to stand erect and still for long periods, e.g. soldiers on parade. It is usually heralded by feelings of dizziness, blurring of vision, nausea, salivation and sweating. During this time, the heart rate rises and peripheral resistance is reflexly increased in an attempt to maintain blood pressure. At the time of fainting, the blood pressure suddenly drops. This is due partly to a sudden fall in peripheral resistance as cholinergic vasodilator fibres dilate muscle blood vessels. It is also due to vagal inhibition of the heart. Though the way in which these effects are triggered by the falling cerebral blood flow is not known, the fainting is referred to as a *vasovagal syndrome* because both vagal inhibition of the heart and vascular dilatation contribute to it. In the absence of an adequate arterial pressure, no amount of vasodilatation in the brain can maintain an adequate blood flow.

### Causes

Postural hypotension can occur in normal people who have to stand for long periods in conditions which favour peripheral vasodilatation, e.g. a warm environment.

Some degree of postural hypotension is seen in most people after they have been confined to bed for some time; with lack of practice, circulatory reflexes seem to become less efficient.

It also occurs when the activity of the sympathetic nervous system is depressed, either by disease or by drugs. Conditions predisposing to peripheral circulatory failure also predispose to postural hypotension.

### Treatment

The long-term treatment of postural hypotension is to avoid or remove the causes which predispose to it. However, the immediate and most important treatment is to remove gravitational stress from the circulation, i.e. let the patient be horizontal. When a person faints or is about to faint, serious damage may be done to his brain by lack of oxygen if he is held or supported in an upright position. For this reason fainting, while being supported in a dentist's chair, is a serious hazard of dental anaesthesia.

Raising the legs of the patient while he is recumbent is useful because it shifts blood from the legs to the central circulation.

## DISORDERS CAUSED BY INCREASED GRAVITY

### Definition

Increased gravitational forces may be experienced by man when he is exposed to severe acceleration or deceleration. It is a problem for passengers in space vehicles, aircraft and fast cars.

### Effects

**Increase in weight.**— A man who weighs 70 kg when subjected to the earth's gravitational force (1 g) weighs 700 kg when subjected to the 10 g force which may be experienced shortly after rocket launch. The body muscles are not powerful enough to deal with these weights so the ability to make manipulations is lost. The increase in weight caused by increased g forces may well limit man's ability to visit the larger planets.

**Circulatory problems.**— The circulatory problems depend on the orientation of the man relative to the gravitational field. If the acceleration occurs in a head-first position (positive g), blood tends to pool in the lower extremities and a "black out" occurs due to reduction in cerebral blood flow. If acceleration occurs in a feet-first position (negative g), cerebral blood flow is maintained and the intracranial vessels do not distend or burst because the CSF pressure rises simultaneously and correspondingly with the intravascular pressure. However, the extracranial tissues are not so protected and temporary blindness ("red out") may occur because of the congestion of blood vessels in and about the eye with blood.

Few circulatory problems are encountered if acceleration occurs with the subject lying transversely to the gravitational force.

**Tissue distortion.**— The increased weight can cause considerable distortion of elastic tissues. This is less severe when the tissues are surrounded by material whose mass per unit volume is similar to that of the tissue. However, in the heart, which is surrounded by air-filled lungs, gross distortion may occur which impedes the heart action.

### Causes

Large g forces are encountered only when man is violently accelerated or decelerated and tend to be of short duration. Amongst the most common and destructive are those seen when a man in a fast moving car is suddenly decelerated by crashing into an oncoming truck; the tissue distortion is usually fatal. Milder g forces of greater duration may be encountered in aircraft and rockets and are the cause of many of the problems of manned flight.

### Treatment

There are several ways to mitigate the problems of increased g. The subject may lie transversely to the plane of gravity so that blood does not tend to pool in his head or feet. He may wear an anti-g suit, i.e. one that can be inflated to provide counter pressure to his lower extremities if he is exposed to positive g.

Finally, he may travel in a tank of water; this supports the circulation by providing counter pressure which is proportional to the g force and facilitates muscular movement by reducing the apparent weight of the tissues by the principle of Archimedes.

## DISORDERS DUE TO DECREASED GRAVITY

### Definition

Man is exposed to a decrease in gravitational force when he leaves the earth's gravitational field and when in orbital flight, where the centrifugal force due to the orbital speed just balances earth's gravitational pull. Such exposures result in weightlessness and, while rare, are of considerable physiological interest.

### Effects

**Bone weakness.**— The structure of bone is regulated to some extent by the mechanical stresses it experiences. Increased stress results in increased bone deposition. Prolonged weightlessness has the opposite effect; demineralization can occur to cause bone weakness and the increased urinary output of calcium increases the risk of calcium deposition as stones in the urinary tract.

**Muscle weakness.**— Muscle power and size increase when muscles are given additional work to do and they waste and become weaker if they are not used. In prolonged weightlessness, the decreased work required of the muscles can result in muscular wasting and weakness and a reduction in exercise tolerance. Heart muscle may be affected similarly; this makes heart failure a hazard on returning to the gravitational status of earth.

**Disturbance of labyrinthine function.**— The otolith organs in the saccule and utricle normally provide information about the position of the head in space relative to the gravitational field. They will not work normally in a weightless state. This may account for the motion sickness that is common on space flights.

**Reduced efficiency of certain circulatory reflexes.**—A condition like postural hypotension is seen in astronauts when they return to earth after a period of weightlessness. The exact mechanism of this is not clear but it resembles the postural hypotension seen in patients who have been confined to bed for long periods. The reflexes that compensate for the effects of gravity on the circulation seem to require practice to maintain their efficiency.

It is of interest that many of the effects of weightlessness are similar to those of being confined to bed.

### Treatment

To prevent bone and muscle atrophy, astronauts should exercise their muscles as much as possible during flight. For prolonged stays in space it may be possible to overcome the problem of weightlessness by having the vehicle rotate at a rate sufficient to generate a suitable g force.

# 16. DISORDERS CAUSED BY LACK OF FOOD

## Definition

MAN DERIVES his energy from the controlled oxidation of certain chemical compounds (food) which he ingests regularly. Though energy is being expended continually to perform the mechanical, electrical, chemical and osmotic work necessary for bodily function, the intake of energy can be periodic. Man is able to store large quantities of energy in the fat depots of his body; this allows him the freedom to do things other than hunt for, and eat food. He can survive for quite long periods without food intake by metabolizing first his energy stores and later the other tissues of the body.

Thus when food intake stops, man *adapts* to the new conditions by altering his source of energy and modifying his energy expenditure. However, he can reduce his energy expenditure to a limited extent only. Complete starvation eventually leads to *death* when the tissue reserves are so depleted that it is no longer possible to provide sufficent energy to support life from the catabolism of body tissues. If an adequate water intake is provided, man can survive for 1-3 months without food intake, depending on the prevailing conditions. However, if there is no water intake, death occurs within about two weeks from water depletion.

## Effects

### Adaptation

Man survives periods of starvation firstly by using his body tissues to provide energy and secondly by reducing his energy expenditure.

(*a*) *Energy production by oxidation of body tissues.*—As described in Chapter 31, the body uses successively its stores of carbohydrates, fats and proteins to provide energy. Body weight falls by about 0.5 kg/day but this figure depends on many factors including the amount of work performed. Though weight loss occurs in all organs, the brain, spinal cord, skeleton and plasma proteins are spared, compared with the other organs such as the muscles, gut, liver, skin, kidney and heart. The smooth muscle coat in the wall of the alimentary tract becomes very thin so that the contents are easily seen through the wall. The bulk of the cardiac muscle may be reduced by half so that heart failure may result if the circulation is loaded with intravenous fluids. The skeletal muscle is reduced in bulk but that which remains retains its contractile power.

Eventually the patient looks as if he is composed of skin and bones, the outlines of the skeleton being seen clearly beneath the skin. The eyes tend to be sunken and the skin is thrown into loose folds. Oedema is common; it may be related to a fall in the tissue pressure as the subcutaneous tissue is metabolized. It cannot be explained entirely by a fall in the level of plasma proteins.

(*b*) *Reduction in energy expenditure.*—In severe starvation, not only is the amount of spontaneous activity reduced, but the activity that remains is performed at a reduced metabolic cost. As the patient's weight reduces, less work is required to maintain the remaining tissues and move the body.

The reduction in spontaneous activity is responsible for most of the energy saving. The starving are listless and apathetic. Despite the fact that there is little reduction in their physical fitness in the early stages as measured by the Harvard Step Tests, starving people tend to avoid unnecessary tasks and carry out essential ones with the minimum expenditure of energy. Psychologically there is a disinterest in the problems of others and an unwillingness to do work on their behalf. Sexual interests are lost quite early in starvation.

### Death

Starvation results in death after 1-3 months. The speed with which starvation leads to death varies with many factors. Death tends to occur early if the environmental temperature is low or if the work load is high because these increase energy expenditure. Young animals tend to die before older animals, perhaps because they carry smaller fat depots. Men tend to die before women during starvation; this again is probably related to the greater fat content of women. Thin people tend to die before fat people, for the same reason.

The cause of death in starvation is not quite certain. It may be that the utilization of tissue protein for energy production finally causes irreparable damage to cellular mechanisms. Another possibility is that the energy production is eventually too small to support life. It has also been suggested that the reduction in metabolism results in a fatal fall in body temperature.

Death in the average person usually occurs when about 40 per cent of the initial body weight has been lost. However, this percentage must depend on the initial state of the fat depots.

## Causes

Acute deprivation of food results usually from a natural or other disaster which separates the subject from food sources. Thus, getting lost in mountains or marooned on life rafts, are possible causes. Voluntary food deprivation may be seen in hunger strikers or in patients with the condition anorexia nervosa. Conditions which interfere with swallowing may also cause acute deprivation of food. Much more common is the semi-starvation due to poverty which exists in many parts of the world.

## Treatment

The treatment of severe starvation involves much more than supplying food. The walls of the alimentary tracts become very thin and the digestive processes are impaired. Giving rough food may cause diarrhoea and death from dehydration. Fat absorption is particularly poor. Skimmed milk, fresh or dried, is the food of choice and it should be given in frequent small feeds. For very weak and apathetic patients, the milk can be given by a tube passed into the stomach. Intravenous feeding is dangerous because the weakened heart may not be able to cope with the increased fluid load.

Good and sympathetic nursing is required because of the mental changes that are seen in starvation. In severe starvation, hunger does not provide an adequate drive to eat.

# 17. DISORDERS CAUSED BY LACK OF WATER

## Definition

THE CELLS OF the body require that the osmolarity of the fluids they contain be maintained within narrow limits. If water intake ceases, the *status quo* is not maintained; water continues to be lost in perspiration, urine etc. and salt is not lost in proportion. The trouble is that man cannot reduce his water output to zero when his water intake is zero. As the body becomes dehydrated, the concentration of solutes in the cells gradually rises so that death occurs from derangement of cellular function within 1-2 weeks of stopping water intake.

Water deprivation occurs whenever water intake is less than the minimal water output (about 1.5 l/day including urine, sweat and respiratory tract loss). The severity of the effects depend on the prevailing conditions. This chapter first describes the effects of complete water deprivation on a healthy adult in a temperate climate and then mentions some factors which may modify the clinical picture.

## Effects

### (a) After 1 day

The subject has depleted his water store by about 1.5 litres. This represents about 2 per cent of his body weight. About 500 ml was lost as insensible perspiration, 500 ml in exhaled air and about 500 ml in the urine. This latter is about the minimum volume (volume obligataire) to which the urine can be concentrated while permitting complete excretion of urinary waste products. Faecal water loss is negligible.

The loss of water concentrates the solutes in extracellular fluid. The increased osmolarity of the blood gives rise to intense thirst and stimulates the osmoreceptors in the hypothalmus. The resulting release of antidiuretic hormone by the posterior pituitary gland causes increased reabsorption of water in the collecting ducts of the kidney and a reduction in urine output.

The increased osmolarity of extracellular fluid draws water from the intracellular to the extracellular compartment.

Because of the mobilization of intracellular water, the fall in extracellular fluid volume is not as great as it would otherwise be. Thus the plasma volume, haematocrit, blood viscosity and arterial pressure are not affected very much in the early stages of water deprivation.

### (b) After 4 days

By this time body water will have been reduced by about 6 litres, causing a reduction in body weight of about 8 per cent. Though the subject is still capable of work and rational thought, he now looks ill and feels weak. Personality changes such as irritability become obvious. As water is shifted from the intracellular to the extracellular compartment, there is little reduction in plasma volume and blood pressure. The subject experiences intense thirst and the mouth is very dry. Salivary secretion has been found to fall in some animals in response to water deprivation and the same may be true for man. About 500 ml of concentrated urine is excreted per day.

### (c) After 7 days

By this stage the subject is very weak and incapable of work. Mentally he is irrational and confused and this state gradually progresses into loss of consciousness. After seven days of water deprivation the body water is reduced by about 10 litres. This is about 14 per cent of the body weight. Death usually occurs when the water loss amounts to 15 per cent of body weight. The cause of death is uncertain but the hypertonicity of the intracellular fluid is thought to interfere with cellular function. This aspect of water deprivation seems of more importance than the reduction of plasma volume and blood pressure.

### (d) Modifying Factors

**Air temperature.**—The sequence of events described above is different if water deprivation occurs in desert conditions where sweating may occur at rates of 10 or more litres per day; death then occurs much more rapidly. Water loss amounting to 15 per cent of body weight would occur in less than two days.

**Renal function.**—The ability of the kidneys to concentrate urine and therefore conserve water is also of importance in water deprivation. Poor renal function would result in an earlier death.

**Drinking sea-water.**—People who are deprived of water by being shipwrecked at sea may be tempted to relieve their thirst by drinking sea-water. This only aggravates the effects of water deprivation. Normal kidneys can only concentrate urine to an osmolarity of about 66 per cent the osmolarity of sea-water. Therefore, sea-water when drunk increases the hypertonicity of extracellular fluid and thus aggravates the intracellular dehydration.

**Age.**—The very young and, to a lesser extent, the very old, show a diminished ability to concentrate urine. Their ability to survive water deprivation is correspondingly reduced.

**Food intake.**—The availability of food and its nature may modify the picture to some extent. This is because food such as animal and plant tissue is about 60 per cent or more water. In addition, water is a metabolic product of the oxidation of food stuffs. However, these favourable points may be countered by the fact that additional urinary output may be required to eliminate the salts contained in the food and the end products of metabolism such as urea.

## Causes

Water deprivation occurs in people lost in deserts or marooned on life rafts at sea. However, it is also a feature in people who are unable to swallow because of obstruction in the upper alimentary tract, loss of consciousness or feebleness.

## Treatment

Water by mouth is the obvious cure for water deprivation. In unconscious people or in those cases where drinking is not possible, water can be given as an intravenous glucose (5 per cent) solution. The glucose is added to make the fluid isotonic with blood; it is eventually metabolized leaving the water. Distilled water given intravenously might cause some haemolysis of the red blood cells.

# 18. DISORDERS CAUSED BY LACK OF SLEEP

## Definition

It is normal to spend about 8 hours (one-third of each day) in sleep. The amount of sleep required by different people seems to vary; children tend to sleep rather more than older people. Severe sleep deprivation is rare, as prolonged wakefulness makes the tendency to fall asleep nearly irresistible. However, factors such as work, prolonged travel, anxiety, pain or discomfort may result in an inability to sleep for long periods. This chapter describes some of the consequences of such lack of sleep.

## Effects

### Sleepiness

Subjectively, lack of sleep causes the sensations of "sleepiness", "tiredness" and "fatigue" which are relieved by the sleep that they invoke. The degree of sleepiness depends on the type of sleep which is lost. It is now recognized that there are two main varieties of sleep. About 80 per cent of sleeping time is characterized by EEG waves of low frequency and high amplitude. The remaining 20 per cent is characterized by rapid eye movements, dreaming, involuntary muscle jerks and EEG waves of high frequency and low amplitude similar to those seen in the alerted state. This type of sleep normally occurs six or seven times for short periods during the course of a night's sleep. It is referred to as REM (rapid eye-ball movement) sleep.

Though REM sleep is associated with apparent restlessness, it appears to be the most valuable part of sleep for the relief of sleepiness. If, under experimental conditions, people are woken up every time their EEG patterns indicate a change from ordinary to REM sleep, the remaining sleep does not seem to relieve the fatigue and sleepiness associated with loss of sleep and the behavioural changes associated with sleep deprivation become evident. Thus, loss of REM sleep can cause the effects of sleep deprivation even though the total time spent in sleep is not greatly reduced.

### Physiological Changes

Though sleep must have considerable biological importance, the physiological purpose of sleep is not really understood. To this extent the physiological consequences of sleep deprivation are not well documented; it may be that the wrong quantities are being measured. If we knew what sleep was supposed to do, we would be in a position to make more apt measurements.

Sleeplessness has not been found to cause consistent changes in the biochemistry of the blood or urine. The normal cyclical changes associated with circadian rhythms persist. The blood glucose level does not fall. Heart rate, blood pressure and ventilation of the lungs are normal. There is some evidence that sensitivity to pain increases but it may be that people who are sleepless and tired are less willing to put up with discomfort. It is interesting, however, that many patients seem to become more aware of their pains and call out their doctors late at night.

### Behavioural Changes

A person deprived of sleep tends to become irritable and cross. Personality tests show that he becomes more hostile and aggressive. Angry letters written late at night are better re-read in the morning before they are posted; things look different in the morning. At times of crisis for politicians, the associated sleeplessness may warp their judgement. Prolonged deprivation of sleep leads to difficulties with perception which may lead to peculiar behaviour. The person may become disoriented in time and make faulty assessments of the passage of time. Reported examples of visual confusion include the refusal by a subject to drink milk which appeared to him to be covered in hairs. Other sleepless subjects have reported seeing changes in size and colour in the familiar objects which surrounded them. Eventually the subject may begin to show some of the signs and symptoms of severe mental illness.

The ability of sleepless persons to perform tasks has received a lot of attention. Surprisingly, moderate sleep deprivation does not greatly impair the ability to carry out short, complex, exciting tasks. Presumably the alerting qualities of such tasks stimulate the cerebral cortex to overcome the effects of sleep loss. The ability to perform routine long dull tasks is progressively impaired with increasing sleeplessness. This impairment is less marked if the subject is carrying out physical exercises or has been given drugs such as amphetamines which stimulate the reticular activating system. The impairment is greater when the subject has been drinking alcohol. These factors are of considerable importance for people working machines or driving when they are tired.

## Causes

People sometimes lose sleep because travel, domestic emergency or the nature of their work limits the opportunity to sleep. More commonly, inability to get to sleep in people who have adequate opportunity is the cause of sleep deprivation. This is known as *insomnia* (L. *somnus* = sleep) and may be due to pain, anxiety, breathlessness or discomfort.

It has been claimed that the sleep induced by hypnotic drugs is poor in REM sleep and does not provide the refreshment of natural sleep.

## Treatment

When a person is permitted to sleep after he has stayed awake for more than about 24 hours, he tends to sleep for about 9-12 hours. This is a longer than average sleep period but it does not necessarily make up the sleep "debt" incurred during the wakeful period. Despite the sleep deficit the subject feels adequately refreshed and he does not exhibit obvious defects in the performance of skilled tasks.

When other factors prevent sleep, sleep can be induced by drugs called hypnotics (Gr. *hypnos* = sleep). These all act by depressing the central nervous system and are dangerous because large doses can cause death by paralyzing the respiratory centre.

# 19. DISORDERS CAUSED BY TRAUMA

## Definition

Trauma (Gr. *trauma* = a wound) is the tissue injury that results when excessive mechanical stress is applied to the body.

The tissues of the body have considerable ability to continue normal function when mechanical forces deform them. This is especially true of the superficial tissues which are most likely to encounter such forces. Normal skin, because of the leathery stratum corneum on the outside and the collagen fibres below, can withstand and survive strong shearing stresses. Skeletal muscle fibres are also very resilient. The fact that so many players walk off the pitch after a hard game of rugby football is a magnificent tribute to tissue resilience. Many of the deeper tissues such as the central nervous system, the liver, the kidneys and the lungs, which are less tolerant of rough treatment, are well protected by bony and other tissue layers. Pain sensation is an important factor protecting the tissues from excessive deformation.

When the mechanical forces applied to the body exceed the tolerance limits of the tissues, trauma or traumatic injury results.

## Effects

### Pain

Tissue damage causes stimulation of pain receptors in the affected area, probably by releasing chemical substances which depolarize the nerve endings. Curiously, the severity of pain is not directly related to the severity of trauma; it seems to be related more to the anxiety and fear associated with the injury. For this reason, the pain of trauma is often better relieved with drugs that produce a tranquil mental state than with drugs which relieve pain sensation specifically.

Pain often gives rise to reflex autonomic activity and increased tone in skeletal muscle and these effects may be seen following trauma. They include slowing or speeding of the heart, increase or a decrease in blood pressure, nausea, salivation, sweating, vomiting and increased intestinal activity.

### Local Tissue Response

When tissue cells are killed by trauma or other agents, they undergo a series of changes which have been carefully documented by pathologists. The dead cells release enzymes, perhaps from lysosomes, which bring about their own dissolution by a process called "*autolysis*". Factors released from the disintegrating cells produce striking changes in local blood flow, vascular permeability and the tissue fluids. These changes are known collectively as "*acute inflammation*". Local vasodilatation and increased capillary permeability raise the local tissue temperature and cause swelling of the part with oedema fluid. Leucocytes are attracted to the area by chemical agents released from the cells (chemotaxis). Initially, neutrophils predominate, but monocytes are eventually the most numerous white cells in the inflamed tissue. The mechanism and function of acute inflammation are not well understood but it results in the phagocytosis and removal of the cellular debris following traumatic injury.

Eventually new blood vessels grow into the damaged area and fibrous tissue is laid down by fibroblasts to heal the wound. The vascular fibrous tissue which grows into the wound caused by trauma is known as *granulation tissue.* As time progresses, it becomes less vascular and the wound is "healed" by tough fibrous *scar tissue*. In some tissues, the original tissue may regenerate to make good the tissue which was lost. The factors controlling tissue repair and regeneration are not understood.

### General Metabolic Response

Trauma induces a complex sequence of metabolic changes whose extent depends on the severity of trauma. Some of these changes are shown diagrammatically in Fig. 22. For the first few days after trauma, protein is broken down and the body is in a state of negative nitrogen balance. This has been called the *catabolic phase.* It cannot be prevented by giving large protein supplements in the diet before or after the injury. It does not occur in animals who have had their protein stores depleted by starvation or who have had their adrenal glands removed; these animals are more easily killed by trauma than normal ones which suggests that the protein catabolism is protective in some way.

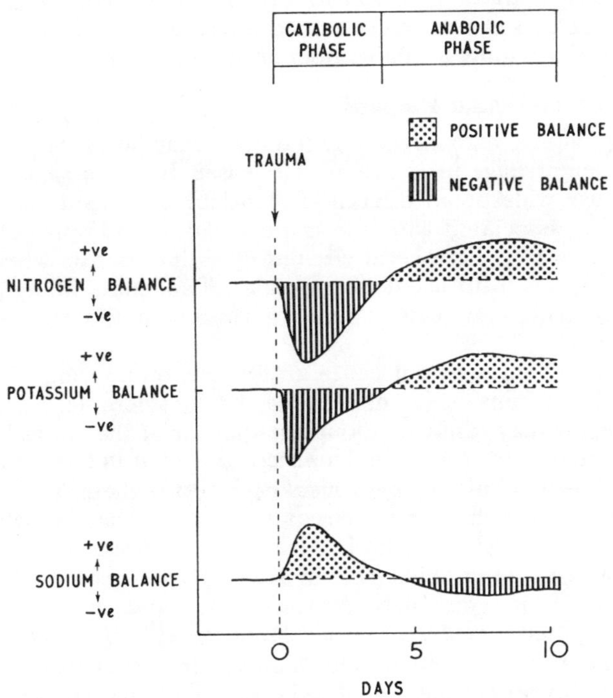


FIG. 22.—Changes in nitrogen, potassium and sodium balance in the catabolic and anabolic phases following trauma.

There is also a negative potassium balance in the catabolic phase which may reflect the renal elimination of potassium released from cells involved in the protein catabolism. There is a fairly constant relationship between the potassium liberated and nitrogen excreted. Sodium balance is positive during this time.

Following the catabolic phase there is an *anabolic* or *convalescence phase.* During this phase, protein is resynthesized, the nitrogen and potassium balance are positive and the sodium balance is slightly negative. The subject replaces the weight that was lost in the catabolic phase.

The mechanism of these changes is probably related, at least in part, to the secretion of adrenal cortical hormones that is known to occur in response to trauma. Some of these hormones such as cortisol, foster gluconeogenesis, i.e. the formation of glucose from protein sources, tubular reabsorption of sodium and the excretion of potassium in the urine. It is possible that other factors, such as starvation and immobilization following injury, contribute to the changes.

It seems that the metabolic response to trauma is protective and fosters survival, but the exact way in which it accomplishes this end remains a mystery. The products of protein breakdown may be made readily available for repair work on the tissues during the catabolic phase but the matter is probably much more complex than that. Patients with adrenal insufficiency have a very poor tolerance for serious trauma; administration of adrenal cortical hormones may be life saving to a person who has sustained severe injuries. It can hardly be all a simple matter of a readily available pool of amino-acids.

**General Cardiovascular Response**

Following trauma, venous return to the heart may fall if the blood volume is reduced by haemorrhage or if products released from damaged tissues, e.g. histamine, cause widespread dilatation of peripheral blood vessels or if the emotional stress associated with the trauma causes widespread dilatation of skeletal muscle vessels. Peripheral circulatory failure occurs when the venous return to the heart is insufficient to maintain an adequate filling pressure and cardiac output. This may lead to underperfusion of the tissues with blood (shock).

Even when trauma does not cause an obvious outpouring of blood to the exterior, blood loss from the circulation may be considerable. A fracture of the shaft of the femur may result in about one-quarter of the entire blood volume bleeding into the tissues. Abdominal injuries may result in large fractions of the blood volume bleeding into the peritoneal cavity. It is these types of blood loss that are mainly responsible for the circulatory disturbance in trauma.

The fall in venous return following trauma may result in a fall in arterial pressure and a decrease in the activity of arterial baroreceptors. This results in reflex activation of the sympathetic nervous system and secretion of catecholamines from the adrenal medulla to cause stimulation of the heart and constriction of peripheral blood vessels. The increase in peripheral resistance and cardiac output plus the contraction of the vascular capacity by venoconstriction help to maintain the arterial pressure and combat the effects of peripheral circulatory failure.

**Fall in Body Temperature**

Following severe trauma which gives rise to peripheral circulatory failure, body temperature falls as the general metabolism of the body is reduced. The mechanism of this is obscure but it may be a consequence of the underperfusion of the tissues with blood.

**Increased Clotting Tendency in Blood**

Following trauma, the platelet count in the blood rises and the coagulation time for the blood is reduced. The surface of the platelets becomes spiky with protoplasmic protrusions. This seems to increase their "stickiness" in that they adhere to one another and tissue surfaces more readily. Though one can see a rationale for such a change as a compensating mechanism to limit blood loss, the mechanism of the change is not understood.

## Causes

Most trauma results from accidents, in cars, at work and in the home. It is also a feature of warfare, and some of the main advances in the understanding and treatment of trauma have been made during times of war.

Controlled trauma is a feature of surgical operations; it may be trivial as when a wart is removed or very extensive as when the colon is excised.

## Investigations

The severity of trauma cannot always be gauged by inspecting the body for wounds; in many cases the trauma causes internal damage. Neither is pain a good index of severity; the sensation of pain and the patient's reaction to it are related more to the patient's feelings of anxiety and fear than to the size of the wound.

Because severe trauma causes peripheral circulatory failure, investigations appropriate for shock, e.g. measurements of heart rate, arterial pressure, central venous pressure and blood volume, are also appropriate for the assessment of trauma.

## Treatment

(1) **Relief of pain and anxiety.**—Traumatic injury causes pain and emotional distress which aggravate the other disturbances. A drug such as morphine, which relieves both pain and anxiety, is most valuable for patients with severe traumatic injury.

(2) **Correction of peripheral circulatory failure.**—Procedures which expand the blood volume or contract the capacity of the circulatory system are helpful in most cases of severe trauma.

(3) **Administration of adrenal cortical homones.**—As mentioned earlier, these hormones seem essential to enable a person to survive severe trauma, though the mechanism by which they aid survival are not known. Injections of the hormones following injury may help people whose adrenal glands are normal and are essential for those patients whose adrenal function is impaired.

**(4) Prevention of intravascular clotting.**—The increased tendency of the blood to clot may result in blood clotting spontaneously in veins where blood flow is sluggish. Such clotting may occur in the veins of the legs and is dangerous because the clot may become mobile in the circulation and travel as an embolus to block a pulmonary artery. Intravascular clotting may be discouraged by procedures which keep venous blood on the move by activating the muscle pump. If evidence of clotting is found, further clotting may be prevented by the use of anticoagulant drugs, provided that they do not lead to further bleeding into the traumatized tissues.

**(5) Promotion of healing.**—Healing occurs more rapidly when (a) the space between the edges of the wound is small, (b) the wound is not infected, (c) the tissues involved are immobilized, (d) the patient is in a satisfactory state of nutrition and (e) the damaged tissues have a satisfactory blood supply. Surgical and other measures should be taken to ensure these ends.

**(6) Restoration of function.**—The process of healing, though it restores the integrity of the tissues may, by producing a surfeit of scarring, reduce the mobility of the injured part. Patients must be encouraged therefore, by physiotherapy and other means, to keep using their muscles and joints during convalescence so that mobility and function are retained.

# 20. DISORDERS CAUSED BY IONIZING RADIATION

## Definition

Rays consisting of certain atomic particles, e.g. helium nuclei (alpha rays or particles), electrons (beta rays or paticles), and certain electromagnetic waves, e.g. X and gamma rays, can cause tissue damage by ionizing tissue molecules that lie in their path. They are thought to do so by displacing electrons from the outer shells of the molecules they strike and so altering the chemical nature of the molecules.

On Earth, man has not needed to develop any physiological apparatus to deal with such radiation. Ionizing radiation travelling towards us from space is mostly trapped in the Van Allen belts of the geomagnetic field that surrounds Earth. Because of this, our environment is virtually free of ionizing radiation and man has not developed physiological mechanisms which can sense it, adapt to it or protect against it.

The problem of ionizing radiation is essentially man-made. It stems from the development of atomic fission and fusion to generate energy for peaceful and military purposes, the development of X-ray technology to diagnose and treat disease, the production of isotopes for use in biological and medical investigation and the wish of man to travel into space where the production of the geomagnetic field is lost.

## Effects

The effect of ionizing radiation can be considered at cellular, individual tissue and whole organism levels.

### Effects on Cells

These effects are conveniently studied on cells grown in tissue culture. The effects are dose dependent. Very small doses have no detectable effect. Higher doses cause a temporary slowing in the rate of mitotic division but the cells ultimately recover. Still higher doses stop mitotic division and produce degenerative changes in the cells which results in their death after a few days. Very high doses kill the cells immediately. Cells which are just going into mitotic division are most vulnerable to ionizing radiation.

The ionization of cellular molecules is thought to disturb cellular function in two ways, though there is little direct evidence on the matter. Firstly, cellular function may be disturbed if the ionization produces chemical change in molecules in the nucleus that cannot be replaced. Thus, if a DNA molecule in a chromosome is altered, the whole future of the cell is altered because that molecule may be essential for the proper functioning of the cell and cannot be replaced. It is to such alteration in cellular genetic material that loss of mitotic ability is attributed. Damage to the chromosomes may also explain the high incidence of genetic mutations and uncontrolled cancerous growth that are seen in tissue exposed to ionizing radiation. Secondly, the products produced by ionization in the cell may be toxic to it. For example, if the radiation ionizes water to produce free hydrogen and hydroxyl radicals, these may continue with molecular oxygen if present, to form hydrogen peroxide and other oxidizing radicals. Hydrogen peroxide ($H_2O_2$) is an oxidizing agent and is toxic to cells; in fact, it is widely used as a disinfectant to kill micro-organisms. In this respect, it is very interesting that the damaging effect of radiation on cells is greatly reduced if the local oxygen pressure is lowered, and vice versa. It is probably the production of toxic products within cells that accounts for the immediate cell death with very high doses of radiation.

### Effects on Tissues

Different tissues differ in their sensitivity to ionizing radiation. This is not surprising in that cells are most sensitive as they begin mitotic division and different tissues vary greatly in their mitotic activity. In general, tissues composed of cells which multiply rapidly are highly sensitive, whereas tissues composed of cells which do not, are relatively insensitive. The former include lymph nodes, the bone marrow, the mucosa of the alimentary tract, the epithelial layers of the skin and the gonads. The latter include skeletal muscle, heart muscle and nerve tissue. The liver, kidneys, bone and other connective tissues, are of intermediate sensitivity.

It should be stressed that radiation can kill all tissues if the dose is great enough, but the tissues which are first affected are those which undergo constant renewal by mitotic cellular division.

This graded sensitivity is important in the *treatment of cancer by ionizing radiation.* Certain types of cancer have an extremely high mitotic rate and hence are sensitive to levels of radiation which have a relatively mild effect on normal body tissues. Natually the radiation is so directed that as little as possible reaches normal tissues.

### Effects on the Body

**(a) Acute effects.**—When man is exposed to high doses of ionizing radiation, he develops *radiation sickness*. The form it takes depends upon the dose and the rate at which it is received. The dose of radiation absorbed is usually measured in *rads.* A rad is the absorbed dose of ionizing radiation which liberates 100 ergs of energy per gram of absorbing material.

With acute massive radiation (more than 5000 rads), many tissue cells are killed immediately. The clinical features reflect mainly the disturbance of nervous system function; the person develops tremors and convulsions and dies within two days from respiratory paralysis and oedema of the brain. With smaller yet lethal doses of about 1000 rads, the story is more prolonged. After a few hours, the patient develops nausea and vomiting. After a few days, damage to the bone marrow results in lack of neutrophils leading to infections, lack of red cells leading to anaemia and lack of platelets leading to spontaneous bleeding. Damage to the mucosal cells of the gut leads to ulceration and haemorrhage. Damage to the epithelium of the skin may cause skin burns, ulcers and loss of hair. The patient pursues a downhill course and dies after some weeks from loss of blood, infection and circulatory collapse.

With small doses in the 100 rad range, some vomiting may ocuur and there may be a transient fall in the white cell count, but recovery is usually rapid and complete.

**(b) Long-term effects.**—If the patient survives the initial exposure, or if he is exposed to low doses of radiation over long periods, he faces certain long-term hazards. He carries and *increased risk of developing cancer.* The mechanism of this is not clear; it may be due to a change in the genetic material that programmes cellular growth or it may be due to a repression of the lymphocytes that maintain an immunological surveillance for cancer cells. In any event, people whose work exposes them to X-rays are more likely to develop cancer of the skin, bone and marrow cells than the rest of the population. Radiation may also exert long-term effects by producing *mutations.* It certainly increases the rate at which mutations occur in micro-organisms and it may also do so in man by an effect on the chromosomes of the germ cells in the ovary and testis.

## Causes

Though an occasional person may encounter large doses of radiation in accidents at nuclear power stations and the like, exposure to low doses is much more common. It may occur at work if the work involves the handling of radioactive material, but exposure is probably most frequent in hospital. Here X-rays are used widely for diagnostic purposes and though the dose for each individual patient is low, the cumulative dose for the personnel who work in these departments has to be kept under surveillance. Radioactive tracer substances are also used widely for diagnositc and research purposes and carelessness in their use may result in exposure.

Ionizing radiation is sometimes given on purpose in the treatment of certain forms of cancer. The principle underlying their use is that active tissues such as cancer which undergo constant mitotic division are more sensitive to radiation than normal cells. For that reason, irradiation, if used in appropriate dose, should kill or suppress the cancer cells and leave the normal cells unscathed. Nevertheless, careful beaming of the rays is needed to limit exposure as far as possible to the cancer area.

Exposure to radiation is likely to be a serious hazard to space travellers because they will not have the protection offered by the earth's magnetic field.

## Investigations

**Detection of radiation.**—Ionizing radiations on entering a gas chamber ionize the gas and convert it from an insulator to a conductor. This principle is used in a number of instruments, e.g. the Geiger Counter, to detect radiation. Such instruments should always be to hand where isotopes are used.

**Estimation of radiation dose.**—People who work in places where exposure to radiation is a danger, e.g. X-ray and radiotherapy departments, should wear a device that estimates the cumulative dose that they have received while working there. A simple device is a piece of photographic film covered by a thin paper which is opaque to light. This is worn by the subject and ionizing radiation produces a fogging of the film which is proportional to dose received.

## Treatment

### (1) Protection Against Radiation

There are a number of general approaches to the problem.

(a) One of the best protections is to make as much use as possible of the *inverse square law* which states that the dose of radiation received is inversely proportional to the square of the distance between the subject and the radiation source (Fig. 23). Thus protection rises geometrically as the subject moves arithmetically away from danger. Put in simple language, one should stay as far away from radiation as possible.

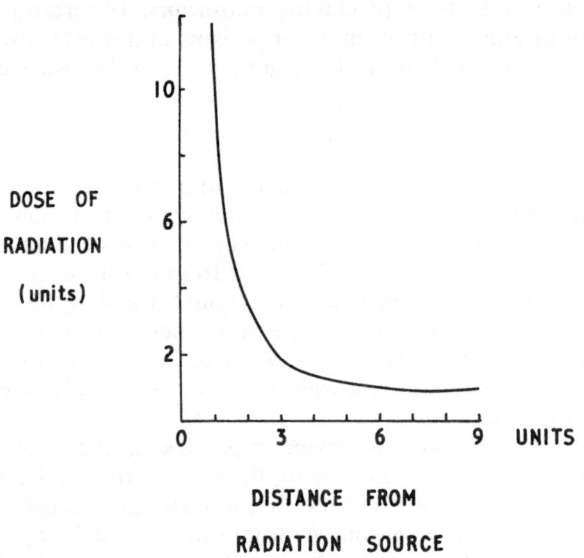


FIG. 23.—The effect of distance from the source of radiation on the dose of radiation received. The dose (d) is inversely related to the square of the distance (D). $d \propto \frac{1}{D^2}$

(b) *Screening* is another approach for people whose work brings them into contact with radiation problems. The type of screening required depends on the type and energy of radiation being encountered. It can range from a simple apron containing a metallic shield to a concrete bunker deep in the earth lined with blocks of lead. In general, the more dense the screening material, the better the protection.

(c) As the effect of radiation is cumulative, protection can also be gained by *limiting the time of exposure* so that the cumulative dose never exceeds a permissible level.

(d) *Hypoxia* is known to protect cells and tissues against the effects of radiation. It is thought to act so by reducing the availability of oxygen for forming oxidizing radicals in the cells. In converse, a raised oxygen tension is sometimes used to potentiate the action of therapeutic radiation on cancer cells.

(e) *Cysteine* and some of its derivatives afford protection against radiation if injected *before* exposure. They increase the amount of radiation that can be tolerated, but it is not clear how they work. They may produce tissue hypoxia but there are many alternative theories.

**(2) Treatment of Radiation Sickness**

If the dose received has been lethal, sedation by drugs is the only appropriate treatment. If there is a chance of survival, several lines of treatment are possible. Drugs which relieve sea sickness are effective in combating the nausea and vomiting of radiation sickness. Blood transfusion can be used to make good the anaemia and bleeding tendency. Antibiotics can be used to make good the lowered resistance to infection due to the leucocyte deficiency. For severe gastrointestinal disturbance, correction of the resulting water and electrolyte disturbances is necessary. If it seems that the bone marrow has been completely destroyed and will not regenerate, the possibility of a transplantation of bone marrow from a donor should be considered.

# 21. DISORDERS CAUSED BY EMOTIONAL STRESS

## Definition

In the course of everyday life, men (and women!) are exposed to countless irritations, doubts, fears, frustrations and threats which are sometimes real and sometimes imaginary. Real or imaginary, they can all generate psychological stress. People adapt to these stresses both mentally and physically, the nature of the adaptation depending on the nature of the stimulus. In general, the adaptations are helpful in preparing the individual to deal with the stimuli and many appear to have considerable value in helping him to survive.

However, if the psychological stress increases above a certain level, the *anxiety state* that arises from it may become incapacitating. Normal bodily function may be so disturbed that the individual cannot work or think effectively. If psychological stress is severe and prolonged, it is thought that the physiological adaptations may actually damage the tissues and cause *psychosomatic disease.* Some people react to severe emotional stress by subconsciously simulating the effects of disease so that they can escape from, or be excused from, facing up to the stressful stimulus. These reactions are called *hysterical reactions.*

Emotional stress and its effects are responsible for a large proportion of the disorders that make patients seek help from their doctors.

## Effects

### Anxiety State

Some anxiety is normal and useful in dealing with the hazards of life. It provides the drive that is necessary to make people take care on motorways, hurry to catch a train, work for examinations and arrive in time for lectures. By alerting the brain, increasing muscle tone, shortening reflex time and accelerating the heart, it prepares the body for the instant action that may be needed if and when the problem causing the anxiety is actually encountered.

However, if the anxiety is disproportionate to the problems, it may handicap rather than help the individual to deal with his problems. Anxiety can manifest itself in nearly every system of the body.

**(a) Mental state.**—Anxiety usually results in restlessness, irritability, inability to relax, sleeplessness and eventually feelings of fatigue. There is an inability to concentrate on any one subject for long and this interferes with ability to work, especially if the work is of an intellectual character.

**(b) Skeletal muscle.**—Muscle tone is increased and this accounts for the feelings of 'tenseness''. Musle tremor, an exaggeration of the normally inapparent "physiological tremor", may be seen in different areas but is usually most easily seen in the outstretched fingers. The increased tone in the muscles at the back of the neck is thought to account for the headache which is common in anxiety.

(c) **Cardiovascular system.**—The heart action tends to be increased in both rate and force and may cause a conscious awareness of the heart beat (*palpitations*). Active dilatation of the blood vessels in skeletal muscle is brought about by sympathetic cholinergic nerves and the drop in peripheral resistance may result in dizziness, light-headedness and in severe cases fainting.

(d) **Urinary system.**—Emotional stress causes the smooth muscle in the bladder to contract and so raise the pressure in the bladder. It seems to do so by causing supraspinal facilitation of the micturition reflex. Frequency of micturition is common in anxious people.

(e) **Alimentary tract.**—The functioning of the alimentary tract may be considerably modified in a number of ways in anxiety. Appetite is usually decreased and loss of weight is common. Nausea and vomiting are seen in some cases. It has been known for a long time that emotion can affect the secretory activity and motility of the stomach. Indigestion is frequently seen in anxiety. The behaviour of the colon is also greatly affected by emotional stress and this may show itself either as constipation or diarrhoea.

(f) **Sweat glands.**—Emotional stress results in an increased sweat rate, especially from the hands and feet though the whole body seems to be involved. The purpose of this response of the eccrine glands is not clear; by causing the hands to drip sweat it may interfere with normal tasks such as writing or examining patients. Emotional stress also increases the secretions of the apocrine glands in the axilla and groin area which give rise to characteristic odours.

(g) **Respiratory system.**—Anxiety can cause a variety of respiratory symptoms. Usually breathing is rapid and there may be hyperventilation. This may be sufficiently marked to lower the carbon dioxide level in the body and hence cause alkalosis and tetany. In certain individuals, bronchoconstriction may occur to cause asthma.

(h) **Pain.**—One of the most important features of anxiety from the medical point of view is the way in which it increases the apparent severity of pain sensation. the unpleasantness of pain is only partly related to the tissue disturbance that gives rise to it; it is greatly modified by the anxiety that accompanies the disturbance. Thus reassurance can greatly reduce pain whereas anxiety can greatly increase it. This explains how injections of water may relieve severe pain in patients who are reassured by thinking that the injection should cause relief. It also explains how many of the pains that bring patients to their doctors may reflect the patient's anxiety rather than any serious disease and how important it is that the doctor should be able to remove the patient's anxiety about his health if there are no grounds for such anxiety.

### Psychosomatic Disease

The borderline between anxiety and psychosomatic disease is very indistinct. As has been mentioned, anxiety can cause vomiting, constipation, diarrhoea, asthma and pain. It is clear that emotional disturbances in the mind can produce changes in the body via either sympathetic, parasympathetic or somatic nerves and endocrine secretions and one can envisage these changes, i.e. prolonged vasoconstriction or bronchoconstriction causing damage to the fabric of the body. However, evnisaging it and proving it are two different matters. It is relatively easy to quantitate the "somatic" changes in disease, e.g. the size of the peptic ulcer and the reduction in respiratory function as measured by respiratory function tests. It is extremely difficult to define or quantitate the psychological stress that is associated with the somatic change. The subtleties of emotional stress are personal matters for the subject and are not easily analyzed by an outside observer. Thus it is difficult to make the correlation between somatic disease and psychological stress that is necessary to prove or disprove the concept of psychosomatic disease. This should be kept in mind when the term "psychosomatic disease" is being bandied about.

In true psychosomatic disease, the onset of symptoms and the course of the disease should be related to the onset and course of the emotional stress. Diseases which are thought to have a psychosomatic component include inflammatory diseases of the skin, e.g. eczema (Gr. *ekzeo* = I bubble up), ulceration of the mucuous membrane of the gut, e.g. peptic ulcer, ulcerative colitis, pain-producing conditions such as migraine and lumbago, endocrine disorders such as hyperthyroidism and circulatory disorders such as hypertension.

It has been suggested that emotional stress is a factor in the disease where degenerative changes in the walls of coronary arteries result in death or disablement from lack of coronary blood flow (ischaemic heart disease). A proposed mechanism for this possibility is as follows. In response to repeated episodes of emotional stress, adrenaline is repeatedly liberated from the adrenal medulla and mobilizes free fatty acids from the fat depots. In a primitive situation, these fatty acids might be metabolized in a fight or flight response to the stress. Many modern stresses do not provide such an effective outlet and it has been suggested that the fat gets laid down in the walls of blood vessels when this is the case. The deposition of lipids in the walls of coronary arteries is certainly a feature of ischaemic heart disease but the evidence that emotional stress is responsible for it is largely circumstantial. Bus drivers whose life is thought to provide more stress and less exercise than that of bus conductors, have a higher incidence of ischaemic heart disease than the latter.

The problem is made more complex by the likelihood that the state of mind can influence the course of disease which arises spontaneously. It is often said that the patient who "does not try" or "gives up hope" deteriorates more rapidly than one whose mental powers are harnessed towards his recovery. This may well be true, but it is extremely difficult to prove one way or the other.

**Hysterical Reactions**

When the psychological stresses of life are greater than a person is prepared or able to tolerate, he may subconsciously simulate the effects of disease, so that he may opt out of the situation which is causing him distress. Thus a student may find that his right arm suddenly becomes paralysed just before an examination he does not wish to sit. A person may develop a loss of memory or "black out" to eliminate a period of emotional distress or embarrassment from his consciousness. When faced with the prospect of work for which he has little inclination, a person may develop attacks of weakness, faintness, breathlessness or pain which make it possible for him to postpone or avoid it.

The boundary line which separates *malingering* from hysteria is hard to draw but malingering implies an awareness of deceit whereas hysterical reactions are brought about subconsciously.

### Causes

Life is full of emotional hazards which cause considerable distress to the majority of people. The slings and arrows of outrageous fortune are not easy to bear and most people carry feelings of guilt, insecurity and inadequacy behind relatively cheerful facades. Even when the problems to be faced are not very great in absolute terms, the mind may inflate them out of all proportion. Unpopularity with one's colleagues, marital discord, failure to fulfil early promise, grievance at being unfairly treated, an unhappy love affair, fear of growing old, fear of dying, family bereavement, inability to measure up to one's job, guilt about disliking a parent, are among the countless problems that most people have to face at some time or another. In addition, there are the minor stresses, the embarrassment when a dental appointment is forgotten, the anger when the driver behind honks his horn belligerently, the guilt at not having written the thank-you letter, and so on.

Emotional stress seems to be more distressing when it involves conflict of ideas and uncertainty. Anxiety is compounded by indecision. If the person can decide on an appropriate course of action to deal with his problem and then carry it out resolutely, he feels much better. Discussing the problem with a sympathetic listener often helps to sort things out and lack of such an opportunity due to social isolation tends to perpetuate anxiety.

Most of the causes of emotional stress cannot be seen and an unwillingness of the patient to talk about his secret inner life plus his ability to mask his facial expression may easily deceive the doctor (and the patient's friends). One of the most common reasons now for a medical student to fail examinations is an unhappy love affair which takes from him (or her) the wish and capacity to work. However, the university staff may never know of this because of natural reticence on all sides and the remedial action may therefore be inappropriate.

### Investigations

Before any disease can be designated as a product of emotional stress, it is necessary to perform such examinations and tests as are required to exclude a physical cause. This may be very difficult in the case of hysterical reactions.

After that, the main task is to get the patient to talk about his emotional problems. Though skill in interview technique is important, it is necessary to win the confidence and trust of the patient, if the interviews are to be productive. It is a time-consuming process and many doctors may not have the time that is necessary to unravel a tangled web of fears.

### Treatment

The best treatment is to have the patient discover the origins and the true nature of the problems that are causing him emotional distress. If he can learn to recognize them, accept them and deal with them, they will cause much less mental and physical incapacity.

If the emotional stress has caused physical disease, the disease has also to be treated in the appropriate fashion. If it is found that the patient has no physical disease, reassurance of the patient is very important. If the patient is frightened about the significance of a symptom arising from a minor bodily disorder, his fear may greatly exaggerate his sense of discomfort. Sympathetic and confident handling of the patient by the doctor does much to relieve this common type of distress. To admit to the patient that one does not know the cause of the discomfort may be honest, but it is not good therapy.

Merely talking about problems with a sympathetic listener or group of listeners seems to alleviate emotional stress. This may be because it helps the patient to formulate his fears in a more coherent fashion or because it gives him a better sense of proportion. In any event, confession is good for the soul.

Unfortunately for many people, life's problems do not disappear, even when their cause is recognized; they just hurt a little less severely. For these people, certain drugs which interfere with the delicate chemistry of the brain may be given. Some of these depress the level of consciousness (sedatives) and others (tranquillisers) interfere with emotional reactions in such a way that the patient develops a more tranquil mental state. The widespread and heavy use of these drugs nowadays is a very clear indication of the difficulties caused by emotional stress in modern life.

Other drugs may be directed more peripherally. Bronchospasm may require the use of bronchial muscle relaxants and palpitations may respond to a beta adrenoceptor blocking drug.

## RECENT LEADING ARTICLES

At what temperature should you keep a baby? *Lancet,* 1970, **2,** 556.
Don't just freeze there: hibernate. *Lancet,* 1973, **2,** 1425.
Life at the top. *Lancet,* 1972, **1,** 674.
Man's future in space. *Brit. med. J.,* 1973, **1,** 321.
Neurohumoral and metabolic aspects of injury. *Lancet,* 1972, **2,** 912.
Severe accidental hypothermia. *Lancet,* 1972, **1,** 237.
Thresholds and low level irradiation effects. *Lancet,* 1973, **2,** 599.
Where to measure temperature. *Lancet,* 1971, **1,** 439.

*Section IV*

# THE CARDIOVASCULAR SYSTEM

# 22. HIGH BLOOD PRESSURE (HYPERTENSION)

## Definition

THE HUMAN circulation works as a constant pressure system. The level of pressure in the arteries is rigorously controlled and local flow to tissues is regulated by varying local peripheral resistance. When an increase in flow to any area is required, vascular resistance in that area is lowered and more blood leaves the arterial system. The tendency for arterial pressure to fall is countered by blood pressure control systems.

There are two main control systems which maintain arterial pressure. The system with the fastest response involves the baroreceptor reflexes. The second and slower system operates through the kidney. There may be other systems which have not yet been identified.

Hypertension is the condition where these control systems are overcome or deranged and arterial blood pressure persistently exceeds normal values. The upper limits of normal are taken arbitrarily as around 150 mm Hg in systole and 90 mm Hg in diastole in a resting young adult (1 mm Hg = 133 pascal). These limits depend on age. The average blood pressure at birth is around 75/40 and it tends to rise gradually throughout life. It also varies throughout the day and falls during sleep. It is worth stressing that the high pressure in "hypertension" is confined to the systemic arteries and arterioles. The pressure in all other vessels, capillaries, veins, pulmonary vessels etc. is usually normal.

## Effects

Hypertension imposes stresses both on the heart and blood vessels and the ill effects of hypertension are due to the changes that occur in these structures. Thus death is usually due to (i) heart failure or (ii) renal or cerebral damage caused by vascular disease.

### *Cardiac Changes*

#### Stage of Hypertrophy

Starling's law of the heart indicates that the work done during contraction is a function of the initial length of the muscle fibres. In other words, the more the muscle fibres are stretched, the more vigorously they contract. The relationship is shown in the ventricular function curve in Fig. 24. *Stroke work* (the amount of work done each beat) is used to indicate the strength of left ventricular contraction. It is obtained by multiplying the volume ejected per stroke (stroke volume) by the mean arterial pressure. The higher the arterial pressure, the more work has to be done to eject the same stroke volume. *End-diastolic pressure* (the pressure in the ventricle at the end of diastole) is used as an index of the amount of stretching of left ventricular fibres. In the normal ventricular function curve, stroke work increases quickly initially as the end-diastolic pressure rises. However, eventually the curve flattens when further stretching produces little further increase in work. To understand the workings of the heart, it is very helpful to think in terms of ventricular function curves.

Let us consider the ventricular function curve of a normal individual who maintains a stroke work of $SW_1$ with an end-diastolic pressure of $P_1$ (Fig. 24). If he becomes hypertensive a greater stroke work (say $SW_2$) is required to eject the same stroke volume against the higher arterial pressure. The normal stroke work $SW_1$, is not sufficient to expel the normal stroke volume so that blood accumulates in the ventricle to raise the end-diastolic pressure to a value of $P_2$ where the required stroke work, $SW_2$, is obtained. Thus the ventricle first compensates for hypertension by dilating slightly and moving to a new point on the normal ventricular function curve (A-B). This slight "physiological" dilatation is not detectable clinically and should not be confused with the pathological dilatation described below under "Stage of Dilatation and Failure".

When given extra work to do over a long period, cardiac muscle, like skeletal and smooth muscle, tends to increase in bulk. This is achieved by an increase in fibre size (hypertrophy) rather than by an increase in fibre number (hyperplasia). Cardiac muscle cells are unable to multiply by mitotic division after the age of about six months. The additional muscle in the left ventricle shifts the ventricular function curve upwards and to the left. Because of this shift, the ventricle can carry out the new work load ($SW_2$) at the original end-diastolic pressure. Thus, the left ventricle adapts to an increased work load, first by increasing end-diastolic pressure and then by hypertrophy which

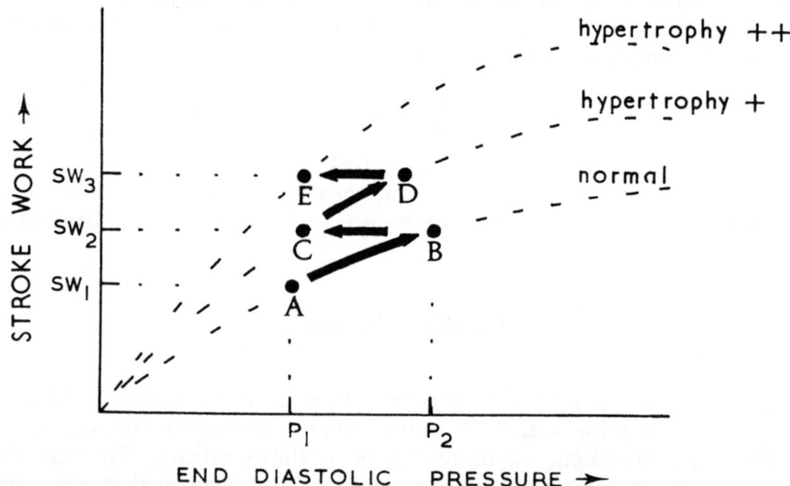


FIG. 24.—Left ventricular function during increasing ventricular hypertrophy in hypertension.

A—B The left ventricle first compensates for the increased work of hypertension by dilating and moving to a new point on the normal ventricular function curve.

B—C Hypertrophy of the ventricle gives rise to a new ventricular function curve and the increased work is now possible at the original end-diastolic pressure.

C—D Progression of the hypertension is compensated by further increase in end-diastolic pressure causing a further increase in stroke work.

D—E Further hypertrophy allows the increased work to be executed at the original end-diastolic pressure.

permits more stroke work to be accomplished at the normal end-diastolic pressure on a new ventricular function curve (B-C).

Unfortunately, hypertension tends to progress and further increases in peripheral resistance lead to further increases in arterial pressure requiring further increments of stroke work. Each of these increments is met initially by an increase in end-diastolic pressure (C-D) and eventually by an increase in muscle bulk by hypertrophy (D-E). In this way the heart compensates for its increased load and the delivery of blood to the tissues may be adequate to permit strenuous exercise. The only clinical evidence of hypertension associated with the heart at this stage may be an enlarged heart, forceful apex beat and characteristic changes in the electrocardiogram indicating hypertrophy of the left ventricle.

**Stage of Dilatation and Failure**

Hypertrophy of the ventricular fibres can cope with the increased work of hypertension to a certain extent only. Eventually the increase in muscle bulk outstrips the ability of the circulation to provide it with nutrition. Because hypertrophy results in an increase in the diameter of the muscle fibres, the distances for diffusion from the capillary wall to the centre of the fibre increase with increasing hypertrophy (Fig. 25). Further, the normal ratio of about one capillary to about one muscle fibre does not increase with hypertrophy. When fibre diameter exceeds about 25 $\mu$m, the distances are too great for diffusion to be an effective means of nutrition and hypertrophy ceases to be adaptive. The situation may be made worse by the degenerative changes in the small arteries that result from or accompany hypertension.

When the blood supply becomes unable to satisfy the metabolic needs of the myocardium, the ventricle starts to fail and its ventricular function curve is shifted down and to the right (Fig. 26). The point labelled (A) represents the situation in compensated hypertension where the new ventricular function curve of the hypertrophied ventricle permits increased stroke work ($SW_1$) to be performed at the normal end-diastolic pressure ($P_1$). (B) represents the point when the ventricle starts to fail and does less stroke work ($SW_2$) at that pressure. With less stroke work, stroke volume falls and blood accumulates in the ventricle until the original work ($SW_1$) can be performed at a higher end-diastolic pressure ($P_2$). As failure progresses, the cycle is repeated and end-diastolic pressure gradually rises to give the necessary stroke work. Eventually a time comes when the rise in end-diastolic pressure brings the ventricle to the plateau of the ventricular function curve (F-G). At this stage, increases in end-diastolic pressures have progressively less effect on stroke work. The stroke work becomes insufficient to maintain an adequate stroke output and death from cardiac failure follows.

Though dilatation of the ventricle helps it to generate more stroke work, it causes the ventricle to work at a mechanical disadvantage because of the law of Laplace. This law applied to a hollow viscus states that the pressure (P) generated in the viscus varies directly with the tension (T) developed in the wall and inversely with the radius (r) of the viscus. This can be written:

$$P = \frac{2T}{r}$$

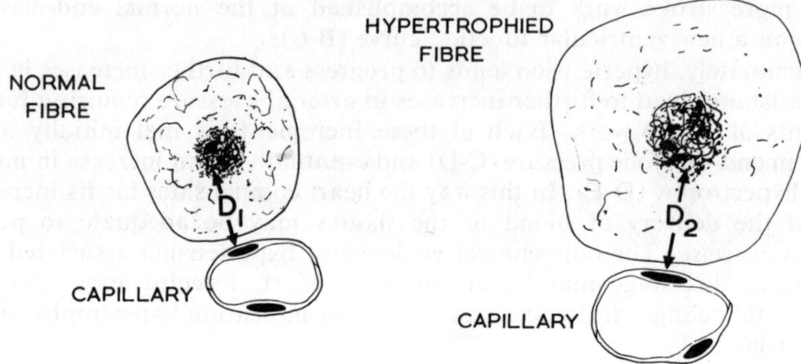


FIG. 25.—Hypertrophy in cardiac muscle. In the hypertrophied state the distance for diffusion from the capillary wall to the centre of the fibre ($D_2$) is greater than in the normal state ($D_1$).

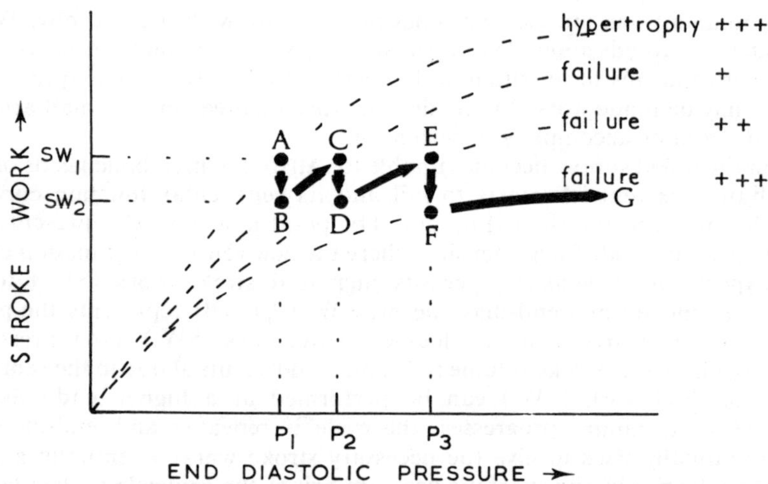


FIG. 26.—Stages in the development of left ventricular failure and dilatation in hypertension.

A—B As the hypertrophied ventricle exceeds its capacity for adequate nutrition, it begins to fail. The stroke work performed at the original end-diastolic pressure, $P_1$, falls.

B—C The stroke work required to evacuate the ventricle is achieved by dilatation of the heart so that the ventricle functions at a higher end-diastolic volume ($P_2$).

C—D Further failure decreases the stroke work attainable at $P_2$.

D—E As blood accumulates in the failing ventricle the end-diastolic pressure rises to $P_3$ to maintain the necessary stroke work.

E—F Further failure decreases the stroke work attainable at $P_3$.

F—G Because of the flattening of the ventricular function curve, a time comes when it is no longer possible to keep up the stroke work by dilatation.

The relationship is shown graphically for different values of the radius in Fig. 27. At the normal radius of the ventricle, the pressure generated in the ventricle is $P_1$ when the tension developed in the wall is $T_1$. When the radius is doubled, the tension developed in the wall has to be doubled to generate the same pressure ($P_1$), i.e. the muscle has to develop twice the tension to do the same amount of work. As it becomes dilated, it becomes progressively more expensive in terms of energy, for the heart to generate a given pressure. This is another factor which limits the ability of the ventricle to compensate for failure by dilatation.

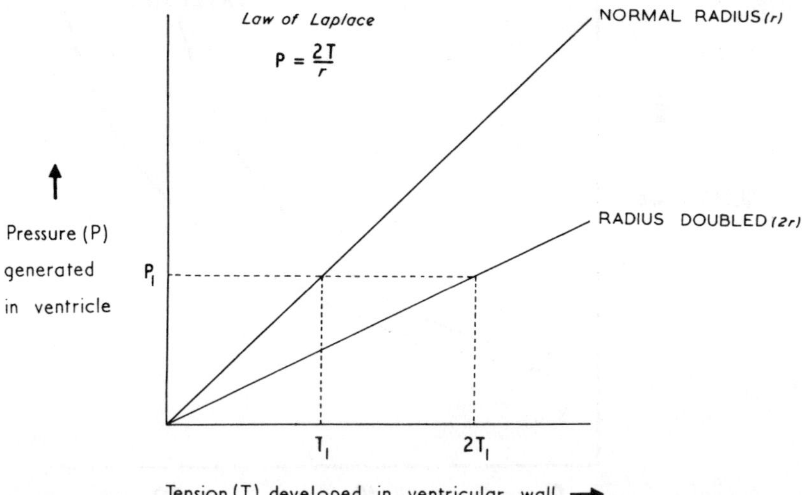


FIG. 27.—The effect of increasing the radius of the ventricle on the relationship between internal pressure and wall tension (from the law of Laplace). When the radius is doubled, the tension developed by the ventricular wall has to be doubled to generate the same pressure.

## *Vascular Changes*

### Wall Hypertrophy

The vascular effects of hypertension are seen mainly in small arteries and arterioles and probably result from the increase in the distending pressures applied to the walls. When vascular smooth muscle is stretched it contracts because of an intrinsic myogenic response. This helps the arteries and arterioles to withstand distension in hypertension. However, the increase in work that the smooth muscles have to do results in their hypertrophy. Hypertrophy of the elastic and subintimal tissues is also seen. In long-standing cases, degenerative changes may be seen in these tissues.

### Vascular Resistance and Reactivity

As a result of these changes, hypertensive vessels offer more resistance than normal, even when fully relaxed, because wall hypertrophy results in a reduction in the lumen. Further, shortening of the muscle in hypertensive vessels causes a greater increase in resistance than the same percentage shortening in normal vessels (Fig. 28). Sympathetic stimulation, by contracting the fibres at the outside of the vessel tends to obliterate the lumen with the hypertrophied inner layers. This would explain why patients with hypertension show exaggerated responses to vasoconstrictor stimuli.

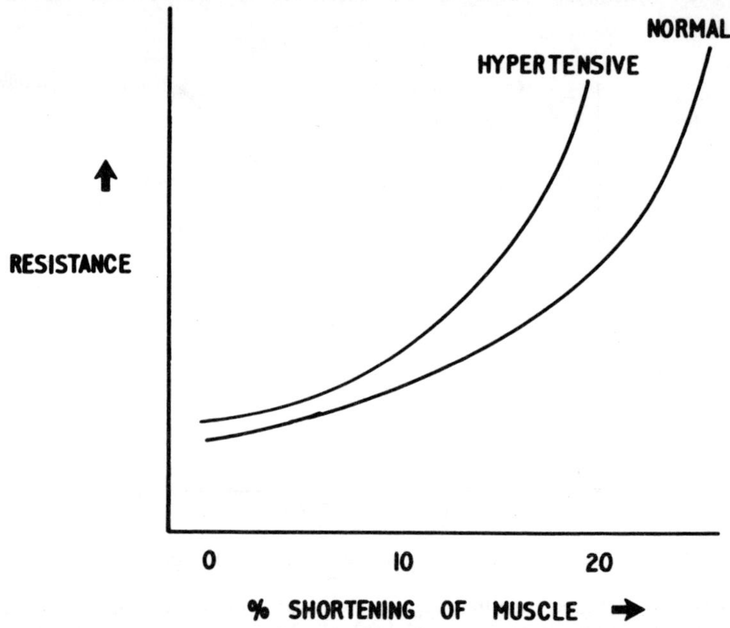


FIG. 28.—The effect of smooth muscle contraction on the resistance offered by normal and hypertensive vessels. Note that the hypertensive vessels offer more resistance to flow than normal vessels *even when fully relaxed.* Contraction of the smooth muscle produces relatively greater increases in resistance in hypertensive vessels.

### Vascular Damage

In severe hypertension, the smooth muscle in the walls of arterioles may be damaged by the high internal pressure. Sometimes the smooth muscle in the wall dies as a result of damage (arteriolar necrosis, Gr. *nekrosis* = death). Sometimes the damage to the smooth muscle irritates it so that the vessels go into spasm. Occasionally the distending force is sufficient to burst the vessels. This is most likely to occur at a place where there is an inborn weakness of the vessel wall.

### Complications Arising from the Vascular Changes

(*i*) *Aggravation of the hypertension.*—The vascular changes in severe hypertension increase peripheral resistance and hence tend to make the hypertension worse. This constitutes a positive feedback system and may account for the progression of the disease.

(*ii*) *Acceleration of degenerative arterial changes.*—With increasing age fatty degenerative changes (***atheroma,*** Gr. ***athere*** = porridge) occur in the larger arteries. Plaques of fatty material are laid down in the subendothelial layers. These plaques may lead to vascular obstruction if thrombi of blood or platelets are laid down on the damaged intimal surface, or if part of the plaque breaks off and travels to block the artery further down stream or if haemorrhage occurs into the soft centre of the plaque causing the wall to swell. The mechanism and cause of the degeneration are not known. However, the degeneration is worse in people with hypertension than in normal people and atheroma occurs in the pulmonary artery system in people who have pulmonary hypertension.

(*iii*) *Cerebrovascular accidents.*—This term covers brain damage due to rupture of a cerebral blood vessel (cerebral haemorrhage), clotting of blood within a cerebral vessel (cerebral thrombosis) or obstruction of a cerebral vessel by a circulating mass in the blood stream (cerebral embolism). The abrupt nature of the onset of symptoms in cerebrovascular accidents has led to the term "*stroke*" being used as a general descriptive term. Cerebrovascular accidents are a common cause of death in hypertension.

(*iv*) *Renal failure.*—Damage to blood vessels in the kidney by hypertension results in patchy destruction of renal tissue. There is a progressive impairment of renal blood flow and glomerular filtration rate. The resulting renal failure is another cause of death in prolonged hypertension.

(*v*) *Loss of vision.*—Many of the changes in blood vessels caused by hypertension can be seen directly through the pupil with an ophthalmoscope. The lens of the patient's eye magnifies the retina, so that the arteries and veins which course over it are easily observed. As hypertension develops, hypertrophy of the walls of the arteries makes them thicker and reflect more light (copper or silver wire effect). Their increased bulk tends to compress and nip the veins where they cross. Dilatations and spasms are sometimes seen. In very severe cases, bursting and leaking vessels give rise to haemorrhages and exudates. The congestion in the area of the optic disc causes it to bulge forward (papilloedema). The damage caused to the retina by these changes results in a progressive loss of vision.

(*vi*) *Spontaneous haemorrhage.*—Spontaneous haemorrhage due to bursting of cerebral vessels by the high intravascular pressure in hypertension has already been mentioned. Spontaneous bleeding can also occur from the nose to cause nose bleeds (epistaxis, Gr. *epi* = upon; *staxo* = I drop), from the respiratory tract to cause coughing up of blood (haemoptysis, Gr. *haima* = blood; *ptyo* = I spit) or into the stomach to cause vomiting of blood (haematemesis, Gr. *haima* = blood; *emeo* = I vomit).

(*vii*) *Hypertensive encephalopathy.*—In severe hypertension the patient may develop headaches, convulsions and eventually coma. The condition is a transient one and full recovery is usual. The condition has been attributed to spasm of cerebral vessels injured by the high intravascular pressures. However, it may be due to local oedema of the brain tissue produced by exudation of fluid from cerebral vessels damaged by the high intravascular pressure.

(*viii*) *Headache.*—This is common in severe hypertension though the mechanism is still obscure. It may be due to distension of vessels by the high intravascular pressure.

The effects of hypertension are summarized in Fig. 29.

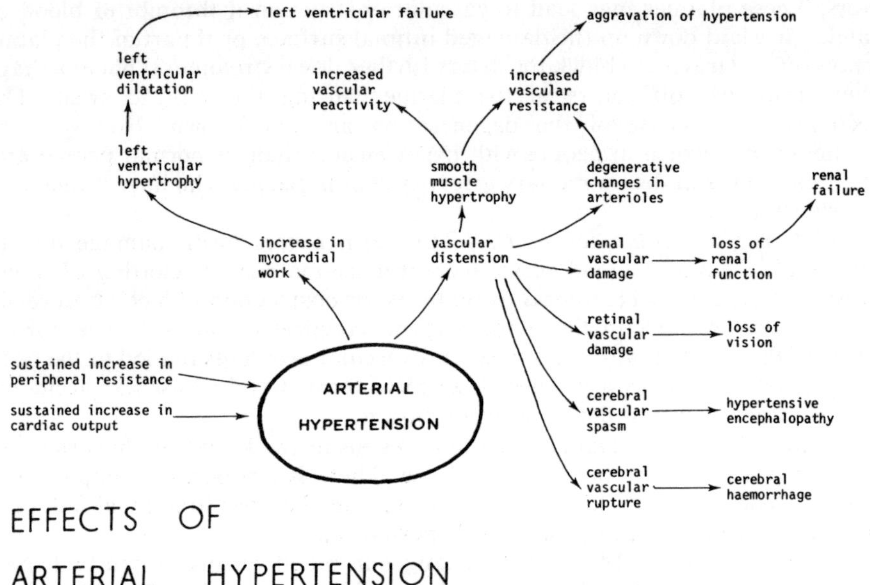


FIG. 29.—The effects of arterial hypertension.

## CAUSES

The level of blood pressure in the arterial system depends on the cardiac output and the peripheral resistance. In hypertension the raised arterial pressure is nearly always due to an increase in peripheral resistance, the cardiac output being normal.

Total peripheral resistance (R) in the systemic circuit may be calculated by dividing the perfusion pressure (ΔP) by the flow around the systemic circuit:

$$R = \frac{\Delta P}{F}$$

The flow around the circuit is measured by measuring cardiac output. The perfusion pressure is measured by subtracting mean central venous pressure from mean arterial pressure. For convenience, central venous pressure is assumed to be zero. Arterial pressure fluctuates with the cardiac cycle but the mean pressure lies closer to diastolic than to systolic pressure. There are many ways to estimate mean arterial pressure but a simple one is to divide the sum of twice the diastolic pressure plus once the systolic pressure by three. In a normal person at rest the total peripheral resistance (R) might be:

$$R = \frac{\Delta P}{F} = \frac{100 \text{ mm Hg}}{5 \text{ litre/min}} = 20 \text{ mm Hg/litre/min}$$

In a patient with hypertension the resistance might be found to be raised e.g.:

$$R = \frac{\Delta P}{F} = \frac{150 \text{ mm Hg}}{5 \text{ litre/min}} = 30 \text{ mm Hg/litre/min}$$

It can be seen that there is a greater energy cost to drive a litre of blood around the systemic circuit in a hypertensive person than in a normal person.

From Poiseuille's equation for describing the flow of fluids through tubes it can be derived that the resistance to flow offered by tubes depends on the length (l) and radius (r) of the tube and the viscosity of the fluid (v). This can be represented:

$$R \propto \frac{lv}{r^4}$$

In most cases of hypertension the increase in resistance is due to a decrease in the radius of the vessels mainly responsible for peripheral resistance. It should be noted that a halving of the radius results in a 16-fold increase in resistance. However, in some conditions such as polycythaemia, the increase in peripheral resistance is due to increased blood viscosity. There is no good evidence that any cases of hypertension are due to elongation of resistance blood vessels.

Hypertension is usually classified into *essential hypertension* where the underlying cause is not known and *secondary hypertension* where it is. Hypertension may also be subdivided into "benign" and "malignant" forms based on the course of the disease. The benign form usually occurs in older people and progresses slowly. The malignant form usually occurs in younger people, and progresses rapidly to death.

### Essential Hypertension

This accounts for about 80 per cent of all cases of hypertension. An increase in the renin content of the blood has not been demonstrated. The baroreceptor reflexes still operate though there may be an alteration in their sensitivity. It has been found that cardiac output in essential hypertensives is normal. This finding indicates that a raised peripheral resistance is the cause of essential hypertension. The viscosity of the blood and the length of the blood vessels are thought to be normal in essential hypertension suggesting that the underlying cause is a decreased internal radius of the resistance blood vessels. Why this narrowing should occur remains a mystery.

### Secondary Hypertension

**Renal hypertension.**—Diseases which restrict blood supply to the kidney result in the release of renin and thus cause hypertension. The raised arterial pressure is due to the vasoconstrictor effects of angiotensin and the salt and water retention caused by aldosterone. The salt and water retention tends to increase cardiac output by raising the filling pressure and hence the end-diastolic pressure in the ventricles. It also increases peripheral resistance by an unknown mechanism. This may involve an increase in the amount of extracellular fluid in the walls of resistance blood vessels which restricts their lumen or an increase in the intrinsic myogenic tone in the vessels.

Restriction of renal blood flow may occur with inflammatory diseases of the kidney which raise intrarenal pressure and cause scarring. Restriction also occurs if the renal arteries or the aorta above the origin of the renal arteries are narrowed by a constriction (stenosed, Gr. *stenos* = narrow). Such narrowings are usually inborn abnormalities.

In some ways the hypertension that occurs with restriction of renal blood flow is a compensatory mechanism to maintain blood flow to these vital organs. However, when only one kidney is supplied by a narrowed artery, the induced hypertension may damage the kidney whose vessels are not narrowed. This may lead to persistent hypertension after correction of the arterial narrowing.

Hypertension is also frequently seen in patients in whom both kidneys have been removed and dialysis with an artificial kidney has been substituted for renal function. Because of this, it has been suggested that the kidneys normally have a lowering effect on blood pressure and that renal damage may cause hypertension (*renoprival hypertension*) by removing this effect. However, the hypertension with kidney removal may be due to disturbances of water and electrolyte balance in the body which are hard to avoid when dialysis is substituted for normal renal function.

**Endocrine hypertension.**— There are three endocrine causes for hypertension, a catecholamine-secreting tumour of the adrenal medulla (phaeochromocytoma), and aldosterone-secreting tumour of the adrenal cortex (Conn's syndrome) and over-secretion of cortisol from the adrenal cortex (Cushing's syndrome).

A *phaeochromocytoma* may release adrenaline and noradrenaline which cause hypertension by peripheral vasoconstriction and stimulation of the heart.

In *Conn's syndrome* excessive secretion of aldosterone leads to excessive reabsorption of sodium from the renal tubules in exchange for potassium or hydrogen ions. The potassium and hydrogen concentrations in the blood are low (hypokalaemic alkalosis). The sodium retention together with the chloride and water which accompany it expand the extracellular fluid volume. The increase in blood volume causes a rise in end-diastolic pressure and an increase in cardiac output. Expansion of extracellular fluid in the walls of blood vessels may increase the resistance they offer.

In *Cushing's syndrome* there is an increased secretion of adrenal glucocorticoids. These cause hypertension because they have some aldosterone-like effects. It has also been suggested that they potentiate the effects of normal vasoconstrictor influences and so increase peripheral resistance.

**Neurogenic hypertension.**— When the blood flow to the brain stem is reduced drastically, a reflex stimulation of the heart and reflex peripheral vasoconstriction occurs which helps to maintain an adequate blood supply to the brain. This is an emergency mechanism that seems to play little part in the day-to-day regulation of the blood pressure. However, the mechanism can explain the hypertension which is often seen when the blood flow to the medulla oblongata is restricted by local disease or injury or by increased intracranial pressure.

**Hypertension of pregnancy.**—*Pre-eclampsia* is a complication of pregnancy in which there is salt and water retention, oedema and hypertension. It disappears when the pregnancy ends or is terminated. The mechanism of the salt and water retention is not known.

**Hypertension with increased blood viscosity.**—The viscosity of blood is related to the haematocrit value. In conditions where the red cell count is raised (polycythaemia), hypertension is often seen.

## INVESTIGATIONS

### To Assess the Severity of the Hypertension

**(i) Measurement of arterial pressure.**—Arterial pressure is greatly affected by emotional factors so it is very important that the patient is in a state of physical and mental rest when measurements of blood pressure are taken. Single observations are of little value and diagnosis should be based only on repeated observations which give similar values. In very obese arms, greater cuff pressures will be needed to compress the tissues and occlude the artery, so that the values of pressure obtained will be erroneously high. This can be overcome to some extent by bearing the problem in mind and using a wider cuff.

**(ii) Chest X-ray.** If the heart is enlarged due to either hypertrophy or dilatation, this is usually detectable as an enlargement of the heart shadow. A rough guide is that the heart is probably enlarged if its diameter, as seen on the standard chest X-ray (postero-anterior view), is more than half the diameter of the thoracic cage (cardio-thoracic ratio greater than 50 per cent).

**(iii) Electrocardiogram.**—Hypertrophy of the left ventricle causes a deviation of the electrical axis to the left (as judged from the standard leads I-III) and by left ventricular preponderance (as judged from the chest leads, $V_{1-6}$).

**(iv) Ophthalmoscopic examination of the retina.**—As mentioned above, the eye provides a way in which blood vessels can be visualized directly. Changes in these vessels in patients with hypertension suggest that other vessels in the body are similarly affected.

### To Establish the Cause of the Hypertension

It is necessary to carry out a variety of tests to see if any of the known causes of hypertension are responsible for the condition. These tests should include X-ray examination of the renal tracts to see if the kidneys are abnormal, measurement of serum electrolytes to see if adrenal cortical hormones are involved, the use of adrenaline and noradrenaline antagonists to see if the adrenal medulla is involved and measurement of the haematocrit to see if polycythaemia is present.

If, as is usually the case, no cause is found for the hypertension it is diagnosed as essential hypertension. Evidence of renal abnormality does not exclude the diagnosis of essential hypertension. Alteration in renal function may be the result as well as the cause of hypertension.

## TREATMENT

Where the cause of the hypertension is diagnosed, as it may be in renal or endocrine disease, the treatment of choice is to treat the cause. In most cases, however, the cause is not known or is not amenable to satisfactory correction and decisions about treatment are harder to make.

In treating cases of essential hypertension one must decide whether or not to give drugs which lower arterial pressure. It could be suggested that the high blood pressure in essential hypertension is a necessary compensatory mechanism ensuring an adequate blood flow to the tissues when the peripheral resistance is pathologically high.

In elderly patients and when hypertension has been severe and long-standing, there is the risk that reducing the blood pressure may seriously reduce flow to vital tissues, such as the kidneys and brain, or, in the case of a pregnant woman, to the fetus. Thus in a hypertensive patient with a raised blood urea level, lowering the blood pressure may cause a further rise in the blood urea. In such cases a compromise position may be the best that can be achieved and the adverse effects of hypertension must be balanced against the adverse effects of inadequate blood flow to the tissues.

However, it has been shown that prolonged hypertension has damaging effects on both the heart and the blood vessels. Certainly treatment which lowers blood pressure has prolonged the lives of patients with very severe (malignant) hypertension and this may also be true for less severe cases. The evidence about the value of such treatment for patients with mild forms of hypertension is less clear cut.

Treatment to lower blood pressure is usually instituted for patients with moderate or severe hypertension especially if there are signs of damage to retinal vessels and of strain on the heart. In most cases it has to be continued throughout life. General measures such as reducing the tempo of life, reducing body weight, bed rest, sedation and a low-salt diet have been used to reduce arterial pressure. Drugs are also used but it is important that their side-effects should not make life miserable. A wide variety of drugs are in use indicating that the perfect drug for hypertension has not been found.

**Drugs Promoting Elimination of Salt and Water**

Diuretic drugs which encourage the urinary excretion of salt and water are used for this purpose. Just as salt and water retention causes hypertension, elimination of salt and water lowers the blood pressure. It may be due to the reduction in the blood volume leading to a fall in cardiac output or a decrease in peripheral resistance due to shrinkage of the cells in the blood vessel wall. A low blood potassium may result from diuretic therapy but this can be countered by giving potassium supplements.

**Drugs which Decrease Sympathetic Vasoconstrictor Tone.**

The sympathetic nervous system exerts a tonic vasoconstrictor effect on peripheral blood vessels. If this effect is blocked, the peripheral blood vessels relax and a fall in peripheral resistance and arterial pressure occurs. One of the difficulties of such treatment is that reflex vasoconstriction by these nerves is a necessary circulatory adjustment during a change in posture from the horizontal to the vertical position and during exercise. During treatment, patients may feel faint during exercise and on standing up due to the fall in arterial pressure that occurs when the sympathetic reflexes are blocked.

The action of sympathetic vasoconstrictor nerves can be interfered with by drugs which depress the vasomotor centre or block autonomic ganglia, sympathetic nerves or noradrenaline synthesis. It can also be affected by surgical section of the sympathetic nerves.

**Drugs which Block the Effect of Sympathetic Nerves on the Heart**

Sympathetic nerves to the heart release noradrenaline which acts on beta receptors on the myocardial cells. If beta receptors are blocked with drugs, the heart is not subjected to sympathetic stimulation in times of stress so that the tendency for blood pressure to rise at these times is reduced. These drugs have little effect on the sympathetic vasoconstrictor nerves so they do not cause fainting on standing up (postural hypotension). However, they should not be used in patients with heart failure or bronchial constriction (asthma). In the former case, the patient depends on beta adrenergic stimulation to maintain an adequate cardiac output. Loss of sympathetic drive may make the heart failure much worse by slowing and weakening the heart and moving its ventricular function curve downwards and to the right. Bronchial constriction in asthma is countered by stimulation of beta receptors by noradrenaline released by sympathetic fibres to bronchial smooth muscle. Loss of sympathetic drive increases the constriction of the bronchi.

# 23. PERIPHERAL CIRCULATORY FAILURE

## Definition

The object of the circulation is to deliver a blood supply to the tissues which meets the needs of the body. This object may fail if the pump breaks down (*central failure*) or if the return of blood from the peripheral circulation is not enough to maintain the necessary cardiac output (*peripheral failure*). Thus, there are two distinct varieties of circulatory failure. In central or cardiac failure, the central venous pressure is normal or greater than normal but cardiac output is reduced because of an impaired ability of the heart to pump. In peripheral failure, the ability of the heart to pump is normal but cardiac output is inadequate because central venous pressure is too low. The changes in ventricular function in these two conditions are illustrated in Fig. 30. In peripheral circulatory failure the ventricular function curve is normal. The stroke work falls because inadequate venous return causes a fall in end-diastolic pressure and function moves down the normal ventricular function curve. In central circulatory failure, stroke work falls because the ventricular function curve of the failing heart lies below that of the normal heart. Thus, for the same end-diastolic pressure, the stroke work is less. In both cases stroke work is decreased, accounting for the inadequate perfusion of the tissues. It is important to note that inadequate perfusion of the tissues or "*shock*" can result

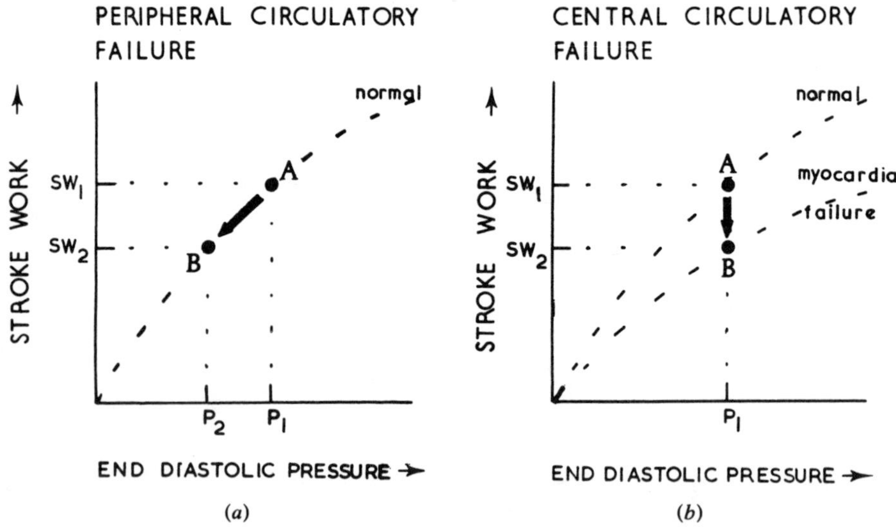


FIG. 30.—Ventricular function in (*a*) peripheral and (*b*) central circulatory failure. In (*a*) the stroke work falls because an inadequate venous return results in a low end-diastolic pressure. The ventricular function curve is normal. In (*b*) the stroke work falls because the ventricular function curve for the failing heart lies below that for the normal heart so that, for the same end-diastolic pressure, the stroke work is less.

from both central and peripheral failure. However, in peripheral failure the end-diastolic pressure is below normal, whereas in central failure it is normal or above normal. The two subjects, peripheral circulatory failure and central circulatory failure, will be considered separately.

It is very easy to confuse "peripheral circulatory failure" which is a *mechanism* of failure with "failure of the circulation at the peripheries" which is an *effect* of circulatory failure caused by any mechanism.

In peripheral circulatory failure, cardiac output is reduced because venous return is insufficient to maintain an adequate filling pressure for the heart; the ability of the heart to pump is normal.

## Effects

In the early stages of peripheral failure most of the effects are manifestations of the compensatory mechanisms which are brought into play to maintain the circulation. At this time compensation ensures that perfusion of the vital tissues is little affected. However, as the failure progresses, the compensatory mechanisms are eventually overwhelmed and the effects of poor tissue perfusion are superimposed.

### Effects of Compensatory Mechanisms

The reduction in cardiac output in peripheral circulatory failure tends to lower the arterial pressure. This brings into play a number of control systems which normally regulate pressure.

As a result of the fall in pressure at the arterial baroreceptors and the decrease in perfusion of the peripheral chemoreceptors, a number of reflex consequences occur which are integrated by centres in the medulla oblongata.

The compensatory effects of peripheral circulatory failure include:

**Increase in heart rate and strength.**—This is brought about by a decrease in vagal activity and an increase in sympathetic activity to the heart. It tends to raise cardiac output. In severe peripheral failure, the pulse is fast but thready because of the small stroke volume.

**Increase in vasoconstrictor tone.**—This is mediated through the sympathetic vasoconstrictor nerves and occurs mainly in the skin, muscle, splanchnic area and kidney. These tissues can survive a temporary restriction in their blood supply. The coronary and cerebral vessels are not richly supplied with vasoconstrictor fibres and escape the general vasoconstriction. Because of this, the smaller output can be channelled preferentially to these vital areas. The vasoconstriction affecting the salivary gland reduces secretion and makes the mouth dry, thereby increasing the sensation of thirst. The vasoconstriction in the mucous membrane in the alimentary tract results in poor secretion, digestion and absorption in the gut. The vasoconstriction in the kidneys causes a reduction in the secretion of urine. The increase in peripheral resistance caused by vasoconstriction helps to maintain arterial pressure. However it offers no long-term solution because the tissues cannot remain for very long in a vasoconstricted state without being damaged. In fact, some of the ill effects of prolonged peripheral circulatory failure are thought to result from such tissue damage.

**Mobilization of tissue fluid.**—When the arterioles supplying the less essential tissues constrict, there is a fall in capillary blood pressure. This alters the balance of forces across the capillary wall in such a way that tissue fluid is mobilized into the capillaries to augment the blood volume. This mobilization of tissue fluid accounts for the pinched appearance of the skin in severe cases of peripheral circulatory failure. By increasing venous return, the mobilization helps to raise the end-diastolic pressure.

**Venoconstriction.**—This is brought about by activity of sympathetic venoconstrictor nerves. In states of shock the veins may be so constricted that it is very difficult to insert a needle into a vein for intravenous therapy. It has been shown in man that venous distensibility is decreased in peripheral circulatory failure (Fig. 31).

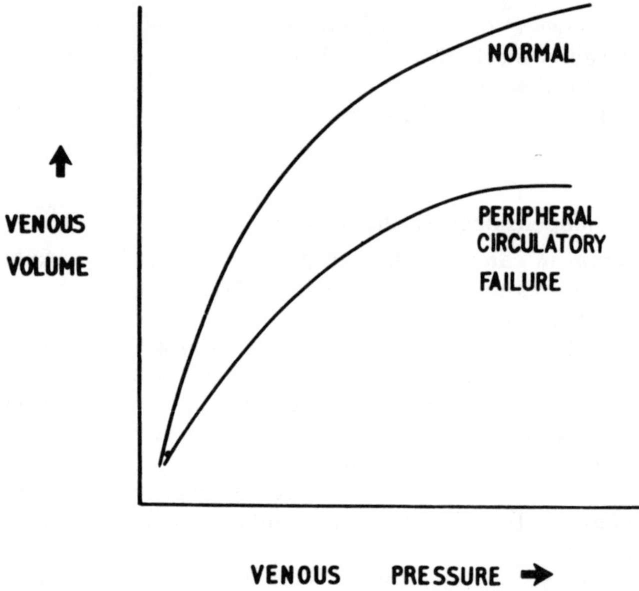


FIG. 31.—The reduction in venous distensibility in patients with peripheral circulatory failure.

**Mobilization of splenic blood.**—In some animals such as the dog, haemorrhage results in a contraction of the smooth muscle in the capsule of the spleen so that its contained blood is mobilized to augment the circulating blood volume. The human spleen does not contain much smooth muscle in its capsule and cannot behave in the same way. However, it has been claimed that the sphincters controlling the outflow from the splenic sinusoids can release the blood contained in the spleen in times of emergency.

**Release of catecholamines from the adrenal medulla.**—Part of the reflex response to a fall in blood pressure is the release of adrenaline and noradrenaline from the adrenal medulla. These increase the peripheral resistance and stimulate the heart, both in rate and in strength. They thus help to maintain arterial pressure.

**Increase in ventilation.**—There is a reflex increase in the respiratory rate which accounts for the rapid shallow breathing in peripheral circulatory failure.

**Increase in the release of renin from kidneys.**—The low pulse pressure in the renal arteries leads to release of renin from the juxtaglomerular cells and the formation of angiotensin I and II in plasma. The resulting increase in the release of aldosterone from the adrenal cortex stimulates sodium reabsorption by the kidney and contributes to the reduction of urinary volume.

**Increase in the output of ADH.**—The fall in central blood volume in peripheral circulatory failure is thought to alter the activity of "volume receptors" in the great veins. As a result of this, the output of antidiuretic hormone from the posterior pituitary is increased. It causes increased water reabsorption by the kidney and a further reduction in urinary output.

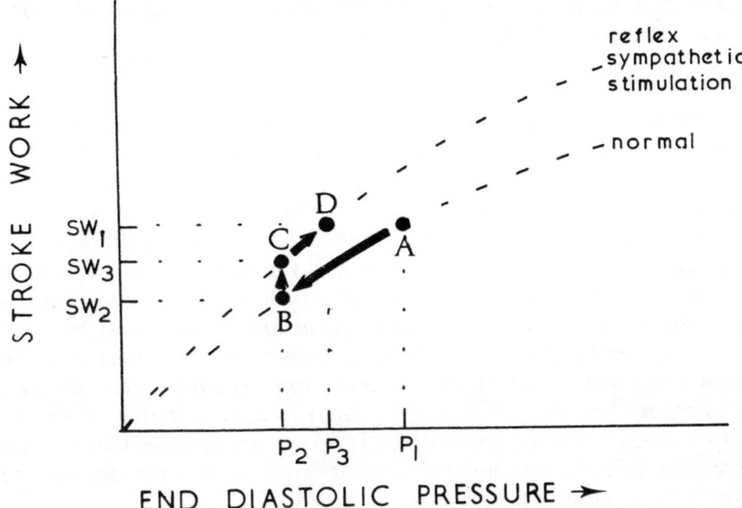


FIG. 32.—Changes in ventricular function in peripheral circulatory failure.

A The normal situation where the normal stroke work ($SW_1$) is performed at the normal end-diastolic pressure $P_1$.

A—B The reduction in stroke work ($SW_1$—$SW_2$) that occurs when end-diastolic pressure falls from $P_1$ to $P_2$ due to a decrease in venous return.

B—C The increase in stroke work accomplished at $P_2$ when reflex stimulation of the heart raises the ventricular function curve.

C—D The further increase in stroke work ($SW_3$—$SW_1$) that results from the rise in end-diastolic pressure produced by venoconstriction, mobilization of tissue fluid and aldosterone secretion.

**Change in ventricular function.**—The course of events in peripheral circulatory failure is illustrated schematically in Fig. 32. 'A' represents the normal situation where the normal stroke work ($SW_1$) is accomplished at the normal end-diastolic pressure ($P_1$). When the venous return falls, the end-diastolic pressure in the ventricle falls to $P_2$ causing a reduction in the stroke work to $SW_2$ and therefore a fall in the cardiac output. The circulatory system compensates for this by (a) attempting to raise the cardiac output and (b) redistributing the smaller cardiac output to the heart and brain by constricting blood vessels in the less vital tissues. Three factors tend to raise the cardiac output. Firstly, reflex sympathetic stimulation of the heart and circulating adrenaline elevate the ventricular function curve upwards and to the left. This changes the situation from B to C where more stroke work is accomplished at the same end-diastolic pressure. Secondly, reflex venoconstriction, mobilization of tissue fluid, and salt and water retention induced by aldosterone secretion increase venous return and thus the end-diastolic pressure. This helps to increase stroke work by moving ventricular function further out the new function curve (C to D). The third factor increasing cardiac output is the reflex increase in heart rate.

The sympathetic nervous system plays a very important role in protecting the individual with peripheral circulatory failure. Animals whose sympathetic nerves have been sectioned are much less able to withstand the effects of haemorrhage than normal animals.

### Effects of Underperfusion

As peripheral circulatory failure progresses, the ability of the compensatory mechanisms to maintain the circulation may be overwhelmed and the direct effects of underperfusion of the tissues are seen. These effects are accentuated in some tissues by the compensatory reflex vasoconstriction already present.

It should be noted that flow in the tissues is not directly proportional to the perfusion pressure (Fig. 33). This means that if the perfusion is reduced by half, the flow need not necessarily fall in the same proportion. In the range around the normal perfusion pressure, good *autoregulation* is seen, that is, the flow is relatively independent of perfusion pressure. However, if the perfusion pressure falls below a certain critical value (*critical closing pressure*), the vessels appear to close and blood flow stops. Critical closing pressure is higher in states where sympathetic vasoconstrictor tone is high. It is important to remember that if the arterial pressure falls to very low values in peripheral circulatory failure, flow through the tissues is not just reduced; it stops.

The effects of underperfusion are described below in relation to each particular tissue.

*The brain.*—The brain is normally well protected against underperfusion by the preferential redistribution of blood to it by selective vasoconstriction and the fact that its blood flow shows good autoregulation. However, severe underperfusion causes clouding of consciousness which may progress into coma. Oxygen lack may cause permanent damage to the brain neurones if the underperfusion is severe or prolonged.

*The heart.*—Myocardial blood flow is normally protected by the same factors as cerebral blood flow. However, a severe fall in blood pressure may so reduce coronary blood flow that myocardial contractility suffers and central or cardiac failure is added to the peripheral circulatory failure.

*The muscles.*—Underperfusion of the muscles may mean that most of the energy production is achieved by anaerobic metabolism. This results in the accumulation of lactic and pyruvic acid. This metabolic acidosis (lactic acidosis) leads to a reduction in plasma bicarbonate which is used to buffer the acids. The acidosis also stimulates respiration and there may be a fall in the $P_{CO_2}$.

*The kidneys.*—The underperfusion of the kidneys together with the vasoconstriction that occurs in these organs causes a decrease in urine output (oliguria, Gr. *oligos* = scanty). If the arterial pressure falls below about 60 mm Hg, urinary output stops. In very severe cases the impairment of blood flow may cause damage or death of the tubular cells in the renal cortex (tubular necrosis). In this case renal function fails to return to normal when the blood pressure is restored to normal (acute renal failure).

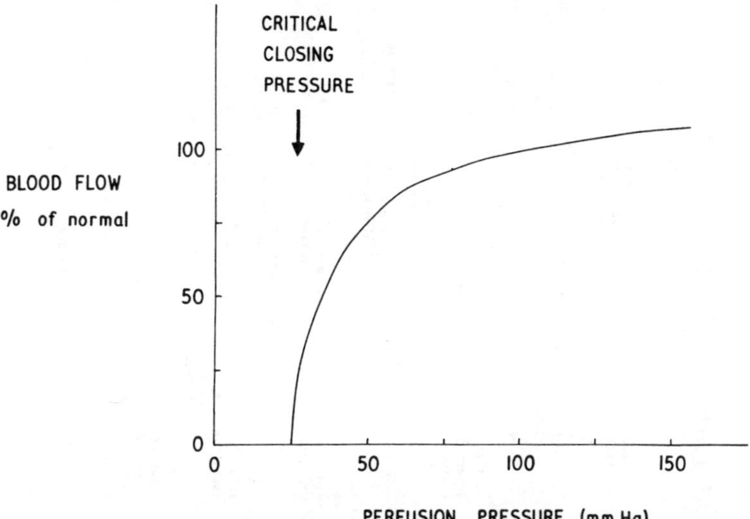


FIG. 33.—The relationship between blood flow and perfusion pressure in vascular beds. Around the normal perfusion pressure (100 mm Hg), the flow is little affected by changes in perfusion pressure (autoregulation). Below a certain critical value for perfusion pressure (critical closing pressure) blood flow ceases abruptly.

*The extremities.*—Underperfusion of the extremities together with the intense vasoconstriction that occurs in them may lead to death (gangrene) of the tips of the fingers and toes.

*The blood vessels.*—In severe and prolonged peripheral circulatory failure, vascular smooth muscle becomes paralysed and does not respond to vasoconstrictor stimuli. As death becomes imminent, erythrocytes form a viscous "sludge" in the capillaries which causes local circulatory arrest. At about this stage the underperfusion of the tissues (shock) becomes *irreversible* in the sense that it will not respond to treatment and death is inevitable.

*The gut.*—The low perfusion pressure and intense vasoconstriction in the walls of the intestine may lead to necrosis of mucosal cells with oedema and small haemorrhages. It has been suggested that absorption of toxic material from damaged alimentary tract may be responsible for circulatory failure becoming "irreversible". Impaired liver function with reduced liver blood flow would accentuate the effects of this because the liver would be less able to deal with the absorbed material by detoxication.

The effects of peripheral circulatory failure are summarized in Fig. 34.

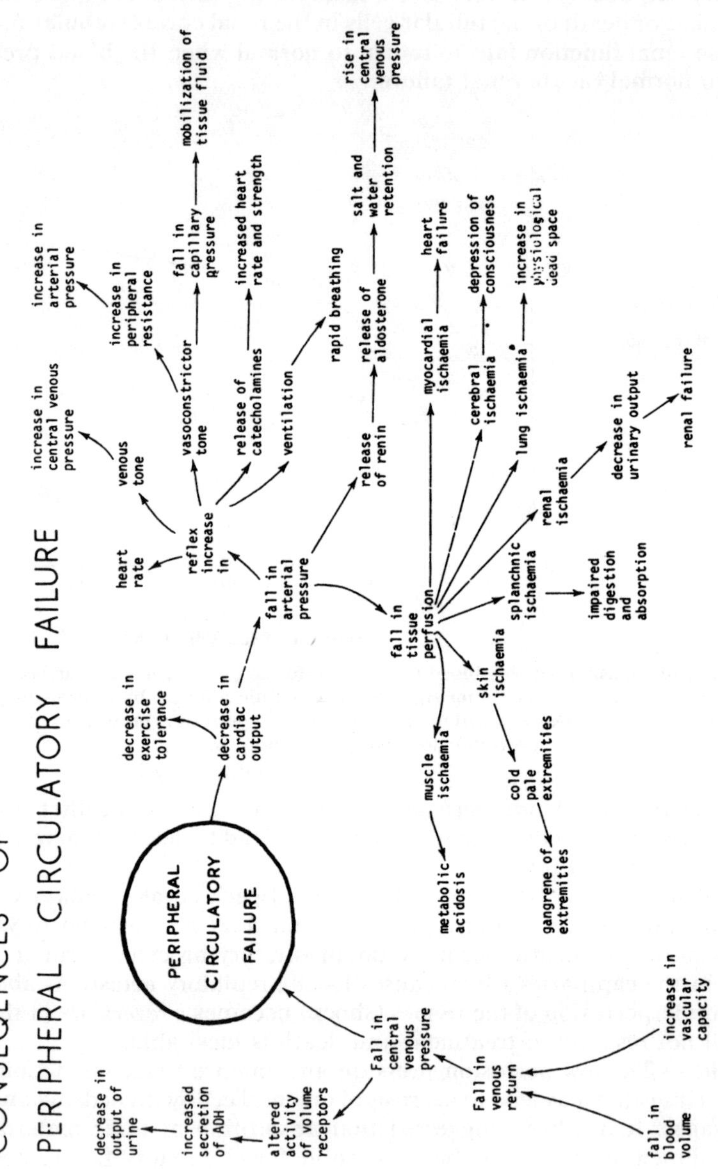


Fig. 34.—The effects of peripheral circulatory failure.

## CAUSES

The main feature in most cases of peripheral circulatory failure is a *disproportion between blood volume and vascular capacity* so that venous return to the heart is reduced. More commonly, blood volume is reduced relative to vascular capacity. This is seen in haemorrhage, where whole blood is lost, in severe burns, where plasma is lost across the walls of damaged capillaries, and in diabetic coma, adrenal cortical failure, severe vomiting, diarrhoea, etc. where the volume of extracellular fluid is reduced.

Less commonly, vascular capacity is increased relative to the blood volume. This is seen with generalized vasodilatation as may occur in histamine poisoning, anaphylactic shock, severe bacterial infections, very hot environments, emotional stress, loss of sympathetic vasoconstrictor tone, etc. The opening up of many peripheral vascular beds simultaneously causes a shift of blood volume into the tissues. The circulation "bleeds" into its own capillaries. The resulting fall in venous return causes a fall in cardiac output and circulatory failure.

*Postural hypotension* is a condition in which peripheral circulatory failure is brought on by adopting the upright posture. In some people, especially after prolonged bed rest, the sympathetic control of peripheral blood vessels is inadequate. On standing up, these people do not constrict the blood vessels in their legs sufficiently to prevent pooling of blood in the lower extremities. The reduction in venous return causes peripheral circulatory failure. Whenever they change from a lying to a standing position, arterial pressure falls and they may feel dizzy or faint because of the decrease in cerebral blood flow.

During emotional stress, cholinergic fibres cause dilatation in blood vessels in skeletal muscle as part of the fight or flight reaction. If very severe, blood may pool in muscle to such an extent that the venous return is inadequate to maintain cardiac output. The subject may then faint. Emotional stress can accentuate the effects of the other varieties of peripheral circulatory failure such as haemorrhage. Excessive warmth and hot drinks may act similarly by causing peripheral vasodilatation.

Severe adrenal failure is a cause of circulatory failure which probably acts both by lowering blood volume and by raising vascular capacity. Inadequate aldosterone means that sodium reabsorption in the tubules is inadequate to maintain the blood volume. Inadequate cortisone activity leads to failure to maintain adequate vascular tone. The reason is not completely clear but it is thought that cortisone potentiates the activity of the sympathetic mediator noradrenaline on vascular smooth muscle.

Thus the causes of peripheral circulatory failure may be classified as:

(a) Fall in blood volume
- — loss of blood
- — loss of plasma
- — loss of water and electrolytes.

(b) Rise in vascular capacity
- — severe infections
- — anaphylactic reactions
- — inappropriate autonomic nerve activity.

(c) Combination of (a) and (b)
- — adrenal failure.

### Investigations

The ideal measurements to assess peripheral circulatory failure would be *blood volume* and *vascular capacity*. Because it is not practical to measure these routinely, measurements are made of quantities which alter when there is a disproportion between blood volume and vascular capacity. Examples of such measurements include *heart rate, arterial pressure* and *central venous pressure.* Central venous pressure is a more sensitive index of disproportion than arterial pressure because the latter may be maintained near normal by compensatory mechanisms even when considerable disproportion has developed.

**Pulse and blood pressure record.**—Early peripheral circulatory failure is suggested by a steadily rising pulse rate combined with a steadily falling blood pressure. Such records are kept routinely in situations where circulatory failure is liable to develop, e.g. after a surgical operation when there is the risk of further blood loss.

**Central venous pressure measurement.**—Central venous pressure is measured by passing a catheter into a large central vein, such as the superior vena cava and connecting it to a saline manometer or other pressure measuring device. It varies between about 0 and 10 cm $H_2O$, the mean being about 5 cm $H_2O$. The superior vena cava, being a distensible structure in the thoracic cavity, expands and contracts slightly during inspiration and expiration. Hence a properly inserted catheter should reveal pressure fluctuations with the respiratory cycle. Central venous pressure tends to fall as peripheral circulatory failure develops and to rise as it is effectively treated. Excessive treatment, such as infusion of too much fluid, raises the central venous pressure above normal. It is thus a very useful measurement to make when assessing the adequacy or otherwise of treatment.

**Blood volume measurement.**—This is not a routine measurement, but may be used in a unit specializing in the treatment of severely ill patients. The measurement is made by an indicator dilution technique. The patient is given an intravenous injection of a radio-active material or dye of known amount which is retained in the circulation. After about 10 minutes, when mixing of the injected material with the patient's blood is more or less complete, a venous sample is taken and the blood volume determined. The result is compared with the patient's predicted blood volume to obtain an estimate of his blood deficit, if any.

### Treatment

It is important that treatment should not be delayed because peripheral circulatory failure becomes irreversible if it is allowed to progress too far. When irreversible, it will not respond to treatment.

There are four principal approaches to the treatment of peripheral circulatory failure.

#### Treatment of the Cause

This is the obvious approach without which other treatment is of temporary benefit only. Thus haemorrhage should be arrested, infection should be treated, pain should be relieved and endocrine deficiencies made good.

### Expansion of the Blood Volume

The aim is to expand the blood volume to meet the vascular capacity. In some cases, such as haemorrhage, this is best done by whole blood transfusions; in others, such as burns, transfusions of blood plasma or one of the plasma expanders are more appropriate. In others, such as the salt and water depletion of severe adrenal insufficiency, vomiting, etc. infusions of saline solutions are given. Expansion of the blood volume raises central venous pressure and thus ventricular end-diastolic pressure to a value where the stroke work is adequate to maintain the circulation.

### Contraction of Vascular Capacity

(*i*) *Raising the foot of the bed.*—When the patient's legs are raised above heart level the veins in the lower extremities collapse and the blood they contain is shifted towards the heart. This simple manoeuvre reduces vascular capacity by lowering the distending pressure in the capacity vessels of the legs. The increased ventricular filling pressure helps to maintain cardiac output.

(*ii*) *Removal of vasodilator stimuli.*—Emotional stress and pain which cause peripheral vasodilatation should be relieved. Morphine is a useful drug for this purpose but care has to be taken with drugs which depress the central nervous system. It is also important to avoid the use of warmth, alcohol or hot drinks in peripheral circulatory failure as they aggravate the condition by the peripheral vasodilatation they cause.

(*iii*) *Vasoconstrictor drugs.*—It is not wise to attempt to reduce the vascular capacity by drugs. First of all there is a very poor absorption of drugs from subcutaneous and intramuscular sites because of the poor circulation. Drugs which constrict the veins and thus raise the cardiac filling pressure also constrict the arterioles and further diminish blood flow to the tissues. Because blood pressure is usually low in proteins with peripheral circulatory failure, it is tempting to give drugs such as noradrenaline which raise blood pressure. However, as a serious problem in circulatory failure is underperfusion of the tissues, a treatment which further reduces tissue perfusion may cause harm rather than improvement.

### Aiding Tissue Nutrition

(*i*) *Blocking sympathetic vasoconstrictor nerves.*—As the reflex vasoconstriction in peripheral circulatory failure is responsible for many of its ill effects, drugs which block the vasoconstrictor (alpha) effects of noradrenaline have been used in treatment. The value of this approach to treatment is not yet established. It implies that the vasoconstriction which is a protective adaptation to low blood pressure may do more harm than good. One possibility is that blood may be trapped in the capillaries in peripheral circulatory failure by a high tone in small venules. Vasodilator drugs, by relaxing tone in these small veins, may allow this blood to be tranfused into the general circulation.

(*ii*) *Oxygen therapy.*—Breathing oxygen may be useful in overcoming the tissue hypoxia.

(*iii*) *Correction of metabolic acidosis.*—Plasma bicarbonate tends to be low due to buffering of the acidic products of anaerobic metabolism. This may be corrected by adding bicarbonate to the infusion fluids. The correction helps tissue metabolism by making the pH of the tissues more favourable for enzyme activity.

(*iv*) *Dextran infusions.*—Intravenous infusions of dextran (colloidal carbohydrate molecules) solutions lower the viscosity of blood and facilitate the flow of blood in the microcirculation. It is claimed that such infusions can cause blood flow to recommence after blood "sludging" has occurred.

# 24. CENTRAL CIRCULATORY FAILURE

## Definition

In central circulatory failure (cardiac failure) cardiac output is inadequate because of an impaired ability of the heart to pump blood; the filling pressure of the heart is normal or above normal.

## Effects

Normally the heart pumps all the blood that returns to it into the arterial system. When its ability to pump is impaired, there is *inadequate perfusion* of tissues in front of the heart and accumulation of blood (*congestion*) in the circulatory system behind the heart. The effects of cardiac failure are therefore best described as (a) those due to inadequate perfusion of the tissues supplied by the failing chamber and (b) those due to congestion behind the failing chamber. The exact nature of these depends on which chamber fails and the severity of the failure.

If the failure is very sudden in onset (acute cardiac failure) the main effects may be due to congestion behind the failure and/or inadequate forward perfusion. If the failure is very slow in onset (chronic or congestive cardiac failure) the main effects are those due to accumulation of blood behind the failing heart.

Congestion due to cardiac failure tends to be localized to the vessels immediately behind the failing chamber (at least in the first instance). When, for example, the left ventricle fails, blood accumulates behind the ventricle because it is unable to clear the volume which reaches it. Blood does not keep on accumulating; a new equilibrium is soon reached where the increased filling pressure due to congestion can maintain an adequate output. Congestion tends to alter the distribution of blood volume; the volume contained in the vessels behind the failing chamber increases whereas the volume contained in those in front decreases. It is important not to confuse the volume of blood contained in vessels with the rate of blood flow through them. Blood flow may be very low in vessels which are full of blood and very high in vessels which are nearly empty.

Inadequate forward perfusion due to heart failure (unlike congestion due to heart failure) tends to be generalized to the entire circulation. As the output of both ventricles must be equal over a period of time, forward perfusion must also be equal over a period of time. If one ventricle fails in forward perfusion, so must the other.'

The heart may fail in many ways and for many reasons. For convenience, the description in this chapter is of the consequences to left *ventricular failure of gradual onset.*

### Effects of Inadequate Ventricular Output (Forward Failure)

*Reduction in exercise tolerance.*—One of the first effects of left ventricular failure is a reduced ability to raise cardiac output in exercise. The effects of this are mitigated to some extent by a high oxygen extraction from the circulating blood. The arteriovenous oxygen difference is greater than normal for a given level of exercise in people with heart failure. Nevertheless the capacity for muscular work of patients with failure is progressively reduced.

*Compensatory responses.*—When the output of the ventricle is insufficient to maintain arterial pressure the altered activity in the arterial baroreceptors and chemoreceptors initiates a variety of reflex responses. There is a reflex stimulation of heart rate and strength of contraction. Sympathetic vasoconstrictor tone is increased in skin, skeletal muscle, kidney, gut and other tissue which can tolerate a restricted blood flow. This permits the reduced output to be distributed preferentially to the brain and myocardium.

However, the vasoconstriction in the gut results in poor secretion of digestive juices and poor digestion and absorption of food. The vasoconstriction in the kidney results in a decreased output of urine. The vasoconstriction in the skin of the extremities makes them pale and cold. The high oxygen extraction from the blood perfusing the skin results in a bluish discolouration of the skin (peripheral cyanosis). In severe cases the tips of the fingers may become gangrenous. The vasoconstriction in muscle may mean that oxygen delivery is unable to meet the metabolic demands of exercise. When this happens, anaerobic metabolism is used for energy production and results in excessive formation of lactic acid. The resulting metabolic acidosis stimulates breathing and may account in part for the breathlessness with exercise.

The arteriolar vasoconstriction also causes a fall in capillary blood pressure so that tissue fluid is mobilized into the blood. This accounts in part for the raised blood volume in heart failure. Another factor tending to raise blood volume is the increased secretion of aldosterone from the adrenal cortex in heart failure. This causes increased sodium reabsorption in the kidneys and a general expansion of the extracellular fluid volume. The aldosterone secretion is thought to be a consequence of increased renin liberation from the underperfused kidneys. Reflex constriction of the veins is also a feature of heart failure. The expansion of the blood volume together with the constriction of the veins raises central venous pressure. The resulting increase in end-diastolic pressure increases the strength of ventricular contraction.

The expansion of the extracellular fluid volume is mainly responsible for the *oedema* which is such a conspicuous feature of congestive cardiac failure.

*Underperfusion of the brain and myocardium.*—When the compensatory mechanisms are insuffienct to maintain an adequate perfusion for the brain, the patient becomes confused and may show mental disturbances (cardiac psychosis). If very severe, loss of consciousness occurs due to lack of oxygen in the brain.

If the perfusion pressure is inadequate to supply the needs of the myocardium, myocardial weakness becomes superimposed on the original failure. By Making failure worse it sets up positive feed-back or a vicious circle and death follows rapidly.

### Effects of Inadequate Ventricular Input (Backward Failure)

When the ventricle fails it is unable to pump away all the blood which is delivered to it. Blood then accumulates in and behind the failing chamber with effects which depend on the chambers or vessels involved.

*Left ventricle.*—When the ventricle fails, its ventricular function curve shifts downward and to the right (Fig. 26). On this curve an adequate stroke work cannot be achieved at the normal end-diastolic pressure. As blood accumulates in the ventricle, the end-diastolic pressure rises and the ventricle contracts more forcibly. Thus the ventricle first compensates for failure by dilating and moving further out is ventricular function curve.

*Left atrium.*—As the left ventricle fails, blood accumulates in and distends the left atrium. The rise in end-diastolic pressure in the atrium causes it to contract with greater vigour. This helps to raise the end-diastolic pressure in the left ventricle enabling it to accomplish more stroke work. Turbulence produced by blood expelled into the ventricle by this atrial contraction may be heard as a sound just before the first heart sound. The additional sound with each cardiac cycle gives rise to a triple sound which is described as a *pre-systolic gallop rhythm.* That this sound is due to atrial contraction is confirmed by the fact that it is not heard when atrial fibrillation is present.

*Pulmonary blood vessels.*—As pressure rises in the left atrium, so does pressure in the pulmonary vessels, and may exceed the colloid osmotic pressure of the plasma proteins. Normally the hydrostatic pressure in pulmonary capillaries is much lower than the osmotic pressure of the plasma proteins. When the osmotic pressure is exceeded, fluid leaves the capillaries to accumulate in the alveoli. If the amount leaving is great, the resulting *pulmonary oedema* may produce severe respiratory embarrassment by interfering with alveolar ventilation. In less severe cases it may account for the wet crackling noise (*crepitations*) heard on auscultation of the bases of the lungs. Sometimes small vessels are ruptured by the high pressure and the blood which escapes is coughed up in the sputum.

Another problem associated with the increased pulmonary blood volume is a reduction in the compliance of the lungs. The vessels are made more rigid by the blood they contain and more work is therefore required to ventilate the lungs. This may give the sensation of difficulty with breathing (dyspnoea) during exercise.

When the patient lies down, blood is shifted from the legs to chest and may precipitate a bad attack of *pulmonary oedema*. As the left ventricle is suddenly overloaded the sudden distension of the pulmonary vessels may give rise to acute respiratory distress (*cardiac asthma or paroxysmal dyspnoea*). The filling of the pulmonary vessels reduces the vital capacity and congestion may irritate the mucous membranes in the respiratory tract giving rise to coughing. The under-ventilation and interference with gaseous diffusion tends to lead to impaired oxygenation of the blood and accumulation of carbon dioxide.

*Right venticle.*—When the pressure rises in the left atrium and pulmonary capillaries, the right ventricle has to generate a greater pressure in the pulmonary artery to pump blood across the pulmonary circuit. This is achieved by the rise in end-diastolic pressure in the right ventricle which moves ventricular function further out its ventricular function curve to generate more stroke work. With increased work, the right ventricle muscle increases in bulk and the pressure in the pulmonary artery becomes persistently raised. Pulmonary hypertension may be made worse if there is an associated respiratory disease which lowers alveolar $Po_2$. Alveolar hypoxia causes local pulmonary vasoconstriction and can, by itself cause pulmonary hypertension.

In the early stages of left heart failure, the right ventricle overcomes the additional work load imposed upon it by ventricular dilatation and hypertrophy. However, for the reasons discussed in the section on hypertension, hypertrophy is not an effective way of increasing work output after the diameter of the muscle fibres exceeds a certain value. After this, the right ventricle begins to fail.

*Right atrium.*—When the right ventricle fails, its function may be helped by the increase in the strength of right atrial contractions as atrial end-diastolic pressure rises. When stretched, the atrium contracts more strongly to cause greater filling and therefore greater stroke work in the ventricle.

*Systemic veins.*—When the right heart fails, blood accumulates in the systemic veins giving many of the classical signs of *congestive heart failure*. The rise in venous pressure is accentuated by the retention of salt and water and by the reflex constriction of the veins which occurs in heart failure. Thus the *jugular venous pressure* is raised. The back pressure in the systemic veins and the expansion of extracellular fluid volume so raise capillary pressure that there is an increased formation of tissue fluid. This causes *oedema* which is most marked in the dependent part of the body where gravitational forces further increase the capillary pressure. In patients who stand, the oedema is most marked in the region of the ankles, whereas in bed-ridden patients the oedema is greatest in the sacral area.

The raised venous pressure tends to distend the *liver* so that it can be felt in the abdomen below the ribs. In long-standing cases of congestive cardiac failure damage to the liver may result in jaundice and impaired liver function. Congestion of the gastric mucosa may cause loss of appetite and vomiting. Fluid may accumulate in the peritoneal cavity (ascites—Gr. *askos* = bag) and in the pleural cavities (pleural effusion) due to the raised capillary pressure in these regions.

The widespread consequences of left ventricular failure and their sequence of development are summarized in Fig. 35.

## Causes

The heart may fail, in part or as a whole, if its muscle is damaged by disease or if it is persistently given excessive work to do over a long period or if its rhythm is seriously disturbed. In many instances of heart failure, all factors are involved.

### Myocardial Damage

Any condition which impairs the delivery of oxygen to the heart muscle is a potential source of damage. Thus obstruction of the coronary arteries as in coronary thrombosis, a reduced oxygen-carrying capacity of the blood as in anaemia or a very low arterial pressure may lead to weakness or death of myocardial tissue.

Some diseases damage the myocardium. For example, rheumatic fever may produce inflammatory changes in the muscle as well as the valves of the heart.

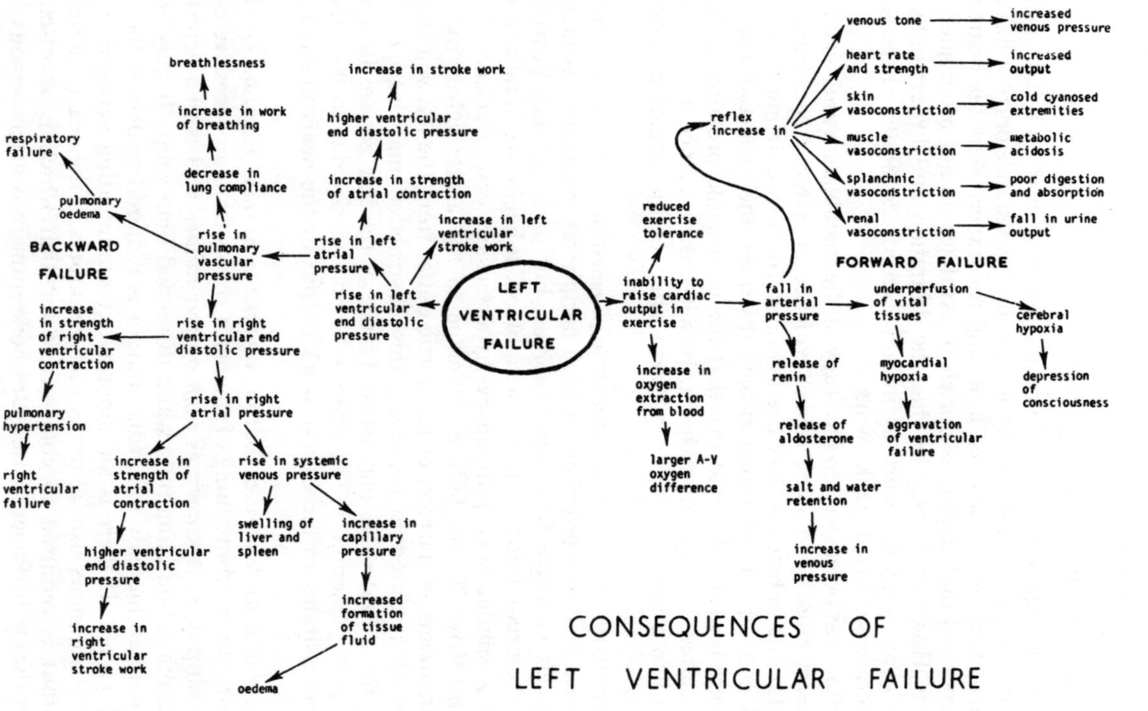


FIG. 35.—The consequences of left ventricular failure.

Bacterial toxins such as those produced by the diphtheria bacillus and certain viral infections may weaken the heart muscle. Sometimes the cause of the myocardial weakness is not obvious. An epidemic of heart failure which occurred in America was found to be due to poisoning by cobalt which had been added to beer during its manufacture.

### Persistently High Work Load

A persistently high work load is imposed on the heart if (a) it has to shift the normal cardiac output against an abnormally high resistance or (b) it has to shift an abnormally high output against a normal resistance. Why the heart should fail with persistent overloading is not clear. Athletes do not often succumb to heart failure. However, it may be that the dilatation and hypertrophy which result from continuous overwork eventually exceed the limits where they are able to generate increased stroke work.

Much effort has been given to the study of the metabolism and biochemistry of failing heart muscle. In some cases, evidence of reduced concentrations of energy containing substrates has been found. In others, it appears that the weakness is due to poor utilization of the substrates, their concentrations being normal. There is also some evidence that the contractile proteins in the myocardium are altered in the failing heart. However the evidence does not add up to anything like a tidy story, and a clarification of the situation must await further work.

(*a*) *Pumping a normal output against high resistance.*—In hypertension, an increased work load is imposed by the high peripheral resistance. As mentioned earlier, the left ventricle first hypertrophies, but as its growth outstrips the capacity of the circulation to supply it with blood, it begins to dilate and go into failure. In a similar way pulmonary hypertension can overload the right ventricle and drive it into failure. Damage to the aortic valves may cause stenosis (Gr. *stenos* = narrow) of the aortic outlet from the left ventricle. The increased work required to overcome this obstruction may lead to failure. Stenosis of the mitral valve may cause failure of the left atrium in a similar fashion. In polycythaemia, the increase in blood viscosity due to the raised red cell count necessitates an increase in work output by the heart to circulate the blood.

(*b*) *Pumping a high cardiac output against a normal resistance.*—In this type of failure the overburdening of the heart comes from an excessive demand for cardiac output. It is known as *high output failure* and is characterized by warm extremities, bounding pulses and distended limb veins. It may occur in anaemia where an increase in cardiac output is required to deliver the normal quantity of oxygen. If there is an abnormal communication between an artery and vein (arteriovenous fistula) or between chambers in the heart, an increased cardiac output is required to compensate for that which is short-circuited through the shunt. In Paget's disease of bone numerous arteriovenous fistulae develop in the bone and may give rise to high output failure. A high cardiac output is also required when a heart valve becomes incompetent. For example, if the valve regulating the exit from the left ventricle becomes incompetent (aortic incompetence), the left ventricle has to increase its stroke output to compensate for the amount of blood which regurgitates back from the aorta to the ventricle during diastole. Excessive thyroxine secretion (thyrotoxicosis), vitamin $B_1$ deficiency (beri-beri) and chronic lung diseases which cause a reduced $Po_2$ in arterial blood may also cause a high output failure.

### Abnormal Rhythms

In conditions where conduction or rhythm in the heart is grossly disordered, the efficiency of heart action may be so reduced that it is unable to maintain an adequate cardiac output. Disorders of rhythm, which can be tolerated by a healthy heart, may precipitate failure in a diseased heart.

Conditions associated with very fast heart rates, such as ventricular tachycardia, may lead to failure by limiting the time available for ventricular filling in diastole. If multiple pacemakers start generating impulses all over the ventricular muscle (ventricular fibrillation), the patchy irregular contractile behaviour that results cannot maintain stroke volume. Cardiac output falls to zero with ventricular fibrillation.

Conditions associated with very slow heart rates such as those which occur with damage to the conducting system of the heart (heart block) or with certain forms of drug therapy may also precipitate failure. At very low rates, the heart may not be able to increase stroke volume enough to maintain cardiac output. In addition, the greater diastolic filling may result in such dilatation of the heart that it cannot operate efficiently.

Conditions associated with abnormal conduction routes in the heart may also result in inefficient heart action. If the impulse does not arise in the sino-atrial node, the advantages of atrial contraction on ventricular filling are lost because the timing sequence is disturbed.

### Aggravating Factors in Heart Failure

No matter what the basic cause of the failure, the situation is always made worse by additional factors which stress the circulation. These include exertion, anxiety, pregnancy, a hot environment and infection.

## INVESTIGATIONS

Though clinical examination is most important in the diagnosis of heart failure, there are some special investigations which may aid diagnosis. They include:

(i) **Electrocardiography.**—This procedure is described fully elsewhere; an electrocardiogram may provide evidence of myocardial damage, abnormal rhythm and conduction and of ventricular hypertrophy.

(ii) **Chest X-ray.**—The X-ray shadow may provide evidence of dilatation or enlargement of the heart and of which chambers are involved. Prominent shadows of blood vessels in the lung fields suggest left ventricular failure.

(iii) **Cardiac catheterization.**—Catheters may be introduced into peripheral veins and advanced until they enter the right side of the heart. Abnormalities of the pressure or oxygen content of blood in the different chambers and vessels may help diagnosis. It is more difficult to study the left side of the heart. The pressure in the left atrium is sometimes estimated from a catheter pushed through the right side of the heart and "wedged" in a small pulmonary artery.

The "wedge" pressure is thought to reflect the left atrial pressure only and not be affected by the pulmonary artery pressure. A catheter may also be introduced into an artery such as the femoral, and pushed back until it enters the left side of the heart. A needle may be inserted into the left atrium through the front wall of the trachea using an instrument which permits visualization of the bronchial tree (a bronchoscope). Radio-opaque fluid introduced through a catheter may be followed by X-ray examination during its passage through the heart; it may up abnormal valve action.
show up abnormal valve action.

(iv) **Blood enzyme studies.**—When myocardial tissue dies some of the enzymes contained in it are released into the blood. Finding raised levels of these enzymes in the blood provides evidence that some myocardial tissue has died. Products released from dead myocardium also cause modest increases in body temperature, white cell count and erythrocyte sedimentation rate.

(v) **Miscellaneous.**—A great variety of tests may be employed to determine the disease which is responsible for the heart failure. Obvious ones would include those to test thyroid function, and to see if anaemia was present.

## Treatment

**Treat the cause.**—Where this is remediable as in anaemia and thyrotoxicosis, treatment of the cause is the obvious first step.

The part played by surgery in the treatment of heart disease is gradually increasing. Surgeons can open up valve orifices which have been narrowed by disease, replace valves which have become incompetent or close abnormal communications between heart chambers or blood vessels. Eventually they may be able to replace hearts whose muscle has been irreparably damaged if the problems of heart transplantation are solved. Where a segment of a coronary artery is blocked, it is sometimes possible to remove the narrowed segment and replace it with a segment of normal vessel.

**Remove aggravating factors.**—Prevention of exertion and anxiety by bed rest and sedation produces great improvement in patients with cardiac failure. Treatment of respiratory and other infections is also helpful.

**Strengthen the heart muscle.**—The digitalis glycosides shift the ventricular function curve upwards and to the left in a similar fashion to sympathetic stimulation. Their precise mode of action on the muscle is not known. However, as shown in Fig. 36, they improve function in the failing ventricle and allow the ventricle to do its work at a lower end-diastolic pressure. This explains how digitalis tends to lower the venous pressure in heart failure. The fall in venous pressure allows mobilization and excretion of oedema fluid. The slowing of the heart rate and the increase in the output of urine that follow treatment with digitalis is partly explained by the increase in the cardiac output with the improvement in ventricular function.

As the sympathetic drive to the heart is a compensating factor helping to maintain cardiac output, the use of beta blocking drugs, e.g. propranolol, may make heart failure worse and should be avoided.

**Improve heart rhythm.**—Abnormal pacemaker activity leading to disordered rhythms may be countered by giving drugs which depress the excitability of the heart. Atrial fibrillation may be converted back to normal rhythm by applying a large voltage to electrodes placed on the chest. If there is great slowing of the heart due to vagal activity as may occur in the early stages after obstruction of a coronary artery, atropine is the drug of choice. If the slowing is permanent and due to some abnormality in the conducting system of the heart, the heart can be driven electrically (paced) at a suitable rate by small electrodes and a power source implanted in the body.

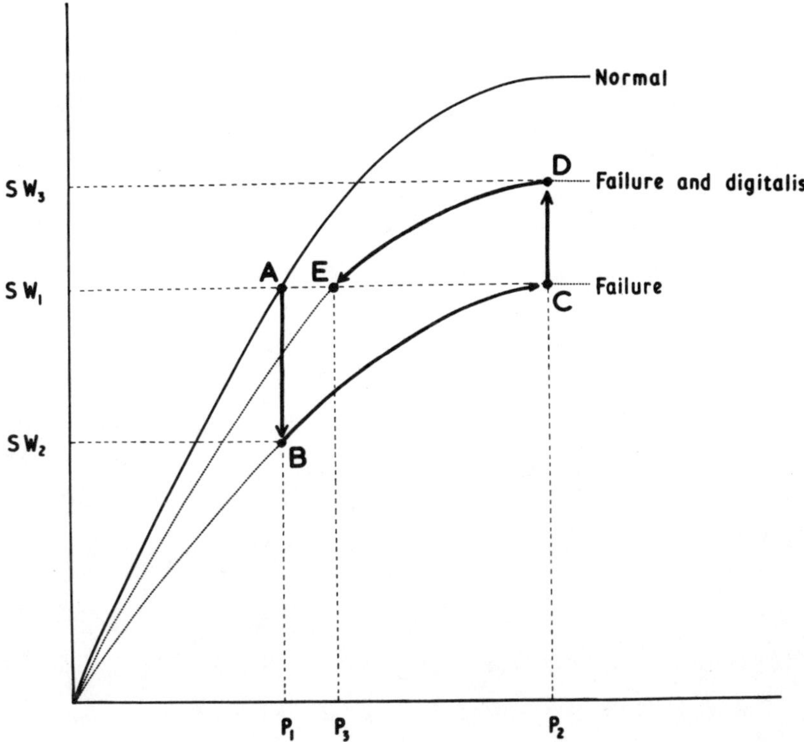


FIG. 36.—The effect of failure and digitalis treatment on ventricular function. When the heart fails, the normal stroke work falls from A to B. The ventricle compensates for this by dilating to C where the original stroke work is accomplished at a higher end-diastolic pressure ($P_2$). With digitalis, the patient increases the stroke work to D at this end-diastolic pressure ($P_2$). This is greater than the stroke work required so the end-diastolic pressure falls to E. At this point the required stroke work is accomplished at a slightly raised end-diastolic pressure ($P_3$).

**Decrease extracellular fluid.**—As explained earlier, salt and water retention is a feature of central circulatory failure. This is a compensatory mechanism which results in an increase in central venous pressure, an increase in diastolic pressure and an increase in stroke work. However, in severe congestive cardiac failure, the oedema and congestion of the lungs and other tissues may become the most incapacitating feature of the disease. In addition, the increase in end-diastolic pressure may overstretch the heart to a point where further stretching causes a reduction, rather than an increase, in stroke work, i.e. to the descending limb of the ventricular function curve. When this is the case, a reduction in blood volume not only relieves congestion but causes stroke work to increase. When congestion is a problem, treatment must aim at reducing extracellular fluid volume. Low sodium diets, drugs which increase urinary output, and removal of a fraction of the blood volume through an intravenous needle are used for this purpose.

# 25. CARDIAC ARREST

## Definition

In cardiac arrest, the output from the heart suddenly falls to about zero to cause acute circulatory failure.

## Effects

With cardiac arrest, delivery of oxygen to the tissues stops. The organ most affected by this is the brain. Consciousness is lost within a few seconds and convulsions often appear within a minute. Depression of the respiratory centre results in cessation of breathing. After about 5-10 minutes of cardiac arrest, the brain cells are irreparably damaged. Restoration of the circulation after this time may permit the body to recover but the mind will never return to its former state.

Following cardiac arrest the skin is first very pale and then becomes blue as the stationary blood in the minute vessels of the skin loses its oxygen. No pulses can be felt in the arteries or heart sounds heard in the chest. The pupils become widely dilated due to cerebral hypoxia and consequent intense sympathetic nervous activity. Evacuation of the bowel and bladder may occur. Fixed dilated pupils indicate that the brain has sustained serious damage.

## Causes

Cardiac arrest may be due to the heart stopping beating (ventricular asystole), or to fibrillation of ventricular muscle due to abnormal pacemaker activity in ventricular muscle (ventricular fibrillation). In both cases the output from the ventricle is effectively zero. These conditions may occur with electrocution, sudden obstruction of a coronary artery (coronary thrombosis), surgical operation under general anaesthesia, injection of a drug to which the patient is hypersensitive, or with sudden block of the conducting system of the heart (heart block).

The way these conditions give rise to fibrillation or asystole is not clear. Following coronary thrombosis asystole may result from reflex inhibition of the heart via the vagus in response to stimulation of sensory nerve endings in the damaged heart. In heart block, asystole may be due to failure of the ventricle to set up its own pacemaker system when the conducting pathway from atrial muscle is blocked. The mechanism of ventricular fibrillation is thought to be as follows: Co-ordinated activation of the ventricles comes to an end as the result of bombardment of premature impulses re-entering previously excited muscle from an "irritable focus", e.g. an area of heart muscle deprived of its blood supply. Repolarization and reactivation become less and less co-ordinated during the resulting series of extrasystoles until the final stage of chaos is reached where activation and repolarization become completely desynchronized.

## Investigations

The time available for instituting investigations is brief in cardiac arrest. Diagnosis of the cause of arrest may be made from the electrocardiogram. The electrocardiogram patterns of asystole and ventricular fibrillation are quite characteristic. In situations where cardiac arrest is likely to occur, as in certain surgical operations and following coronary thrombosis, continuous monitoring of the electrocardiogram permits treatment to begin immediately the cardiac arrest occurs.

## Treatment

Cardiac arrest presents a critical medical emergency where prompt effective treatment is essential.

The aim of treatment is to provide some cardiac output by massaging the heart and some ventilation of the lungs by mouth to mouth artificial respiration until such time as the heart begins to beat again or the ventricular fibrillation is corrected. To get some cardiac output, the heart is compressed between the sternum and the vertebral column by applying strong pressure rhythmically to the front of the chest, at a rate of about 60 presses per minute. Artificial respiration is carried out by blowing expired air into the patient's mouth to inflate his lungs. The patient expires this gas by the passive recoil of his lungs and thoracic cage.

If it is established by electrocardiography that the patient has ventricular fibrillation, this may be treated by applying a strong voltage through electrodes placed on either side of the chest. This usually causes asystole and when the heart starts beating again it does so in normal rhythm.

In cases of asystole in which the cardiac tone is relatively normal, it has been found possible to "pace" the heart by a series of firm precordial blows. This activity produces a QRS complex after each blow and a powerful cardiac contraction. Injections of noradrenaline may be used to restore cardiac tone because it rapidly produces an arterial blood pressure high enough to provide good coronary perfusion, something which cannot be provided during external cardiac massage. Also it is less likely to produce abnormal rhythms than adrenaline. In the presence of coronary artery disease, external cardiac massage, which produces an arterial blood pressure of about 120/40, does not appear to provide adequate oxygenation of the myocardium. Thus patients with myocardial infarction who suffer cardiac arrest due to ventricular fibrillation will die if the abnormal rhythm is not corrected within 20 or at most 30 minutes, even with optimal massage and artificial respiration.

# 26. LOCAL CIRCULATORY FAILURE

## Definition

THE CONDITION when local blood supply becomes inadequate for tissue needs is known as *ischaemia* (G. *ischein* = I suppress; *haima* = blood). Tissues are well protected against this. If blood flow falls, products of metabolism (metabolites) accumulate in the tissues. Many of these products such as carbon dioxide, oxygen lack, hydrogen ions and lactic acid have vasodilator properties. By relaxing vascular smooth muscle they permit blood flow to rise until the excess metabolites are washed away. This local regulatory mechanism which ensures that local blood flow to a tissue is matched to its metabolic needs does not depend on vasomotor nerves or other outside influences. In addition, if a blood vessel becomes blocked, collateral and anastomotic vessels open up so that the block may be by-passed. If this is insufficient for the metabolic needs of the tissues, new blood vessels will develop.

Ischaemia occurs when the vascular reserves and compensatory mechanisms have been exhausted or overwhelmed.

## Effects

When blood flow to a tissue is inadequate there is insufficient delivery of oxygen and nutrient material to the tissue and inadequate clearance of waste products from the tissues. Such conditions interfere with local metabolism and result in *impairment of function, degenerative changes in structure* and eventually *death of the tissue.* Accumulation of waste products is thought to be responsible for the *pain* that is a feature of some types of ischaemia.

The specific effects of ischaemia vary with the tissue involved and the speed of onset of the ischaemia. Some tissues such as the brain, which have a high metabolic rate, are much more affected by ischaemia than those such as skin, whose metabolic requirements are low. Ischaemia causes more disorder when its onset is rapid because there is less time for compensatory mechanisms to come into play.

### Myocardial Ischaemia

(a) *Acute ischaemia.*—This occurs when the lumen of coronary arteries is narrowed or obliterated by pathological changes in their walls. If a large vessel is suddenly blocked, the muscle supplied by that artery dies. The area of dead tissue becomes pale and swollen and is called and *infarct* (L. *infarcire* = to stuff in). Though collateral and anastomotic vessels do exist, they cannot open up quickly enough when a major vessel is occluded to prevent infarction. For this reason, coronary arteries are called *functional end arteries.*

Infarction can cause the patient to die in one of several ways. If the *damage* is widespread in the left ventricle, the heart will be so weakened that it is not able to maintain an adequate cardiac output to support life and death results from acute heart failure.

*Very rapid heart rates* may upset the ratio of myocardial work to coronary flow and so extend the zone of irreversible myocardial damage. Increased sympathetic tone may lead to an increase in heart rate (*sinus tachycardia*) which increases myocardial work yet reduces the time available for perfusion of the coronary arteries. Death of myocardial tissue often sets up abnormal pacemaker activity (ectopic foci) in adjacent living muscle which shows an inflammatory response to products released from the infarcted area. A very rapid rhythm may be set up by an ectopic focus and be heralded by ectopic beats originating in the focus. If a number of these abnormal pacemakers are set up randomly in the muscle, the ventricle may fibrillate. In this state the ventricle cannot give regular rhythmic co-ordinated contractions; its muscle ripples ineffectively like "a bag of worms". Cardiac output is virtually zero and the patient collapses and dies.

Another cause of death in the early stages after myocardial infarction is intense reflex inhibition of the heart mediated through the vagus nerves. This causes extreme slowing of the heart (*sinus bradycardia*) or block of the conducting system (atrioventricular block). The afferent limb of this reflex is thought to be the sensory nerves whose receptors are stimulated by chemical changes in the myocardium. Like sinus tachycardia, sinus bradycardia can also upset the ratio of myocardial work to coronary blood flow so that myocardial damage is extended. In this case the fall in coronary blood flow due to the fall in blood pressure may be relatively greater than the reduction in myocardial work. In addition, the heart is more vulnerable to the effects of abnormal pacemaker activity at very slow heart rates and there is an increased risk of ventricular fibrillation.

In the early stages after infarction, products released from the damaged tissue cause a rise in body temperature and an increase in the white cell count. The blood level of certain enzymes normally found in the heart increases.

Most deaths from acute myocardial ischaemia occur in the first few hours after the event. However those who survive a week or so are still at risk. The infarcted area of the ventricular wall may rupture and allow blood to escape into the pericardial sac. This produces "*cardiac tamponade*". The ventricle can no longer fill effectively and the patient dies of acute heart failure. If the infarcted area includes the endocardial layer of the heart, a thrombus may form on the damaged area. If it detaches it may block a major artery such as the carotid.

However, if all goes well and none of these catastrophes occur, the infarcted area is organized and replaced with a fibrous scar. Hearts can work quite efficiently with such scars.

(b) *Chronic ischaemia.*—If the narrowing of the coronary arteries develops gradually, the effects are less dramatic as time is permitted for the development of collateral vessels. However, degenerative changes in the myocardium may be seen at postmortem, in addition to the new anastomotic channels. Small areas of tissue which die from local hypoxia are replaced with fibrous tissue. During exercise or excitement, the work done by the heart may exceed the functional capacity of the coronary circulation to clear metabolites from the myocardium and cause pain in the chest (*angina pectoris*). Rest decreases myocardial metabolism and relieves the pain. The nature of the metabolite producing pain is not known. It may be potassium ion, because potassium is released from myocardial cells when they are excited and it is known to cause pain when applied to the base of a blister on the skin.

Pain arising in the heart is felt in the chest and neck and may radiate down the left arm. This is because these areas are served by the same spinal cord segments as the heart.

### Brain Ischaemia

(a)*Acute ischaemia.*—If the blood flow to the brain stops, consciousness is lost in 5-10 seconds and the brain tissue is irreparably damaged within about 5-10 minutes. With its high rate of metabolism and absence of oxygen stores, brain tissue cannot tolerate hypoxia.

If a major cerabral artery is suddenly blocked by an embolus, the part supplied by that artery dies to form a cerebral infarct. This gives rise to a neurological condition whose features depend on the area affected. The area of infarction is usually much smaller than the normal distribution of the affected artery because of good anastomosis between cerebral vessels, notably at the circle of Willis.

(b) *Chronic ischaemia.*—This is usually due to degenerative changes in the small cerebral arteries. It is associated with gradual mental deterioration (*dementia*) as patchy destruction of neurones occurs over a wide area from lack of blood supply. Damage to the pyramidal tract or basal ganglia can also occur to cause disorders of movement and muscle tone.

### Renal Ischaemia

This may occur in an acute form when a renal artery is suddenly blocked by an embolus or in the extreme vasoconstriction which occurs in severe circulatory failure. In both cases, death occurs in the cortical parts of the kidney (*tubular necrosis*). This is because the convoluted tubules in the cortex have higher metabolic needs than the ducts in the medulla.

A less severe reduction in blood flow causes release of renin from the juxta-glomerular apparatus which leads to a compensatory rise in arterial pressure. If this is inadequate to maintain glomerular filtration, renal failure will occur.

### Lung Ischaemia

As the lungs are supplied by both pulmonary and bronchial arteries, obstruction of a pulmonary artery does not cause death of the tissue supplied by it unless there is some pre-existing cause of circulatory failure. However, there is a local decrease in the ventilation of the underperfused area of lung which is probably due to the fall in $Pco_2$ and rise in $Po_2$ of the air which ventilates it. The affected part of the lung tends to collapse. This is a compensatory mechanism which prevents wasteful ventilation of underperfused alveoli. Collapse is thought to be due to a decrease in the alveolar surfactant. A normal blood supply seems to be necessary for its manufacture. Lung surfactant normally prevents collapse of the alveoli by reducing the surface forces of the fluid which coats the alveolar walls.

**Liver Ischaemia**

Ischaemia of liver is rarely sufficient to cause tissue death unless the hepatic artery is completely and suddenly blocked. However, impairment of liver function can be caused by factors which reduce hepatic blood flow, such as the splanchnic vasoconstriction in peripheral circulatory failure.

**Limb Ischaemia**

(a) *Acute ischaemia.*—The limbs have a good collateral circulation so that even when a major artery in a healthy individual is completely blocked, opening of collaterals is adequate to maintain the life of the tissues. There is a reduction in his reserve capacity to supply the limb with blood in exercise. However, growth and development of new collateral vessels may make this good.

When the arterial vessels show degenerative changes, good compensation is not possible and occlusion of an artery results in the limb becoming cold and pale. Eventually the peripheral tissues die, becoming black and gangrenous. A line of demarcation develops between the living and the dead tissue which later separates from the rest of the limb. There is no attempt by the body to regrow the part of the limb which dies.

Gangrene (Gr. *graino* = I gnaw) is used to describe death of part of the body large enough to be seen. Necrosis (Gr. *nekrosis* = *death*) is normally used to describe death of internal tissues. There are many exceptions to this usage.

(b) *Chronic ischaemia.*—With slow obliteration of limb blood vessels by degenerative arterial disease, ischaemia produces a variety of effects. Though muscle blood flow may be adequate for resting conditions, it may be indequate to support exercise. After walking for some distance the patient has to stop because of pain in the calf. This is due to the accumulation of a pain-producing metabolite in the muscle. The pain disappears with rest as the metabolite concentration falls. Thus, walking has to be interrupted periodically because of pain. The condition is called *intermittent claudication* (L. *claudicatio* = I limp). Degenerative changes occur in the tissues. The skin becomes shiny and loses its hairs. Cuts and abrasions are very slow to heal and may develop into persistent skin ulcers. Eventually the extremities of the limbs may become gangrenous.

**Gut Ischaemia**

(a) *Acute ischaemia.*—This may occur if a segment of the gut becomes twisted or constrained in such a way that its blood supply is obstructed. It causes death of the cells in the gut wall and the results may be disastrous. As the affected gut cannot propel food material, intestinal obstruction occurs. Even more dangerous is the likelihood of rupture of the gangrenous wall and the release of intestinal contents into the peritoneal cavity. This leads to severe infection and is often fatal. Severe bleeding can also occur from disintegrating blood vessels in the gangrenous area.

(b) *Chronic ischaemia.*—This is rare but may result from widespread degenerative arterial disease. Intestinal function is depressed in that the digestion and absorption of food is impaired. Increased intestinal activity may result in intestinal discomfort ("gut angina").

## CAUSES

Tissue ischaemia occurs with a wide variety of conditions which lower the blood pressure or obstruct blood vessels. Factors which lower blood pressure have been considered in the chapter on peripheral circulatory failure.

Vascular obstruction can occur with:

**External pressure.**—When an elderly or unconscious patient lies without moving in bed for some time, the pressure on skin over bony prominences may so restrict blood flow that the skin and subcutaneous tissues die to cause local ulceration (*bed sores*). A tightly fitting plaster cast applied to a fractured limb may cause severe tissue damage by vascular compression. In the forearm this may cause some necrosis of skeletal muscle. The replacement of the necrotic tissue with a fibrous scar which subsequently contracts may produce disablement and deformity known as *Volkmann's ischaemic contracture.*

**Vascular spasm.**—In some people the blood vessels of the extremities are unusually sensitive to cold. When the extremities are exposed to cold, the arteries in the digits go into spasm and the skin becomes white and cold. Later the fingers become painful and blue. Usually the spasm wears off spontaneously but necrosis of the skin of the finger-tips can occur. The condition is known as *Raynaud's phenomenon.* The cause of this condition is not known. Vascular spasm due to damage to cerebral blood vessels by severe hypertension is thought to be responsible for the convulsions and transient paralysis that occur in hypertensive encephalopathy. Spasm also may occur accidentally as when the anaesthetic agent pentothal, which is a powerful vasoconstrictor in high concentration, is inadvertently injected into an artery.

**Embolism.**—An embolus is an abnormal mass of material which travels in the circulation. If it arises in the systemic veins it tends to lodge in the pulmonary artery and its branches. If it arises in the pulmonary veins or the left side of the heart it lodges in some branch of the systemic arteries.

The embolus may consist of blood clot formed on a damaged area of vascular endothelium which breaks free. It may consist of air sucked into an incision in a vein whose intraluminal pressure is subatmospheric (air embolism). Such an embolus is converted into froth by contractions in the right side of the heart. The high viscosity of froth can stop the circulation. Fat droplets released from a fractured bone can also act as emboli (fat embolism). Many of these tiny fat emboli lodge in the lungs and may cause disturbance of respiratory function. Others pass through the lungs to lodge in the systemic circulation where they may cause disturbance of cerebral and renal function.

**Thrombosis.**—When an area of the endothelial lining of the heart or blood vessels is damaged, platelets adhere and may provoke the local formation of a clot or thrombus which blocks the vessel. This will occur more readily if the blood shows an increased propensity for clotting. Thrombus formation on arterial lining damged by degenerative change is responsible for many cases of myocardial and cerebral ischaemia.

**Arterial disease.**— If the arterial wall is affected by disease, the lumen of the vessels may be narrowed or obliterated. In addition the vessels may lose their ability to dilate in response to metabolic activity.

## INVESTIGATIONS

### Assessment of Regional Flow

Many techniques of varying complexity are available for estimating regional blood flow in different parts of the body.

*Skin temperature measurement.*—In the extremities, skin temperature is determined almost exclusively by the amount of blood flow. The heat produced by local metabolism is negligible. Thus measurements of skin temperature are a useful index of flow at these sites. In proximal parts of the body, skin temperature depends more on the underlying tissue temperature and does not vary much with local changes in blood flow.

*Venous occlusion plethysmography.*—With this technique, the volume of part of an extremity is measured with a plethysmograph. When the veins draining that part are occluded by a cuff inflated to a pressure somewhat below arterial diastolic pressure, blood can enter by the arteries but cannot leave by the veins. The part then swells at a rate equal to the rate of arterial inflow.

*Fick principle.*—When an organ takes up or eliminates a substance from the blood, the rate of blood flow through the organ can be calculated if the rate of uptake or excretion and the concentrations of the substance in the blood entering and leaving the organ are known. Use of this principle has been made to measure blood flow in the kidney (using PAH excretion), brain (using nitrous oxide uptake) and heart (using nitrous oxide uptake).

*Indicator clearance techniques.*—When an indicator such as an isotope is injected into a tissue, its rate of clearance from that tissue is a function of the blood flow at the site of injection. The rate of clearance of radioactive sodium when injected into skeletal muscle has been used as an index of skeletal muscle blood flow. These techniques are really further applications of the Fick principle.

### Angiography

The patency or otherwise of the vessels in a vascular bed may be visualized by taking X-ray pictures of tissues following injection of radio-opaque material into the artery. Blocks in the arterial system can be accurately located by this technique.

### Sympathetic Release Tests

These tests are performed to see how much improvement in blood flow can be expected following therapeutic section of the sympathetic vasoconstrictor nerves. If the area in question is the hand, blood flow to the hand is estimated (e.g. by skin temperature) before and after release of sympathetic vasoconstrictor tone. Tone may be released by infiltrating the appropriate nerves with local anaesthetic solution. More commonly tone is released reflexly by putting the subject's legs in warm water (indirect heating). The ensuing rise in body temperature normally causes a reflex increase in hand blood flow. Failure of flow to increase with indirect heating suggests that there is local obstruction of the blood vessels and that surgical section of the sympathetic nerves to the hand would not help local blood flow.

**Miscellaneous**

When tissues die from ischaemia, they may release some of their enzymes into the circulation. Thus, raised blood levels of certain enzymes are found in the blood after infarction of the heart. Cell degradation products are probably responsible for the moderate rise in body temperature and white cell count that occur after death of cardiac tissue.

## Treatment

The two basic objectives of treatment of local ischaemia are to *increase local blood flow* and to *reduce the metabolic needs of the organ.*

**Increasing Local Blood Flow**

(*a*) *Removal of obstructions.*—If there is a constriction or block localized to a small segment in an artery, this may be treated by surgical removal of the segment and its replacement with a substitute vessel. For example if disease results in obstruction of only one segment of the coronary arteries, the offending segment may be replaced with a transplanted segment of vein. If a major artery is blocked by an embolus, surgical removal of the embolus may save the tissue supplied by that artery.

(*b*) *Vasodilator drugs.*—There are many drugs which dilate blood vessels but they are of limited use in the treatment of local ischaemia. Some act by a direct relaxant effect on the vascular smooth muscle, others by blocking adrenergic transmission at sympathetic vasoconstrictor nerves. When they are given by mouth or intravenously, a dilator action capable of opening the diseased vessels would also dilate all the other vessels in the body. The resulting drop in peripheral resistance would make the patient liable to faint whenever he stood up (*postural hypotension*). Another problem with the use of vasodilator drugs is that the dilatation of vessels in healthy tissue may "steal" blood from the ischaemic tissue.

(*c*) *Section of sympathetic nerves* (*sympathectomy*).—The blood vessels in the skin, and to a lesser extent, the muscles of the extremities, are normally held constricted by the action of sympathetic nerves. Lumbar or cervical sympathectomy thus increases blood flow to the leg and arm respectively. Its value is limited because of a progressive return of intrinsic tone to the sympathectomized vessels which occurs progressively in the weeks following operation. At least part of the return of tone seems to be due to a steep increase in the sensitivity of the denervated vessels to circulating adrenaline and noradrenaline (*denervation hypersensitivity*).

(*d*) *Anticoagulant therapy.*—Where ischaemia is thought to be due to widespread clotting occurring in blood vessels, drugs which inhibit clotting have been tried without much success.

**Decreasing Metabolic Requirements**

This can be achieved by reducing the work carried out by the ischaemic tissue. *Physical rest* can thus help patients with myocardial or skeletal muscle ischaemia. Great *care of the skin* is required in ischaemic extremities so that injuries and infections, which might give rise to inflammatory responses, are prevented. Ischaemic limbs should not be heated in an attempt to increase local blood flow. This is because the increase in tissue temperature may increase local metabolism more than local blood flow.

*Vasodilator drugs* such as glyceryl trinitrate are used in cases of myocardial ischaemia. They act by decreasing peripheral resistance and so lowering the blood pressure. This reduces the work the heart has to do in ejecting blood. It is unlikely that these drugs do anything for the coronary vessels in coronary ischaemia that is not being done already by the accumulation of metabolites.

*Hypothermia* is a potential method for reducing metabolic requirements. It can be used for the whole body or for parts of the body when local circulation is restricted. The induction of whole body hypothermia is a very complex affair and its value in ischaemia is therefore very limited.

*Beta-blocking drugs* block the action of sympathetic nerves on the heart and thus protect the heart against excessive sympathetic drive during emotional stress or exercise.

# 27. ELECTROCARDIOGRAPHY

THE AIM OF this chapter is to provide an understanding of the normal electrocardiogram (ECG) and some of the commoner and more definite abnormalities to be recognized in the ECG. Note that the machine which produces the record is the electrocardio*graph* and the record produced, the electrocardio*gram*.

## BASIS

Initial understanding of the ECG will be helped by understanding four basic facts:

(i) *The ECG is a recording of induced electrical activity.*—It is not a direct recording of the electrical activity at the membrane of the cardiac muscle cell (the cardiac action potential), or even of the electrical changes to be recorded from the surface of the heart. Instead it is a recording of the electrical activity induced in the body fluids by the electrical impulse which spreads through the heart during each cardiac cycle.

(ii) *The ECG consists of three main deflections from the baseline*—the P wave, the QRS complex and the T wave (Fig. 37). The P wave is due to atrial depolarization, the QRS complex to ventricular depolarization and the T wave to

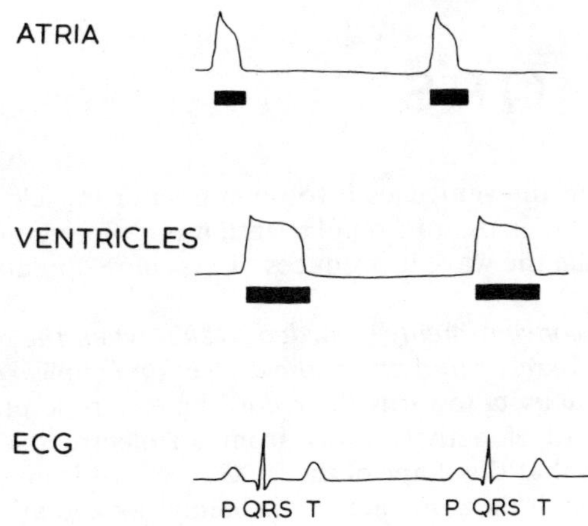


FIG. 37.—Correlation of the ECG with electrical activity in a typical cell in the atria and in the ventricles. The correspondence is:
P — atrial depolarization
QRS — ventricular depolarization
T — ventricular repolarization
Note that systole (black rectangle) in the atria corresponds to the period P-QRS. The longer ventricular systole, due to the longer ventricular action potential, corresponds to the period QRS-T. The time taken for the impulse to spread through atria and ventricles (duration of P, QRS respectively) is small, compared with action potential lengths (PR, RT).

ventricular repolarization. Atrial repolarization does not usually produce any detectable effect. Firstly, any deflection produced is much smaller than the P wave; secondly, it occurs about the time of ventricular depolarization.

Although generally referred to as the QRS complex, the wave due to ventricular *de*polarization often has only one or two components (Fig. 38), e.g. QR, RS etc. Occasionally ventricular *re*polarization produces a second, smaller wave after the T wave and in the same direction as the T wave; this is known as the U wave.

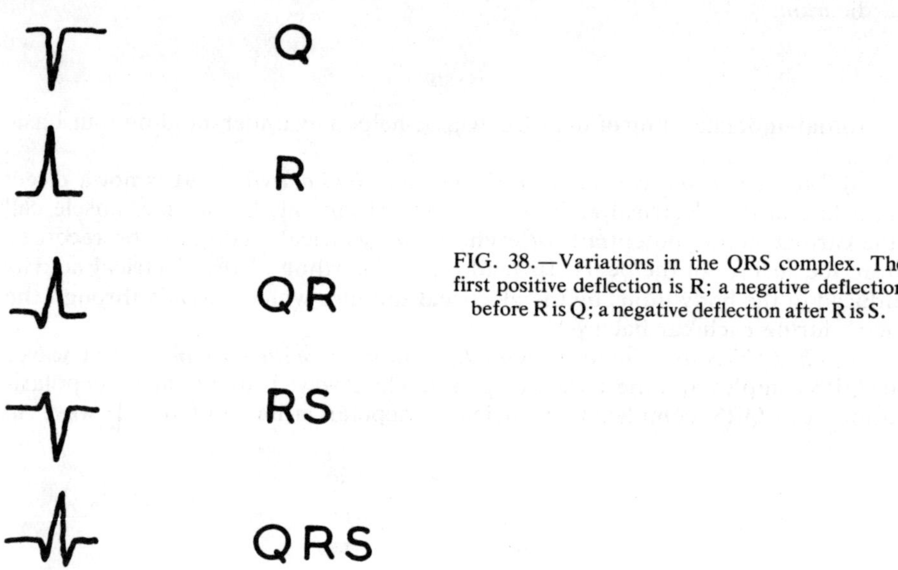


FIG. 38.—Variations in the QRS complex. The first positive deflection is R; a negative deflection before R is Q; a negative deflection after R is S.

Depolarization in the ventricles involves a greater muscle mass than in the atria; hence QRS is taller than P; repolarization is a more gradual process than depolarization; hence the wave it produces, T, is more spread out and less tall than QRS.

(iii) *The ECG is conventionally recorded so that when the cardiac impulse is spreading towards the recording point, the deflection is upwards.*—Because the spread of electrical activity towards the recording electrode produces a positive deflection and spread of activity away from an electrode causes a negative deflection, it follows that the shape of the ECG recorded from different sites will vary widely even though the *same* activity is being "looked at" from the various points (Fig. 39).

(iv) *The basis of the ECG is extremely complex.*—The ECG is the end result of electrical events whose pattern in space and time is complex. It is not at present possible to predict in detail what the ECG of a given heart should look like. Even with the excellent equipment now available it is not possible to construct a detailed map of the electrical events occurring in the heart. Some concepts such as the electrical axis of the heart and the mean cardiac vector, while useful in practice, are somewhat removed from reality, because they summarize in simple form what is inherently complex. Despite all this, the ECG is, of course, a useful diagnostic tool because it provides a clear pattern of the overall electrical events which can then be correlated with other signs of disease.

## The Standard Leads

The three standard leads (or ECG tracings) are bipolar, that is, they consist of the record obtained by recording from two electrodes on the surface of the body. In practice an electrode is applied to both wrists and both ankles, electrode jelly being used to aid conduction from skin to electrode. The electrode on the right ankle is for earthing only and the 3 leads are recorded between the other 3 electrodes:

Lead I records between the left and right arms
Lead II records between the left leg and the right arm.
Lead III records between the left leg and the left arm.

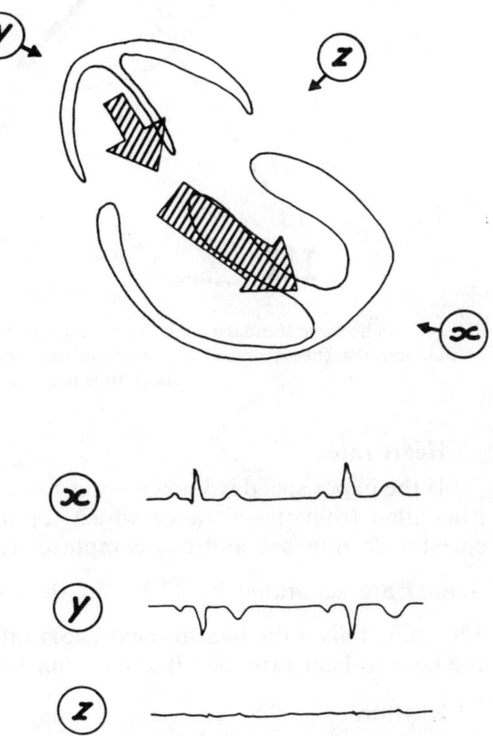


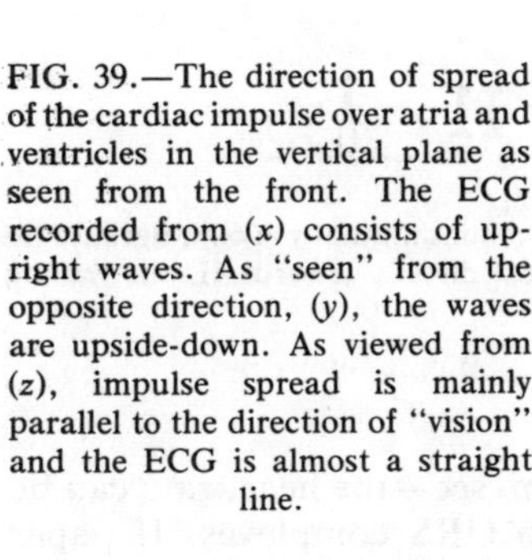

FIG. 39.—The direction of spread of the cardiac impulse over atria and ventricles in the vertical plane as seen from the front. The ECG recorded from ($x$) consists of upright waves. As "seen" from the opposite direction, ($y$), the waves are upside-down. As viewed from ($z$), impulse spread is mainly parallel to the direction of "vision" and the ECG is almost a straight line.

The standard leads are *bipolar leads* and hence cannot strictly be said to *view* the heart from a particular direction. However, their leads are so connected to the ECG machine that in the average normal heart, the P wave, QRS complex and T wave in all three standard leads are predominantly above the baseline. This is a matter of convention. By reversing the lead connections to the ECG machine the complexes would be reversed. Since conventionally an upright deflection implies spread towards the point of reference, the three standard leads may be said to *view* the heart as shown in Fig. 40.

In any patient, the standard leads enable deductions to be drawn about *heart rate, rhythm* and *electrical axis.* In addition, various *abnormal patterns* may be present.

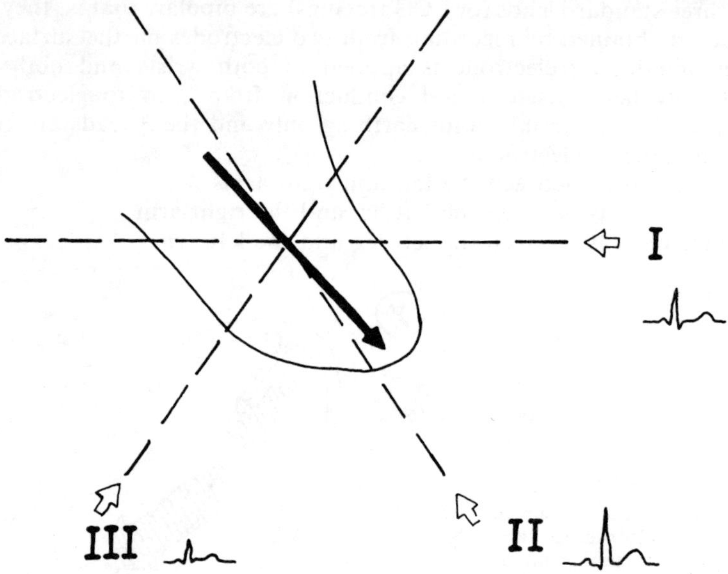


FIG. 40.—The three standard leads "view" the heart from 3 directions mutually at 60°. Because the cardiac impulse (heavy arrow) in the above case spreads almost directly towards II, the tracing amplitude is greatest in lead II.

**Heart rate**

If the paper speed is known—it is usually 25 mm/sec—the heart rate can be calculated from the distance which separates two QRS complexes. If paper speed is 25 mm/sec and two complexes (tips of R waves are useful reference points) are separated by 12.5 mm then the heart rate is $60 \times \frac{25}{12.5}$, i.e. 120/min. This is the beat-to-beat heart rate, if the cardiac rhythm is irregular, the beat-to-beat rate may fluctuate markedly.

**Rhythm**

As with the heart rate, only one lead is required. Lead 2 is often used, because the various waves P, QRS, T are usually clearly recognizable here. Use of the ECG is the most satisfactory way of diagnosing the precise nature of an abnormal cardiac rhythm. The commoner rhythms will now be considered briefly. Some examples are shown diagrammatically in Fig. 41.

*Sinus rhythm.*—This is the normal heart rhythm which is due to impulses that originate in the sinu-atrial (SA) node and follow the normal pathway through atria, atrioventricular (AV) node and the Purkinje tissue in the ventricles (bundle of His, its left and right branches and subsequent smaller branches). The rhythm is diagnosed by the presence of regular QRS and T waves of constant shape in any one lead, each QRS being preceded at a constant interval by a normal P wave.

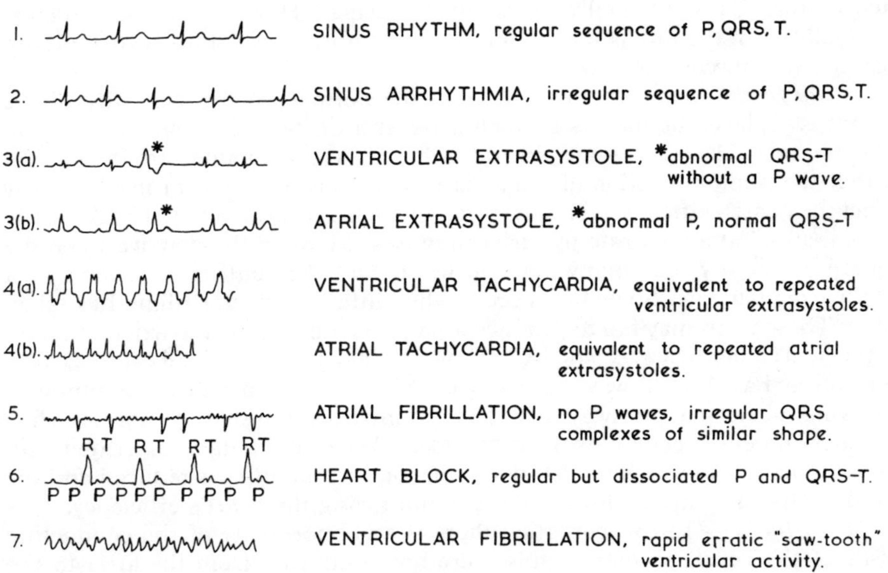


**FIG. 41.—Diagrams of the commoner varieties of cardiac rhythm as shown by the ECG.**

*Sinus arrhythmia.*—This is a normal variation of sinus rhythm, the only difference being that the impulses arise irregularly at the sinus node. The rate increases in inspiration and decreases in expiration, probably because activity in the respiratory centre affects activity in adjoining regions of the medulla which control heart rate. Beat-to-beat rate may vary, for example, from 80-90/min early in inspiration to 50-60/min in expiration. Each QRS-T is preceded by a normal P wave. This rhythm is common in children and young adults; its presence does not indicate any abnormality.

*Extrasystoles.*—These are beats which arise from some region of the heart other than the SA node. Ventricular extrasystoles are usually very obvious. Because they arise in the ventricles there is no P wave. As the impulse spreads through the ventricles by an abnormal pathway, the shape of the QRS wave is abnormal and, because the rapidly conducting Purkinje system is not made use of in the normal manner, QRS tends to be prolonged. Atrial extrasystoles are not so easy to spot as QRS is normal. However, there is no normal P wave and, as with ventricular extrasystoles, if the heart is in sinus rhythm, the regularity of the QRS complexes is disturbed. Occasional extrasystoles occur in normal people, especially if they smoke.

*Tachycardias.*—A rapid heart rate, e.g. 120-180/min at rest is due to stimulation of the heart by abnormally frequent impulses. If these arise in the SA node, the condition is sinus tachycardia, such as may be seen in people with normal hearts during a fever. Other tachycardias are due to impulses arising at an abnormal site. The effect then is of a rapid series of extrasystoles. Thus in atrial tachycardia, the QRS complex resembles the normal QRS; in ventricular tachycardia, QRS is usually markedly abnormal. However, in some cases, especially when the rate is very rapid, it may require considerable experience to distinguish between the two.

*Atrial fibrillation.*—In this condition the normal electrical activity of the atrium is replaced by impulses which arise at an extremely rapid rate—up to 500-600/min. The resultant rapid and feeble atrial contractions are ineffective, so that the pumping action of the atria is lost. This does not, of itself, greatly affect the heart's efficiency. However the ventricle is also affected. The AV node is bombarded at a high rate by atrial impulses. Many of these arrive when the node is refractory, but many activate it, so that the ventricle contracts at a totally irregular rate. Some beats occur when little ventricular filling has taken place. These beats may fail to produce a palpable impulse at the wrist, and add to the heart's inefficiency. The ECG shows atrial activity as small rapid oscillations of the base line. P waves are absent. Since QRS is produced by impulses following the normal pathway through the ventricles, the shape is normal while the interval between complexes varies widely. When the patient is treated with digoxin, the rate at which the AV node can conduct impulses is reduced and the rapid ventricular rate is slowed, thereby improving the heart's efficiency.

*Heart block.*—In this condition there is an abnormality of the conducting tissue of the heart, so that impulses are not conducted from the atria to the ventricles. The effect is similar to that produced by tying the second Stannius ligature between the atria and ventricle of the frog's heart. The atria beat as before, but the ventricles must generate their own impulse, generally at a slower rate than that of the atria and sometimes at a rate which is so slow that heart failure results. The ECG shows regular P waves and also regular QRS and T waves, but the ventricular QRS-T activity is slower than and completely independent of the atrial P activity.

*Ventricular fibrillation.*—This is a condition similar to atrial fibrillation in that electrical impulses arise at a rapid rate in ventricular muscle and are associated with completely ineffective mechanical activity. The condition is rapidly fatal unless promptly terminated, either spontaneously or by an electric shock which may be followed by a return to a more normal rhythm. The ECG shows rapid large waves produced by ventricular activity and having a "saw-tooth" appearance.

## Electrical Axis

The general direction of spread of the electrical impulse through the heart is its electrical axis. The electrical axis does not correspond precisely to the anatomical axis of the heart. The direction of this electrical axis in the frontal plane may be deduced from the standard leads. The *cardiac vector* is the diagrammatic representation of the direction and size of the cardiac impulse and may be plotted as shown in Fig. 42. Lines representing the three standard leads are drawn mutually at 60°. The areas under the QRS complexes are then obtained, e.g. by counting the small squares of the ECG paper. The areas are then marked along the appropriate lines. The cardiac vector passes through the point where verticals drawn through the three axes meet as shown in Fig. 42.

In practice, it is sufficient to draw two of the verticals, and, with experience, the axis may be deduced by merely inspecting the three standard leads. Referring to Fig. 43, it can be seen that if (a) all 3 leads have positive QRS complexes, with that in lead II biggest, the electrical axis is average normal. If QRS in lead I is positive and in lead III strongly negative, *left axis deviation* is present. The reverse holds for *right axis deviation.*

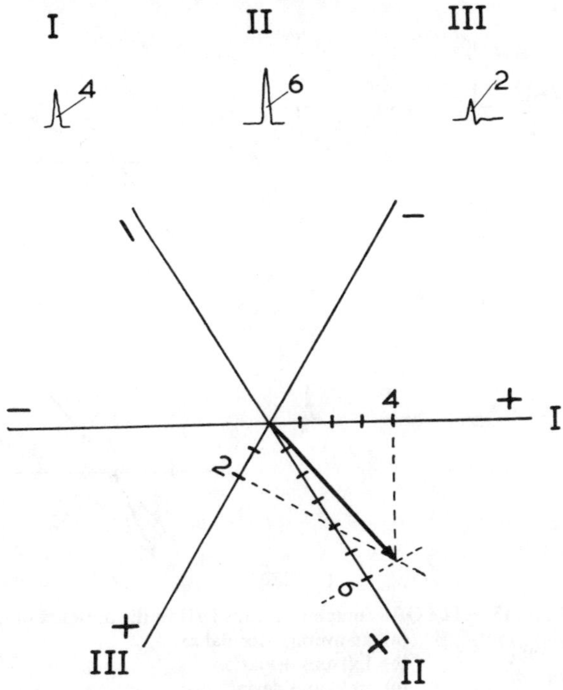


**FIG. 42.—Method of plotting the cardiac vector using the QRS complex in the standard leads, I-III. The figures 4, 6, 2 refer to the areas under the QRS complex.**

Two factors which influence the electrical axis are *body build* and *ventricular hypertrophy.* In a tall, thin person, the heart tends to lie in the chest more vertically than average and in such a person the electrical axis tends to move to the right of the average. A short stocky person tends to have an axis to the left of the average. *Left ventricular hypertrophy* tends to "pull" the electrical axis to the left; *right ventricular hypertrophy* to the right.

A vector can also be obtained from the T waves. This T vector usually points in a similar but not identical direction to the QRS vector. Myocardial damage, e.g. due to infarction, can lead to wide separation of the directions of the QRS and T vectors.

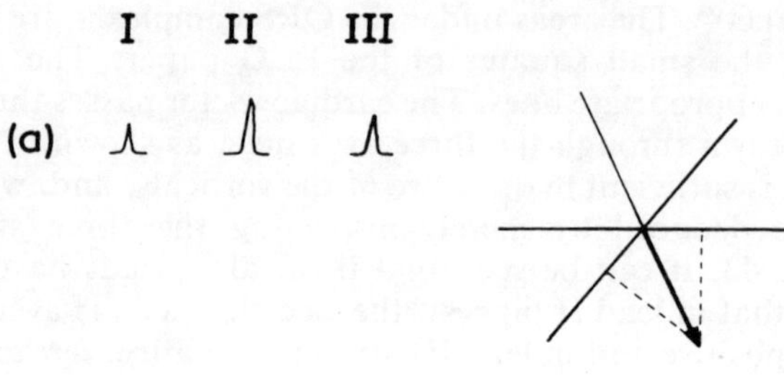


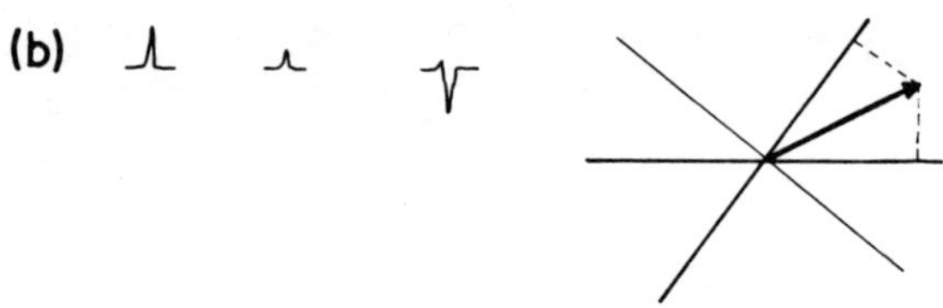


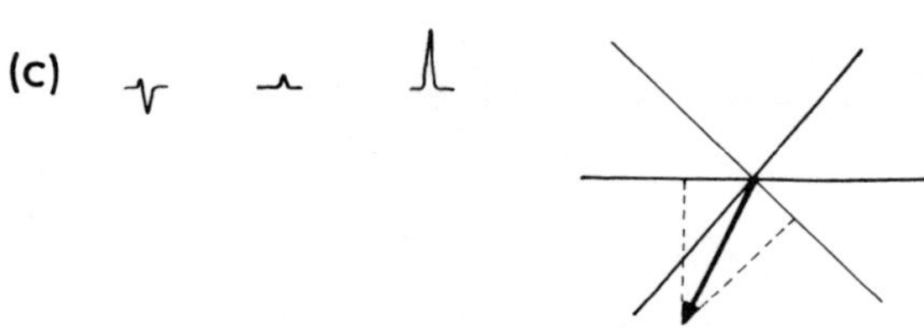


FIG. 43.—The QRS complex in leads I-III in the presence of
(a) an average normal axis
(b) left axis deviation
(c) right axis deviation.

**Abnormal Patterns**

A variety of conditions, such as damage to heart muscle, abnormal body levels of potassium and calcium and the presence of digoxin may modify the shape of the PQRST sequence. In general, these changes cannot at present be explained on a physiological basis and will not be considered here. However, one of the commonest reasons for recording the ECG is to look for evidence of interference with myocardial function due to an inadequate blood supply (ischaemic heart disease). The changes produced by such damage to heart muscle will be outlined briefly.

Because such damage generally affects the left ventricle, where blood supply is the most precarious, the P wave is normal and it is the QRST region of the

ECG which is affected. As damage to the myocardium is followed by a gradual healing process, whereby the damaged heart muscle is replaced by fibrous scar tissue, it is not surprising that the ECG changes gradually alter following an acute heart attack which has led to destruction of a region of heart muscle (*myocardial infarction*).

Typical early changes are the presence of an abnormally large Q wave and elevation of the ST segment between the QRS and T waves (Fig. 44). The Q wave is attributed to the fact that an area of heart muscle has been destroyed; the electrical axis may become markedly abnormal, e.g. pointing up and to the right. The S-T changes are attributed to abnormal currents produced by regions of heart muscle which have been injured but not destroyed. The S-T elevation tends to disappear in a few weeks, to be replaced by T wave inversion. The T wave may later return to normal, but the Q wave tends to be permanent.

These changes may be present in one or more of the standard leads or may be detected only by recording other leads.

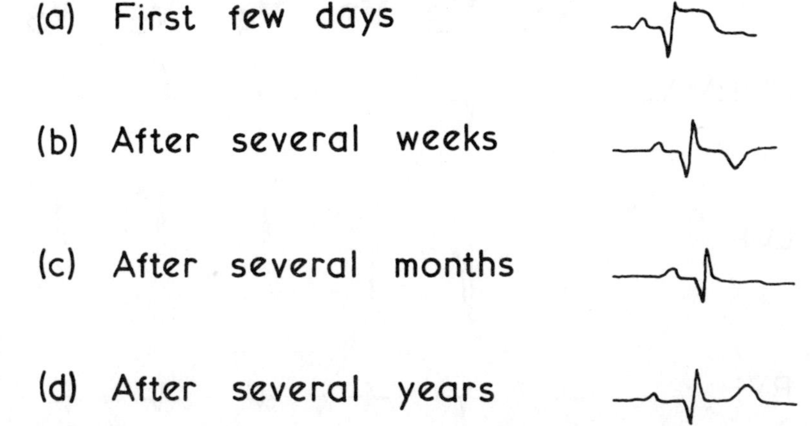


FIG. 44.—Diagrammatic representation of the evolution with time of the ECG changes associated with damage to heart muscle by loss of its blood supply.

## Other Leads

One of the main reasons for recording leads other than the standard leads is to detect evidence of a localized area of myocardial damage. Another reason is that the anterior chest leads ($V_{1-6}$) may provide evidence of left or right ventricular hypertrophy.

ECG recordings may be made from any part of the body, including the inside of the heart (during cardiac catheterization) and the inside of the oesophagus behind the heart (the lead is swallowed). However, the commoner leads recorded are (a) the augmented limb leads—aVR, aVL, aVF and (b) the chest leads, $V_{1-6}$. These are unipolar recordings—recordings of electrical activity at one point, as opposed to the standard leads which record potential difference between two parts of the body. The letter V in these leads denotes voltage recorded by a unipolar lead. R stands for the right arm, L, left arm and F, foot. The 'a' refers to the fact that the leads are electrically augmented to produce a larger record than would otherwise be the case. Leads $V_{1-6}$ are recorded from:

$V_1$ 4th intercostal space to right of sternum
$V_2$ 4th intercostal space to left of sternum
$V_3$ midway between $V_2$ and $V_4$
$V_4$ 5th intercostal space, mid-clavicular line
$V_5$ 5th intercostal space, anterior axillary line
$V_6$ 5th intercostal space, mid-axillary line

The activity recorded in leads $V_{1-6}$ is due mainly to the left ventricle. This produces mainly downward deflections (S waves) in $V_1$ and $V_2$, and mainly upward deflections (R waves) in $V_5$ and $V_6$. $V_3$ and $V_4$ usually are the transition leads between a predominant S wave and a predominant R wave (Fig. 45).

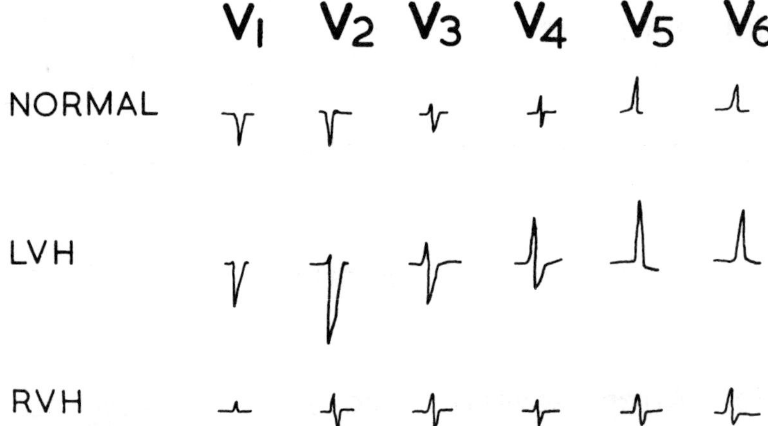


FIG. 45.—The QRS complex in leads $V_{1-6}$ in the presence of *left ventricular hypertrophy* (LVH) and *right ventricular hypertrophy* (RVH) compared with normal.

### Ventricular Hypertrophy

*When the left ventricular muscle increases in bulk*—e.g. due to aortic valvular disease or high blood pressure—the general pattern in $V_{1-6}$ is unchanged but exaggerated. The S waves in $V_{1-2}$ increase in size, as do the R waves in $V_{5-6}$. A rough guide is that when the voltage of the biggest S plus the biggest R exceeds 4 mV (40 mm, assuming the usual sensitivity 10 mm = 1 mV), then left ventricular hypertrophy is likely to be present.

*When the right ventricle is hypertrophied* (pulmonary valvular disease, pulmonary hypertension), its activity subtracts from that of the normal left ventricle as far as leads $V_{1-6}$ are concerned. Thus the S wave in $V_{1-2}$ over the right ventricle is reduced and may be converted into an R wave; the R waves over the left ventricle ($V_{5-6}$) are reduced in size. These changes are exemplified in Fig. 45.

# 28. MURMURS

## Definition

Murmurs are the noises which can be heard over areas of the cardiovascular system where blood flow is turbulent. The turbulence causes the vibrations of surrounding structures which generate the sound waves.

Murmurs vary in intensity and frequency. If the intensity is great, the murmur may be heard at some distance from the patient and may be felt as a palpable "thrill" over the area involved. Most murmurs are less intense than this and require a stethoscope, if not an audio-amplifier and display unit, for detection. High and low frequency vibrations give rise to high and low pitched murmurs respectively.

## Effects

Murmurs are not really harmful in themselves; their importance in medicine is mainly derived from the help they give in diagnosing faults in the heart and circulation.

However, it is more expensive in terms of energy to transport blood by turbulent than by laminar flow because energy is dissipated by the turbulence. It has also been suggested that constant vibrations of vascular walls due to local turbulence may lead to structural weakness and result in localized dilatations (aneurysms, Gr. *aneurysma* = dilatation) in the blood vessels. A localized dilatation of an artery is usually seen just distal to a constricted segment which sets up turbulence and this has been attributed to a "stress fatigue" analogous to that seen in metals exposed to continuous vibration.

## Causes

The cardiovascular system is so designed that the flow around it is usually laminar or streamline and murmurs are not normally heard.

The tendency for laminar flow to become turbulent is increased by an increase in the velocity of flow, the diameter of the vessel and the density of the fluid and by a decrease in the viscosity of the fluid (Fig. 46). Turbulence also occurs when rough areas on walls of blood vessels or valve cusps set up eddy currents in the blood as it flows past.

Thus murmurs may be heard in conditions where there is:

### Increased Blood Velocity

The velocity of blood in a blood vessel depends on the volume of blood flow and the cross-sectional area of the vessel. Thus the velocity rises if (a) blood flow increases or (b) the cross-sectional area decreases.

Blood flow is increased in the circulation of normal people taking exercise or who are excited. This explains the murmurs that may be heard over the arteries or heart of such people. In disease states characterized by high cardiac outputs such as anaemia or excessive thyroxine secretion, circulatory murmurs are common. For similar reasons, murmurs may be heard in the heart or vessels of pregnant women.

Cross-sectional area is decreased when the lumen of an artery is partly occluded by a blood pressure cuff (Korotkow sounds) or by disease causing narrowing (stenosis) of the vessel. Narrowing of the exit orifice of a heart chamber due to valve disease leads to an increase in the velocity of blood leaving the chamber and causes a murmur best heard in systole (systolic murmur).

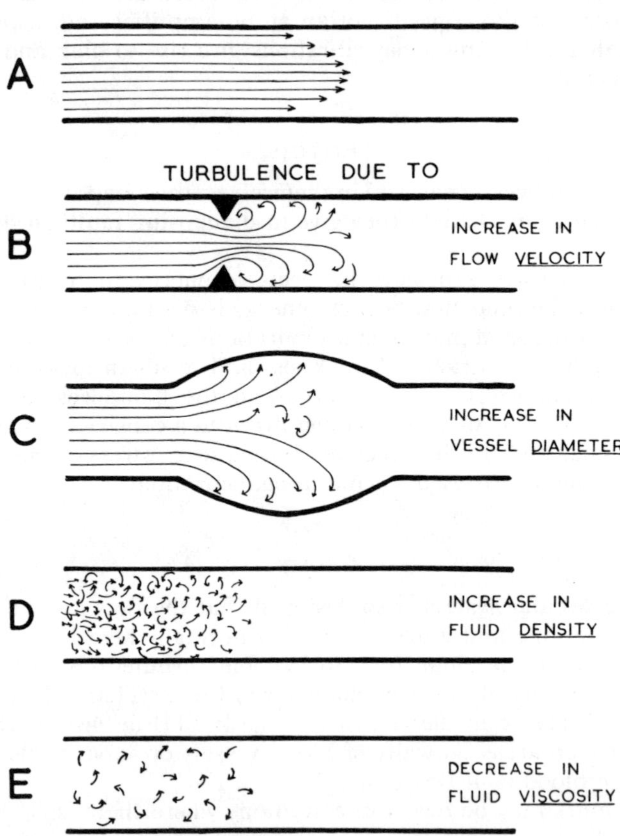


FIG. 46.—Factors promoting turbulent flow.

A. Laminar flow where fluid moves in streamline fashion in concentric laminae, the velocity of the different laminae increasing towards the centre of the vessel.
B. Turbulence occurring where flow *velocity* is increased when passing through a narrowed segment of vessel.
C. Turbulence occurring when the vessel *diameter* is increased.
D. Turbulence occurring when the *density* of the fluid is increased.
E. Turbulence occurring when fluid *viscosity* is decreased.

### Increased Vessel Diameter

This accounts for the murmur heard over a dilated segment of artery (aneurysm). It also contributes to the murmur that is heard when the aortic valves are incompetent and blood regurgitates from the aorta into the left ventricle during diastole.

### Decreased Blood Viscosity

This contributes to the murmurs that may be heard in patients with anaemia. Such patients have not only a high cardiac output but also a reduced blood viscosity. An increase in the red cell concentration (polycythaemia) reduces the possibility of turbulence or murmurs.

Changes in blood density are not usually great enough to influence the nature of flow in the circulation.

## INVESTIGATIONS

Phonocardiography is a technique for recording the amplitude and frequency of the vibrations generated by the heart using a sensitive microphone applied to the skin. It may be used for confirming the timing and characteristics of sounds heard using a stethoscope.

## RECENT LEADING ARTICLES

Blood gas tensions in acute pulmonary oedema. *Lancet,* 1972, **1,** 1006.
Cardiac transplantation today. *Brit. med. J.,* 1971, **4,** 377.
Cardioversion in chronic atrial fibrillation. *Lancet,* 1972, **1,** 1002.
Carotid-sinus-nerve electro-stimulation in angina pectoris. *Lancet,* 1973, **1,** 304.
Chest thump in ventricular tachycardia. *Lancet,* 1971, **1,** 488.
Cold and the heart. *Brit. med. J.,* 1974, **3,** 430.
Ectopic beats. *Lancet,* 1970, **1,** 603.
Fat embolism. *Lancet,* 1973, **1,** 672.
Glucose and the heart. *Lancet,* 1972, **2,** 1295.
Hypertension and cerebral blood flow. *Lancet,* 1973, **1,** 526.
Intermediate coronary syndrome. *Brit. med. J.,* 1973, **3,** 601.
Intra-aortic balloon pumping. *Lancet,* 1972, **2,** 1238.
Long term use of implanted pacemakers. *Lancet,* 1971, **1,** 635.
Mitral stenosis and the crescendo presystolic murmur. *Lancet,* 1972, **1,** 1007.
New thoughts on essential hypertension. *Brit. med. J.,* 1972, **2,** 121.
New vasodilator drugs for hypertension. *Brit. med. J.,* 1973, **4,** 185.
Pacemakers for heart block. *Brit. med. J.,* 1971, **4,** 442.
Postural hypotension in the elderly. *Brit. med. J.,* 1973, **4,** 246.
Predicting response to sympathetic ablation. *Lancet,* 1974, **1,** 441.
Problems in renovascular hypertension. *Brit. med. J.,* 1973, **4,** 566.
Pulmonary oedema. *Brit. med. J.,* 1973, **2,** 437.
Renal hypertension. *Lancet,* 1971, **1,** 483.
Results of treatment of hypertension. *Lancet,* 1971, **1,** 217.
Resuscitation after electric shock. *Lancet,* 1973, **1,** 244.
Risko. *Lancet,* 1973, **2,** 243.
Significance of ectopic beats. *Brit. med. J.,* 1973, **2,** 191.
Surgery for coronary occlusion. *Brit. med. J.,* 1973, **3,** 420.
The heart and the bicycle. *Lancet,* 1972, **1,** 1223.
The left ventricle in chronic bronchitis. *Lancet,* 1971, **2,** 1019.
The odds on getting a coronary. *Brit. med. J.,* 1973, **2,** 375.

The sick sinus syndrome. *Brit. med. J.*, 1973, **2,** 677.
Tourniquet techniques. *Brit. med. J.*, 1972, **2,** 247.
Treatment of ventricular arrhythmias. *Lancet,* 1971, **2,** 857.
Ultrasounding the heart. *Brit. med. J.*, 1974, **1,** 83.
When and where should hypertension be treated? *Brit. med. J.*, 1972, **2,** 1.

*Section V*

# DISORDERS OF THE RESPIRATORY SYSTEM

# 29. DISTURBANCE OF UPPER RESPIRATORY TRACT FUNCTION

## Definition

THE UPPER respiratory tract extends from the nostrils to where the lower border of the larynx adjoins the upper end of the trachea. The effects considered in this section are those of upper respiratory tract infections (URTI) and allergies (which produce similar effects and may pave the way for subsequent infection).

Upper respiratory tract function is lost when the region is bypassed by a tracheostomy—a direct opening (Gr. *stoma* = mouth) from the outside into the trachea. Such effects are considered in the treatment of respiratory disease by tracheostomy.

## Effects

The regions which may be involved are the nose, throat, sinuses and larynx. As the infecting agent—commonly a virus—is generally breathed in, the infection usually begins in the area where the infective droplets are most likely to land, namely the back of the nose or the throat. The initial symptoms are thus those of the common cold or sore throat. It appears that the common cold has been common for many centuries since the Greeks had a word for it—*koryza*—which gave rise to the medical term coryza, which simply means a cold in the head. Rhinitis (Gr. *rhis* = nose) is a parallel term referring specifically to inflammation in the nose.

As with all infection, effects are local and general (systemic). The main local effects are irritation of mucus cells causing production of excess mucus, swelling of the mucosa due to local oedema and congestion with blood and in some areas, especially throat, larynx and trachea, irritation of pain endings. Irritation of receptors in the larynx, trachea and bronchi, and the accumulation of excess mucus at these sites, may activate the cough reflex. The nasal passages tend to become blocked by mucosal swelling and excess secretions. The general effects are varying degrees of loss of energy, raised body temperature and lack of a feeling of well-being.

When the mucosal cells are damaged by viral invasion, there is commonly a secondary invasion by bacteria, so that the initial clear mucous discharge becomes greenish or yellowish in colour due to the presence of pus (dead neutrophil granulocytes).

These infections tend to spread, some more than others. They spread along the respiratory pathways—into the sinuses (L. *sinus* = a cavity) to cause sinusitis; downwards into the larynx, to cause laryngitis, and further down to cause a lower respiratory tract infection—bronchitis or pneumonia. Spread into the sinuses and down into the larynx, trachea and bronchi is against the normal flow of the mucus caused by beating of the cilia and is more likely to occur in patients in whom previous infections or irritating factors have damaged the cilia-bearing cells of the mucosa.

Complications of the common cold due to local spread of the infection are summarized in Fig. 47. Spread of the infection up the nasolachrymal duct to the conjunctiva (*conjunctivitis*) and obstruction of the duct lead to the pink and watering eyes commonly associated with a cold in the head.

Spread to one or more sinuses leads to *sinusitis* which is associated with headache and increased systemic upset. In its chronic form, sinusitis is a not uncommon cause of morning misery in which the trapped sinus secretions raise the pressure in the sinus and cause a persistent headache. As the secretions drain, pain is relieved, but there is much coughing and spluttering. Because the normal ciliated mucosa is damaged, drainage of the sinuses depends on the less effective method of emptying under the influence of gravity. The sinuses then empty best with the head in the upright position, which explains why they tend to fill up at night. Surgical operations which aim to improve sinus drainage by increasing the size of the sinus openings, are often unsuccessful because they cannot replace the normal action of the cilia.

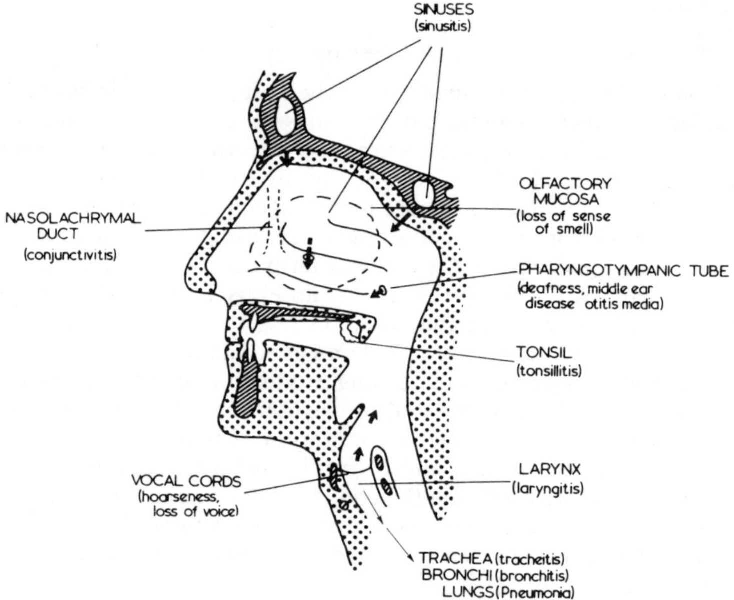


FIG. 47.—How the common cold infection may spread (complications in brackets). The direction of mucus by cilia is shown by the heavy arrows.

Loss of the sense of smell indicates involvement of the olfactory epithelium and, because smell plays an important part in the appreciation of flavour, food may lose its taste. Repeated onslaughts of this kind on the olfactory mucosa may account for the impairment of the sense of smell which accompanies ageing.

If the mucous membrane around the opening of the pharyngo-tympanic tube becomes swollen by oedema, the tube may not open as usual with swallowing. In this case the air in the inner ear may be partially absorbed into the surrounding tissues with the result that the eardrum is pulled inwards and fails to transmit sounds normally, causing partial deafness. Skin-diving is inadvisable in such circumstances because, during descent, as external pressure on the eardrum increases, further strain is placed on the drum, and it may rupture (giving the sensation of a painful bang).

The tonsil is part of a protective ring of lymphoid tissue around the pharynx. This ring consists of the adenoid (pharyngeal tonsil), the lingual and the faucial tonsils. This tissue appears to have a protective function of trapping and providing immunity against invading organisms. In the process, the tissue may become swollen and painful (*tonsillitis*) and in certain cases the tonsils or adenoids may be so damaged by recurrent infections that they harbour rather than destroy micro-organisms. In addition, swelling of the tissue may obstruct nasal breathing. In such cases these lymphoid organs are more of a liability than an asset and are better removed. Deciding when they should be removed is a difficult problem and the criteria for doing so have changed markedly from time to time so that tonsillectomy exemplifies one of the great swings of medical fashion. At one time vast numbers of tonsils were removed on very little pretext; now the approach is more cautious. As with a number of operations which remove a part of the body, the problem is to decide whether, in the long term, the damaged organ will be more of a liability than an asset, bearing in mind also the risks of surgery.

*Laryngitis* tends to cause local discomfort, and involvement of the vocal cords leads to voice abnormalities. When the cords are oedematous the frequency of vibration tends to fall, causing hoarseness; severe inflammation can cause almost complete loss of vibration and hence of voice. Occasionally, usually in infance or childhood, laryngeal inflammation may cause severe obstruction to the airway and death may result unless the obstruction is bypassed by making an opening (tracheostomy) directly into the trachea.

Spread of infection below the larynx is relatively rare unless the individual has lower airways already damaged by disease. *Tracheitis* causes a burning central chest pain especially on coughing. Further spread of infection into the bronchi (*bronchitis*) will be considered in the next main section (obstructive disease of the lower airways).

### Causes

As mentioned above, viruses and less commonly bacteria generally initiate upper respiratory tract infections, and bacteria tend to attack as secondary invaders in the presence of viral infection. At this point it may be mentioned that foreign materials such as pollens may, in certain people, provoke an allergic response which resembles the response to infections. The commonest form of this is hay fever (allergic rhinitis). The number of cases of this condition can be correlated with the pollen count in the local atmosphere. Less rarely, but much more dramatically, the larynx may be obstructed by severe oedema as part of the generalized severe allergic response (anaphylactic shock) to an antigen to which the patient is highly sensitive.

Whatever the initial cause, the symptoms and signs of upper respiratory tract infection are related to increased local blood flow (causing visible redness) and increased capillary permeability (causing oedema) with stimulation of pain endings and of mucus and serous glands to cause excessive mucous and serous secretions. Toxic materials, from damaged cells and invading organisms, cause the systemic symptoms.

### Investigations

Most mild upper respiratory tract infections do not require any special investigation. In certain patients, for whom the condition constitutes a special risk, a throat swab may be taken for bacteriological examination.

### Treatment

As the infecting virus cannot at present be specifically treated the condition must "run its course" until the body defence systems inactivate the invading organisms and repair the damaged tissue. Symptoms such as headache and fever can be relieved by drugs such as aspirin or paracetamol. Inhalations of a drug which mimics noradrenaline's vasoconstrictor action may relieve nasal obstruction. Bed rest is necessary in more severe infections such as influenza. In some cases antibiotic drugs may reduce the risk of serious secondary bacterial invasion.

In view of the very large total amount of disability produced by upper respiratory tract infections, strenuous efforts have been made to develop vaccines to prevent these infections. Apart from limited success in the prevention of influenza, these efforts have met with little reward. This is probably due to two main reasons. Firstly, a very large number of antigenically different organisms are involved, and mutations of the organisms, e.g. influenza viruses, lead to further antigenic changes. Secondly, many of the organisms do not appear to activate the immune system very effectively. These effects which hinder the development of effective vaccines also prevent the acquisition of natural immunity to upper respiratory infections, with the well-known consequence that frequent recurrences are the rule.

# 30. DISTURBANCE OF LOWER RESPIRATORY TRACT FUNCTION

## DEFINITION

DISTURBANCE of respiratory function occurs in conditions which impair the flow of air into and out of the alveoli, the flow of blood through the pulmonary capillaries, or gas exchange between the alveoli and the pulmonary capillary blood.

## EFFECTS

### Respiratory Impairment, Insufficiency and Failure

In the progression of respiratory disease there are three phases, *respiratory impairment, respiratory insufficiency,* and *respiratory failure.* These phases are illustrated in Fig. 48.

The lungs of a healthy young adult have a very large functional *reserve.* Maximal exercise causes a smaller increase in minute volume than does maximal voluntary ventilation. Thus in the stage of *respiratory impairment,* the patient may be unaware that anything is wrong though respiratory function tests show results which fall progressively below the accepted normal values.

In the stage of *respiratory insufficiency,* the patient becomes aware of respiratory discomfort (dyspnoea), during exertion. This leads to a progressive limitation of exercise tolerance.

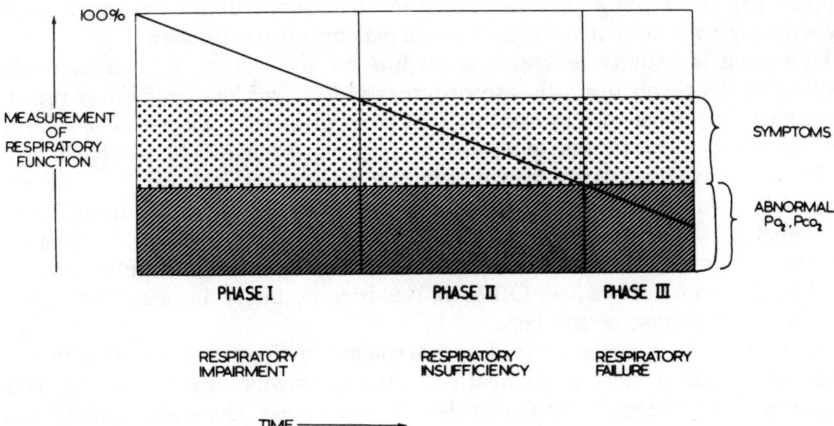


FIG. 48.—Three phases in the progression of diseases which disturb respiratory function (indicated on an arbitrary scale). In the initial phase (*respiratory impairment*) there are no symptoms (clear zone). Symptoms appear in phase II (*respiratory insufficiency*). In phase III, *respiratory failure* is present. Phase III is terminated by death from respiratory failure.

In the stage of *respiratory failure,* the lungs lose their ability to maintain normal arterial gas tensions at rest. There is a fall in the resting pressure of oxygen ($Po_2$) and a rise in the pressure of carbon dioxide ($Pco_2$) in arterial blood. There are differences of opinion about the definition of respiratory failure. Some authors include a fall in $Po_2$ without a rise in $Pco_2$; others require that both $Po_2$ and $Pco_2$ be abnormal. The stage of respiratory failure is terminated by death when tissue gas exchange falls below the level necessary to maintain life.

### Dyspnoea and Limitation of Exercise Tolerance

Dyspnoea is an unpleasant awareness of breathing (Gr. *dys* = difficult, painful, bad; *pnoia* = breathing). There is the sensation of something abnormal in the breathing process. This sensation of dyspnoea, or breathlessness, presents in many varieties and has many causes, just as the sensation of something wrong with the legs when walking may be due to painful joints, poor circulation, muscular weakness, cerebellar disease, etc.

Thus a patient may complain of difficulty in breathing because of a painful condition of the chest wall; or because narrowing of the airways or loss of pulmonary compliance increase the work of breathing; or because inefficiency of the lungs increases the ventilatory minute volume required to maintain gas exchange. During pregnancy, ventilation increases to meet the needs of the fetus and the pattern of respiration changes; due to abdominal distension by the uterus, expansion of the rib cage plays an increased role in ventilation. These changes in the pattern of respiration may also be experienced as dyspnoea. At high altitudes and during strenuous exercise, increased ventilation is required and this may be described as dyspnoea, even though the respiratory system is entirely normal.

Thus dyspnoea may be experienced when sensory information reaching the brain indicates an involuntary disturbance of the normal respiratory pattern required for the resting state or exertion. Not surprisingly dyspnoea can be relieved only by a return towards the normal breathing pattern.

Dyspnoea is usually assessed according to the degree of exertion which produces it. As respiratory disease progresses, less and less exertion is required to produce dyspnoea, and eventually it is present at rest. Decreasing tolerance of exercise is a characteristic feature of respiratory disease. The patient's feeling that he must stop his activity is probably due largely to dyspnoea. In some cases abnormal blood gas tensions may stop exertion by impairing muscular efficiency. Psychological factors may exaggerate the reduction in exercise tolerance. It must be remembered that respiratory disease is only one of a number of causes of reduced exercise tolerance. Other causes include heart disease, and vascular and muscular disease of the legs.

In *asthma,* dyspnoea is caused by narrowing of the airways and is typically worse in expiration than in inspiration. This is because the act of inspiration expands not only the lung alveoli but also all the airways, these also having elastic walls. In asthma, inspiratory expansion of the airways aids inflow of air but the reduction in airway lumen diameter during expiration may cause a severe rise in airway resistance and in the work of breathing. Expiration becomes difficult, prolonged and wheezy. The fact that expiration in such circumstances requires greater effort than inspiration is the reverse of the normal situation where inspiration is active and expiration passive. Wheezing, which may or may not accompany dyspnoea, is associated with increased velocity of air flow due to narrowed airways. This increased velocity leads to air turbulence and hence vibrations which may be felt by a hand on the chest or heard as rhonchi (Gr. *rhonchos* = a snoring sound) using a stethoscope.

Breathing is more efficient in the upright position. In the lying position, movements of the diaphragm are hindered by the abdominal contents. Thus, when a patient experiences marked dyspnoea at rest, he usually sits up. This preference for breathing in the upright position is referred to as *orthopnoea* (Gr. *orthos* = straight). In this condition, the accessory muscles of respiration (e.g. sternomastoids) are commonly used. Orthopnoea is also associated with the dyspnoea of pulmonary oedema due to left heart failure. In this case, sitting up tends to reduce venous return to the right atrium and lungs and hence pulmonary capillary pressure falls, especially in the upper parts of the lungs.

When breathing is excessively difficult, or if it is temporarily suspended by, e.g. laryngospasm, the sensation of *suffocation* is experienced. This is the feeling that life is threatened by the stoppage of respiration. This sensation is associated with considerable fear and panic and greatly adds to the distress of a patient with, e.g. severe bronchospasm in an attack of asthma.

**Reduction in Arterial Oxygen Pressure ($Po_2$)**

*Hypoxia* implies lack of oxygen by the cells of the body. It is used interchangeably with *anoxia* which, strictly speaking, means no oxygen. Hypoxia has been classified as being of four varieties—*hypoxic* (when the arterial $Po_2$ is low), *anaemic* (when the oxygen-carrying capacity of the blood is inadequate), *stagnant* (when the hypoxia is due to poor circulation of blood) and *histotoxic* (when the cells are unable to use oxygen due to poisoning). Hypoxia due to respiratory failure is hypoxic hypoxia and is characterised by a reduced arterial oxygen tension. Although the $Po_2$ tends to fall with age, the lower limit of normal may be taken as around 80 mm Hg.

While measuring arterial $Po_2$ is the only precise method of diagnosing hypoxia, the condition may be suspected in a patient with respiratory disease who shows (a) *central cyanosis,* (b) *abnormal behaviour,* (c) *pulmonary hypertension* and (d) *polycythaemia.*

*Cyanosis* denotes bluish discolouration of the skin (Gr. *kyanos* = blue). It is usually due to hypoxia, although abnormal blood pigments may also cause it. When due to hypoxia, cyanosis is caused by the presence of a moderate amount of reduced haemoglobin in the blood; 5 g/100 ml, or about a third of the normal total haemoglobin concentration is generallly regarded as the level at which most people can just detect cyanosis. Cyanosis is of two varieties — peripheral and central. Peripheral cyanosis is due to local stagnant hypoxia. It occurs in normal people in cold conditions—e.g. hands, feet, ears "blue with cold"—and is not in itself a sign of disease. In contrast, central cyanosis is associated with hypoxic hypoxia. It is due to the presence of an increased amount of reduced haemoglobin in arterial blood and is abnormal. It may be diagnosed by observing cyanosis in parts of the body, such as the tongue and inside of the mouth, where there is adequate blood flow and hence no stagnant hypoxia. Usually a bluish facial hue indicates central cyanosis provided the face is reasonably warm. If the peripheries are markedly warm, indicating a high rate of blood flow, then blueness of the hands supports the impression of central cyanosis.

Cyanosis is often a difficult sign to be sure about. Detection of slight cyanosis requires good light and well-developed colour vision. If a patient has little blood in superficial vessels due to vasoconstriction, cyanosis may be hard to detect. Comparison of a patient's tongue, face or warm palms with those of a normal person is often helpful.

As cyanosis depends on the presence of around 5 g/100 ml of reduced haemoglobin, it is not an accurate indicator of hypoxia if the haemoglobin level is abnormal. An anaemic patient with a haemoglobin level of 5 g/100 ml cannot be cyanosed and remain alive; a polycythaemic patient with a haemoglobin level of 20 g/100 ml may be cyanosed and yet have a satisfactory blood oxygen content. Patients with chronic hypoxia often have an increased haematocrit and peripheral vasodilatation due to an associated increase in $Pco_2$. Hence they may have very severe cyanosis.

***Abnormal behaviour.***—A large variety of conditions can lead to abnormal behaviour by interfering with the function of brain cells. The conditions include a low $Po_2$, a high $Pco_2$, an altered pH, liver failure, renal failure, a low blood sugar and an abnormal body temperature. All these conditions impair brain cell function and can cause behaviour resembling in many ways intoxication with alcohol or other drugs. Thus a patient suffering from any one of these conditions may be noisy, difficult, unpleasant, unco-operative, drowsy, apathetic, merry or unduly talkative. With any of these conditions, as with alcoholic intoxication, increasing severity interferes progressively with more brain cells, passing more or less from the frontal cortex to the vital centres in the medulla and leading progressively to unconsciousness and death. It is important to remember that, when a patient with hypoxia is abusive and unco-operative, the correct treatment is not a lecture on how to behave, but measures to improve the blood gases.

*Pulmonary hypertension.*—Long-standing (chronic) hypoxia can produce progressive damage to cells in the brain, liver, kidney etc., but one of its most important effects is to cause heart failure by raising pulmonary arterial blood pressure. A reduced alveolar oxygen level affects pulmonary arterioles in the opposite way to systemic arterioles—it leads to vasoconstriction. In the normal lung this probably has a beneficial effect by diverting blood from regions of the lung where there is inadequate oxygenation to regions where oxygenation is satisfactory. However, when alveolar air *throughout* the lungs is poorly oxygenated and all pulmonary arterioles constrict, the effect is to raise progressively pulmonary vascular resistance and hence pulmonary blood pressure. This raised pulmonary blood pressure (pulmonary hypertension) requires increased work from the right ventricle. The ventricular wall thickens (right ventricular hypertrophy) and the ventricle may eventually fail, leading to the signs of right ventricular failure—raised jugular venous pressure, enlarged liver and oedema. This state of heart failure due to lung disease is known as *cor pulmonale.*

*Polycythaemia.*—Chronic hypoxia of the hypoxic variety (reduced arterial $Po_2$ due to respiratory failure, shunting of venous blood into the arterial circulation or the low $Po_2$ of inspired air at high altitudes) leads typically to *secondary polycythaemia.* The reduced arterial oxygen content causes increased release from the kidney of the hormone erythropoietin, which in turn stimulates the bone marrow to increase its rate of production and release of red blood cells. This increased circulating red cell mass usually manifests itself in a raised haemoglobin level, red cell count and packed cell volume; although in some cases there is also a rise in plasma volume which at least in part counteracts the rise in the blood indices. An increased packed cell volume leads to a steep rise in blood viscosity and this may contribute to the development of cor pulmonale.

**Increase in Arterial Carbon Dioxide Pressure ($Pco_2$)**

*Hypercapnia* denotes the presence in the body of excess carbon dioxide (Gr. *kapnos* = smoke). It has a number of effects on the body. It causes *peripheral vasodilatation, respiratory acidosis, stimulation of ventilation* and *cerebral depression.*

*Peripheral vasodilatation.*—Carbon dioxide causes dilatation of systemic arterioles and hypercapnia is associated with warm peripheries making the detection of cyanosis rather easier. Dilatation of cerebral blood vessels can lead to a marked increase in cerebral blood flow;the rise in cerebral capillary pressure may cause cerebral oedema and raised cerebrospinal fluid pressure. This may be recognized by a bulging of the optic nerve head into the vitreous humour (papilloedema).

*Respiratory acidosis.*—One of the functions of the lungs is to help maintain a normal arterial pH of 7.4 by regulating the $Pco_2$ in arterial blood. When the lungs cannot excrete adequate amounts of $CO_2$, the blood $Pco_2$ and hydrogen ion concentration rise because the reactions

$$CO_2 + H_2O \xrightarrow[\text{anhydrase}]{\text{carbonic}} H_2CO_3 \longrightarrow HCO_3^- + H^+$$

are driven in the direction shown with the formation of excess hydrogen ions.

This respiratory acidosis is compensated, at least partly, by the renal tubular mechanism which, under the influence of carbonic anhydrase, synthesizes $H_2CO_3$ and excretes hydrogen ions into the urine and bicarbonate ions into the blood. The hydrogen ion concentration in the blood depends on the ratio of carbonic acid to bicarbonate ion which is normally 1 : 20. When a rise in the carbonic acid concentration gives rise to respiratory acidosis, the kidneys attempt to compensate by manufacturing bicarbonate ion to restore the ratio of carbonic acid to bicarbonate to 1 : 20 (Fig. 49). Excess $CO_2$ in the blood is there-

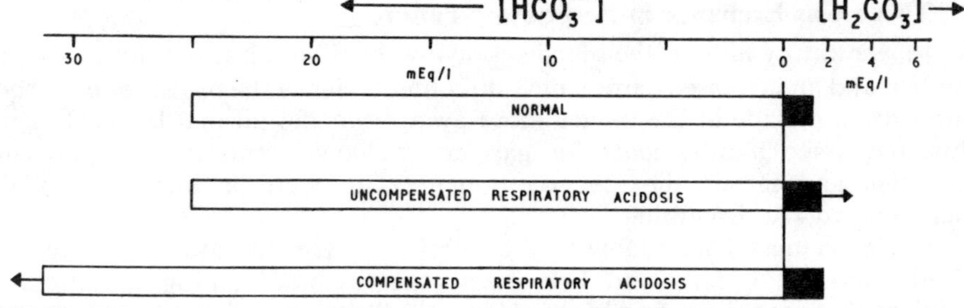


FIG. 49.—Changes in carbonic acid and bicarbonate concentrations in uncompensated and compensated respiratory acidosis.

fore associated with a rise in the bicarbonate level. While measurement of the $Pco_2$ is the most definite indicator of hypercapnia, the presence of a raised bicarbonate level from the normal value of around 25 mmol/l to 30-35 mmol/l in someone with respiratory disease makes it fairly likely that $CO_2$ is being retained by the body.

***Stimulation of ventilation.***—A moderate rise in the $Pco_2$ stimulates ventilation directly by its action on the respiratory centre and indirectly by its action on the peripheral chemoreceptors. *Severe rises in $Pco_2$, however, depress ventilation and can lead to coma and death.*

Some patients suffering from severe obstructive airway disease fall into one of two distinct groups and the differences between the groups seem to stem from *different sensitivities to carbon dioxide.* The first group of patients known as *"pink puffers"* have a normal sensitivity to $CO_2$. As the $Pco_2$ tends to rise due to their lung disease, these patients hyperventilate markedly and thereby maintain fairly normal blood gas tensions and are not cyanosed.

The second group of patients known as "*blue bloaters*" have an abnormally low sensitivity to $CO_2$. They show much less tendency to hyperventilate and in fact do not ventilate sufficiently to maintain normal blood gas tensions, and hypoxia results. The hypoxia is evident as cyanosis and in the lungs tends to cause a rise in pulmonary vascular resistance and eventual right heart failure (cor pulmonale).

The sequence of events in the two groups is shown in Figs. 50 & 51. Each group contains patients with severe obstructive airway disease and patients do not usually move from one group to the other. Blue bloaters can become pinker by voluntarily breathing more deeply but they seem incapable of keeping this up, and lapse into the relative comfort of inadequate ventilation and chronic respiratory failure. It has been suggested that they are by personality more lethargic than the "pugnacious pink puffers" who show a greater tendency to "fight" the disease! As with most classifications of disease, many patients with obstructive disease are neither typical pink puffers nor blue bloaters, but lie in the spectrum between the two extremes.

***Cerebral depression.***—When carbon dioxide is breathed in high concentrations it acts as a general anaesthetic. The high $Pco_2$ in hypercapnia can cause depression of consciousness and abnormalities of behaviour similar to those caused by hypoxia. Eventually accumulation of carbon dioxide can cause loss of consciousness, coma, and death from respiratory paralysis.

### Tissue Gas Exchange in Respiratory Failure

In respiratory failure, the blood supplying the tissues has a reduced oxygen content and an increased carbon dioxide content. Hence the pressures of oxygen and carbon dioxide in the tissues move away from the normal levels. Despite this, the tissue requirements for gaseous exchange continue. Oxygen consumption and carbon dioxide excretion may, in fact, be increased by the increased work of breathing.

Gas exchange is maintained by a number of compensatory mechanisms. Firstly, tissue $Po_2$ falls and $Pco_2$ rises, thereby maintaining the pressure gradients between tissue and blood. However, the abnormal gas tensions in the tissues may impair tissue function. In early respiratory failure, the arterial $Po_2$

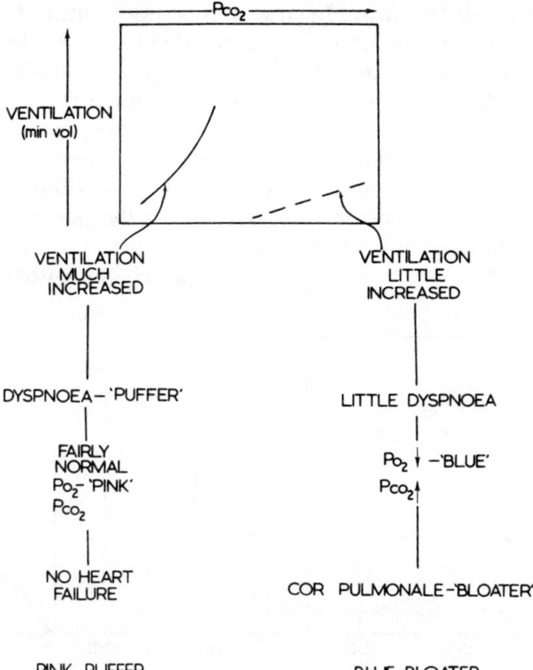


FIG. 50.—Whether a patient with severe obstructive airway disease becomes a "pink puffer" or a "blue bloater" depends on his sensitivity to a raised arterial $Pco_2$ as indicated in the rectangle.

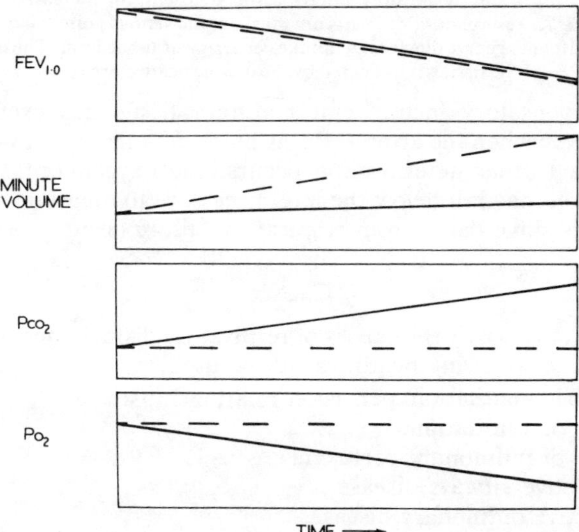


FIG. 51.—Comparison of the progression of lung disease in the case of a ***blue bloater*** (continuous line) and a *pink puffer* (dashed line). It is assumed that lung function (indicated by forced expiratory volume in one second — $FEV_{1\cdot 0}$, for example) deteriorates at the same rate in the two patients. The pink puffer ventilates at a higher rate than the blue bloater and avoids respiratory failure.

may fall, e.g. from 80-100 mm Hg (normal) to 60-70 mm Hg. Because of the shape of the oxygen dissociation curve of blood (Fig. 52 *b*), the oxygen content of arterial blood is little affected and tissue oxygen supply is well maintained. The raised blood $Pco_2$ and lowered pH tend to move the oxygen dissociation curve to the right and downwards, thereby aiding oxygen release at any given $Po_2$. As arterial $Po_2$ falls below 50 mm Hg, the steep part of the oxygen dissociation curve is reached, and the saturation of arterial blood with oxygen rapidly falls (Fig. 52 *c*). However, the increase in the haemoglobin stimulated by the low $Po_2$ allows the oxygen *content* of arterial blood to be maintained at near normal values despite the fall in oxygen saturation.

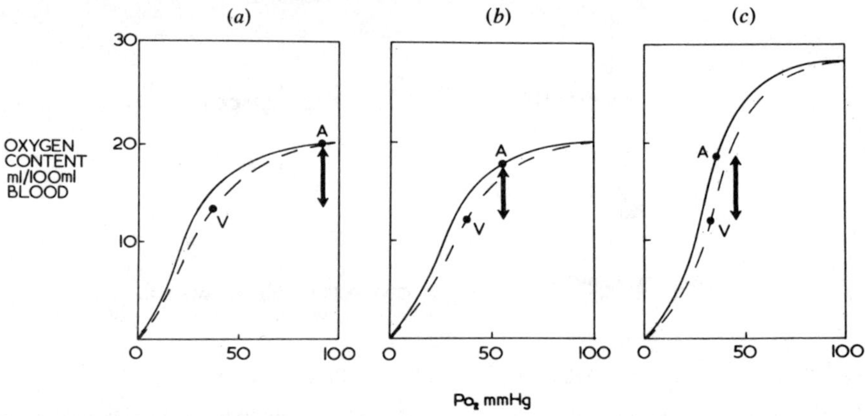


FIG. 52.—Diagrammatic representation of oxygen uptake (*a*) under normal conditions, (*b*) in early respiratory failure, (*c*) in late respiratory failure associated with an increased haemoglobin concentration. Arterial (A) and venous (V) points are marked, the venous point lying on the dotted line, representing the shift in the curve due to $CO_2$ uptake occurring at tissue level. The oxygen delivered to the tissues is indicated by the double headed arrow.

These compensatory factors can maintain tissue gas exchange at near normal levels even when the arterial $Po_2$ is below 50 mm Hg. However, if at this critical point a further deterioration occurs, the oxygen uptake and carbon dioxide excretion may fall below the level necessary to maintain minimal tissue function. A this stage death from respiratory failure occurs.

## Causes

It is difficult to classify the causes of respiratory disturbance on a completely logical basis. The following headings will be used.

I Disturbed ventilation/perfusion relationships.
II Failure of ventilation.
III Failure of pulmonary perfusion.
IV Obstructive airways disease.
V Restrictive pulmonary disease.

Section I describes the fundamental disturbances which can occur in pulmonary disease. Sections II and III describe more or less "pure" disturbance of ventilation and of perfusion, respectively. The remaining sections describe recognizable types of pulmonary disease in which the functional disturbance is complex.

## I *Disturbed ventilation/perfusion relationships*

The function of the lung is to permit gas exchange between the mixed venous blood in the pulmonary capillaries and the alveolar air in the alveoli, the composition of this air being maintained by appropriate alveolar ventilation. Two main abnormalities are theoretically possible—ventilation may be either inadequate or excessive relative to perfusion (Fig. 53).

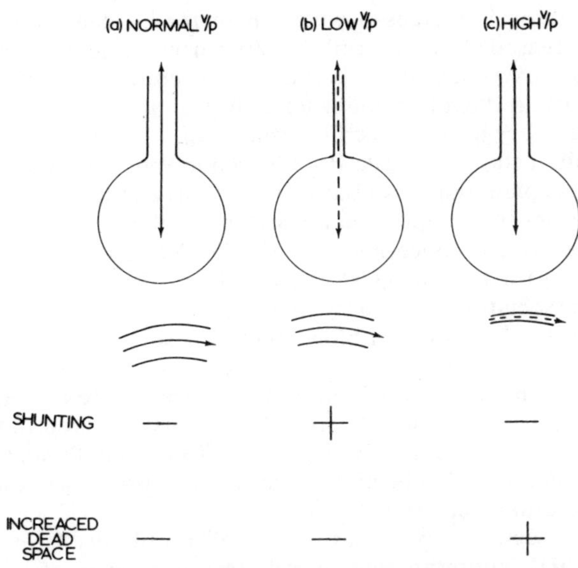


FIG. 53.—A diagram of various ventilation/perfusion (V/P) relationships: (a) normal V/P ratio, (b) low V/P ratio — ventilation is inadequate; gas exchange is inadequate and the physiological shunt is increased, (c) high V/P ratio — ventilation is excessive; this tends to increase physiological dead space.

When an alveolus receives *inadequate ventilation,* e.g. due to bronchial narrowing, alveolar $Po_2$ falls and $Pco_2$ rises. The blood passing through capillaries in the wall of such an alveolus is not adequately oxygenated and does not give off the normal amount of carbon dioxide. This situation is referred to as "*venous admixture*". It has the same effect as adding venous blood from the right side of the heart to arterial blood. The passage of blood directly from the venous to the arterial circulation, e.g. through a septal defect between the atria or ventricles, is referred to as an anatomical shunt—the blood is shunted past the pulmonary circulation. Venous admixture due to alveolar underventilation has the same effect. Total shunting is referred to as the physiological shunt.

Physiological shunt = anatomical shunt + venous admixture.

Shunting tends to lower arterial $Po_2$ and raise arterial $Pco_2$.

*Excessive ventilation* may result from increased ventilation of a normally perfused alveolus or from normal ventilation of an underperfused alveolus. The effect in this case is to raise the alveolar $Po_2$ and lower $Pco_2$. Because the pulmonary capillary blood is almost completely saturated with oxygen at the normal alveolar $Po_2$, an increased $Po_2$ confers little benefit and does not materially improve oxygenation. As far as oxygen transfer is concerned, the extra ventilation is wasted, and the effect is similar to an increased dead space. The increase does not affect the anatomical dead space of the airways and the increased dead space is thus an ***increased physiological dead space.***

Physiological dead space = anatomical dead space + alveolar dead space.

The work of breathing is increased when an increased dead space is ventilated. While an increased alveolar ventilation/perfusion ratio has little effect on pulmonary capillary oxygenation, carbon dioxide excretion is facilitated and a lowered arterial carbon dioxide content tends to result.

In diffuse pulmonary fibrosis, the disturbance of V/P relationships is often such that in the initial stages, arterial $Po_2$ falls while $Pco_2$ is normal or even low (***hypocapnia***). The explanation is probably that in this condition over-ventilated alveoli compensate or over-compensate for under-ventilated alveoli as far as the $CO_2$ content of the blood is concerned because the $CO_2$ dissociation curve rises in a linear fashion in the physiological range of $Pco_2$. However, because of the plateau of the $O_2$ dissociation curve, a rise in $Po_2$ in over-ventilated alveoli causes a negligible rise in the $O_2$ content of blood perfusing such alveoli. Thus the low $O_2$ content of blood from under-ventilated alveoli is not compensated for.

In certain conditions, ventilation and perfusion are reduced in parallel, e.g. when part or all of a lung is removed surgically. In this case, ventilation/perfusion relationships in the remaining lung are not affected—both ventilation and perfusion increase in parallel. The effect is to reduce respiratory reserve rather than to upset ventilation/perfusion relationships.

In the next two sections, examples of "pure" ventilatory (section II) and perfusion (section III) failure are considered. However, in most cases of lung disease (sections IV and V), a mixture of alterations occurs, the V/P ratio being increased in some regions and decreased in others. In such diseases, there is a combination of the effects of a decreased V/P ratio (shunting) and a raised V/P ratio (increased dead space ventilation).

## *II Failure of ventilation*

In this condition, the lungs may be normal, but there is inadequate ventilation due to failure of the mechanical system which expands the chest. Expansion of the chest depends on the system shown in Fig. 54.

Ventilatory failure may be caused by depression of the respiratory centre in the medulla oblongata by drugs, e.g. in attempted suicide, and by other conditions which interfere with brain function.

Disease of the nerves supplying the respiratory muscles may reduce respiratory reserve and may eventually cause respiratory failure.

Muscle disease causing inability to ventilate adequately may be due to abnormality at the neuromuscular junction (myasthenia gravis; Gr. *mys* = muscle, *astheneia* = weakness; L. *gravis* = severe) or in the muscle proper. Potential respiratory failure from muscle paralysis is a side-effect of the muscle relaxants used in many abdominal operations. Artificial ventilation must then be used to maintain gas exchange.

Abnormalities of the bones of the thorax, either of long-standing as in the hunchback deformities, or due to injury, e.g. multiple rib fractures, can seriously interfere with expansion of the chest. Severe obesity may be associated with hypoventilation, possibly because the excess fat interferes with the mechanics of ventilation.

Ventilation may also be inadequate because the work required to expand the lungs is greatly increased. This occurs when there is a lack of *pulmonary surfactant.* Such a lack is quite common among premature infants, because adequate production of surfactant begins late in fetal life.

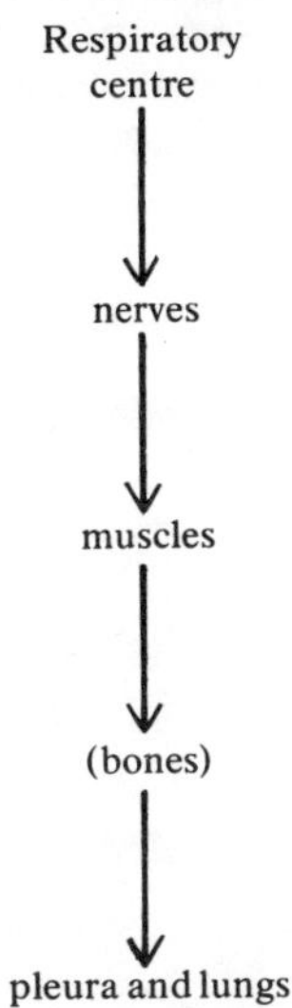


FIG. 54.—The system responsible for expansion of the chest and hence of the lungs (assuming a normal pleural cavity). Ventilatory failure can result from a failure of this system at any point.

The work required to expand the lungs may also be increased by partial obstruction of the respiratory airways.

Thus, chronic obstructive lung disease (see below) may result in *failure of alveolar ventilation.* Whereas hypoxia in lung disease may be due to disturbed ventilation/perfusion ratios, hypercapnia is always due to inadequate total alveolar ventilation.

### *III Failure of pulmonary perfusion*

Failure of pulmonary perfusion in the absence of disease of the lungs occurs in pulmonary embolism in which a blood clot travels to and blocks a pulmonary arterial vessel. Perfusion of the corresponding region of the lung is impaired and the V/P ratio rises. Dead space ventilation increases and the patient may complain of shortness of breath. Central cyanosis is not a feature.

### *IV Obstructive airways disease*

This heading covers a large group of diseases which form the great bulk of chronic lung disease in affluent societies where tuberculosis has largely disappeared but where atmospheric pollution and cigarette smoking are major problems. The group includes a variety of diseases, including asthma, chronic bronchitis and emphysema, which have in common an *obstructive* pattern of respiratory function. The *rate* at which air can be forcibly expired from the lungs is reduced, due to narrowing of the airways.

Obstructive disease of the lower airways may be caused by:

(i) spasm of the smooth muscle in the wall of the airways
(ii) oedema of the wall of the airways
(iii) mucus in the lumen of the airways
(iv) loss of elastic supporting tissue which keeps smaller airways patent.

These abnormalities are shown diagrammatically in Fig. 55. Usually two or more of these abnormalities are present together in the same patient; irritation of the bronchial wall may lead to muscle spasm, oedema and excess secretion of mucus.

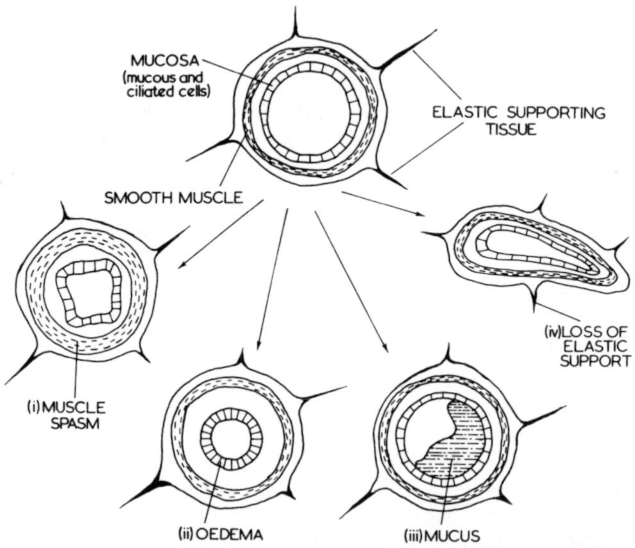


FIG. 55.—Four ways in which the normal airway lumen (top) may be reduced.

Muscle spasm and oedema tend to occur in intermittent attacks, giving rise to the clinical condition known as bronchial asthma (Gr. *asthma* = panting). Asthma may be defined as recurrent attacks of wheezy dyspnoea due to airway narrowing. There is often excessive secretion of mucus and inadequate drainage of these secretions; the combination of airway narrowing and lumen blockage by mucus may cause severe distress and, in the very severe case, death.

### (i) Muscle Spasm

The lower airways, other than the trachea, are approximately circular. The resistance to flow in a circular tube is inversely proportional to the fourth power of the radius. Hence if the radius of an airway is reduced to ⅓ of its original size the resistance of the airway will increase 81-fold. In this way the lower airways, which normally offer less resistance to airflow at rest than the tortuous upper airways, may offer such resistance to airflow that, even at rest, all the patient's energies are required to maintain an adequate airflow to and from his lungs.

Despite the fact that muscle spasm in the airways is a common condition, the reason for its occurrence in a particular patient may be difficult or impossible to discover. One well-recognized cause is Type I hypersensitivity or *allergy*. When a patient inhales a material to which he is allergic—plant and animal proteins such as pollen and cast-off skin particles are common offenders—the local reaction of antigen with antibody seems to liberate a material or materials such as histamine, which act locally to cause the airway muscle to contract, thereby reducing lumen radius. The fact that antihistamine drugs do not usually help asthma suggests that histamine is not the major substance causing the trouble.

As well as causing spasm of the bronchial smooth muscle, the products of antigen-antibody reactions cause dilatation of local blood vessels and an increase in capillary permeability, so that mucosal swelling due to oedema adds to bronchial narrowing.

Airway muscle spasm may also be caused by *psychological factors*. It is well recognized that asthmatic patients are often rather tense individuals and attacks are often brought on by emotional upsets and stressful situations. Possibly such factors cause over-activity of the constrictor parasympathetic nerves to the airways; however, again this does not appear to be a major factor because division of the nerves to the airways has not been found to relieve the spasms.

*Irritant materials* such as smoke, dusts or the products of local infection (bronchitis) may also directly cause bronchospasm in a susceptible individual.

#### (ii) Oedema

Oedema of the bronchial wall tends to be produced by many factors which cause muscle spasm. In addition, pulmonary oedema is caused by *left heart failure* which causes the hydrostatic pressure in the pulmonary capillaries to rise from its normal value of 5-10 mm Hg to exceed the colloid osmotic pressure of about 25 mm Hg. Fluid passes into the alveoli and hence into the airways. In severe cases this pulmonary oedema may kill the patient by a process resembling drowning in that fluid prevents the passage of gases along the airways and into the alveoli.

In some cases the left heart failure leads to severe bronchospasm (*cardiac asthma*). It is not clear why some patients should present mainly bronchospasm and others mainly pulmonary oedema.

#### (iii) Mucus

This also tends to be produced by the conditions which lead to bronchospasm. It is, however, particularly associated with ***chronic bronchitis,*** a disease which seems to be due mainly to cigarette smoking and atmospheric pollution. Chronic bronchitis is characterised histologically by the presence of excessive numbers of mucus-secreting cells in the airways. Normally, mucus is wafted upwards by the cilia and passes unnoticed into the alimentary tract. In chronic bronchitis the cilia are damaged and the excess mucus must be coughed up. The regular coughing up of excess mucus secretions is the hallmark of chronic bronchitis. Blobs of mucus vibrating in the airways give rise to characteristic wheezy noises (rhonchi) audible with or without a stethoscope.

The blue bloaters, referred to under the effects of respiratory disease, tend to come from the large group of patients with chronic bronchitis, but they form only a small minority of the group.

A condition associated with the production of large quantities of sputum and which affects children, is the fairly rare disease known as *mucoviscidosis* or *fibrocystic disease.* The first name indicates the cause of the trouble—viscid mucus which cannot be cleared normally from the lung by the cilia. The other name describes the resulting damage to the lung—obstruction to air passages tends to be associated with cystic dilatation of the bronchi, infection, damage to lung tissue and replacement by fibrous scar tissue. Such children generally suffer from a similar disorder in their pancreas and this leads to severe impairment of digestion.

*Bronchiectasis,* or dilatation of the bronchi (Gr. *ektasis* = dilatation) is a condition in some respects intermediate between chronic bronchitis and mucoviscidosis. It tends to affect people in early adult life (chronic bronchitis occurs more in middle age and the elderly). Bronchiectasis is usually associated with even more copious sputum production than is chronic bronchitis.

The dilated bronchi may result from an illness causing obstruction in the bronchi with lung collapse and infection beyond the obstruction. Such bronchi have their normal ciliated mucosa replaced by squamous epithelium so that the means of keeping the smaller bronchi clear of secretions is lost. Despite the fact that the name of the condition refers to the characteristic regions of dilated bronchi, the main problem is bronchial obstruction by the mucus which accumulates in the lungs due to loss of the normal ciliary transport mechanisms. This mucus often becomes infected and hence purulent. Cough-up of material from bronchiectatic cavities is hampered by the fact that the large cross-sectional area in these regions leads to low air velocity and hence difficulty in moving the sputum. Because gravity aids drainage of the upper lobe bronchi, the most troublesome bronchiectasis is that affecting the lower lobes.

### (iv) Loss of Elastic Supporting Tissue

This cause of obstructive airways disease is rather different from the preceding three in that the previous types of obstruction tend to respond to treatment, whereas destruction of elastic supporting tissue is permanent. Such disease is referred to as "irreversible airways obstruction". Further, in the preceding conditions, the trouble is in the wall of the airways; in the present condition the trouble is outside. In forced expiration during exercise, intrapleural pressure is positive (that is, greater than atmospheric). This pressure is transmitted throughout the lung tissue and tends to close the smaller non-rigid airways which have no cartilage in their walls. In normal lungs, the tendency for these airways to close is successfully resisted by elastic tissue attached to the airways (Fig. 56). When this elastic tissue is deficient, the small airways collapse and trap air in the alveoli. This condition is referred to as emphysema (Gr. *emphysan* = to inflate) and patients who suffer from it cannot empty their lungs adequately. As a result their chests are constantly in the expanded position (barrel chest). Many of these patients prefer to breathe out against the resistance of pursed lips. This may possibly help to maintain airway patency by raising luminal pressure during expiration.

It must be realised that the term emphysema, although a useful general concept, is a rather loose one. To the clinician it means a condition characterized by increasing dyspnoea of pulmonary origin with or without associated chronic bronchitis and sputum. Lung function tests show that there is obstruction to expiratory airflow; the chest X-ray typically shows little of note other than some reduction in the density of lung markings. To the pathologist, emphysema is a condition where there is generalized overdistension of alveoli and rupture of their partition walls. The two conditions do not always correspond, hence the concept of emphysema is a rather vague one.

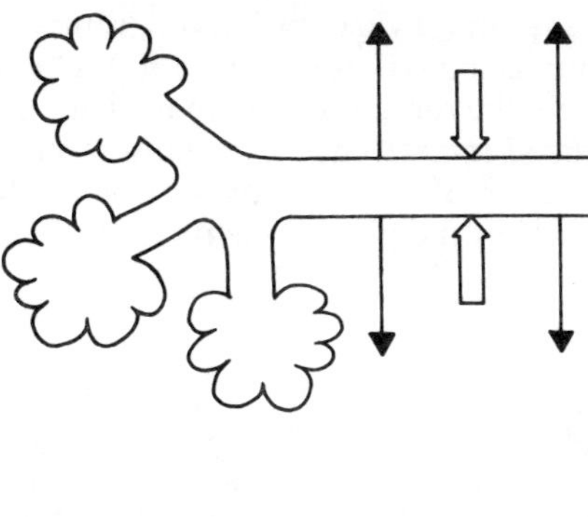

FIG. 56.—During forced expiration, as in exercise, positive intrapleural pressure (open arrows) tends to close the smaller bronchi. In normal lungs (top) elastic supporting tissue keeps these airways open. When the supporting tissue is damaged (bottom) the airways close during expiration and trap air in the alveoli.

Patients suffering mainly from emphysema rather than from bronchitis, tend to become pink puffers as opposed to blue bloaters; the reason for this is not clear.

The cause of emphysema is uncertain, but the disease is frequently associated with chronic bronchitis and may then have a similar cause—the inhalation of irritant gases from cigarettes or a polluted industrial atmosphere. Some patients with emphysema have been shown to have a deficiency of circulating antitrypsin activity. Possibly in such patients excessive proteolytic activity in the circulation causes structural lung damage and hence emphysema.

### *V Restrictive pulmonary disease*

In restrictive pulmonary disease there is a fall in total lung capacity and its subdivisions, vital capacity and residual volume. The larger airways are typically normal, but gas exchange across the alveolar-capillary membrane is limited by a fall in alveolar volume and surface area. Dyspnoea in these conditions is often associated with rapid shallow breathing rather than the slow wheezy breathing of obstructive disease. This may be because a reduction in the number of functioning alveoli results in a correspondingly greater expansion of the remaining alveoli for the same tidal volume. The increased expansion of these alveoli may cause premature inhibition of inspiration through the Hering Breuer reflex.

For the physiological point of view, two main groups of restrictive pulmonary diseases can be distinguished—those in which the restriction is limited to certain parts of the lungs, the remainder being normal, and those in which there is more or less generalized interference with lung structure—pulmonary fibrosis. Restriction of the volume of normal lung may be produced by consolidation, by collapse, by fluid or air in the pleural cavity and by surgical removal of lung tissue (Fig. 57).

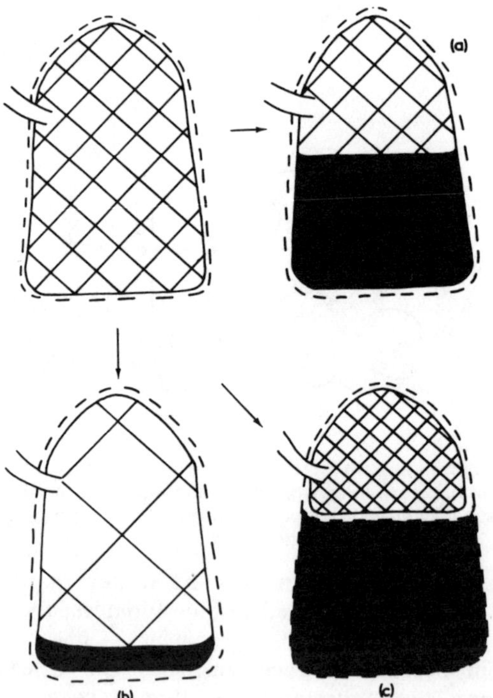


FIG. 57.—Three ways in which the ventilation of the lung (normal, top left) may be restricted:
(a) consolidation of part of lung
(b) collapse of part of lung — remainder is overdistended
(c) fluid etc. in the pleural space — the lung cannot be fully expanded.

**Consolidation**

This refers to the presence in the alveoli of material which excludes the entry of air and in which the lung loses its sponge-like appearance when seen after death. The alveolar surface area available for gas diffusion is thus reduced. In consolidation produced by pneumonia, inflammatory oedema, bacteria and leucocytes fill the alveoli. A somewhat similar condition is produced when some of the lung tissue dies from sudden obstruction of its blood supply. This is known as lung infarction (L. *infarcire* = to stuff in). It is due to obstruction of a pulmonary artery, often by a clot from a pelvic or leg vein; the alveoli contain large numbers of red cells and blood is often present in the sputum.

**Collapse**

Collapse of the lung (atelectasis; Gr. *ateles* = imperfect, *ektasis* = expansion) occurs when a moderately large bronchus is completely obstructed, by thick mucus, for example. The gas beyond the obstruction is then absorbed into the local blood vessels where total gas pressure is considerably less than atmospheric, e.g. 700 compared with 750 mm Hg. This lowered gas pressure in capillary and venous blood explains why gas trapped in the tissues anywhere tends to be absorbed, and is due to the fact that the $O_2$ which passes from blood to tissues causes the $Po_2$ to drop by about 55 mm Hg from around 95 to 40 mm Hg., whereas the $CO_2$ which enters the blood raises the $Pco_2$ only from around 40 to 45 mm Hg.

Collapse of part of the lung reduces the alveolar surface area available for gaseous exchange, and tends to overdistend the rest of the lung. Pleural pressure tends to become more negative on the side involved so that the mediastinum shifts to that side and the other lung tends to share in the excess distension.

The bronchial blockage which leads to collapse may be due to sticky secretions which cannot be coughed up. This is a well-recognized post-operative complication. Its frequency can be reduced if patients deliberately breathe deeply and cough in the post-operative period. More rarely, there may be a tumour of the bronchial wall, pressure from without the wall, or an inhaled foreign body, such as a tooth which was extracted under general anaesthesia or an item of food which was choked over, especially by a child.

**Fluid or Air in the Pleural Cavity**

This interferes with lung function by causing the lung to collapse, so reducing alveolar volume. Fliud in the pleural cavity (*pleural effusion*) may be due to local pleural irritation, e.g. due to infection or cancer, or to heart failure with its associated raised venous pressure. Air in the pleural cavity (*pneumothorax*; Gr. *pneuma* = air) may be due to an injury which penetrates the chest wall, or a surgical operation on the chest. Not uncommonly, however, the air enters from the lung and this may occur in otherwise healthy people. It appears that a leak develops between the pleural cavity and an airway or air-containing cavity connected with the airways. As pleural pressure is on average subatmospheric and air in the lungs is on average at atmospheric pressure, air is sucked into the pleural cavity. The rise in pleural pressure tends to cause the mediastinum to move towards the other side—the opposite to what happens with lung collapse.

Occasionally a large amount of air enters the pleural space when there is a valve-like situation at the point of entry so that air can enter more easily than leave. In this case, the pressure of the air in the pleural cavity rapidly exceeds atmospheric pressure (tension pneumothorax), the lung is severely compressed and death may result in a short time from grossly impaired gas exchange unless appropriate action is taken to remove the air from the pleural cavity.

### Surgical Removal of Lung Tissue

Surgical removal of lung tissue may be necessary in the treatment of lung cancer and other diseases. When lung tissue is removed, the remaining lung expands to occupy the available space, but pulmonary function is impaired because of the reduction in alveolar surface area available for gaseous exchange.

### Pulmonary Fibrosis

This heading covers a variety of conditions in which there is more fibrous tissue than normal in the lungs. The fibrosis may follow lung damage, for example by infection, and be relatively localized—coarse local fibrosis—or it may be fine fibrosis which involves lung tissue generally—fine diffuse fibrosis. The scarring reduces alveolar volume and the surface area available for gas diffusion.

*Coarse local fibrosis* is the scar tissue which follows damage to the lungs, usually by a prolonged infection such as tuberculosis, suppurative pneumonia or a lung abscess. If the amount of lung tissue destroyed is small, this type of fibrosis has little effect on lung function. If much lung has been replaced by fibrous tissue, the effect is similar to that of lung collapse, with over-inflation of the remaining lung. Damage to bronchi may lead to associated bronchiectasis.

*Fine diffuse fibrosis* is a condition where the fibrosis is present throughout the whole lung, and it tends progressively to impair lung function. It may be caused by the irritant effects of inhaled dusts, by deep X-ray therapy to the lung (vastly more than is received from routine chest X-rays) and by a number of diseases which affect collagen. When due to accumulation in the lungs of inhaled dust, the condition is referred to as pneumoconiosis (Gr. *pneumon* = lung, *konis* = dust). Pneumoconiosis may be caused by many industrial dusts and each tends to produce its own characteristic features. Each pneumoconiosis is named after its causative agent—e.g. silicosis due to silica inhalation, asbestosis due to asbestos inhalation. Similar fibrosis can result from repeated inhalation of materials which cause allergic alveolar disease, although the offending particles may not accumulate in the lungs. An example is farmer's lung due to the inhalation of fungal particles from mouldy hay.

Some agents produce fibrosis which is quite obvious on X-ray examination of the chest but which causes little interference with lung function, at least in the early stages of the disease. In other cases, there is progressive and serious impairment of lung function. The condition produced causes impairment of transfer (or diffusion) between alveoli and capillaries. This condition used to be called *alveolar-capillary block* because it was thought that there was such thickening of the alveolar-capillary membrane that passage of gases was slowed. It is now thought that the cause is a disturbance of ventilation/perfusion relationships rather than a simple uniform thickening. As well as interfering with gas transfer, such diseases make the lungs much stiffer than normal, thereby adding to the work of breathing.

## INVESTIGATIONS

The investigation of respiratory disorders involves (i) the demonstration of localized disease, e.g. lung cancer, tuberculosis, pneumothorax, and (ii) the demonstration of diffuse disease, e.g. obstructive airways disease, diffuse pulmonary fibrosis. The former are demonstrated mainly by *X-ray examination* and the latter by *respiratory function tests.*

### X-RAY EXAMINATION

The demonstration of localized disease relies mainly on X-rays. In addition to routine postero-anterior and lateral films, a *bronchogram* may be undertaken to demonstrate bronchial dilatation. For this test, the patient's throat and airways are treated by the inhalation of a local anaesthetic mist which suppresses the cough reflex. A tube can then be passed into the appropriate bronchus and radio-opaque material introduced. The inside of the trachea and the larger bronchi can be examined directly through a tube (bronchoscope) passed into the larger airways. A small sample of tissue can be removed for examination if required.

The demonstration of generalized disease, in contrast, relies mainly on tests of respiratory function. An exception is diffuse pulmonary fibrosis which often presents characteristic X-ray changes.

### RESPIRATORY FUNCTION TESTS

This subject is considered under four headings.

(a) *Classification of respiratory function tests.* A brief description of the tests is given.
(b) *Variation between individuals.*
(c) *Questions which may be answered by respiratory function tests.*
(d) *Summary of the place of respiratory function tests in the management of patients with respiratory disease.*

#### *(a) Classification of respiratory function tests*

Seven groups of tests related to respiratory function are considered:

1. static lung volumes
2. dynamic lung volumes
3. efficiency of gas mixing
4. gas transfer
5. mechanical properties of the lungs
6. blood analysis
7. exercise tests.

The first group is concerned with lung size, groups 2, 3 and 5 with ventilation of the alveoli, group 4 with transfer of gases across the alveolar-capillary membrane and group 6 with the end-product of lung function—arterial gas tensions and pH. Tests in groups 1,2 and 6 are the most commonly used in practice.

**Static lung volumes.**—This group includes the vital capacity (VC) and the residual volume (RV), which, when added together, give the total lung capacity (TLC). VC is typically reduced in both obstructive and restrictive lung disease. Patients with obstructive airways disease cannot exhale as completely as is normal because their small airways close during expiration. This gives rise to a barrel chested appearance and can be detected by measuring the volume of air left in the lungs at the end of a complete expiration—the residual volume (RV). When there is air trapping RV is raised and VC lowered, so that the ratio RV to TLC (total lung capacity = RV + VC) may be greatly increased.

RV is measured by asking the patient to exhale completely until only RV is left in his lungs. He is then connected to a known volume of gas containing an indicator such as helium which he breathes for about 5 min until mixing is complete. Dilution of the helium indicates the final volume of gas, including the patient's RV. By subtracting the initial gas volume, his RV is obtained. Alternatively, the patient can be connected at end expiration to a volume of oxygen and the dilution of the nitrogen in the patient's lungs used to calculate RV. In this case the initial $N_2$ concentration in RV is the alveolar $N_2$ concentration.

Lung volumes are measured, saturated with water vapour, in a spirometer at room temperature. Conventionally the results are converted to what they would be in the body, saturated with water vapour at 37°C. Variations in barometric pressure make relatively little difference to the result, and are not usually corrected for.

Measurements of the other subdivisions of TLC, such as inspiratory and expiratory reserve volume, are of little practical importance in the management of respiratory disease.

**Dynamic lung volumes.**—These tests measure volume/unit time. As obstructive airways disease, which by definition reduces maximal air flow rate, is likely to cause more difficulty in expiration than in inspiration, it is usual to measure expiratory rather than inspiratory rates. Forced expiratory volume, the volume in litres which can be expired in one second ($FEV_{1\cdot0}$) and peak flow rate, the maximum rate in litres/min at which air can be expired (PFR), are useful examples of such tests.

In the measurement of $FEV_{1\cdot0}$, the patient is asked to fill his lungs completely and then to breathe out as rapidly as possible, keeping on until he cannot expire any further. The volume expired in one second—after the initial slower expiration of the first 200 ml—is either recorded automatically or obtained from a tracing. The fast vital capacity (FVC) is also noted and the ratio $FEV_{1\cdot0}$/FVC calculated. As the test takes only a few seconds, it is usually repeated five times and the mean of the last three measurements recorded. If the patient has carried out the test properly, these three results should be very similar, e.g. differing by only 2-5 per cent.

PFR is measured in a similar way, except that complete expiration is unnecessary. Respiratory minute volume is not generally a particularly helpful measurement to make because it may be within the normal range despite advanced lung disease.

Airways obstruction may be classified as reversible (bronchial asthma for example) or irreversible (emphysema). If one of the above tests of airflow reveals airways obstruction the patient may be given a large dose of a drug which relaxes smooth muscle, such as an adrenaline aerosol inhalation. If the $FEV_{1\cdot 0}$ or PFR show an improvement, then the airways obstruction can be said to be at least partially reversible.

There are several simple ways of making a rough estimate of dynamic lung volumes. One is to see whether a patient with his mouth wide open can blow out a lighted match held 15 cm from his mouth. If he succeeds, his $FEV_{1\cdot 0}$ is probably at least one litre. Another simple test is to measure forced expiratory time by listening with a stethoscope over the trachea while the patient expires his vital capacity as quickly as possible. A time of more than 6 seconds suggests that obstructive airways disease is present. A whistle with an adjustable leak can also be used to estimate airways patency. The greater the patient's $FEV_{1\cdot 0}$, the greater the leak in whose presence a sound can still be produced. Such tests may with practice be useful in a situation where other equipment is lacking, but they are a poor substitute for more precise measurements if the patient's condition is to be accurately assessed and his response to treatment observed.

**Efficiency of gas mixing.**—This can be assessed by the ***nitrogen washout test.*** The patient makes a maximal expiration and then breathes in pure oxygen and breathes out to air for a standard period, e.g. 7 minutes. The nitrogen originally in the RV should by then have been almost completely washed out if mixing in the lungs is normal. Alternatively, the rate of wash-in of helium can be measured, e.g. during measurement of RV by helium dilution. Gas mixing is inefficient in obstructive airways disease, especially in emphysema.

**Gas transfer.**—Gas transfer is a measure of the capacity of the alveolar-capillary membrane to transfer gas between alveolar air and blood. Because gas transfer is entirely passive and depends on the difference in partial pressure of the gas between alveolar air and blood, rate of gas transfer is related to the pressure gradient in mm Hg. The units are therefore ml/min/mm Hg.

For convenience, carbon monoxide (CO) is often used as the test gas. Its pressure gradient can be estimated readily because CO combines so readily with haemoglobin. With the low concentration of CO used, it can be assumed that the amount of dissolved CO in the pulmonary capillaries is virtually zero, hence the Pco here is zero and the gradient is taken as the alveolar Pco. The amount of CO taken into the body in such a test is too small to have an adverse effect on the patient.

At the beginning of the test, the patient breathes out completely and then takes in a deep breath of air + helium + CO. After holding his breath for about 10 secs, during which gas transfer occurs, he breathes out again and a sample is taken. Dilution of the helium indicates the alveolar volume during breath-holding. Subtraction of the volume of expired CO from that inspired gives the volume of gas transferred. This test requires fairly elaborate equipment; full co-operation on the part of the patient is essential.

A reduction in the rate of gas transfer is sometimes an early sign of diffuse pulmonary fibrosis. The result of a gas transfer test is given in terms of the particular gas used, e.g. $T_{LCO}$—lung transfer factor for CO. Sometimes the result is referred to as the diffusion coefficient—$D_{LCO}$.

**Mechanical properties of the lungs.**—This group of tests includes measurements of airway resistance and lung compliance. The tests are somewhat complicated and are usually reserved for use in specialized laboratories, largely for research purposes.

Resistance of the airways (R) in the lung is analogous to the resistance in an electrical circuit (Resistance $= \frac{\text{potential difference}}{\text{current}}$) and the resistance of blood vessels (Resistance = pressure gradient along vessel/flow). Thus airways resistance, R = pressure gradient along airways/airflow, the units being cm $H_2O$/litre/min. Airflow can easily be measured, but the pressure gradient along the airways requires an indirect method of estimation.

*Compliance* of the lungs is a measure of their expansibility. It is the change in lung volume produced by unit change in pressure gradient. The units are litres/cm $H_2O$. The measurements required are volume change and the pressure gradient which produced the change in volume. The pressure gradient used is that between the pleural cavity and the atmosphere. Intrapleural pressure is estimated from oesophageal pressure, using a swallowed small balloon (some patients have difficulty is swallowing the balloon). Oesophageal pressure is similar to pleural pressure because both the oesophagus and the pleura are elastic sacs containing only a small quantity of fluid and themselves contained in the same expanding and contracting chamber, the thorax.

A variety of diseases decrease lung compliance and hence increase the work of breathing. Such diseases as diffuse pulmonary fibrosis make the lungs "stiffer", i.e. less compliant. There is a severe fall in compliance when pulmonary surfactant is deficient. In heart failure the congestion of the pulmonary capillaries with blood makes the lungs more rigid and reduces their compliance.

**Blood analysis.**—Measurement of blood $Po_2$, $Pco_2$ and pH indicates the presence or absence of respiratory failure and its severity if present, provided other body systems—e.g. cardiovascular and renal—are normal.

As respiratory failure is defined in terms of abnormal *arterial* $Po_2$ and $Pco_2$ levels, the most direct method of diagnosing and assessing this condition is to obtain a sample of arterial blood and measure the $Po_2$ and $Pco_2$. With practice, it is not difficult to insert a needle into the brachial artery in the antecubital fossa, or into some other superficial artery. The blood sample is then analysed as soon as possible lest metabolic activity in the blood alters the gas pressures. The acid-base status of the blood should also be assessed.

*The effect of breathing 100 per cent oxygen* for about 5 minutes serves to distinguish between patients with an abnormal arterial $Po_2$ due to respiratory disease and those in whom the reduced $Po_2$ is due to a shunt between venous and arterial blood due, e.g. to an atrial septal defect. If the low $Po_2$ is due to respiratory disease, breathing 100 per cent $O_2$ will correct it; if it is due to a shunt, breathing 100 per cent $O_2$ will not correct it.

**Exercise tests.**—Certain abnormalities of respiratory function may be obvious only when the patient is submitted to the stress of exercise. For example, a patient may have a more or less normal arterial oxygen content at rest but may show marked desaturation when he carries out moderate exercise. In some cases, bronchospasm is precipitated by exercise. The exercise may consist of stepping on and off a low platform or walking on a moving treadmill. Care needs to be taken to avoid causing distress to patients, especially those suffering also from heart disease.

### *(b) Variation between individuals*

While values such as the blood pH or calcium level vary relatively little with individual size, sex and age, this is not the case with lung function tests, particularly those involving lung volume. The values are highly dependent on height, decrease with age and are less in females than in males. A tall man (1.9 metres, 6 ft 3 in) of 20 should have a vital capacity around 6 litres, whereas a small elderly woman (1.4 metres, 4 ft 7 in, age 60) would be expected to have a VC of under 2 litres.

It is common practice to express results of such tests as a percentage of the predicted value which may be obtained from formulae and nomograms. However, because of the wide normal variation, a range of about 20 per cent would be necessary to include about 95 per cent of normal people in the case of the vital capacity. Thus a person whose vital capacity is 4 litres, or 80 per cent of that predicted, could be a normal individual with smaller than average lungs or someone whose VC ought to have been 6 litres and who has had it markedly reduced by disease. This may be clarified by considering the result alongside other symptoms and findings. The most satisfactory way of using such tests, however, is to carry out *serial observations*. One can then tell fairly precisely whether the patient's condition is static, improving or deteriorating.

### *(c) Questions which may be answered by respiratory function tests*

1. Is lung function normal or abnormal?
2. Is the condition improving or deteriorating?
3. What kind of abnormality is present?
4. Can this patient survive removal of part or all of a lung?

*Is lung function normal or abnormal?*—In some cases of obstructive respiratory disease of the emphysema type, the patient may complain of moderate shortness of breath but general examination and chest X-ray may reveal little or nothing of note. In contrast, measurement of the FVC and $FEV_{1\cdot0}$ or PFR, may show definitely that obstructive respiratory disease is present.

*Is the condition improving or deteriorating?*—As already mentioned, a single assessment of lung function can give only an approximate indication of the severity of disease present, because the original normal values are uncertain. Serial testing overcomes this difficulty and is very useful in observing the progress of disease, including the effects of treatment.

*What kind of abnormality is present?*—Many varieties of lung abnormality have been described, but most common lung diseases can be grouped as either obstructive or restrictive disease. Obstructive disease implies an impaired maximal rate of airflow, so that $FEV_{1\cdot0}$ and PFR are reduced whereas VC is relatively normal. Thus the ratio $FEV_{1\cdot0}/VC$ is reduced, e.g. from 75 per cent to 50 per cent or less. Due to air trapping, the RV is increased.

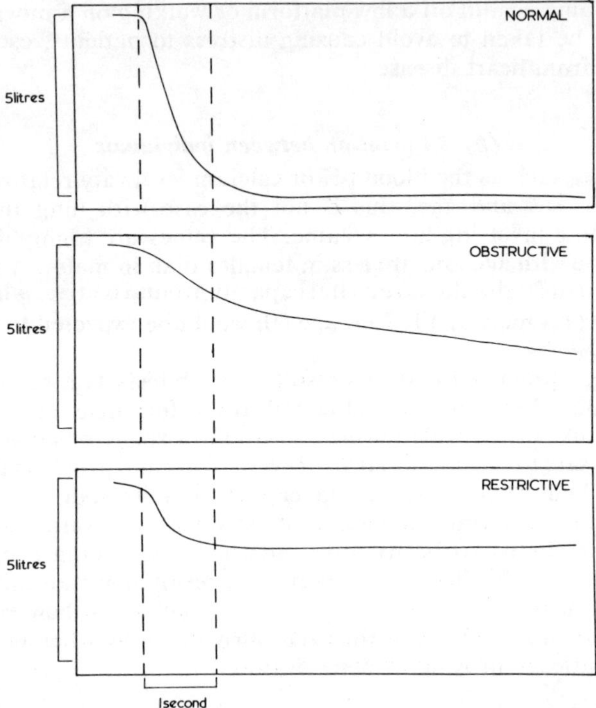


FIG. 58.—Tracings of a forced expiration in obstructive and restrictive respiratory disease compared with normal. The trace starts with the patient holding his lungs fully inflated, he then expires as rapidly and completely as possible. The dotted lines indicate the first second of expiration. In obstructive disease the trace is elongated in time; in restrictive disease it is a miniature of the normal.

TABLE III

| TEST | OBSTRUCTIVE DISEASE | RESTRICTIVE DISEASE |
|---|---|---|
| VC | ↓ | ↓ |
| $FEV_{1\cdot0}$ | ↓ ↓ | ↓ |
| $FEV_{1\cdot0}/VC$ | ↓ | — |
| PFR | ↓ ↓ | — |
| RV | ↑ | ↓ |

**Contrasting respiratory function results in obstructive and in restrictive disease. One arrow indicates moderate reduction; two arrows severe reduction. Compare with Fig. 58.**

In restrictive disease VC is reduced. $FEV_{1\cdot 0}$ is reduced in proportion, so that $FEV_{1\cdot 0}$/VC is normal or increased. PFR is usually also normal. RV is decreased in parallel with the decrease in VC. Obstructive and restrictive disease are contrasted in Fig. 58 and Table III.

Diffuse pulmonary fibrosis often causes restrictive lung disease. However, in some cases there may be no restriction, whereas measurement of the transfer factor for CO may reveal a very low value and show that the type of lung disease present is diffuse fibrosis.

*Can this patient survive removal of part or all of a lung* (e.g. for cancer)?—This problem is somewhat similar to that of whether to carry out artificial ventilation. In both cases the procedure is potentially life-saving but if the final respiratory function is inadequate to maintain life then the procedure causes only unnecessary hardship to patient, relatives and hospital staff. A rough guide is to measure the $FEV_{1\cdot 0}$ and deduct 30 per cent for removal of a *functioning* lobe; 55 per cent for removal of a functioning lung. If the result is less than 25 per cent of the *predicted* $FEV_{1\cdot 0}$, then the operation should not be carried out.

There is no particular rule for other operations. The worse the lung disease, the lower are the patient's chances of survival. Thus, possibly helpful but non-essential operations may be deferred until the patient's condition is optimal, e.g. in summer as opposed to winter. Life-saving operations should be proceeded with as usual.

### *(d) Summary of the place of respiratory function tests in the management of patients with respiratory disease*

Respiratory function tests range from such measurements as PFR, $FEV_{1\cdot 0}$ and FVC, which require only a few minutes and relatively inexpensive equipment to carry out, to elaborate time-consuming tests which require considerable co-operation by the patient, experience and skill on the part of the operator and expensive equipment. As the time required is considerable, it is impracticable to repeat them, say, five times in succession as can be done with the simpler tests. Hence their accuracy is reduced. They may be of considerable value in certain situations, but generally in clinical practice they add little, if anything, to the information that can be obtained from simpler tests (Fig. 59). Measurement of $FEV_{1\cdot 0}$ and FVC gives near maximum information in most cases at relatively little "cost".

Respiratory function tests are used in the diagnosis and assessment of diffuse lung disease or ventilatory insufficiency, when X-ray studies may supply little information. They supply objective quantitative information and are most informative when serial tests are done.

Measurements of $FEV_{1\cdot 0}$, VC and PFR are relatively easy to make and may be compared with the measurement of the haemoglobin level in anaemia. The measurement of arterial $Po_2$, $Pco_2$ and pH in the treatment of respiratory failure may be compared with the measurement of blood sugar levels in diabetic ketosis. The more elaborate tests are used in the elucidation of diagnostic problems and in research into the mechanisms of respiratory disease.

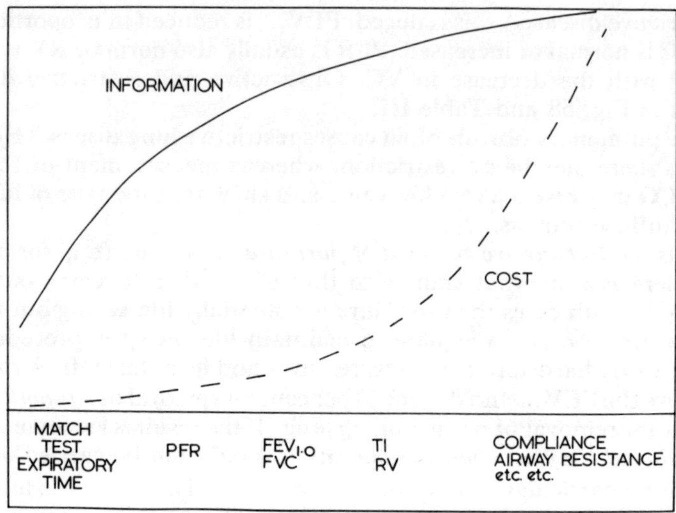


FIG. 59.—An attempt to relate the "cost" in terms of time, equipment and expertise etc. of respiratory function tests of increasing complexity to the increased information obtained by carrying out such tests.

## TREATMENT

This is described under three headings.
(a) obstructive airways disease
(b) restrictive pulmonary disease
(c) respiratory failure.

### *(a) Obstructive airways disease*

Treatment will be considered under the same headings as were used in describing the causes of airways obstruction—spasm, mucus, oedema and loss of elasticity.

**Spasm**

Treatment may be related to the cause of the trouble, or may be non-specific. Bronchial spasm due to hypersensitivity (allergy) is *ideally* treated by removing the offending allergen or allergens. This is not always possible, but in some cases dramatic improvement follows the elimination of certain plants from the environment or the departure of a pet budgerigar, cat, rabbit, etc. Even though no specific allergen can be identified, reduction in the inhalation of house dust (containing protein particles), substitution of cotton etc. for woollen blankets and covering a hair mattress by a polythene bag may all help the situation.

Drugs which interfere with antigen-antibody reactions are often of value. Of these, glucocorticoids can be very effective, but in view of their serious adverse side-effects, some potentially fatal, they should be used only when other treatments have failed and the condition is extremely disabling. In these circumstances they seem to help most varieties of airways obstruction due to muscle spasm. They tend also to reduce oedema and mucus secretion. Disodium cromoglycate is a more recently introduced drug which is thought to interfere with the release of spasm-producing chemicals from mast cells in the bronchial wall, these being mediators of the bronchospasm caused by antigen-antibody reactions.

Psychological factors causing asthma may be helped in some cases by discussion with the patient and appropriate social action. Tranquillizing drugs are generally of rather limited value.

Direct bronchial irritants, such as dusts inhaled in certain occupations, may in some cases be removed by a change of job where this is possible. Included in the group of bronchial irritants is cigarette smoking. Giving up smoking is probably the one most important factor in the prevention and treatment of obstructive airways disease. Infections causing bronchospasm require treatment with an appropriate antibiotic.

In addition to these measures aimed at the *cause* of bronchial spasm, drugs which relax bronchial muscle directly are often of benefit. Bronchial muscle is relaxed by adrenergic nerves, the effect being mediated at adrenoceptors of the $\beta_2$ variety. Sympathomimetic drugs such as adrenaline tend to relax bronchial muscle; isoprenaline, a more marked $\beta$-stimulator being more specific. However, such drugs stimulate the heart to beat faster and more powerfully and this side-effect may, in extreme cases, be fatal. It was found recently that the death rate in young asthmatics in particular was rising rather than falling and it has been postulated that this was due to panic-stricken asthmatic patients taking excessive doses of isoprenaline from their portable inhalers. Thus caution must be exercised in using $\beta$-stimulating drugs and efforts are now being made to overcome this difficulty by using selective $\beta_2$-stimulator drugs which have relatively little effect on the heart. Conversely $\beta$-blocking drugs which remove the bronchial relaxing effects of the sympathetic nervous system in maintaining bronchial patency may make bronchial asthma worse.

Cholinergic fibres to the bronchi tend to constrict them, so it seems logical to use anticholinergic drugs, e.g. atropine, in the treatment of asthma. However, this treatment is of limited value, suggesting that vagal tone to the bronchi is not generally of great importance in the causation of asthma. Drugs which mimic the action of acetylcholine (parasympathomimetic drugs) may, however, make asthma worse.

Other drugs, such as aminophylline, relax bronchial smooth muscle by an effect which seems to be independent of adreno- or cholinergic receptors.

**Mucus**

Although the anticholinergic drug atropine, given routinely before surgical operations, is used to cause a short-term reduction in airways secretion of mucus and hence in the risk of obstruction when the cough reflex is suppressed by anaesthesia, the long-term treatment of excess mucus is generally aimed at removing it from the respiratory tract, rather than preventing it from being formed. The means of removal are physical rather than chemical and consist of the relatively simple procedures of postural drainage and persuading the patient to cough. These are used mainly in chronic bronchitis, bronchiectasis and mucoviscidosis.

Adequate coughing by the patient is one of the most important parts of the treatment of obstructive disease associated with marked retention of secretions. It may be necessary to use severe verbal force to *make* a lethargic patient cough repeatedly and vigorously, but there is no other way whereby his airways may be adequately cleared.

Because most patients spend most of their time upright or in the head-up position in bed, excess bronchial secretions tend to accumulate in the lower parts of the lungs. Loss of the normal action of the cilia favours this. During *postural drainage* the position must be reversed, with the patient's head well below his lung bases. Rhythmical slapping on the patient's chest may help to dislodge the rather sticky mucus and the patient is encouraged to cough frequently and vigorously. In cases where the patient is drowsy as a result of respiratory failure, the exhortations may have to be very forceful, verging on the bullying. If the patient is unconscious, a large dose of a drug which stimulates the nervous system may arouse him sufficiently to cough and thereby clear his airways and improve gas transport to and from the alveoli. If bronchospasm is present, this should be treated before the effort to drain the mucus is begun.

In cases where the mucus is particularly thick and sticky, as in mucovisicidosis, inhalation of a fluid aerosol may help, as may drugs aimed at liquefying secretions. Conversely administration of dry oxygen may make sputum much more difficult to cough up.

### Oedema

Oedema due to bronchial irritation is helped by treatment directed towards the removal of such irritation as described in the treatment of bronchospasm. Oedema of cardiac origin requires treatment of the heart failure. It is important not to mistake cardiac asthma for bronchial asthma and give adrenergic drugs which would impose a greater work load on the heart. Aminophylline, which strengthens the heart's action as well as dilating the bronchi, is a safe drug to give. In very severe cases of cardiac asthma, reduction of the blood volume by the ancient procedure of venesection, or blood-letting, may lead to rapid improvement.

### Loss of Elasticity

There is at present no treatment for this condition. Avoidance of smoking may help to prevent the disease from progressing.

### Surgery

Very rarely surgical removal of part of a lung may be carried out on a patient with bronchiectasis who has a fairly small region of lung seriously affected or on a patient who has a very large useless air cavity which is compressing other parts of the lung.

## *(b) Restrictive pulmonary disease*

Treatment of the restrictive pulmonary disease due to infection consists mainly in combating the organisms by antibiotic drugs, with general support of the patient's condition by nursing care.

Where pleural fluid is seriously interfering with lung ventilation it should be removed so that the lung can expand more fully. Similarly a large pneumothorax which interferes with adequate ventilation should be treated by removal of air. This is done by connecting the pleural cavity to tubing whose end is under water (*under-water seal*) as shown in Fig. 60. When intrapleural pressure rises above atmospheric pressure, air bubbles out through the water; when intrapleural pressure is below atmospheric pressure, fluid is sucked up the tubing to a certain extent but air cannot re-enter it. This drainage method provides, incidentally, an excellent visual indication of the changes in pleural pressure associated with breathing and coughing. If respiratory swings are not present, then the tubing has become blocked—or the patient has stopped breathing!

When a portion of lung has collapsed due to blockage of a major airway it may be possible to examine the area directly through a tube (bronchoscope) passed into the bronchial system. If a foreign body is seen it may be removed to cure the condition.

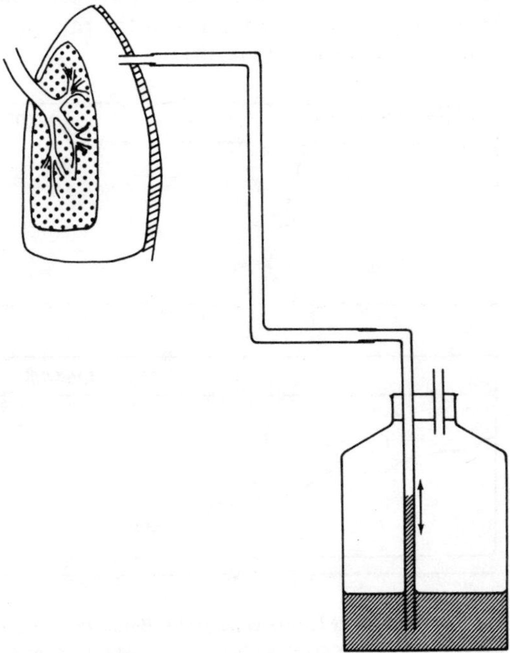

FIG. 60.—The system used to remove air from the pleural cavity. During coughing, a deep expiration, or constantly in the case of a tension pneumothorax, air bubbles out through the water in the bottle. During quiet breathing, the fluid level in the tube indicates a negative intrapleural pressure which varies during the breathing cycle, as shown by the arrows.

**Pulmonary fibrosis** does not usually respond to treatment, although occasionally where the fibrous tissue is in the process of being laid down, possibly as a result of an antigen-antibody reaction, treatment with glucocorticoids may hinder progress of the condition. Acute attacks of allergic alveolitis, e.g. in farmer's lung, may apparently resolve completely, provided that the patient is removed from exposure to the allergen; the rate of resolution may be increased by treatment with glucocorticoids. Continued or repeated exposure to the allergen may lead to permanent irreversible fibrosis.

With regard to industrial pulmonary fibrosis, the main "treatment" is prevention, by ensuring working conditions where there is the least possible exposure to dust. This may be achieved by enclosing and by damping dusty processes, and, less effectively, by the use of masks—the problem being that the masks tend to be so uncomfortable that they are not worn.

### (c) *Respiratory failure*

Patients with respiratory failure fall into two distinct categories, depending on whether the respiratory failure is acute or chronic, although occasional patients may have some features of both groups. Treatment for these two groups is quite distinct. The two groups differ mainly in the rate of onset of respiratory failure and its potential reversibility (Fig. 61).

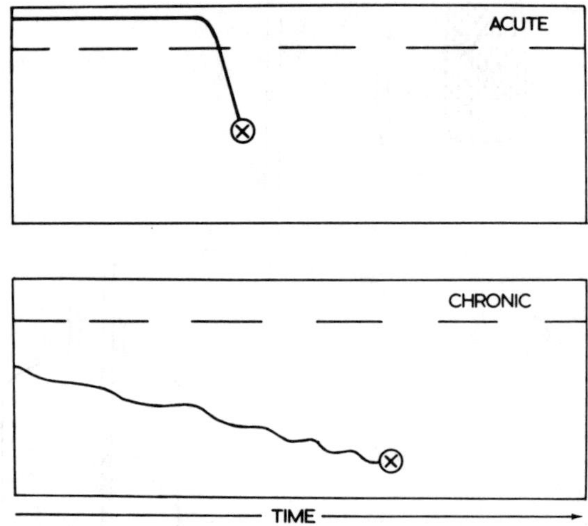


FIG. 61.—Acute and chronic respiratory failure contrasted. Below the dotted line respiratory failure is present. Acute respiratory failure may arise in the course of days or weeks in an otherwise healthy person. If the condition is curable, artificial ventilation should be given. A patient with chronic respiratory failure usually takes years to reach a similar point; in this case, artificial ventilation is not given because the patient has no prospect of subsequently doing without artificial ventilation.

#### Acute Respiratory Failure

This develops over the course of days or weeks. Provided there is a reasonable chance that the condition will eventually improve sufficiently to allow adequate unassisted ventilation, the treatment is artificial ventilation. If, however, there is little or no subsequent improvement, the patient must receive artificial ventilation for the rest of his life. This is not generally regarded as a situation in which one should place a patient, hence the aim to reserve artificial ventilation for those in whom it will be needed only temporarily, i.e. for a matter of days, weeks or months. In the case of the occasional patient with chronic chest disease who suddenly develops respiratory failure due to a severe infection, artificial ventilation may be justified because considerable improvement can be expected when the infection has been treated. A useful guide in such cases is the "hundred yards rule". If the patient before the acute illness was able to walk 100 yards on the level without stopping to take breath, then artificial ventilation may be considered.

*Anticipation* of the need for artificial ventilation is possible in certain conditions. In these conditions, such as polyneuritis affecting the respiratory nerves, the patient's ventilatory ability may steadily fall. However, as long as he can maintain a minute volume adequate for his needs at rest, respiratory failure will not develop. Once he falls below the necessary minute volume, respiratory failure may develop rapidly and threaten life. In order that the patient may be transferred to a special unit for treatment of his respiratory failure *before* the emergency arises, it is desirable to anticipate the onset of respiratory failure. This may usually be done by making serial measurements of the $FEV_{1\cdot0}$ or VC.

The commonest method of ***artificial ventilation*** used at present is intermittent inflation of the lungs using a machine which provides a modest positive (above atmospheric) pressure to drive the gas in. Expiration occurs by passive recoil of the lungs, or a slight negative pressure may be applied by the machine. The gas is usually passed directly into the patient's trachea through a tracheostomy. This avoids the risk of damage to the larynx by tubing and makes much easier the removal by suction of secretions in the trachea and bronchi. As the upper respiratory tract with its humidifying mucosa has been by-passed, it is essential to humidify artificially the gases entering the lungs to prevent drying of and damage to the alveoli. Also, because the larynx has been by-passed, a person with a tracheostomy cannot speak unless the tracheosotomy opening is temporarily covered.

With artificial ventilation, the regulating effects of the chemoreceptors are absent, so it is necessary to replace their function by regular monitoring of the chemical levels which they monitor in the body—the $Po_2$, $Pco_2$ and pH. On the one hand, inadequate ventilation fails to provide adequate oxygenation; on the other hand, excessive ventilation blows off too much $CO_2$ and causes a harmful alkalosis.

### Chronic Respiratory Failure

This is usually due to chronic obstructive respiratory disease, and tends to progress in a series of undulations whereby the patient develops episodes of worsening of his condition, generally precipitated by bronchial irritation due to infection or chemical irritants (e.g. smog). During these episodes he suffers a fall in $Po_2$ and a rise in $Pco_2$. He then requires appropriate therapy for his underlying condition, e.g. antibiotics, bronchial relaxants, physiotherapy and treatment for heart failure (cor pulmonale) if present. In addition however, he requires specific treatment for his respiratory failure. The mainstay of this treatment is *controlled oxygen therapy.* This phrase implies that the rate of oxygen administration to the patient is carefully regulated. This must be done because too little or too much oxygen may be equally harmful. Too little oxygen in the body is likely to cause generalized tissue damage, and probably plays a role in the causation of cor pulmonale because oxygen lack causes constriction of pulmonary arterioles. It is impossible to state a precise figure at which the arterial $Po_2$ is dangerously low, but it is generally considered that below a $Po_2$ of 50 mm Hg, death from serious oxygen lack becomes increasingly likely.

Too much oxygen in the body is liable to harm anyone eventually, but the typical patient with chronic respiratory failure may be at risk if his arterial $Po_2$ is raised merely to the normal value of around 100 mm Hg. The reason is as follows. Such patients have a low sensitivity to a raised $Pco_2$ and their $Pco_2$ is typically raised. Thus oxygen lack, which stimulates ventilation at a $Po_2$ of around three quarters normal, i.e. 75 mm Hg, is important in maintaining adequate ventilation. If the oxygen lack is suddenly removed, their rate of ventilation may fall. Their $Pco_2$ will then rapidly rise, and, because at high levels it depresses the function of cerebral cells, including those of the respiratory centre, there may be a vicious circle, with ventilation falling off and the patient passing into coma and death.

If the $Pco_2$ can be measured regularly, oxygen can be administered at a rate which raises the $Po_2$ from, say 30-40 mm Hg to 50-60 mm Hg. As these oxygen tensions correspond to the steep part of the oxygen dissociation curve, a small rise in alveolar $Po_2$ will lead to a relatively large rise in the oxygen uptake by blood passing through the pulmonary capillaries. If the $Po_2$ cannot be measured, therapy cannot be so precise, but it has been found in practice that breathing 25-30 per cent oxygen will usually lead to improvement in such patients. Increased drowsiness in the patient is an indication for reducing the rate of oxygen administration. Special devices are available for administering oxygen in this controlled manner. Humidification is important.

"Respiratory" stimulant drugs have enjoyed considerable popularity in the treatment of respiratory failure. However, although a substance which would stimulate the respiratory centre without having any other effects appears desirable, such a substance is not at present available. Drugs sometimes described as respiratory stimulants seem to be generalised cerebral stimulants, or else to act by causing an acidosis (e.g. inhibitors of carbonic anhydrase, which interfere with acid excretion by the kidney). It is doubtful whether such drugs have a place in the long-term management of respiratory failure, because the adverse effects of the drugs may outweigh the beneficial effects of improved gas tensions. However, they may be helpful in the acute exacerbation of respiratory failure by rousing the patient from his stupor so that he can actively help to break the vicious cycle of his disease by coughing up sputum blocking his airways and by co-operating in general nursing procedures such as the maintenance of an adequate intake of food and fluids. Injected drugs, such as nikethamide, may produce a very dramatic improvement in the level of consciousness—from coma to a fair level of alertness. In contrast, the use of drugs which depress cerebral function—e.g. morphine—may lead to coma, deterioration and death.

If the patient does not respond to controlled oxygen therapy, there remains the possibility of carrying out a tracheostomy to reduce dead space and aid in removing airways secretions; finally artificial ventilation may be considered. As mentioned previously, these procedures should not be carried out unless there is evidence that the patient's underlying condition is capable of moderate improvement.

## RECENT LEADING ARTICLES

Analgesics and asthma. *Brit. med. J.*, 1973, **3**, 419.
A new stimulant for ventilatory failure. *Lancet*, 1973, **1**, 753.
Asbestos hazard. *Brit. med. J.*, 1973, **4**, 312.
Asthma and wheezy bronchitis in childhood. *Brit. med. J.*, 1973, **4**, 749.
Asthma deaths: a question answered. *Brit. med. J.*, 1972, **4**, 443.
Bronchial mucocoele. *Brit. med. J.*, 1971, **3**, 721.
Closing volume. *Lancet*, 1972, **2**, 908.
Drowning. *Lancet*, 1972, **2**, 691.
Drug-induced respiratory disorders. *Brit. med. J.*, 1973, **2**, 320.
Enzymes and emphysema. *Brit. med. J.*, 1973, **1**, 1.
"First in, last out" in the lung. *Brit. med. J.*, 1973, **3**, 119.
Hypercapnia in bronchial asthma. *Lancet*, 1973, **1**, 140.
Increasing the transpulmonary pressure in respiratory-distress syndrome. *Lancet*, 1973, **2**, 244.
Lung damage by oxygen. *Lancet*, 1970, **2**, 1292.
Oxygen therapy. *Lancet*, 1970, **1**, 130.
Pathogenesis of emphysema. *Brit. med. J.*, 1974, **1**, 527.
Pneumoconiosis redefined. *Brit. med. J.*, 1972, **2**, 552.
Respiratory complications of obesitiy. *Brit. med. J.*, 1974, **2**, 519.
Reversibility of asthma. *Lancet*, 1974, **1**, 1327.
Spread of colds. *Brit. med. J.*, 1973, **4**, 123.
The carotid bodies in oxygen regulation. *Lancet*, 1972, **1**, 79.
Treatment of acute exacerbation of chronic bronchitis. *Brit. med. J.*, 1973, **4**, 437.
Types of emphysema. *Brit. med. J.*, 1974, **2**, 571.

*Section VI*

# ALIMENTARY SYSTEM

# 31. DEFICIENT NUTRITION

## DEFINITION

THIS CHAPTER includes conditions in which body tissues are malnourished because of an inadequate supply of necessary nutrients. Malnutrition may arise from the *malabsorption* of food as well as from an inadequate food supply (*starvation*).

## EFFECTS

Generalized malnutrition varies in severity from sudden complete starvation to prolonged mild deficiency. The effects of complete starvation have been studied in some detail. Lesser degrees of food deprivation produce a similar but less severe effect; if the deficiency is not unduly severe, the individual may survive indefinitely by reducing his activity to reach a state of equilibrium below the optimal state.

### Starvation

Intake of food is required to provide for the energy requirements of the body. Even when food intake ceases, energy expenditure can only be reduced to a certain minimal level. In these circumstances the body has to metabolize its own tissues to provide the basal energy requirements to sustain life. The problems of food deprivation are those associated with the breakdown of the body tissues to provide energy.

Complete starvation (an adequate supply of water is assumed) is likely to cause death in the average healthy individual within 1-3 months. Three stages may be distinguished. The carbohydrate, fat and protein stores are successively metabolized to provide energy.

*Utilization of carbohydrate stores.*—The carbohydrate stores of liver and muscle glycogen last for only 1-2 days.

*Utilization of fat stores.*—This stage lasts for about a month in the average individual; its duration depends on the size of the fat stores. The fat utilized comes from the adipose tissue stores, not from essential cellular lipids. This stage can be recognized by a fall in the respiratory quotient (RQ) to approximately that of fat (around 0.7). The blood glucose level is maintained within normal limits. The liver produces the glucose from amino-acids, lipids and lactic acid. Fat metabolism, however, is increased relative to carbohydrate metabolism and, as in diabetes mellitus, leads to ketosis. The associated acidosis must contribute to the individual's feeling of loss of energy and well-being. Vitamin deficiencies develop, especially of the water-soluble vitamins B and C which cannot be stored for long in the body. In complete starvation, but not in partial food deprivation, the sensation of hunger is usually lost after the first few days.

Body protein is slowly depleted because protein lost in the normal turnover processes cannot be replaced. Just as the vasoconstriction which compensates for a fall in blood volume does not involve the cerebral circulation, so in starvation the protein content of the brain is largely preserved at the expense of other body proteins. The mechanisms of this effect is unknown. The plasma proteins and the protein in the skeleton are also fairly well preserved. The oedema seen in starvation has been attributed, at least in part, to the increasing laxity of the tissues as the fat stores and other tissues are reduced in bulk.

Hormonal levels tend to fall. A fall in the level of gonadotropins leads to cessation of menstruation (amenorrhoea) in the female. In both sexes, thyroid activity decreases, so that the basal metabolic rate falls. Loss of body weight is also associated with a fall in the basal metabolic rate because the mass of metabolizing tissues has decreased. As there is generally a feeling of lethargy, the energy expenditure above the basal level is small. This lethargy and reduction in metabolic rate makes starvation in some ways analogous to hibernation, although body temperature tends to remain normal. In this way, the body may compensate for an indequate diet and maintain life at a markedly sub-optimal level.

At the end of this stage the fat stores are exhausted and body weight has fallen markedly, e.g. by 25 per cent, the exact amount depending on the extent of the original fat stores.

*Utilization of body proteins.*—In this stage the body proteins, which have been gradually depleted in the earlier stages, are used as the main source of energy. This is indicated by a rise in the RQ to that of protein (around 0.8) and a rise in the excretion of nitrogen in the urine. This stage is associated with rapidly increasing weakness and falling blood pressure; death is likely to occur within a week or two of the onset of this stage of rapid protein breakdown.

When generalized malnutrition occurs more *gradually* than in sudden complete starvation, the effects are similar to those described above, but are more drawn out. However, in partial starvation, a persistent and aggravated sensation of hunger is one of the main causes of misery.

Chronically starved individuals tend to be lethargic and weak; sustained strenuous exertion is impossible. The classical "skin and bones" picture, due to absence of fat stores and wasting of muscles, may be modified by swelling of the legs due to oedema. Amenorrhoea may be present in the female. In the case of children, stunting of growth and, in more severe cases, weight loss will occur.

### Malabsorption

When malnutrition is due to malabsorption, the faeces are abnormal. In many cases, there is a marked excess of fat in the faeces—*steatorrhoea* (Gr. *stear* = fat, *rhoia* = flow). This reduces the specific gravity of the faeces, so that they float readily, and they also tend to be pale and greasy looking. In cases where disaccharides cannot be digested, the presence of these small molecules in the gut exerts an osmotic force which prevents adequate fluid absorption and an *osmotic diarrhoea* results (compare the *osmotic diuresis* due to the presence of glucose in the renal tubules in diabetes mellitus). The abnormally large residue of undigested food leads to increased bacterial *putrefaction* with gas formation. This can cause abdominal distension and give the faeces an extremely unpleasant smell.

## Causes

### Inadequate Availability of Food

This is a major cause of malnutrition in the less privileged countries and is determined by the relationship between food production and population size. In more affluent societies it is a smaller problem but occurs to some extent in people with low incomes, particularly elderly people and children. Children aged 10 and over require approximately the same calorie intake as the average adult. Pregnancy and high energy expenditure increase the likelihood of deficient nutrition.

### Impairment of Appetite

Impairment of appetite commonly accompanies chronic disease, e.g. cancer, tuberculosis, renal failure. The mechanisms are obscure, but toxic products of the diseased state may act on the appetite regulating centres in some way, as do certain drugs used to suppress appetite when weight loss is desired.

*Anorexia nervosa* is a condition in which loss of appetite is a prominent feature. The loss of appetite can lead to marked weight loss, and in severe cases, death from starvation. As the term nervosa suggests, the cause of the loss of appetite appears to be psychological. These patients, mainly young women, have a morbid fear of becoming fat and tend to visualize themselves as being much plumper than in fact they are. The reasons for this distorted body image and fear of obesity are obscure. It is possible that both the loss of appetite and the psychological abnormalities could spring from hypothalamic dysfunction.

Local conditions in the upper alimentary tract, e.g. painful ulcers of the mouth, oesophagus and stomach also tend to suppress the appetite because of the pain or discomfort produced by eating.

The effects of *gastrectomy* (removal of part of the stomach) may be considered here. When a considerable portion of the stomach is removed, the reservoir function of the stomach is interfered with. Normally the capacity of the stomach is such that the individual can take in an amount of food which is slowly and steadily released into the intestine in the course of around 3-4 hours. When a person takes 3 or 4 meals per day, his intestine is thus digesting and absorbing food for most of his waking hours. If gastric capacity is severely curtailed, this cannot occur. The alternative would be frequent small meals. In practice this cannot fully compensate, and a degree of malnutrition and weight loss must be accepted as a consequence of gastrectomy in most patients.

### Malabsorption of Food

Malabsorption is a condition in which malnutrition is present despite an adequate diet. The nutrients are lost in the faeces instead of being absorbed into the body.

Malabsorption results from three main causes—*impairment of digestion, impairment of absorption* and *diversion of nutrients to intestinal parasites.*

*Impairment of digestion* is due mainly to lack of one or more of three essential items—the pancreatic secretions, bile, and the small intestinal disaccharidases. The *pancreatic secretions* can produce adequate protein digestion even in the absence of gastric pepsin and hydrochloric acid. However, when pancreatic secretions are seriously reduced, severe malabsorption results. Pancreatic secretions are seriously reduced in *fibrocystic disease* (mucoviscidosis) in which viscid mucus blocks the pancreatic ducts so that the pancreas becomes cystic and degenerated. Fibrocystic disease also damages the lungs in a somewhat similar way. It is believed to be of genetic origin.

An essential contribution of bile to digestion is the emulsification of fat by the *bile salts*. This action aids the digestion and absorption of fat and in the absence of bile there is serious malabsorption of fat. Bile ceases to reach the intestine when the common bile duct is obstructed by a gall stone or by tumour and in some cases of liver disease.

Despite adequate entry of bile salts into the duodenum, their action may be interfered with when surgery has led to the formation of a *"blind loop"* of gut whereby the duodenum is bypassed by a direct connection between stomach and jejunum. Relative stagnation of flow in this region permits bacterial multiplication and the bacteria deconjugate the bile acids by splitting off the water-soluble portion so that they cease to be water-soluble and are ineffective for fat digestion. In addition, the residual bile acids are irritant to the intestine and tend to cause diarrhoea.

While pancreatic secretions can break down most nutrients to the level at which they can be absorbed, this is not the case with carbohydrates. The end product of pancreatic digestion of carbohydrates consists of disaccharides and these cannot be absorbed by the intestine in adequate amounts. When one or more of the intestinal *disaccharidases* is absent—typically in infants—the disaccharides are lost in the faeces. Water is retained in the faeces by the osmotic effect of the disaccharide molecules, leading to "osmotic diarrhoea". The commonest disaccharide to be missing is lactase. The lactase content of intestinal secretions is related to the milk intake. An increased amount of lactose requiring digestion tends to induce a rise in lactase activity, an example of enzyme induction. However, when severe lactase deficiency is present, milk may have to be omitted from the diet.

*Impairment of absorption.*— This is caused by damage to or reduction of the surface area of normal intestine available for absorption. The intestine may be damaged by certain diseases, or the available length may be reduced by surgical removal or by a bypass operation between the stomach and the jejunum. One of the commonest causes of widespread disease of the intestinal mucosa is damage by certain foodstuffs to which the individual intestine is abnormally sensitive. *Coeliac disease* is an example (Gr. *koilia* = belly). In this condition, gluten (a protein constituent of wheat and some other grains) appears to damage the intestinal villi so that the mucosa becomes flattened by the withering away of the villi which play an essential role in the absorption of all nutrients. It is not clear how gluten causes this damage. In coeliac disease there is an increased rate of mucosal cell turnover. This leads to excessive loss of cell proteins which are not reabsorbed due to the impaired absorption.

*Diversion of nutrients to intestinal parasites.*—Parasites, such as a tapeworm, may cause generalized malnutrition, or specific deficiencies. In certain circumstances bacteria may do the same thing, e.g. in the blind loop syndrome, when the stomach is joined to the jejunum, leaving the duodenum as a blind appendage. Bacteria are particularly liable to deprive the body of vitamin $B_{12}$.

### Abnormal Utilization of Nutrients

In certain diseases, such as widespread cancer, body nutrients are diverted to the abnormal tissue at the expense of normal body structures. Extreme wasting of the body (*cachexia,* Gr. *kakos* = bad; *hexis* = condition) may then be seen.

## INVESTIGATIONS

Investigations are required mainly in the malnourished individual with impaired appetite where a cancer of the intestine is suspected or where food intake appears adequate and malabsorption is suspected. The main investigations include:

**X-ray studies** using ingested barium may reveal the presence of cancer or of a peptic ulcer. They may show that a bypass operation has been performed; a blind loop may be present. In addition, when malabsorption is present, the pattern of the barium in the intestine often shows characteristic changes. The barium tends to clump (*flocculation*; L. *flocculus* = a tuft of wool) possibly due to excess secretion of mucus.

**Examination of the faeces** may show the presence of abnormally large amounts of unabsorbed food, especially fat. Normally a small amount of fat is lost in the faeces. In the presence of malabsorption, the total fat excretion and the percentage of fat in the faeces are increased, often markedly. A 72-hour collection of faeces is generally used. Naked eye inspection of the faeces may reveal tape-worm segments.

**Blood levels.**—Low blood levels of a combination of substances such as iron, folic acid and vitamin A, suggest the possibility of malabsorption.

**Tests of absorption.**—In these, the absorption of a particular substance is assessed by administering a standard amount by mouth and estimating its rate of absorption into the body by measuring blood or urine levels. One example is the *glucose tolerance test,* as used in the diagnosis of diabetes mellitus. In malabsorption, the blood glucose levels do not rise as much after glucose ingestion as in a normal person (a "flat" result). A more accurate test involves the use of the sugar *d-xylose.* This substance is absorbed mainly passively, is metabolized more slowly than glucose and its excretion does not depend on a threshold. Hence, its rate of absorption can be estimated from the rate of urinary excretion and indicates the ability of the intestine to absorb substances, e.g. water, passively.

Fat containing *radio-active carbon* may be used as an indicator of fat absorption. The fat is given by mouth and several hours later the radio-activity of expired air is assessed. When fat is absorbed and metabolized, the radio-active carbon appears in $CO_2$ which is excreted by the lungs.

*The Schilling test for vitamin* $B_{12}$ absorption is another example of this type of study. If a person is given a very large injection of vitamin $B_{12}$, followed by a dose of radio-active $B_{12}$ by mouth, then any of the radio-active $B_{12}$ absorbed from the intestine will be surplus to requirements and will be excreted in the urine. Thus the urinary radio-activity will be an index of the efficiency of $B_{12}$ absorption. Should $B_{12}$ excretion be low, a further oral dose of $B_{12}$ is given, this time accompanied by intrinsic factor. If excretion of $B_{12}$ is now much increased, then the cause of its malabsorption was probably lack of intrinsic factor due to disease of the gastric mucosa. If excretion is not increased by the administration of intrinsic factor, then there is probably disease of the terminal ileum where $B_{12}$ is normally absorbed.

**Tests of pancreatic secretion.**—If pancreatic insufficiency is suspected, it is possible to sample the *pancreatic secretions* directly from a tube passed into the duodenum. The end of the tube is placed in the lower duodenum, as judged by X-ray of its X-ray-opaque tip. The tube is double-lumened as far as the stomach, the second lumen being used to aspirate gastric contents to prevent contamination of the pancreatic aspirate. Samples are analysed for bicarbonate and enzyme content after stimulation of the pancreas by injected secretin and pancreozymin.

**Examination of sweat.**—Where fibrocystic disease of the pancreas is suspected, estimation of the *sodium chloride* content of the sweat is useful. In fibrocystic disease, the sweat glands share in the generalized abnormality of exocrine glands and produce sweat with a much higher than normal concentration of sodium chloride. The reason for this is not known.

**Jejunal biopsy.**—In some cases examination of a jejunal biopsy may be useful. A tube is passed into the jejunum, a small amount of mucosa is sucked into the end of the tube and sliced off. If microscopic examination shows loss of the normal villi (flattened mucosa), the cause of the malabsorption is damage to the absorptive surface, e.g. due to coeliac disease.

## Treatment

### Treatment of Starvation

When a patient is severely malnourished as a result of starvation, or any other cause, it is essential to institute adequate feeding gradually. Otherwise the stress placed upon the enfeebled digestive system and body generally may lead to collapse and death. A system rather like that used for infants is required. Sweetened water is given first. When this is tolerated, dilute milk is given, then whole milk and then a light semi-solid diet.

The treatment of *anorexia nervosa* is a special case of the treatment of starvation. Psychiatric treatment is aimed at removing the aversion for food. Great patience may be required to persuade the patient to eat adequately. Despite this, the food may be subsequently vomited. Tube feeding to the stomach and intravenous feeding may help in some cases. Despite treatment, around 5 per cent of patients die, so powerful is the aversion to food.

### Treatment of Malabsorption

When malabsorption is due to disease of the intestinal villi caused by sensitivity to certain food, e.g. gluten, the offending item is carefully excluded from the diet and recovery usually occurs.

The absence of pancreatic secretions can be partly compensated for by feeding extracts of animal pancreas with meals.

In malabsorption, fat digestion and absorption are often particularly affected and the retention of fat in the alimentary tract tends to impair absorption of other nutrients. It is thus desirable to restrict fat intake. Protein, vitamin and mineral intake often require to be increased.

When bacteria in a blind loop are causing malabsorption, antibiotic treatment may help.

**Intravenous Feeding**

Although theoretically simple, prolonged intravenous feeding presents serious problems and is used only to treat certain conditions from which eventual recovery is expected; adequate facilities for supervision are essential. The administered fluid must contain adequate amounts of minerals and vitamins and in addition must meet the patient's requirements for calories and for amino-acids. Fat emulsions can provide adequate calories (concentrated glucose is too irritant to veins to be given for long). Plasma or a solution of amino-acids may be used to meet protein requirements. For long term treatment it is preferable to use a widebore catheter with its tip placed in a major vein because smaller veins tend to become irritated and eventually occluded as a result of such infusions.

# 32. EXCESSIVE NUTRITION

## OBESITY

### Definition

Obesity is the excessive accumulation of fat in the body. Thus men with a fat content much above 15 per cent of body weight (women 25 per cent) can be regarded as obese. As accurate measurements of fat content cannot be made at present with sufficient convenience for general use, an individual is normally considered obese if his or her weight is 10 per cent or more above the ideal weight for a person of corresponding age, sex, height, race and body build. Methods of estimating ideal weight are discussed under "investigations".

### Effects

The effects of obesity can be harmful or beneficial depending on the environmental situation. The *beneficial effects* of increased thermal insulation and energy storage are useful in a very cold environment or in famine following a period of plenty. Following a disaster at sea in cold waters, one of the main causes of death is the fall in body temperature (hypothermia). In such circumstances, an obese person may survive much longer than a thin person.

The *adverse effects* of obesity are seen in affluent societies where food supply exceeds requirements. Obesity increases the risk of developing diseases which between them reduce life expectancy. As a very rough approximation, it can be taken that the life expectancy falls by about a year for every kilogram by which the individual exceeds his ideal weight. The ill effects of obesity include:

**Increased incidence of degenerative arterial disease.**—This may result in obstruction of coronary or cerebral blood vessels to cause heart disease and strokes, two of the major killers of obese patients. The mechanism by which obesity predisposes to arterial degeneration is not known.

**Increased incidence of diabetes mellitus.**—Diabetes mellitus is more common in obese than in normal people and weight reduction alone often cures mild diabetes in elderly patients. Again, the reason for this is obscure. Diabetes further increases the tendency to develop degenerative arterial disease.

**Increased incidence of gall-bladder disease.**—For an unknown reason gall-stones develop more readily in obese patients. Gall-stones can give rise to irritation, obstruction and infection in the gall-bladder and its ducts.

**Increased laxity of body tissues.**—Accumulation of fat in the tissues means that the collagenous tissues offer less support to the structures that run through them. Thus the veins in the lower limbs are liable to become excessively dilated (varicose). Structures in the peritoneal cavity can more easily push through weaknesses and orifices in the abdominal wall. These protrusions (ruptures or hernias) occur most commonly in the region of the umbilicus (umbilical hernia), where the oesophagus enters the abdomen (diaphragmatic hernia), and where the inguinal cord enters the inguinal canal (inguinal hernia).

**Increased incidence of degenerative changes in joints.**—Stress on weight-bearing parts increases the chances of developing wear-and-tear changes (osteoarthritis) in the joints of the spine and legs.

**Increased body work load.**—As the obese patient may carry excess weight equivalent to one or two heavy suitcases, his movements tend to be slow and clumsy. There is consequently an increased risk of accidental injury or death in the home, in traffic, and at work. Irrespective of any lung or heart disease, the obese patient's exercise tolerance is limited by the fact that he must do more work than the average person during physical activity, especially when climbing stairs or hills. This throws an additional strain on his cardiovascular and respiratory systems. Excessive shortness of breath and tiredness results and the obese patient then tends to reduce his physical activity below normal. Excess fat in and around the muscles may account for the inadequate lung ventilation sometimes seen in very obese patients. In extreme cases, this can lead to respiratory failure (Pickwickian syndrome, named after the fat boy in *Pickwick Papers*) and pneumonia. Fat people are more liable to develop complications and die during and after surgical operation than normal people. Finally, in a society where obesity is regarded as undesirable and unattractive, a sensitive obese individual may suffer considerable distress and may tend to avoid social contacts.

## Causes

At the simple level, the cause of obesity is clear. It is due to an *excess of energy intake over energy expenditure,* the excess nutrients being converted into fat.

At the more fundamental level, the cause is complex and obscure. It is necessary to ask why some individuals have an excessive food intake relative to their metabolic needs, while others maintain perfect long-term balance. To answer this question completely would require much more knowledge than is at present available. However, the available knowledge indicates a number of possibilities. Figure 62 indicates some factors which can influence fat deposition. The adipose tissue cell is indicated by the diagram resembling a television set. Despite its large fat content, it is an actively metabolizing cell with mitochondria and the usual subcellular structures in its cytoplasm. The tissue represented by this cell takes up and releases fatty material and fat precursors in accordance with their level in the blood stream. It is also influenced by *nerves and hormones.* The details of these influences are incompletely known, but it appears that adrenergic nerves can stimulate release of fatty acids (a beta receptor effect) and that insulin favours fat deposition.

Another factor which may influence the extent of individual fat stores is the *number of fat cells* in the body. On this basis, individuals with more than the average number of fat cells would tend to store more fat than the average person. The number of fat cells is probably controlled partly by hereditary factors but it may also be related to dietary intake early in life, so that excessive intake would lead to a greater eventual total number of fat cells.

Thus excess fat deposition may be related primarily to an abnormality of adipose tissue and its control, or to an imbalance between food intake and

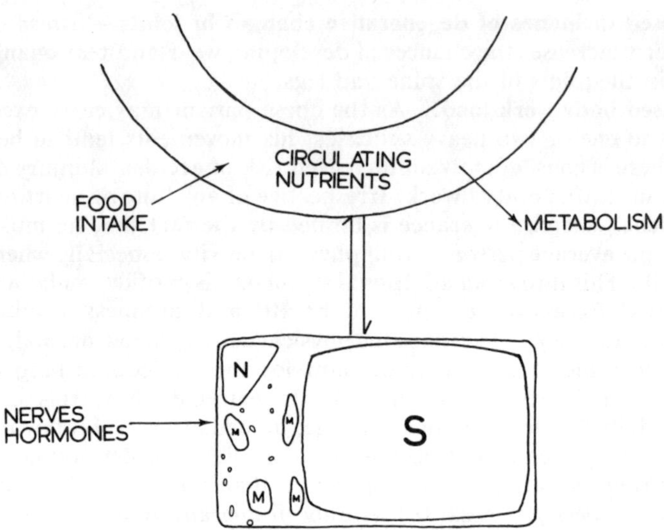


FIG. 62.—Regulation of adipose tissue. The fat cell is represented with a large fat store (S), eccentric nucleus (N), mitochondria (M) and various other vesicles and granules. Fat stores are related to food intake and metabolism and are also subject to nervous and hormonal influences.

metabolism. Excess fat deposition could lead to a secondary increase in food intake, or vice versa. Because primary abnormalities of fat deposition have not yet been clearly identified, attention has been focused on the imbalance between food intake and metabolism. Food intake can be adjusted and so provides a means of treating obesity.

Figure 63 shows factors which influence food intake. Basically there are four: availability of food, the sensation of hunger, the sensation of satiety and

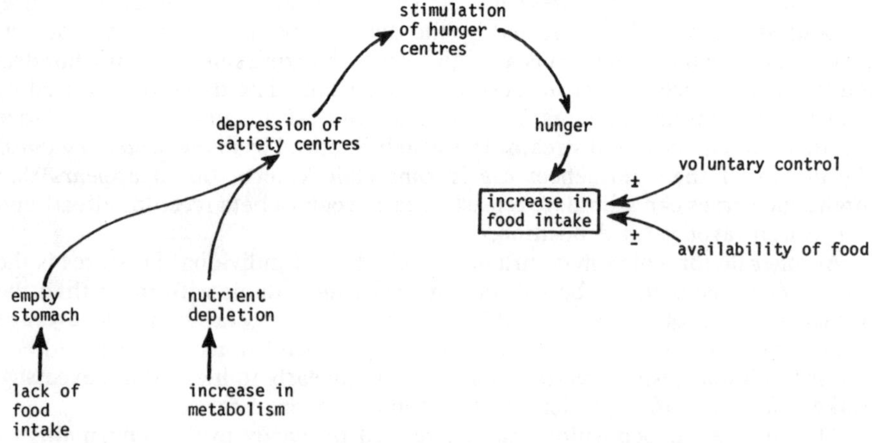


FIG. 63.—Factors which influence food intake.

the influence of voluntary control. *Hunger and satiety centres* probably exist in the hypothalmus and possibly elsewhere; stimulation of these areas in animals produces compulsive eating and voluntary starvation respectively. In societies where obesity is a problem, food tends to be taken "by the clock", so that eating is initiated by habit rather than by hunger. This in itself may lead to obesity in that body nutrients (circulating and temporarily stored) are kept at a higher level than would be the case if hunger were the only factor initiating eating. In these circumstances, the sense of satiety is critical in limiting the amount eaten and there is some evidence that obese people have a weaker than average sense of satiety.

Thus the precise cause of obesity in a given case is generally obscure. However, a number of factors are known to be associated with the tendency to become obese.

(i) *Availability of food.*—This is of crucial importance, because in undernourished communities, obesity is virtually unknown. Very few, if any, concentration camp victims emerged with excess fat deposits. Thus free availability of food may unmask a tendency to obesity which would be of no significance where food is in short supply.

(ii) *Type of food eaten.*—There are suggestions that consumption of refined carbohydrates, especially sugar, is especially likely to cause obesity. Such foods are relatively cheap and this could explain why the less well-off inhabitants of affluent societies have a higher incidence of obesity than their richer fellow citizens.

(iii) *Physical activity.*—It is possible that when a person who has become accustomed to eating a certain amount of food daily reduces his level of physical activity he may thereby start to gain weight. The obesity which rapidly develops after a surgical operation, pregnancy, or other period of enforced rest, may be related to this. In addition, the resting state itself may induce fat deposition at an increased rate.

(iv) *Cessation of smoking.*—Many people who give up smoking begin to gain weight rapidly. Thus the avoidance of one major health hazard can lead to another major hazard.

(v) *Mental state.*—In certain cases of relatively mild loneliness and depression, food may be resorted to as a form of distraction or consolation.

(vi) *Family tendencies.*—Obesity tends to run in families. It is likely that both genetic and enviromental factors, such as eating habits, operate. Fat parents tend to have fat children who grow into fat adults.

Finally, in relation to possible genetic tendencies, it seems possible that variations in the tendency to obesity are part of the genetic variation which has enabled the human race to survive in a great variety of conditions. Perhaps the tendency to obesity should be regarded as part of the normal variation between individuals, requiring compensation by higher centres, rather than as a disease due to a specific cause.

## INVESTIGATIONS

The main investigation is to find out whether there is excess body fat and if so to determine its extent. Total body fat cannot be measured directly by any practicable method but it can be estimated by a variety of methods of varying complexity, e.g. weighing the patient, using skin fold calipers and measuring the patient's specific gravity.

**(a) Weighing the patient.**—The total mass of a body is made up of the *lean body mass* (muscles, bones, viscera, etc) whose mass remains relatively constant in health and *adipose tissue* which can vary greatly in quantity. For this reason, changes in body weight are usually due to changes in the amount of adipose tissue in the body.

The standard procedure is to weigh the patient, preferably without clothes, or at least without shoes and heavy outer garments, on an accurate balance, and compare his weight with the predicted "ideal" weight. Data is available relating average weight to age in the case of children. The "ideal" weight in adults is obtained rather arbitrarily by referring to tables based on surveys by life insurance investigators. Some tables allow for a gradual weight gain between the ages of 25 and 55; others assume that weight should not increase after the age of 25; it is debatable which is the better approach.

**(b) Skin fold calipers.**—Measurement of skin fold thickness, in which the skin at various sites is pinched up and the thickness of the double skin layer recorded, is another means of estimating total body fat.

**(c) Body specific gravity.**—A more precise, but technically difficult measure of total body fat can be made from the individual's specific gravity, obtained by weighing him in water and in air. The greater the proportion of fat, the lower the specific gravity.

It should be remembered that *excessive weight can be due to factors other than obesity*. Large bones and well-developed muscles (e.g. due to physical training) increase weight relative to height. Some weight tables distinguish between "large frame" and "small frame", adding to or subtracting an arbitrary 10 per cent from the average weight. The weight gain due to pregnancy can be ascribed to obesity if the condition is not borne in mind—an embarrassing mistake! Oedema due to heart failure, liver or kidney disease etc., can produce marked weight gain; in early cases of fluid retention there may be no obvious oedema on clinical examination.

*Knowledge of the patient's previous weight can be a useful guide in assessing obesity*. Many people are at approximately their ideal weight around age 20-25. Excess weight above this level, especially if gained fairly rapidly after a period of enforced rest is likely to be due to obesity if the conditions referred to above can be excluded. A record of body weight over the years is a useful aid to diagnosis in cases both of weight gain and weight loss.

## TREATMENT

It has been shown that weight reduction by overweight people is associated with an increase in their expected life span. However, unless or until a specific remediable cause of obesity is discovered, treatment of obesity must consist of a voluntary effort by the patient to take over the control of eating from the hunger and satiety centres. The sensory feed-back must be via the weighing machine or other indicator of body fat and the effort must be maintained in most cases indefinitely (Fig. 64). Although the treatment of obesity appears a simple matter of reducing energy input below the level of energy output, in practice it consists in modifying established habits and fundamental drives, and the chances of complete success are small, unless the individual is strongly motivated to lose weight. In many cases, partial success is the best that can be achieved. Attempts to treat obesity are, in general, time-consuming and often discouraging. One problem is that, when some patients reduce their food intake, there is a matching reduction in their energy expenditure. They become listless and apathetic and may not incur the calorie deficit necessary to lose weight. A number of practical points may help the patient who wishes to lose weight:

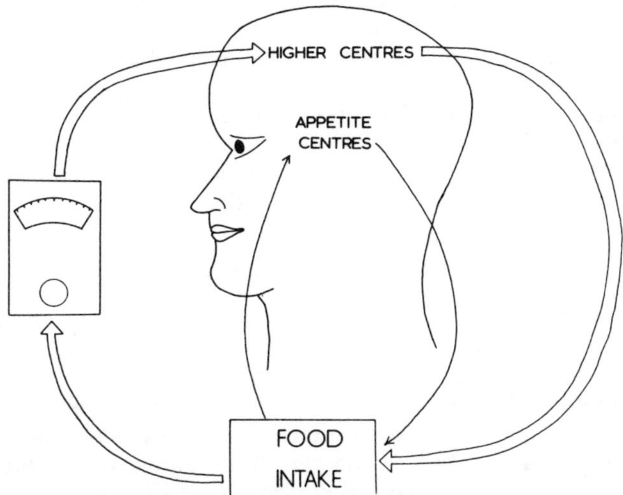


FIG. 64.—In obese people, the sensory feed-back regulating food intake must operate via the weighing machine and the higher centres.

**Adequate explanation.**—Success is most likely when the patient has some understanding of the causes of obesity, the benefits of weight loss and the calorific value of the commoner items of diet. For this reason, success is more often achieved with the more intelligent patient.

**Appropriate diet.**—The diet should be one designed by an adequately trained dietician so that it combines a limited energy content with an adequate protein, mineral and vitamin content. Ideally it should be varied and attractive but not more expensive than the patient can afford. Total calorie content is more important than detailed composition. The various highly original diets sometimes offered probably gain any success they may have by supporting the patient's morale and determination. Rather surprisingly, some patients feel that they should lose weight if they consume a "slimming" diet in addition to their ordinary food intake.

*High bulk diets* containing non-energy producing materials such as cellulose and gelatine can suppress appetite temporarily by filling the stomach and activating the satiety centre. However, the effect is only temporary, because a low calorie intake eventually leads to stimulation of the hunger centres.

**Regular weighing.**—Weight measurements at 1-2 week intervals provides the feed-back information necessary to control food intake properly.

**Increased exercise.**—Increased exercise is generally desirable but is not a very successful way of treating obesity because increased calorie utilization tends to stimulate the hunger centres in proportion to the increase in metabolic rate.

**Drugs.**—The continued problem of obesity, despite the use of a variety of drugs with widely differing modes of action, suggests that these drugs do not modify the mechanisms causing obesity in a satisfactory way. Some of these drugs suppress the appetite; others are thought to modify body metabolism so that fat deposits are depleted. However, their effects tend to be temporary only. Some may produce unpleasant adverse effects and lead to addiction. They may have a limited place in certain cases as a psychological aid, but should be used only with adequate supervision.

**Supervision.**—The treatment of obesity should ideally have careful medical supervision at least in the early stages. In particular, it should be borne in mind that patients who eat excessively for psychological reasons may become more seriously ill if their source of solace is removed.

## VITAMIN TOXICITY

In general, vitamin deficiency is a more serious problem than vitamin toxicity. However, in the case of vitamins A and D, an excess can cause serious illness. Of the two, vitamin D excess is the commoner.

Chronic vitamin A excess leads to fatigue, joint pains and skin changes; acute vitamin A poisoning seems to interfere with brain function, producing headache and drowsiness. The mechanisms of these toxic effects are not known and they are reversed in a few days when the excess vitamin intake is stopped.

Vitamin D toxicity is related to the effect of vitamin D on calcium metabolism. Calcium absorption from the gut is increased and calcium tends to be removed from mature bone and deposited in growing bone and in soft tissues, especially the kidney. The raised blood calcium level is associated with increased calcium excretion and a tendency to renal stone formation. The raised blood calcium level is probably the cause of the general ill-health with impaired appetite. The renal calcification leads eventually to renal failure.

Thus in the case of vitamins A and D, particularly the latter, it is as important to avoid excess as it is to avoid deficiency. This poses a problem in the use of dietary supplements of vitamin D, because vitamin D requirements vary markedly from individual to individual. It seems that the range of individual variation in vitamin D requirements is so large that no fixed dose of a supplement will completely prevent rickets without producing hypercalcaemia in some children. Unless or until individual requirements can be precisely determined, a compromise intake must be advised which will prevent rickets in the vast majority of cases while holding only a very small risk of producing toxic effects.

# 33. VOMITING

## Definition

Vomiting is a reflex action, co-ordinated in the medulla oblongata, whereby the contents of the stomach are forcibly expelled through the mouth.

The *mechanism of vomiting* is complex, but consists of two main components—activity by alimentary tract muscle, mainly smooth muscle, and activity by skeletal muscle of the chest and abdomen, mainly the muscles connected with inspiration and expiration.

The muscle of the alimentary tract sets the stage for the emptying of the stomach by closing the pylorus and by relaxing at the cardiac end of the stomach and in the oesophagus. Reverse peristalsis probably plays only a minor role, if any, in emptying the stomach. It may account for the regurgitation of intestinal contents into the stomach prior to their being vomited. Sometimes when the colon is obstructed by disease, faecal material passes backwards up the intestine to the stomach and is vomited (faecal vomiting).

The motive force for emptying the stomach forcefully comes from the thoracic muscles, the diaphragm and the abdominal muscles. A deep breath is taken, the glottis closes, and the muscles of the anterior abdominal wall contract strongly. The abdominal cavity is thus compressed between the diaphragm and the anterior abdominal wall. Intra-abdominal pressure rises, compresses the stomach, and squeezes out its contents into the oesophagus and mouth. The diaphragm then relaxes, so that increased pressure is transmitted to the thoracic cavity and oesophageal contents are squeezed into the mouth.

During vomiting, the palate is elevated to close off the nasal cavity. In patients suffering from paralysis of the palate, some of the vomited material may pass out through the nose.

The vomiting cycle is often repeated several times until the stomach has been emptied. *Retching* is a fragment of the vomiting reflex and often precedes the fully developed action.

Another phenomenon related to the vomiting reflex is *nausea,* the sensation of impending vomiting which often precedes the actual vomiting (Gr. *naus* = ship; *nausia* = seasickness). Nausea may be accompanied by salivation (water-brash), a part of the vomiting reflex whereby gastric acid is buffered as it passes through the mouth. Nausea may also be associated with other autonomic disturbances which accompany vomiting, such as sweating and pallor. The sensation of nausea may be caused by distension of the stomach; the sensation of satiety following a heavy meal tends to merge into one of nausea. In some normal people nausea is experienced at a considerable interval after a meal and is relieved by a meal. This may be because the food encourages a more normal peristaltic pattern.

## Effects

Vomiting is an essentially benign reflex built into the neural control system of the alimentary tract. It enables noxious material to be expelled from the stomach rather than be retained and absorbed. Most people have benefited in this way from vomiting, often nowadays due to food poisoning.

**Associated Autonomic Disturbances**

Both nausea and vomiting can be associated with autonomic disturbance. Peripheral sympathetic overactivity leads to a cold, pale, moist skin. Lower intestinal peristalsis may be marked, leading to diarrhoea. The heart rate may rise or fall, as may the blood pressure. Such disturbances of pulse and blood pressure, together with pallor and sweating, may falsely suggest that the patient has suffered a heart attack.

**Mechanical Injury**

Although vomiting involves powerful muscular contraction which is unpleasant and at times painful, mechanical injury is uncommon as a result. Occasionally a mucosal tear or complete rupture of the wall of the stomach or oesophagus results from vomiting.

One circumstance in which vomiting is particularly liable to damage the stomach or oesophagus is when a patient has swallowed a corrosive fluid. No attempt should be made to induce such a patient to vomit, nor should a tube be passed into the stomach to empty it. Also, following surgery of the upper alimentary tract, vomiting may cause rupture of the regions weakened by incisions.

**Upset of Body Fluids**

Repeated vomiting causes loss of intestinal secretions as well as preventing fluid intake by mouth. The *loss of water and electrolytes* can lead rapidly to a serious fall in extracellular fluid volume including blood volume. Renal circulation tends to be reduced early in dehydration as a compensatory mechanism to maintain blood pressure. Glomerular filtration is impaired and excretion of substances such as urea reduced so that they accumulate in the blood (*uraemia*). Blood pressure eventually tends to fall and death from circulatory failure can follow rapidly, especially in infants and elderly patients. The vomited gastric contents contain a considerable quantity of *potassium* from gastric secretions and from the cells which are constantly being desquamated from the intestinal mucosa; potassium deficiency tends to occur. As the vomited material is generally *acid,* an alkalosis occurs. This makes the potassium depletion worse by promoting renal excretion of potassium. In summary, there is a loss of water and of electrolytes in general; in addition depletion of potassium and of hydrogen ions (hypokalaemic alkalosis; L. *kalium* = potassium) tends to be particularly severe.

**Disordered Vomiting Reflex**

Reference has already been made to the fact that palatal paralysis can lead to vomited material passing down the nose. Much more serious is failure of the glottis to close during vomiting or less violent regurgitation of gastric contents. This is likely to occur in unconscious patients. Particularly if the patient's head is higher than his abdomen, regurgitated material then passes down the trachea into the lungs. The acid and pepsin of the gastric secretions attack the lung tissue and can cause very severe inflammation of the lungs.

## CAUSES

The vomiting reflex, nausea and retching, can be initiated by afferent impulses from many structures in the body and by direct stimulation of the vomiting centres (Fig. 65).

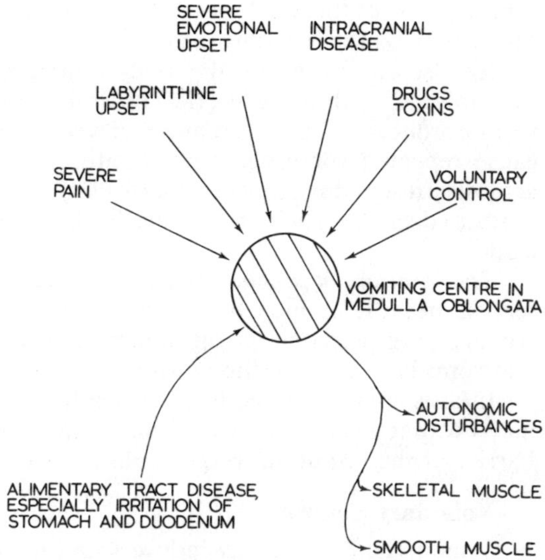

FIG. 65.—The vomiting reflex and some factors which can initiate it. The extra-alimentary factors generally act indirectly on the vomiting centre.

### Alimentary Tract Disease

*Irritation* of the mucosa of the stomach and duodenum by local infection, irritant materials in the food or by blood, is a common cause of vomiting. Often when the irritant material has been got rid of, the patient starts to improve. *Intestinal obstruction* at any level, from the pylorus down, tends to cause vomiting. The lower down the obstruction, the longer is vomiting likely to be postponed. Other intestinal diseases, such as perforation of the intestine and appendicitis can produce vomiting, probably as a result of the severe pain produced.

### Severe Pain, Labyrinthine Upset and Emotional Upset

These are examples of unusual and severe afferent stimulation which can stimulate the vomiting centre to produce nausea, retching and vomiting. The pain may arise in a variety of organs, e.g. the heart when its blood supply is suddenly impaired, the ureter when a renal stone is passing.

As well as causing single episodes where the vomiting is obviously related to emotional upset, *emotional factors can cause repeated vomiting episodes.* In such cases the precipitating cause may be difficult to uncover.

### Intracranial Disease

Raised intracranial pressure due, e.g. to a tumour or an abscess or a cerebral haemorrhage or thrombosis, typically leads to vomiting, probably by interfering with the cerebral circulation generally. Vomiting sometimes accompanies migraine headaches and again a disturbance of cerebral circulation may be responsible.

### Drugs and Toxins

Many drugs and poisons can initiate vomiting. It has been suggested that some at least of these act on chemoreceptors in the floor of the fourth ventricle rather than on the vomiting centre itself.

Any severe upset of the body's metabolism can precipitate vomiting, probably also by the adverse effect on brain cells of the abnormal internal environment produced. The disturbances of severe *renal failure* ("uraemia") typically cause repeated vomiting. This vomiting may to a certain extent relieve the acidosis present due to loss of hydrochloric acid from the stomach. The vomiting tends to decrease when alkalis (e.g. sodium bicarbonate) are given to treat the acidosis.

The *vomiting of pregnancy* is of uncertain cause, but may be included in this section because it may be caused by hormonal and/or metabolic changes. About 50 per cent of pregnant women are troubled in varying degrees by nausea and vomiting, worst in the morning ("morning sickness").

Infants vomit very readily, especially when suffering from an infection. It may be that toxic products of infection crossing the immature blood/brain barrier of the infant are responsible.

### Voluntary Control

Some patients regularly induce vomiting in themselves. This may be done because vomiting relieves abdominal pain, or discomfort, especially first thing in the morning. In some cases the vomiting reflex is induced by stimulating the back of the throat with the fingers. Early morning vomiting may become a habit in such patients, so that the vomiting is more or less a conditioned reflex.

## Investigations

Investigations in vomiting are required for two main reasons—to determine the effects of vomiting on body fluids and to determine the underlying cause of the vomiting.

The effects of vomiting on the body fluids are indicated by blood electrolyte and acid-base measurements. These investigations must be repeated during treatment to ensure that all values are returning towards normal.

The cause of vomiting is often clear. Barium studies may be necessary to confirm or exclude disease of the alimentary tract. The possibility of renal failure must be borne in mind and may require investigation. Other less obvious causes of vomiting which may require investigation include heart, cerebral and labyrinthine disease.

## Treatment

The treatment of vomiting is concerned with the underlying cause where possible. In severe cases, correction of water, electrolyte and acid-base disturbances may be urgently required. Otherwise the tendency to vomit can be reduced, more in some conditions than in others. Treatment aimed at preventing vomiting may be by mechanical means or by means of drugs.

Emptying of the stomach by sucking out its contents through a tube usually cures or prevents vomiting due to gastric or duodenal distension or irritation by removing the stimulus to the sensory side of the vomiting reflex arc. This procedure is widely used in patients who are admitted to hospital with persistent vomiting due, e.g. to pyloric or intestinal obstruction. It is also used routinely after operations on the upper alimentary tract. Firstly it prevents the vomiting which would be likely to follow interference with the gut. Secondly, if the gut wall has been opened during the operation, post-operative vomiting would endanger the site of repair.

Several kinds of drug are used to prevent vomiting. One type of anti-emetic (Gr. *emein* = to vomit) may help vomiting from one group of causes, another type may not be of much value in that group but may help vomiting from a second group of causes. Probably the various groups of drug act at various sites in the brain. Some probably act directly on the vomiting centres, others such as sedatives and tranquillizers presumably act by reducing emotional responses. In general, anti-emetics are more effective in preventing vomiting (e.g. due to travel sickness) than in terminating it once is has started. When repeated vomiting is occurring, an orally administered drug is unlikely to be absorbed effectively due to either its being vomited or failing to pass a closed pyloric sphincter; the drug must be given by injection.

## Hiccups and Flatulence

Despite their everyday occurrence, these two phenomena are rather poorly understood.

### Hiccup

A hiccup is a sudden brief inspiration (due to a spasmodic contraction of the diaphragm) which is abruptly terminated by closure of the glottis. The sudden closure of the glottis is responsible for the characteristic sound.

Although in the vast majority of cases hiccuping is of brief duration, it occasionally persists for days, causing serious distress.

Hiccuping is a reflex action. Afferent impulses from the diaphragm and from the regions above and below it can initiate the reflex. Gastric irritation is an example of this type of cause. The condition, like vomiting, can also be precipitated by disease of the central nervous system such as encephalitis. The metabolic disturbances of renal failure are another recognized cause of persistent hiccuping, presumably by an action on the central nervous system. Most cases of hiccuping are of brief duration and their cause is completely obscure.

It is traditional to treat hiccuping, a disordered respiratory action, by modifying respiration by breath-holding, swallowing or breathing $CO_2$ (rebreathing from a bag is a simple method). Whether or not such procedures terminate brief hiccuping attacks, they do not usually end persistent hiccuping. In such cases certain anti-emetic drugs may help. In the very rare cases where persistent hiccuping causes severe distress, e.g. to an already very ill patient, an injection to block conduction in one of the phrenic nerves may be considered. This procedure usually reduces the violence of the hiccup.

### Flatulence

Flatulence is a condition in which a person suffers from distension of the stomach or intestines with gas. Excess gas (flatus, L. = gas in stomach, intestines) may be brought up from the stomach or passed rectally.

Gas is normally present in the stomach and large intestine, the total amount being around 50-200 ml. The small intestine is normally free of undissolved gas; when gas enters it from the stomach (e.g. when the stomach is empty but contracting as due to hunger contractions) it leads to the characteristic palpable and sometimes audible rumbling described by the appropriate Latin term, *borborygmus.* Free gas in the small intestine is rapidly absorbed so that the small intestine does not normally contain obvious free gas, e.g. when viewed by X-ray.

Gas enters the intestinal tract by being swallowed with food or drink, and by chemical processes in the intestine. It leaves it largely by absorption into intestinal capillaries (as elsewhere total capillary gas tensions are less than atmospheric pressure) and also as flatus. Passage of flatus from the stomach occurs when the individual is more or less erect; the gastric air bubble then finds its way readily towards the oesophagus and passes upwards. The location and movements of gas in the alimentary tract are shown in Fig. 66.

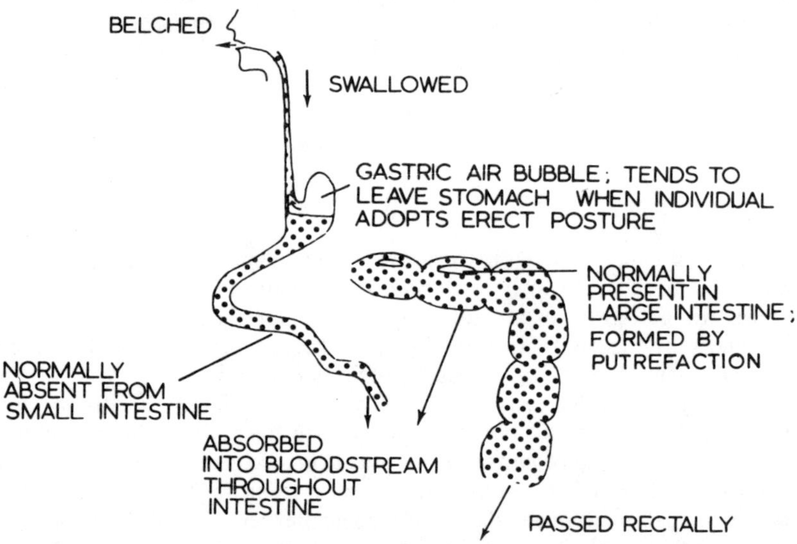


**FIG. 66.—Showing the sources, distribution and elimination of gas in the alimentary tract. X-ray studies normally show obvious gas only in the stomach and large intestine.**

Excessive gas in the alimentary tract can be due to excessive air swallowing, *aerophagy* (Gr. *aer* = air; *phagein* = to eat) or excessive production. Excessive production is due to putrefaction. This can occur in the stomach when pyloric obstruction is present and in the large intestine when malabsorption leaves large amounts of carbohydrate and protein for bacteria to act on. Very occasionally such putrefaction leads to production of an inflammable mixture which may ignite when belched up or cause a minor explosion during an intestinal operation in which heat coagulation of small blood vessels is used to arrest bleeding. Certain foods, e.g. beans, increase gas production; they contain small polysaccharide molecules which cannot be absorbed but which can be attacked by colonic bacteria.

The bringing up of wind by infants after a meal is a ritual which appears to afford much relief to the child as well as the mother. It may be that the swallowing mechanism in infants is relatively inefficient in that considerable quantities of air are swallowed. Similarly patients who complain of excessive flatulence may habitually swallow excess air; alternatively they may generate more intestinal gas or may be more sensitive to the gas present.

The treatment of flatulence due to intestinal disease is that of the disease responsible. Many patients complain of flatulence for no obvious cause. Treatment then consists largely of reassuring the patient of the normality of the presence of gas in the alimentary tract.

# 34. DIARRHOEA

## Definition

Diarrhoea is the passage of abnormally fluid faeces; often the frequency of the faecal discharges is increased.

## Effects

As with vomiting, the effects of diarrhoea can be both beneficial and harmful. Diarrhoea is beneficial when rapid emptying of the intestine speeds the elimination of harmful material. The harmful effects are due to excessive loss of intestinal contents. This results in fluid and electrolyte disturbances and loss of nutrient material.

Again as with vomiting, considerable amounts of *fluid and electrolytes* may be lost (in cholera where the walls of the lower intestine become infected with invading micro-organisms, there is profuse watery diarrhoea which may measure as much as 10-20 litres in 24 hours). The fluid lost in diarrhoea is rich in *potassium* from intestinal secretion and from desquamated mucosal cells. Absorption of nutrients is impaired. In contrast to vomited fluid, diarrhoea fluid has an alkaline pH due to its *bicarbonate* content and diarrhoea thus tends to produce acidosis. Despite the tendency to acidosis, loss of *calcium* in severe diarrhoea can cause tetany.

If fluid loss is severe, there is a rapid fall in extracellular fluid volume. Blood volume and blood pressure fall as a result. Especially in infants and in elderly people, death from *peripheral circulatory failure* may occur in a few days.

## Causes

The abnormal liquidity of the faeces in diarrhoea may be due to excessive fluid entering the lumen of the intestine, poor absorption of the 7-10 l of fluid that are normally presented to the intestinal mucosa every day or both. Of the 1-2 litres of fluid that are ingested daily and the 6-8 litres that are secreted in the digestive juices, 95 per cent is normally reabsorbed by the small intestine and 5 per cent by the colon. The rapid transit of intestinal contents may be the cause or the effect of the condition. A condition *causing* rapid transit by stimulating peristaltic movements may not allow sufficient time for adequate absorption of fluid to occur so that the faeces are fluid. However, a condition which interferes primarily with the absorption of intestinal contents *results* in rapid transit because the large volume of the intestinal contents stimulates peristalsis.

Diarrhoea may be caused by the following conditions:

### Local Irritation of the Intestine

This is the commonest cause of diarrhoea. The irritation is thought to act by stimulating the smooth muscle in the gut wall to increase peristalsis. This tends to limit absorption of fluids as may the impairment of the absorbing mechanisms in the irritated mucosal cells. It may also allow fluid to escape from the dilated capillaries in the mucosa into the intestinal lumen.

The commonest cause of the irritation is mild *food poisoning,* often due to bacterial or viral infection of the intestines. Colonic inflammation (colitis) from any cause leads to diarrhoea, but the volume of the faeces tends to be small compared with the diarrhoea where the small intestine is affected. Unexplained persistent diarrhoea, especially in the elderly, raises the possibility of a *cancer* of the colon or rectum, leading to increased bowel motility, or, in the case of the rectum, directly stimulating the rectal stretch reflex.

Many drugs used in the treatment of constipation (*purgatives*) work by irritating the colon, and overdosage can cause diarrhoea.

When a moderately large haemorrhage from stomach or duodenum has occurred, the *blood* irritates the intestine by an unknown mechanism and diarrhoea results. If the bleeding is very profuse, it may pass through the intestine in a few minutes and be passed as bright red blood. More commonly, the passage of the blood is slower so that it is altered and colours the faeces black. This loose black stool, known as *melaena* (Gr. *melas* = black) typically follows a moderate gastro-intestinal haemorrhage.

### Emotional Stress

A not uncommon cause of persistent mild diarrhoea is *psychological stress.* Anxiety, e.g. before an interview or oral examination, often leads to the passage of a loose bowel motion. Prolonged stress can have a similar effect presumably by its effect on the extrinsic autonomic nerve supply of the intestine. Measurements of colonic movement have shown that increased activity is associated with constipation and decreased activity with diarrhoea. The explanation could be that decreased mixing movements lead to impaired absorption of water and hence to diarrhoea.

### Malabsorption

Malabsorption, particularly of carbohydrate, leaves osmotically active particles in the gut. These retain water and increase the bulk and liquidity of the faeces. Osmotically active substances which cannot be absorbed from the alimentary tract, e.g. magnesium sulphate, may be used as purgatives.

### Miscellaneous Causes

Rapid gastric emptying may be due to a surgical drainage procedure to the stomach such as a connection between the greater curvature and the intestine, bypassing the pylorus and duodenum. Rarely a connection develops between stomach and colon causing diarrhoea (undigested food is passed shortly after eating). Cutting the vagus nerves (*vagotomy*) to treat duodenal ulcers may lead to diarrhoea. The mechanism is obscure.

When an artificial opening is made surgically into the ileum (ileostomy) or ascending colon (colostomy) effects for the patient are somewhat similar to the effects of diarrhoea, because the water and electrolyte-absorbing colon is bypassed. Faeces have more or less reached their final state in the descending colon so a colostomy here produces fairly normal faeces.

Hyperthyroidism causes mild diarrhoea by increasing intestinal motility. Excessive gastrin can also increase intestinal activity excessively to cause diarrhoea.

### Investigations

Many cases of mild transient diarrhoea require no special investigations. In severe diarrhoea, investigations may be required to monitor fluid and electrolyte balance.

Chronic diarrhoea is an important sign of cancer of the large intestine. When diarrhoea is persistent, the possibility of a cancer of colon or rectum requires investigation, e.g. by digital examination of the rectum, visual inspection by proctoscope (Gr. *proktos* = rectum) and sigmoidoscope (for visualizing the sigmoid colon) and X-ray study after a barium enema (barium solution is injected into the rectum and passes into the colon.)

Examination of the faeces for bacteria and parasites may be required when an infective agent is suspected and the diarrhoea persistent.

### Treatment

The vast majority of cases of mild transient diarrhoea require no treatment other than rest in bed, with a fairly high fluid intake.

When fluid loss is marked, extra fluid may have to be given intravenously. In severe cases such treatment is urgently required to save life. Administered fluid should contain saline and glucose, together with appropriate amounts of potassium. The rate of administration is related to the rate of fluid loss. Certain drugs acting on the smooth muscle of the intestine can reduce peristalsis and help moderately severe diarrhoea.

In general, treatment of diarrhoea consists of maintaining fluid and electrolyte balance in more severe cases and dealing with the underlying cause if one is present.

## ***Abnormal Perineal Sensations.***

Two abnormal sensations initiated in the perineal region are tenesmus and proctalgia fugax.

**Tenesmus** (Gr. *teinesmos* = straining) is the sensation that the rectum requires emptying even though it is empty of faeces. It is due to the presence in the rectum of an irritating condition (e.g. tumour) which stimulates the stretch reflex and produces the desire to defaecate.

**Proctalgia fugax** is a brief (e.g. 5-10 minutes) sensation of pain in the region of the rectum (Gr. *proktos* = rectum, *algos* = pain; L. *fugax* = apt to flee). The condition is probably a cramp of the pelvic floor striated muscles and is often relieved somewhat by applying pressure in the anal region. As with muscle cramps elsewhere, it is not associated with serious disease, its causation is obscure and, while it may well spontaneously cease to occur, there is no effective preventive treatment.

# 35. CONSTIPATION

## Definition

Constipation is infrequent or difficult defaecation. The condition is interpreted by patients in relation to their normal pattern of defaecation, whose frequency may range from several times per day to several times per week.

Constipation may be due to slow passage of faeces through the colon or to abnormality of the mechanism of defaecation.

## Effects

Constipation (L. *con* = together; *stipio* = I cram) of itself seems to have little effect on the patient's general health. Effects attributed to constipation include foul breath, poor appetite, headache and irritability; they may be due to distension of the colon by the retained faeces or material absorbed from the faeces. Evacuation of the bowel after a period of voluntary suppression of defaecation cures the symptoms within an hour or two.

Prolonged constipation may play a role in the causation of hernias, and varicose veins due to the raised intra-abdominal pressure associated with straining to defaecate. Very rarely, usually in an elderly bedridden patient, prolonged severe constipation can produce partial intestinal obstruction. Liquid faeces accumulate above the dried faecal mass and trickle past it causing *"spurious diarrhoea"*.

## Causes

Constipation arises in three main ways (Fig. 67):

### Little Food Residue Enters the Large Intestine

A diet low in roughage tends to cause constipation. Restriction of fluid in the diet has the same effect. Clearly the less material remains unabsorbed in the colon, the smaller and less frequent will be the faecal discharges. When, during a feverish illness, food intake consists mainly of fluids, it is not surprising that the faeces should be small in bulk. This is the inevitable consequence of the absence of dietary residue in the colon and should not be the occasion for a dose of laxative. Faeces passed in these conditions are derived from cast-off cells from intestinal mucosa and bacteria.

Obstruction to the passage of intestinal contents into the large intestine produces constipation in the same way.

### Passage Through the Large Intestine is Slowed

If faecal material does not arrive in the rectum, defaecation cannot occur. All grades of obstruction may occur, from complete obstruction with complete absence of defaecation after the bowel beyond the obstruction has been emptied, to partial obstruction, where the passage of faeces through colon and rectum is slowed.

Obstruction may be due to obliteration of the lumen of the large bowel by a tumour. It may also be caused by anticholinergic and ganglion-blocking drugs which cause loss of tone and movement in the large intestine. Some drugs such as morphine increase tone in the large bowel to such an extent that faeces cannot be propelled along it. Similar increases in colonic tone may be produced by emotional stress and by food ingestion in certain patients with the "irritable colon syndrome"; such patients are prone to constipation. *Hirschsprung's disease*, or *megacolon*, is a condition where the nerve plexus in a segment of the rectal wall is congenitally absent. This segment is unable to propel faeces and faeces accumulate in the dilated colon proximal to the abnormal segment.

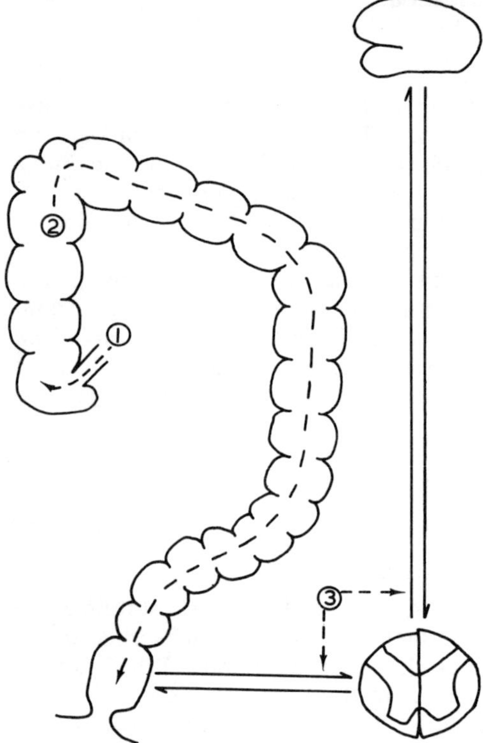


FIG. 67.—Three fundamental causes of constipation:
1. little food residue enters the large intestine;
2. passage through the large intestine is slowed;
3. the defaecation reflex is impaired due to local factors or disturbances affecting higher centres.

When laxatives are used regularly, their withdrawal may lead to constipation. Initially, absence of defaecation is due to the previous emptying of the bowel by the purgative. Subsequently the return to normal defaecation may be delayed by the bowel having become adjusted in some way to an excessive level of stimulation by the laxative.

Hypothyroidism tends to produce constipation by slowing down intestinal activity generally.

### The Defaecation Reflex is Impaired

The defaecation reflex is mediated by stretch receptors in the rectum and parasympathetic cholinergic motor fibres to the rectum plus somatic nerves to adjoining skeletal muscle. The co-ordination centre is in the sacral region of the spinal cord. Any interruption of the reflex arc, such as damage to sensory nerves supplying the rectum, or to the parasympathetic and somatic motor nerves to the rectum and adjoining skeletal muscle must interfere with defaecation. Following upper spinal cord injury, the defaecation reflex, along with other spinal stretch reflexes, is lost during the stage of spinal shock. It later returns, but is not, of course, under normal voluntary control.

Anticholinergic drugs and autonomic ganglion-blocking drugs tend to cause constipation by interfering with the motor side of the arc.

Higher centre control of the defaecation reflex can lead to constipation in a number of ways. Most obviously, defaecation can be voluntarily suppressed, despite the sensation which indicates that the rectum requires emptying. Suppression of defaecation may be due to lack of a suitable opportunity or to some condition around the anus which makes defaecation painful. Less directly, defaecation can be inhibited (or stimulated) subconsciously in anxiety states. It also tends to be inhibited by a depressive illness. In at least some cases of anxiety-induced constipation, the cause seems to be disturbance of autonomic nerve activity in the colon so that faeces are slowed in their passage through it to the rectum.

### Investigations

Investigations of constipation are concerned mainly with identifying or eliminating disease of the large intestine. In particular, a persistent change of bowel habit (towards either constipation or diarrhoea) is an important early warning sign in many cases of large intestinal cancer. Examination of the large intestine by X-rays and visually (by sigmoidoscope) is required as in the investigation of diarrhoea.

### Treatment

Self-medication for constipation is widespread and largely unnecessary in that many such cases of "constipation" are normal variations of bowel habit or the natural consequence of a change in diet or of mild psychological disturbance. Increased dietary roughage and reassurance are the appropriate treatments in most cases.

Where constipation is due to an underlying disease, the treatment is that of the underlying disease as with diarrhoea. In some cases where constipation and its treatment by drugs is of long standing, it may be necessary to continue treatment with, e.g. a material which adds to the normal non-absorbable food residue, or a drug which spurs the intestine into greater activity by mildly irritating it.

An *enema* (fluid injected into the rectum) is used to empty the colon and rectum rapidly, e.g. during X-ray examination of the colon or before surgery to the colon. It may also stimulate reflex defaecation in a person in whom control of defaecation by the higher centres has been lost due, e.g. to an upper spinal cord injury.

# 36. ALIMENTARY TRACT OBSTRUCTION

## Definition

Obstruction of the alimentary tract implies an abnormal hindrance to the passage through it of its contents. The obstruction may be a mechanical hold-up or a paralysis of peristalsis.

In practice, the condition does not usually cause trouble unless the obstruction is fairly severe or complete. The obstruction may be present at any level in the alimentary tract.

## Effects

The effects of obstruction differ, depending on the level of obstruction but certain features are common to most cases and include *pain*, *distension*, *vomiting*, *constipation*, *dehydration* and *impaired nutrition*.

### Stage of Increased Peristalsis

The increased peristalsis is a compensatory effort by the mildly distended gut to overcome the obstruction. Each time a powerful peristaltic wave spreads along the gastro-intestinal tract the patient experiences severe pain, due possibly to stretching of the mesentery. Synchronously with the pain, loud bowel sounds (borborygmi) may be heard by means of a stethoscope, or with the unaided ear. In some thin patients, the outline of the intestine may be visible, with the peristaltic wave passing along. These painful peristaltic waves tend to cause episodes of pain lasting for around 3-5 minutes with intervening pain-free periods of around 5-15 minutes. Reflex vomiting may or may not accompany these intermittent attacks of pain.

### Stage of Intestinal Paralysis and Distension

Distension of the gut normally leads to secretion and peristalsis leading to movement of contents along the alimentary tract. When peristalsis is ineffective, the secretion leads to distension which, in turn, leads to more secretion, so that a vicious circle results:

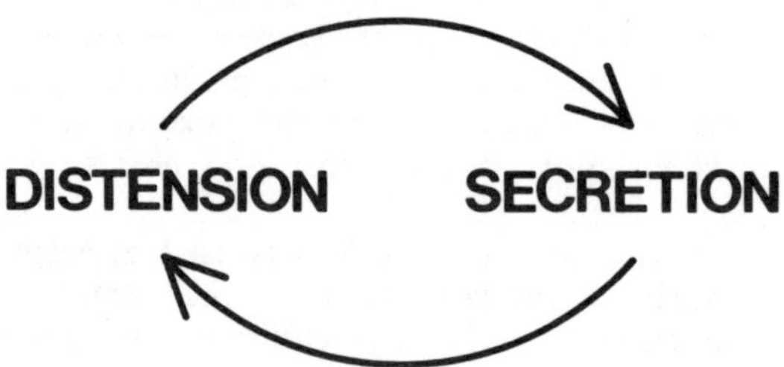


The intestine becomes greatly distended and its over-stretched muscle ceases to contract. With the cessation of peristalsis, the episodes of pain cease, and listening to the abdomen with a stethoscope reveals an ominous silence. Even the normal periodic bowel sounds are now absent.

As fluid continues to accumulate, the intestinal contents pass back to the stomach and are periodically vomited. After the gut beyond the obstruction has emptied, it becomes collapsed and in the absence of stimulation by intestinal contents, quiescent.

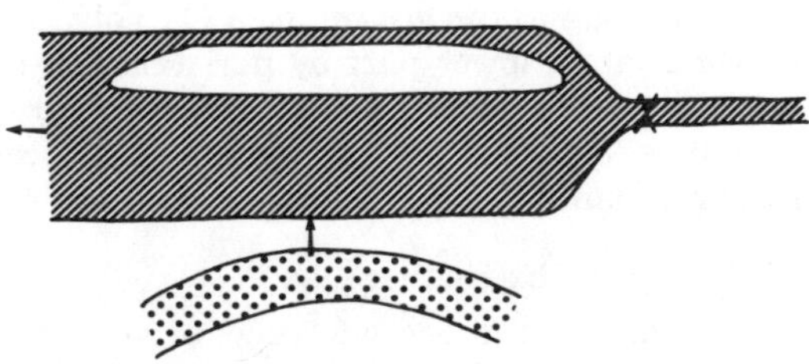


FIG. 68.—Two stages in the progression of intestinal obstruction.
(a) Increased peristalsis above the obstruction. normal bowel beyond.
(b) Intestinal paralysis with massive transfer of fluid from circulation to intestine due to failure to reabsorb excessive secretions; beyond the obstruction the bowel is empty and quiescent.

The severely distended and immobile gut is incapable of absorbing the large volume of fluid present, possibly due to lack of normal mixing movements. As this fluid has been withdrawn from the extracellular space, severe dehydration occurs with the typical features of a falling blood volume—a compensatory rise in the pulse rate and a progressively falling blood pressure. Loss of potassium into the intestinal lumen contributes to the inability of the intestinal smooth muscle to contract effectively.

### Effect of Site of Obstruction

The higher in the alimentary tract the obstruction, the earlier does vomiting occur and the less obvious is distension. Thus when the oesophagus is obstructed, distension is apparent only on X-ray examination. When the lower ileum or colon are obstructed, the abdomen is markedly distended by the distended loops of intestine.

The character of the fluid vomited is influenced by the site of the obstruction. With obstruction at the pylorus, the vomitus is acid and does not contain bile; undigested food may be present. When the large intestine is obstructed, the vomitus tends to progress from undigested food to intestinal mucous and serous secretions and possibly after several days to faecal material.

In all cases of gastro-intestinal hypersecretion and vomiting, there is loss of water, sodium chloride and potassium (from desquamated cells). In pyloric stenosis the fluid lost is acid, so that *alkalosis* develops. When intestinal secretions are also lost, the vomitus tends to be neutral or even alkaline and *acidosis* may result.

## CAUSES

Alimentary tract obstruction may be mechanical or paralytic.

### Mechanical Obstruction

The lumen of the gut may be obstructed when the intestine becomes *kinked,* due for example to the nipping of a portion of gut which has protruded through a gap in the wall of the peritoneal cavity (a hernia). Kinking may also occur when, usually following abdominal surgery, a portion of intestine becomes stuck to a healing scar. Pyloric obstruction is generally due to narrowing by scar tissue resulting from repeated activity and healing of a long-standing ulcer in the duodenum or pylorus.

A tumour of the intestine, generally the large intestine, may also cause obstruction. Much more rarely, food or other ingested materials may do so, as may a twisting of the gut on itself (L. volvulus) or the passing of one part of the intestine into a lower part by peristalsis (intussusception—L. *intus* = within, *suscipere* = to receive). The intussusception may begin when a projecting tumour is "caught" by the peristaltic process. Possibly in children enlarged intestinal lymphoid masses may act in the same way.

### Paralytic Obstruction

This may occur as a late stage of *mechanical obstruction.* It also arises as a general rule after *abdominal surgery.* For some reason which is obscure, handling and otherwise disturbing the gut leads to cessation of its peristaltic activity. In most cases the condition subsides rapidly, but occasionally it persists for many days and endangers life.

*Disturbances of the extracellular environment,* such as a reduction in serum potassium (hypokalaemia) and uraemia, interfere with intestinal smooth muscle contraction, as do drugs which interfere with parasympathetic transmission.

Partial obstruction can also occur when the intrinsic nerves of the gut are absent, e.g. in the lower oesophagus (achalasia of the oesophagus—Gr. *chalasis* = relaxation) and colon (megacolon). Food and its products pass these sites only with difficulty, leading to distension above the paralysed segment.

## INVESTIGATIONS

X-ray studies generally confirm the presence of alimentary tract obstruction. In the case of intestinal obstruction, the presence of excessive gas is a cardinal sign. Because of the intestinal distension, gas is not absorbed at the usual rate and shows up on X-ray either outlining a distended portion of gut or in the form of abnormal fluid levels. It is normal to see a fluid level in the stomach between the gastric air bubble and the gastric contents, but such levels are not normally apparent in the small intestine and indicate severe distension.

Fluid and electrolyte status must be accurately assessed.

## Treatment

### Removal of Secretions by Suction

As the excess secretions cannot be absorbed and are responsible for further secretions and for vomiting, they are removed through a tube passed into the stomach. Removal of the distending secretions is also necessary before the return of peristalsis can be hoped for. In the case of paralytic obstruction, removal of secretions and the correction of fluid and electrolyte levels provide the treatment of the obstruction. When the distension is relieved, the gut activity tends to return spontaneously. Drugs do not appear to help appreciably; as with the heart, their main effect is to *modify* activity which is already present.

### Intravenous Adminstration of Fluid and Electrolytes

This is essential to replace fluid lost, largely from the extracellular compartment into the lumen of the gut. The fluid must contain electrolytes required to replace the deficit—sodium chloride and potassium mainly. Glucose is usually given in a concentration of 5 per cent which is isotonic with the plasma and not unduly irritant to veins as are higher concentrations of glucose. In cases of prolonged paralytic obstruction, fat suspensions and amino-acid solutions may also be required to maintain nutrition.

### Surgical Correction of Mechanical Obstruction

When a mechanical obstruction is present, surgery is required urgently to correct it. If the obstruction cannot easily be removed, it is bypassed, in the case of the colon by an opening to the surface of the body (colostomy).

An indication that gastric suction can be stopped and feeding by mouth cautiously begun is the reappearance of normal bowel sounds and the passage of rectal flatus by the patient.

# 37. DISEASES CAUSING DESTRUCTION OF THE WALL OF THE GUT

## DEFINITION

THIS SECTION is concerned with diseases which damage the wall of the alimentary tract, producing effects such as ulceration, haemorrhage and perforation.

## EFFECTS

Some diseases of the wall of the gut, such as an abscess or a cancer, produce severe general effects on the body. However, in this section, only the local effects of damage to the integrity of the gut wall are considered.

The damage to the gut wall may be diffuse or local. Local damage tends to produce an *ulcer,* which may be defined as an area of destruction in an epithelial surface. The effects of diffuse damage and of ulceration tend to differ (Fig. 69), although in some cases diffuse damage and ulceration may coexist.

**Diffuse damage** must affect the local *secretion* of digestive fluids and the *absorption* of nutrients. For example, when the mucosa of the stomach is diffusely damaged (atrophic gastritis in pernicious anaemia) the secretion of HCI, pepsinogen and intrinsic factor is progressively abolished. Pancreatic secretions can compensate for the loss of HCI and pepsinogen required for protein digestion, but the absence of intrinsic factor prevents normal absorption of vitamin $B_{12}$ in the terminal ileum.

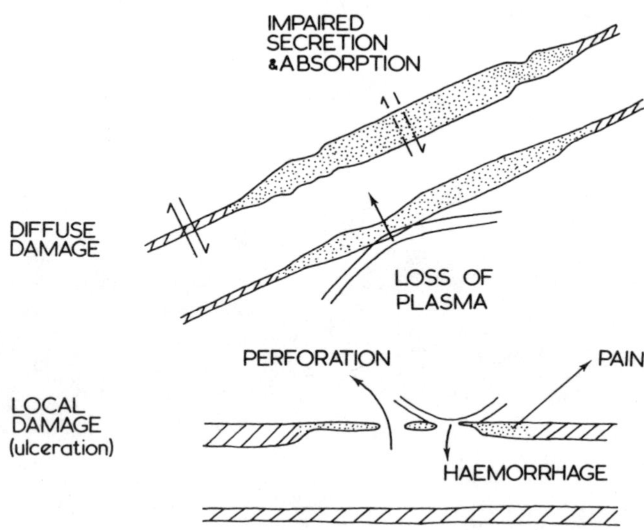


FIG. 69.—Some effects of diffuse and local damage to the wall of the alimentary tract.

Diffuse damage in the small intestine can cause severe generalized malabsorption of nutrients, e.g. when the mucosa is damaged by gluten in coeliac disease. When the large intestine is affected, the absorption of fluids is reduced and diarrhoea results.

*Loss of plasma* occurs when the disease of the intestinal wall leads to capillary damage and increased capillary permeability. The situation is similar to that in which a moderately large area of skin has been damaged by burning. In both cases fluid, electrolytes and plasma proteins are lost from the body. If the loss occurs fairly high in the alimentary tract, subsequent reabsorption may compensate, but if the colon is diffusely affected (ulcerative colitis), reabsorption cannot occur.

**Local damage** often, but not invariably, causes *pain* which is poorly localized and generally referred to the midline in the body segments from which the corresponding gut region has developed in fetal life. Pain from the oesophagus is usually referred to the chest; pain from the stomach and duodenum is referred to the upper abdomen; pain from the small intestine, including vermiform appendix, to the middle of the abdomen and pain from the large intestine to the lower abdomen.

*Pain* is a particularly prominent feature in *peptic ulceration.* This is ulceration in which the ulcer is bathed in acid-pepsin secretions. It can thus occur in the lower oesophagus, stomach and duodenum. It may also occur in the small intestine around the orifice of a direct connection (usually surgically produced) between the stomach and the intestine. The acid apparently irritates local pain endings, because neutralization of the acid by an alkali rapidly relieves the pain.

*Haemorrhage* frequently complicates ulceration at any site. It may be slight but repeated—*occult* (hidden) *bleeding*—and may produce little effect on the appearance of the faeces; biochemical tests can detect it. More severe bleeding may lead to vomiting of blood (*haematemesis*) and/or passage of partly digested blood in the faeces (*melaena,* Gr. *melas* = black). Occasionally massive bleeding so irritates the gut that rapid peristalsis leads to its evacuation within a few minutes. The general effects of bleeding from the alimentary tract are those of haemorrhage from any site.

*Perforation* may complicate ulceration of the gastro-intestinal tract at any site, especially in the stomach, duodenum, vermiform appendix and colon. The wall of the gut breaks down completely so that intestinal contents escape into the peritoneal cavity. Effects differ somewhat when the upper and lower regions of the alimentary tract are perforated. Perforation of the stomach or duodeum leads to sudden very severe pain due to intense irritation of pain endings by the corrosive effect of the gastric and duodenal contents. Pain in the peripheral peritoneum is well localized and the pain is referred to the corresponding region of abdominal wall. Movements of the peritoneum aggravate the pain so that the patient lies as still as possible and resents pressure on the abdominal wall and sudden removal of that pressure which causes "rebound" pain as the abdominal wall springs back into shape. Protective ("guarding") reflexes cause steady contraction of the abdominal muscles and suppress deep breathing, thereby tending to "splint" the painful region.

Following perforation of the appendix or colon, the pain is rather less dramatic because the intestinal contents are less corrosive, but local tenderness and guarding occur. Infection of the escaped fluid is a greater risk at this site because the contents of the large intestine, unlike those of the stomach and duodenum, are heavily infected with bacteria.

If untreated, the eventual widespread inflammation in the peritoneal cavity tends to abolish peristalsis, causing a paralytic type of intestinal obstruction.

## Causes

In some cases, the cause of damage to the gut wall is clear, e.g. a penetrating abdominal injury, infection of the gut wall, serious interference with the blood supply of the intestine, a cancer which has itself become ulcerated. In many cases, however, the cause is obscure. For some reason the mucosa is unable to maintain its integrity in the face of the corrosive and abrasive contents which it normally resists successfully.

Impaired mucosal resistance may be an important cause of disease of the gut wall. Thus in the case of peptic ulcers, while excess acidity seems to be to blame in some cases, in others there is no such excess. Certain items of diet and cigarette smoking may aggravate peptic ulceration. A variety of drugs have an irritant action on the mucosa and may cause bleeding; aspirin is an example of many such drugs.

Psychological stress appears at least partly to blame in some cases, possibly due to increased secretion of glucocorticoids which would delay the healing of minor areas of damage in the gut. Certainly the adminstration of large amounts of glucocorticoid hormones favours the development of ulcers and consequent haemorrhage or perforation in the gastro-intestinal tract.

## Investigations

As with alimentary tract disease in general, X-ray studies using ingested barium are the main means of diagnosing abnormality of the wall of the gastro-intestinal tract, especially ulceration and associated scarring.

*Acid secretion* by the stomach is measured by passing a tube into the stomach and collecting its contents by continuous suction. Resting secretion rate and response to pentagastrin (which contains the four terminal (active) amino-acids of gastrin) are measured.

If a patient has had his vagus nerve supply to the stomach divided, the *insulin test for vagotomy* is applied. This measures the response in terms of acid secretion to a dose of insulin large enough to cause hypoglycaemia. Hypoglycaemia causes a reflex increase in acid secretion, the vagus supply to the stomach forming an essential part of the reflex. The purpose of the reflex is not clear.

Direct examination of the mucosa of the gut through instruments passed through the mouth or rectum is often useful, as is histological examination of a small sample of mucosa obtained during direct examination or from the jejunum by means of a swallowed sampling device at the end of a long tube.

## Treatment

As the cause of disease of the gut wall is often obscure, treatment often fails to cure, but rather minimizes symptoms and complications. Thus in the case of peptic ulceration, elimination of aggravating dietary constituents, and the reduction of gastric acidity usually relieve symptoms but rarely lead to a permanent cure; on the other hand, healing sometimes occurs spontaneously.

*Reduction of gastric acidity* may be achieved by means of drugs or surgery.

**Drugs.**—Although blockade of the parasympathetic nerve supply to the stomach reduces acid secretion, the drugs available tend to produce unpleasant side-effects by interfering with other parasympathetic actions. Treatment is often directed towards the neutralization by alkalis of the acid after it has been secreted. The most suitable alkalis are those which are not absorbed through the intestinal mucosa. An absorbable alkali such as sodium bicarbonate when taken in large quantities tends to produce alkalosis.

One of the simplest means of buffering gastric acidity and relieving peptic ulcer pain is to take a drink of milk. However, even this is not without its dangers, because the consumption of very large quantities of milk, particularly by elderly people, leads to excessive calcium absorption. When both excessive milk and soluble alkalis are consumed, the combination of hypercalcaemia and alkalosis causes loss of appetite, nausea and vomiting, headaches and weakness —a generalised metabolic upset known as the ***milk-alkali syndrome.***

**Surgery.**—When it is found that acid secretion is occurring at a persistently high rate, the branches of the vagus nerves to the stomach may be cut (vagotomy). As this procedure delays gastric emptying (which is facilitated by vagal activity) it is generally accompanied by a procedure to aid drainage from the stomach, e.g. widening of the pylorus or making a connection between stomach and jejunum (gastrojejunostomy). Alternatively, part of the acid- and gastrin-secreting lower region of the stomach may be removed.

Not surprisingly, interference with the normal mechanisms of gastric emptying can create problems. Gastric emptying may be delayed. This is helped by a cholinergic drug. Alternatively, emptying may be excessively fast—the *"dumping syndrome"*. This is a feeling of weakness usually within about half an hour of a meal. It is related to relatively sudden emptying of the stomach with distension of the jejunum. Several mechanisms probably operate. Firstly, reflex effects may be produced by the distension. Secondly, there may be a fall in blood volume produced by massive intestinal secretion due to marked intestinal stimulation. These two mechanisms cause faintness shortly after a meal.

Thirdly, and later, the blood glucose level may fall. This is a consequence of the rapid glucose absorption which leads to a high level of insulin in the blood. Glucose absorption which commences abruptly also ceases abruptly, before the insulin level has had time to return to normal. The consequent hypoglycaemia accounts for the later symptoms in the dumping syndrome.

Both delayed emptying and "dumping" tend to improve with time as the alimentary tract adjusts to the altered conditions.

Complications of ulceration, such as severe haemorrhage or perforation require emergency surgical treatment to stop the bleeding or repair the perforation.

# 38. IMPAIRED LIVER FUNCTION

## Definition

THIS SECTION is concerned with diffuse disease of liver cells, which interferes with the many functions of these cells. As the liver, like kidneys, lungs, etc., has a large functional reserve, disordered liver function becomes apparent only when damage to the liver is widespread and moderately severe. Obvious disorder of liver function is referred to as liver failure.

## Effects

Because the functions of the liver are many and varied, the effects of failure of liver function are correspondingly widespread. A number of these effects seriously alter the composition of the extracellular fluid, the environment of the body cells. The brain is particularly sensitive to these changes and advanced liver failure is often associated with progressive failure of brain function—behaviour disorders, impairment of consciousness and death.

The effects of liver disease derive largely from loss of liver cell function. These include *impaired detoxication*, *impaired protein synthesis*, *disturbed carbohydrate metabolism* and *diminished bile secretion.* In addition, the disease process often obstructs the circulation through the liver, so that pressure in the portal circulation rises (*portal hypertension*).

These various effects tend to advance in parallel, but one or other may predominate in an individual patient. Each main effect give rise to various secondary effects. Together the effects impair function throughout the body.

### Impaired Detoxication

One of the most important functions of the liver is to modify and make harmless (detoxicate) chemicals which would otherwise accumulate and poison the body. Chemicals which are thus treated include intestinal toxins, hormones and certain drugs.

The liver is particularly well situated to deal with intestinal toxins, because blood from the gut passes through the liver before entering the general circulation. Though their exact nature is unknown, these intestinal toxins are thought to be derived from dietary protein, and bacteria in the large intestine are thought to be important in their formation. As in renal failure, no one substance has been identified as the main culprit. In severe liver failure, some of these protein breakdown products may be excreted in the breath, giving a somewhat sickly odour of tissue decomposition— *fetor hepaticus.*

These intestinal toxins can seriously interfere with brain function. Initially the patient may show mild peculiarities of behaviour and have difficulty in carrying out familiar tasks such as dressing. Episodes of grossly abnormal behaviour resembling alcoholic intoxication characteristically follow a meal rich in protein. In advanced liver failure these toxins lead to unconsciousness and eventually death. Irregular tremor of the outstretched hands is presumably due to the effect of the disturbed internal environment on the nervous system (*metabolic flap*).

Haemorrhage into the gut is not uncommon in liver failure; bleeding occurs from dilated blood vessels in the oesophagus, which are used by the circulation to bypass the diseased liver. This provides a high-protein "meal" and may precipitate coma.

Other features of liver failure, such as increased peripheral circulation and clubbing of the fingers, may be related to unidentified circulating substances. Failure of breakdown of oestrogens may be responsible for feminization in the male. The depression of secondary sexual characteristics and sexual activity in both sexes has also been attributed to depression of pituitary activity.

Because some drugs are inactivated in the liver, patients with hepatic failure are much more sensitive than normal to such drugs.

The effects of failure of detoxication are summarized in Fig. 70.

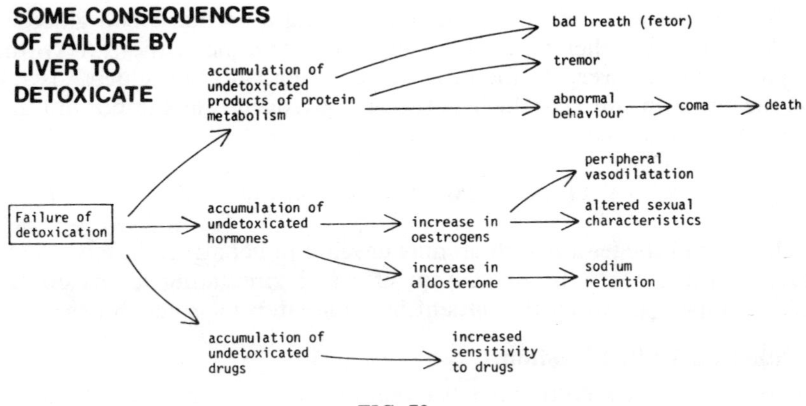


FIG. 70.

### Impaired Protein Synthesis

The liver is the sole significant source of plasma albumin. Relatively early in liver disease, the plasma albumin level starts to fall; the more severe the liver disease, the lower the albumin level. For some reason which is unknown, the globulin level tends to rise, so that the albumin/globulin ratio falls markedly in liver disease.

The fall in the plasma albumin level is probably an important contributory factor to the development of oedema and ascites (abdominal swelling due to the collection of fluid in the peritoneal cavity; Gr. *askos* = bag). Oedema and ascites are considered further in the section on portal hypertension.

A low level of prothrombin tends to occur and interfere with clotting; spontaneous bleeding and excessive blood loss following injury tend to occur.

Since the immunoglobulins are manufactured by lymphoid tissue, these plasma proteins are not depressed in liver disease.

The effects of impaired protein synthesis are summarised in Fig. 71.

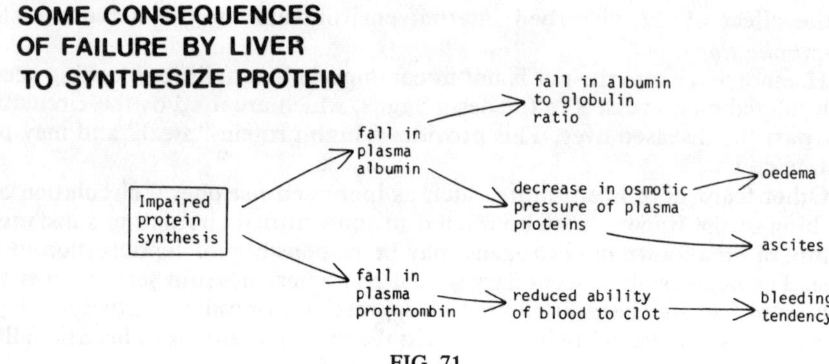


FIG. 71

### Disturbed Carbohydrate Metabolism

The liver has an essential role in maintaining a normal blood glucose level. When glucose and other sugars are plentiful after a meal, they are synthesised to glycogen in the liver. As the blood glucose tends to fall, glucose is released from the glycogen stored in the liver (muscle glycogen is less important in maintaining blood glucose):

LIVER GLYCOGEN $\leftrightharpoons$ BLOOD GLUCOSE

Thus one of the features which may develop in hepatic failure is a low blood sugar. This augments the depressant effect of circulating toxins on cerebral function, especially when the patient has not eaten for some hours.

### Diminished Bile Secretion

Diminished bile secretion has two main effects, one harmful, the other not. The harmful effect is malabsorption of fat and the other effect is jaundice which, in itself, is not harmful to adults.

The bile salts are essential for the digestion and absorption of fat because they emulsify fat into small, water-soluble particles which provide a large interface for the action of lipase and facilitate absorption. In liver failure, fat absorption is seriously impaired and consequently deficiencies of the fat-soluble vitamins A, D and K tend to arise. As vitamin K is required in the synthesis of prothrombin, there are two possible causes of prothrombin deficiency in liver disease: impaired vitamin K absorption and impaired hepatic synthesis of protein.

### Portal Hypertension

Conditions which damage the liver tend to cause scarring which interferes with the free flow of blood through the portal circulation. The rate of blood flow through the portal circulation is determined by the rate of blood flow through the spleen and gastro-intestinal tract. As this flow must be maintained, a rise in resistance in the portal vascular bed leads to a rise in pressure in the portal vein above its normal value of 10-20 mm Hg.

The problems caused by portal hypertension (Fig. 72) include:

*Varicosities.*—Connecting channels exist between portal venous tributaries and systemic veins at a number of sites, including the gastro-oesophageal junction region, the anterior abdominal wall around the umbilicus and the anorectal junction region. When portal venous pressure rises markedly above systemic venous pressure (e.g. to 30-40 mm Hg), portal blood bypasses the liver and is returned to the inferior vena cava by these channels which become greatly distended and tortuous. These varices (L. *varix* = an enlarged and tortuous vessel) bleed profusely if damaged. This is particularly likely to happen in the case of gastro-oesophageal varices and the blood loss in such cases is often severe.

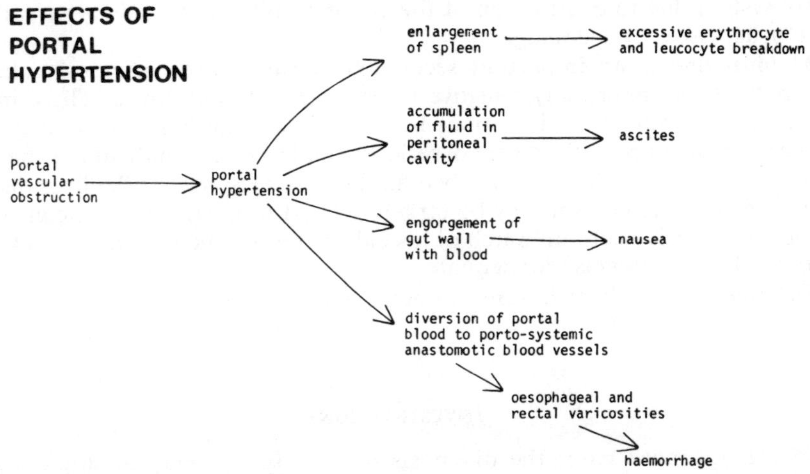


FIG. 72

*Ascites.*—Ascites is a condition in which the abdomen is distended by fluid in the peritoneal cavity. When marked, the condition causes considerable distress. Although ascites does not occur in the absence of portal hypertension, other factors probably play a role in its causation, because portal hypertension does not always cause ascites. The other factors may include obstruction to lymphatic vessels in the liver, increased fluid loss from capillaries due to reduced plasma albumin (and hence osmotic pressure) and increased extracellular fluid volume due to increased aldosterone activity.

*Enlarged spleen.*—The increased pressure in the splenic vessels causes swelling of the spleen which is often easily palpable. Splenic function appears to increase with size, and increased trapping and destruction of red and white blood cells and platelets occurs so that the numbers of these cellular elements in the blood tends to fall.

*Alimentary tract congestion.*—Congestion of the wall of the gastro-intestinal tract tends to mimic the sensation produced by a full stomach. Loss of appetite, nausea and even vomiting may result.

### Causes

In view of its high rate of metabolism, the liver is one of the more sensitive of the body organs to adverse *changes in the internal environment* of its cells. In this way, its function may be temporarily upset by hepatic *venous congestion* (due, e.g. to heart failure), by *disturbed blood gas tension* (as in respiratory failure) and by various *toxins,* including alcohol and certain drugs to which the individual may be exceptionally sensitive.

*Infection* of the liver (hepatitis) generally produces a reversible impairment of liver function in otherwise healthy persons.

*Replacement* of normally-functioning liver can occur due to scarring following previous liver damage, and due to widespread cancer. Distension of the biliary system due to obstruction of the common bile duct tends eventually to cause progressive liver damage.

*Malnutrition* is an important secondary cause of liver disease in that it renders the liver abnormally sensitive to damage from any cause. Thus, in the well-nourished individual, infectious hepatitis is usually followed by complete recovery, whereas up to 50 per cent of seriously malnourished individuals may die from it, due to liver failure. Similarly it has been suggested that the liver disease associated with alcoholism may be largely related to nutritional deficiencies—the alcoholic tends to obtain much of his caloric requirement from alcohol, and often his diet is otherwise inadequate.

In some cases of liver failure, no obvious cause may be found.

### Investigations

In the advanced case, the diagnosis of liver failure may be apparent on clinical examination. Diagnosis in earlier cases and assessment generally depend on a variety of chemical tests which mainly demonstrate the inability of the liver to carry out one or other of its many metabolic functions. Thus most of these tests indicate early liver failure; when liver damage is present but the remaining healthy liver cells can compensate, the tests generally yield normal results.

*Abnormal protein metabolism* is indicated by a fall in plasma albumin and rise in globulin.

*Abnormal biliary metabolism* is shown by a raised level of bilirubin in the blood.

*Release of intracellular enzymes* occurs in liver cell damage. Detection of these escaped enzymes (e.g. glutamic-pyruvic transaminase) in the circulation provides a sensitive index of liver disease.

The existence of gastro-oesophageal varices may be demonstrated by *X-ray* in some cases.

*Scanning* studies reveal the liver outline. Radio-active gold is administered and is concentrated in the liver and to a certain extent in the spleen by phagocytic cells.

The cause of obscure liver disease may be discovered in some cases by microscopic examination of a sample of liver tissue (liver biopsy).

## Treatment

Treatment of liver disease varies somewhat, depending on which feature is most troublesome.

*Bed rest,* which reduces the metabolic demands on the liver, is advisable when the liver is responding to some noxious agent, such as a chemical toxin or the virus of infective hepatitis. When using *drugs,* the possibility must be borne in mind that the patient may be unduly sensitive; drugs which carry the risk of damaging the liver should be avoided if possible.

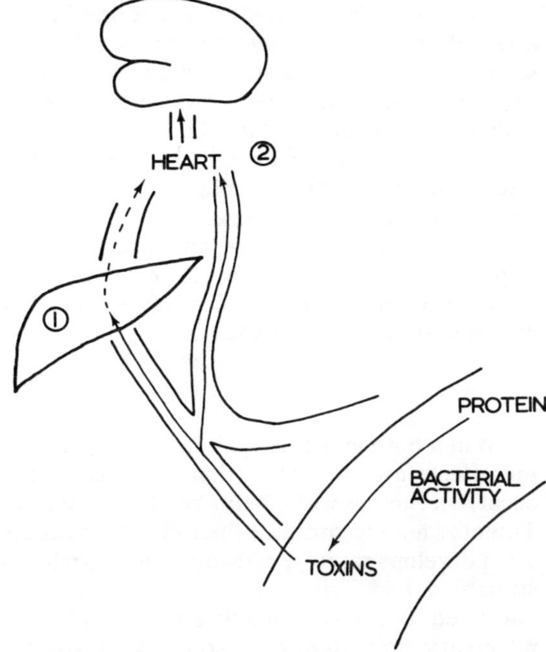


FIG. 73.—Effect of shunting portal venous blood into a systemic vein. Instead of being detoxicated in the liver (1), blood from the gut passes directly to the heart (2). Toxins are then circulated to the brain and may depress its function.

*Ascites and oedema* may respond to a low sodium intake or to diuretic drugs which eliminate sodium from the body and reduce the excessive extracellular fluid volume. Occasionally excessive ascitic fluid may have to be withdrawn directly from the abdomen but this depletes the patient of protein, carries the risk of infection and can generally be avoided by reducing extracellular sodium content.

*Cerebral malfunction* due to failure to detoxicate protein products is treated by a low protein diet. Appropriate non-absorbable antibiotics reduce toxin formation by bacteria in the colon.

*Portal hypertension,* as well as contributing to the development of ascites, carries the risk of severe and recurrent haemorrhage which endangers life. In such cases, portal pressure can be lowered by making a deliberate large anastomosis with a systemic vein (e.g. portal vein to inferior vena cava; splenic vein to renal vein). The risk of such a procedure is that it may precipitate features of cerebral dysfunction due to diversion of toxins past the liver (Fig. 73). This risk is in proportion to the degree of liver failure present.

# 39. PANCREATIC DISEASE

## Definition

THE PANCREAS is a double gland, with both exocrine and endocrine secretions. Disease of the pancreas may involve the exocrine secretions alone, as in fibrocystic disease of the pancreas which causes impairment of digestion and absorption. Alternatively, the endocrine function of the pancreas may alone be affected as in diabetes mellitus or the relatively rare condition of excess insulin secretion. However, despite its common occurrence, it is by no means clear whether diabetes mellitus is in most cases primarily a disease of the pancreas or is due to inactivation of insulin which leads to failure of an otherwise normal organ. Fibrocystic disease of the pancreas is described in the chapter on deficient nutrition. Endocrine disturbances related to the pancreas are described in the section on endocrine disease.

Finally, both endocrine and exocrine functions of the pancreas may be impaired simultaneously by a non-specific invasive disease such as cancer, infection or inflammation for any cause (pancreatitis). Such conditions are considered briefly in this section.

## Effects

When both exocrine and endocrine pancreatic tissue are affected together, the effects are a combination of malabsorption and diabetes mellitus, their occurrence and severity being related to the extent of pancreatic tissue damage. Tumours tend to produce their effects gradually. Inflammation of the pancreas often develops rapidly, causing severe upper abdominal pain related to capsular stretching and irritation of adjoining peritoneum. Nearby intestine may be paralysed as a result, leading to features of intestinal obstruction. Normally pancreatic protein- and fat-splitting enzymes are not active until they reach the intestine and are activated by enterokinase and bile. In this way, the pancreas is protected from autodigestion. However, when the pancreatic ducts are obstructed, the activation of these enzymes can occur when they are still in the pancreas. In at least some cases, the entry of bile salts into the pancreatic ducts may activate pancreatic lipase. Destruction of fat deposits (fat necrosis) around the pancreas and elsewhere is a feature of pancreatitis. As the more powerful pancreatic enzymes act on pancreatic tissue and are absorbed into the blood stream, widespread tissue damage and toxin production occur. In severe cases, all these combined effects may lead rapidly to circulatory failure and death.

Jaundice may occur in addition if the common bile duct is closed off by oedema of surrounding tissues as it passes through the head of the pancreas.

## Causes

Disease of the pancreas often seems to develop from obstruction to the pancreatic ducts (Fig. 74). The obstruction may be due to inflammation or to spasm caused by a gall-stone lodged in the region of the common termination of the bile and pancreatic ducts. The powerful pancreatic enzymes in the ducts are then activated and attack pancreatic tissues and surrounding tissues. They may also spread through the circulation.

Pancreatic tissue may also be damaged by various infecting agents and chemicals, including alcohol and some drugs.

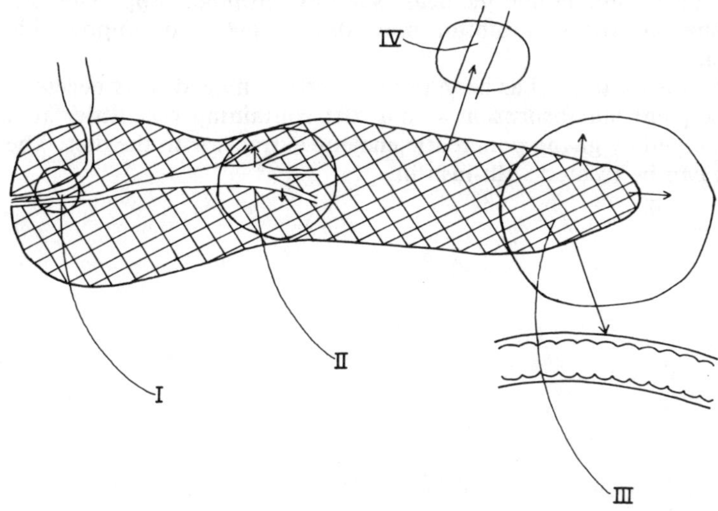


FIG. 74.—Mechanisms in the development of pancreatitis.

- I Pancreatic (and in some cases bile) ducts are obstructed (by oedema etc.), especially in the head of the pancreas.
- II Pancreatic enzymes escape from ducts into the pancreatic tissues, thereby causing tissue damage.
- III Inflammation spreads to adjoining tissues, including the intestine which may become paralysed.
- IV Pancreatic digestive enzymes escape into the blood stream.

## INVESTIGATIONS

The coincidence of evidence for both malabsorption and insulin deficiency suggests the presence of pancreatic disease. In cases where pancreatitis develops rapidly, the leak of enzymes into the circulation can often be detected. For example, the serum amylase level may be several times the normal and in exceptional cases up to 100 times the normal. The serum lipase level may also be raised.

*Scanning* studies reveal the outline of the pancreas. The patient is given a radio-active amino-acid solution, e.g. methionine labelled with radio-active selenium. This material is used by the pancreas in the synthesis of its enzymes and can be detected in the pancreas at an appropriate interval after its administration to the patient.

### Treatment

During an attack of pancreatitis, measures are taken to reduce pancreatic secretion. The measures include stopping of oral feeding, which causes reflex stimulation of pancreatic secretion, and removal, by suction, of gastric secretions. In this way acid and food do not enter the duodenum to cause release into the blood stream of secretin and pancreozymin, the hormones which stimulate the pancreas to secrete. Pancreatic secretion is also favoured by vagal activity and an anticholinergic drug, such as atropine, suppresses this activity. A carbonic anhydrase inhibitor may also be given to suppress bicarbonate secretion.

When the pancreas has been permanently damaged, it is necessary to treat the consequent malabsorption with a diet containing very little fat. Digestion may be helped by giving pancreatic enzyme extracts with the diet. The diabetes mellitus can be treated with insulin.

# 40. INCREASED CIRCULATORY BILIRUBIN (JAUNDICE)

## DEFINITION

JAUNDICE is a condition in which the bilirubin content of the blood is raised and the patient appears orange, yellow or greenish yellow as a result; the colour being most easily seen in the whites of the eyes.

The upper limit of normal for serum bilirubin is around 1 mg/100 ml. Discolouration may be detected when the level rises to around 2-4 mg/100 ml, the exact level depending on the lighting conditions (yellowish artificial light makes detection difficult), and the observer's colour vision and experience.

## EFFECTS

As far as is known, excess bilirubin does no harm to the adult patient. In the newborn it is associated with brain damage. Bilirubin crosses the blood-brain barrier of the infant much more easily than that of the adult.

Troublesome itching sometimes occurs in jaundiced patients. This is thought to be due to an excess of circulating bile salts, rather than excess bilirubin.

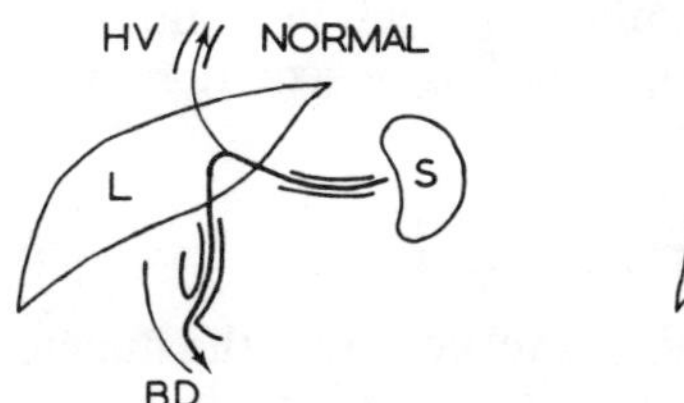

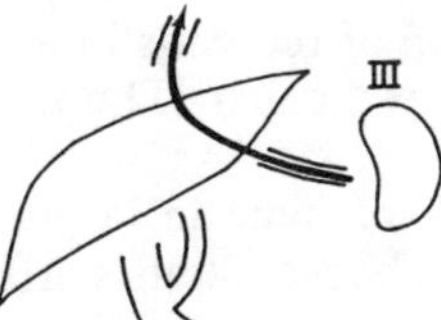


**FIG. 75.—Three main varieties of jaundice. Normally, bilirubin (mainly from the spleen, S) is largely conjugated in the liver (L) and excreted in the bile ducts (BD). A small fraction is returned to the circulation in the hepatic venous blood (HV).**

**In pre-hepatic jaundice (I) the load of bilirubin presented to the liver is excessive. An increased amount of bilirubin is excreted in the bile but excess is also retained in the blood.**

**In hepatic jaundice (II) liver function is so depressed that it is unable to excrete the normal load of bilirubin. The amount of bilirubin excreted in the bile is decreased and the remainder is retained in the blood.**

**In post-hepatic jaundice (III) bile excretion is blocked and bile pigments are retained in the blood.**

## Causes

There are three main causes of jaundice (Fig. 75):

**Pre-hepatic** or **haemolytic jaundice** is due to excessive breakdown of red cells so that the liver is presented with more bilirubin than it can excrete. An increased amount of bilirubin is conjugated to form a water-soluble complex with glucuronic acid and excreted in the bile. However, there remains an excessive amount of bilirubin in the hepatic venous blood. The excess bilirubin is unconjugated with glucuronic acid but is bound to plasma protein so that it cannot pass the glomerular filter and the urine remains free of bile (acholuric jaundice).

**Hepatic jaundice** is due to disease of the liver cells. The liver is unable to deal with the normal load of bilirubin. There may be impairment of bilirubin uptake, impairment of conjugation or impairment of excretion of the conjugated bilirubin.

**Post-hepatic** or **obstructive jaundice** is due to obstruction of the excretory pathway, e.g. obstruction of the common bile duct by a gall-stone or by a tumour. As the bile pigments have been split from their protein carriers and conjugated with glucuronic acid, they can pass the glomerular filter and appear in the urine.

The diagnosis of the cause of jaundice in a given patient is complicated by the fact that more than one reason for jaundice may be present. Thus in hepatitis liver function is impaired but the oedema of the liver tends to produce obstruction of the small intrahepatic bile canaliculi in addition. Again, one type of jaundice may lead to another—long-standing obstruction of the bile duct can lead to back-pressure and liver damage.

## Investigations

Measurement of the serum bilirubin level confirms the diagnosis of jaundice and indicates its severity and how much of the bilirubin is in the conjugated (glucuronide) form.

The hallmark of pre-hepatic (haemolytic) jaundice is evidence of haemolysis in the form of anaemia with an increased number of reticulocytes (the presence of these immature red cells in increased numbers indicates increased bone marrow formation of red cells in an effort to compensate for the increased rate of breakdown of red cells). The urine is free of bile; the faeces contain excess bile pigments.

The hallmark of hepatic jaundice is evidence of hepatic failure including evidence of disturbed protein synthesis and of release into the circulation of hepatic intracellular enzymes.

The hallmark of post-hepatic (obstructive) jaundice is the presence of excess bile pigments and bile salts in the urine, and reduction or absence of bile pigments in the faeces, in the absence of signs of hepatic failure.

## Treatment

As bile pigments may cause cerebral damage in infants, excess bile pigments must be removed in infants with severe jaundice. This may be done by exchange transfusion—small quantities of blood are withdrawn repeatedly and replaced by fresh blood.

In adults, only the underlying condition requires treatment. Treatment of haemolytic anaemia and of liver disease have been discussed elsewhere. Treatment of obstructive jaundice is surgical—a stone in the common bile duct may require removal; in some cases (e.g. cancer obstructing the lower end of the bile duct) a fresh opening may be made between the gall-bladder or biliary ducts and the duodenum or jejunum.

## RECENT LEADING ARTICLES

Achalasia of the cardia. *Brit. med. J.*, 1974, **2,** 515.
Anorexia nervosa. *Brit. med. J.*, 1971, **4,** 183.
Childhood obesity and carbohydrate intolerance. *Brit. med. J.*, 1973, **3,** 122.
Complications of vagotomy. *Brit. med. J.*, 1973, **2,** 256.
Diagnosis of cystic fibrosis. *Brit. med. J.*, 1971, **3,** 489.
Drastic cures for obesity. *Lancet,* 1970, **1,** 1094.
Drop-outs' diarrhoea. *Brit. med. J.*, 1974, **3,** 373.
Drugs causing weight gain. *Brit. med. J.*, 1974, **1,** 168.
Drugs for gastric ulceration. *Brit. med. J.*, 1974, **2,** 186.
Effects of cholecystectomy. *Brit. med. J.*, 1974, **2,** 72.
Electrical activity of the stomach. *Brit. med. J.*, 1971, **3,** 596.
Enterogastrone(s). *Lancet,* 1971, **1,** 1225.
Erosive gastritis. *Brit. med. J.*, 1974, **3,** 211.
Fashions in duodenal ulcer surgery. *Brit. med. J.*, 1973, **1,** 563.
Flatulence. *Brit. med. J.*, 1971, **3,** 595.
Gastric acid, ulcers, and $H_2$ — receptors. *Lancet,* 1974, **1,** 666.
Gastric decompression after abdominal surgery. *Brit. med. J.*, 1973, **1,** 189.
Gastrin and gastro-intestinal disease. *Brit. med. J.*, 1972, **3,** 604.
Haemoperfusion for jaundice. *Brit. med. J.*, 1974, **3,** 486.
Ileus; paralytic or sympathetic. *Lancet,* 1971, **1,** 329.
Infant and adult obesity. *Lancet,* 1974, **1,** 17.
Management of coeliac disease. *Brit. med. J.*, 1973, **2,** 130.
More about chenodeoxycholic acid. *Brit. med. J.*, 1973, **4,** 629.
Obesity and coronary heart disease. *Brit. med. J.*, 1973, **1,** 566.
Operations for obesity. *Brit. med. J.*, 1971, **4,** 247.
Perforation. *Lancet,* 1971, **2,** 749.
Prolonged intravenous feeding. *Brit. med. J.*, 1974, **3,** 431.
Reflux and hernia. *Brit. med. J.*, 1971, **3,** 205.
Starving children before operation. *Brit. med. J.*, 1974, **3,** 213.
Steatorrhoea and the blind loop syndrome. *Lancet,* 1973, **1,** 528.
Success and failure in anorexia nervosa. *Lancet,* 1971, **1,** 900.
Surgery for severe obesity. *Brit. med. J.*, 1974, **2,** 575.
Tests for the pancreas. *Lancet,* 1974, **1,** 1023.
The fibreoptic revolution. *Lancet,* 1974, **3,** 131.
Trasylol for pancreatitis. *Lancet,* 1974, **3,** 133.
Treatment of liver failure. *Brit. med. J.*, 1973, **4,** 126.
Vagotomy for duodenal ulcer. *Brit. med. J.*, 1972, **2,** 2.
Zollinger-Ellison syndrome revisited. *Brit. med. J.*, 1973, **1,** 2.

[illegible] small quantities of blood are [illegible] and given repeatedly and replaced [illegible]

Ideally, only the underlying condition requires treatment. Treatment of [illegible] have been discussed elsewhere. [illegible] [illegible] the lower end of the [illegible]

[illegible]

FURTHER READING

[illegible]

*Section VII*

# DISORDERS OF THE NERVOUS SYSTEM

# 41. DISTURBANCE OF LOWER MOTOR NEURONE FUNCTION

## Definition

THE LOWER motor neurone is the term used to describe the *final common pathway* by which somatic impulses travel to skeletal muscle from the central nervous system. The cell bodies of these neurones lie in the anterior horns of the grey matter in the spinal cord (Fig. 76) Many impulses impinge on these cell bodies from different parts of the nervous system; some tend to depolarize and excite and some tend to hyperpolarize and inhibit the synaptic membranes on the cell. The pattern of impulses in the lower motor neurone depends on the algebraic sum of these excitatory and inhibitory influences. The axons of these neurones, which are thick and heavily myelinated, leave the spinal cord via the anterior roots and conduct impulses at high velocity to the skeletal muscle. In the muscle, the neurone loses its myelin sheath and divides into a number of filaments. Each filament forms a neuromuscular junction with a single muscle fibre. The motor nerve and the muscle fibres it supplies are known collectively as a *motor unit.* In coarse muscles such as those supporting the spinal vertebrae, the motor units are large, one motor neurone supplying several hundred muscle fibres. In muscles used for precise movements such as those which rotate the eye, a motor neurone may supply only two to three fibres. Impulses reaching the neuromuscular junctions result in the release of acetylcholine which results in depolarization and contraction of the muscle cell.

This chapter describes conditions which interfere with the final common pathway from the anterior horn cell to the neuromuscular junction.

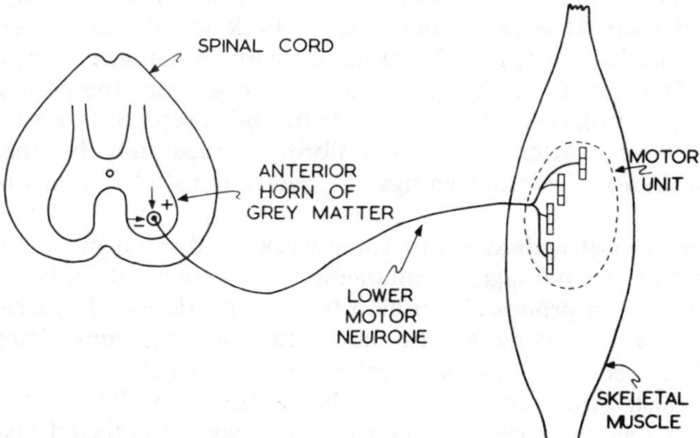


FIG. 76.—The lower motor neurone. It travels from the anterior horn of spinal cord grey matter to innervate a motor unit in skeletal muscle. It acts as the final common pathway to muscle, its activity being the algebraic sum of the excitatory and inhibitory influences which impinge upon the cell body.

## EFFECTS

Damage to lower motor neurones results in muscular weakness. Muscle weakness may occur with damage in many parts of the nervous system, but that associated with lower motor neurone disease has certain characteristics which permit it to be differentiated from the others.

**Weakness of the muscles.**—This may range from slight weakness to complete paralysis depending on the number of neurones involved and the degree of damage. When the damage is in the spinal cord the weakness tends to be segmental in distribution. When peripheral nerves are affected, the weakness lies in the distribution of that nerve and is usually associated with sensory loss.

All movements of the affected muscles tend to be weak, no matter in which way the muscles are stimulated to move. Thus voluntary movements, e.g. clenching the fist, involuntary movements, e.g. tremor of the fingers, reflex movements, e.g. the knee jerk, and synergistic movements, e.g. the contraction of the muscles which extend the wrist when the fist is clenched, are all weak or absent. Death may occur if the muscles responsible for ventilating the lungs or for swallowing become paralysed.

**Loss of muscle tone.**—When an observer passively flexes and extends the limb of a patient, he encounters a certain resistance in the muscles. This resistance is known as "muscle tone" and is thought to result from contraction of individual muscle fibres here and there throughout the muscle. It is a reflex phenomenon and disappears in animals if the posterior roots are cut. The afferent limb of the reflex arc is from the stretch receptors in the muscle spindles and the lower motor neurone forms the efferent limb. Damage to the lower motor neurone, by interfering with the integrity of this reflex arc, reduces muscle tone. This state is sometimes referred to as *flaccidity.*

**Wasting of the muscles.**—When a muscle is exercised regularly it tends to increase in bulk or hypertrophy (Gr. = *hyper,* over; *trophe* = nutrition). The mechanism underlying this phenomenon is not known. Similarly, when a muscle is not used, it tends to decrease in bulk or *atrophy* (Fig. 77). The reduction in muscle activity after lower motor neurone damage is responsible for the atrophy. It begins to be obvious about two weeks after the damage and, in severe cases, may progress until the muscle has been replaced by a thin fibrous strap. Progressive shortening of such fibrous straps may fix the limbs in abnormal positions. The shortenings are referred to as *contractures* of the muscles.

**Spontaneous contractions.**—For some weeks and often much longer after lower motor neurone damage, spontaneous contractions of individual muscle fibres (*fibrillation*) or groups of muscle fibres (*fasciculation,* L. *fasciculus* = a small bundle) may be detected. In the former case the contractions can be detected only by recording the electrical manifestations of the contraction near the fibre by electromyography, because the contractions give rise to no visible movement. In the latter case the contractions can be detected visually as a fleeting puckering of the skin here and there in the area overlying the muscle.

Though there is no general agreement on the mechanism of these phenomena, they may result from the hypersensitivity to acetylcholine which develops in skeletal muscle when it is denervated. A denervated skeletal muscle is many times more sensitive to acetylcholine than a normally innervated muscle. A similar hypersensitivity to neurotransmitter substances is seen when organs supplied by the autonomic nerves are denervated. It would seem that with constant exposure to neurotransmitter, the receptors gradually adapt to it. When traffic along the nerve stops, the target organ sees much less of the transmitter and the adaptation is lost. The target organ then becomes hypersensitive to the transmitter (*denervation hypersensitivity*).

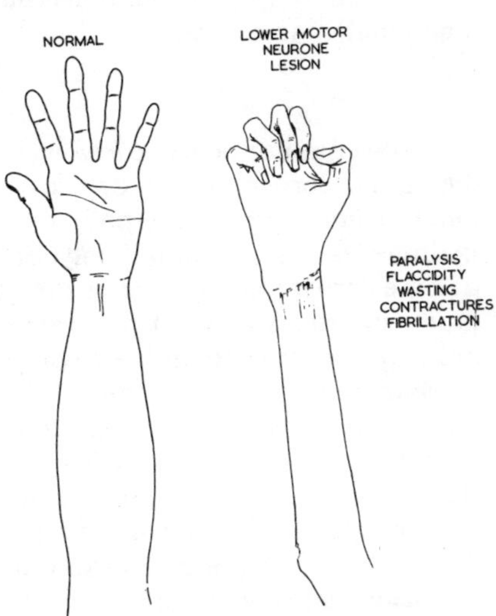


FIG. 77.—Features of lower motor neurone damage in the arm.

At motor end-plates, even in the absence of nerve impulses, a synaptic vesicle occasionally releases its acetylcholine spontaneously in a random fashion. The amount released is not normally enough to depolarize the membrane to threshold and cause the muscle fibre to contract but can be detected by the transient small depolarization that it causes in the post-synaptic membrane. When the muscle becomes hypersensitive to acetylcholine, the quantity released becomes enough to cause contraction. The random nature of the rupture explains the random distribution of the contractions over the muscle. However, fibrillation can persist after nerve degeneration, so spontaneous release of acetylcholine cannot be the whole story of the cause of fibrillation.

**Nerve degeneration.**—When a neurone (motor or sensory) is damaged or cut, the part which is separated from the cell body gradually degenerates. In lower motor neurone damage, therefore, the nerve endings supplying the muscle eventually degenerate. Following injury, the distal portion of the axon which is separated from the cell body shows characteristic changes. The axons and the myelin that surrounds them degenerate and the products are carried away by phagocytic cells. The Schwann cell sheathes proliferate so that each former nerve fibre becomes a solid cord of cells

**Nerve regeneration.**—Following injury, the proximal portions of the damaged axons attached to the cell bodies do not degenerate but send out sprouts which bridge the gap between the proximal and distal portions of the damaged nerve fibres. The sprouts grow down the distal cords of Schwann cells at about 3-4 mm/day. When they reach the target muscle they form neuromuscular junctions with it. Initially, the axons are small and slow-conducting but size and speed of conduction increase as they mature.

Return of function in a peripheral nerve may eventually be almost complete, provided the injury is not extensive and the two portions of the severed nerve are in close proximity. Nerve regeneration with useful return of function appears to be confined to the peripheral nervous system. Severed spinal cords do not regain functional continuity.

## Causes

A lower motor neurone *lesion* (L. *laesio* = I hurt) may affect the cell body in the spinal cord or the axon as it travels to the muscle. The destruction of *anterior horn cells* by the poliomyelitis virus is responsible for the lower motor neurone lesions seen in poliomyelitis. In the condition known as motor neurone disease (progressive muscular atrophy) the anterior horn cells (and some upper motor neurones as well) progressively die off for some unknown reason. The *axons in motor roots* may be compressed by protruding intervertebral discs or tumours in the spinal column. *Axons in the peripheral nerves* may be severed or torn in a traumatic injury. Inflammatory or degenerative changes occur in peripheral nerves in a great variety of conditions such as heavy metal poisoning, vitamin B deficiency, diabetes mellitus, certain allergies and infections and in a number of inherited disorders. The "peripheral neuritis" in these and other conditions is a common cause of lower motor neurone lesions.

Drugs, such as curare, cause temporary paralysis due to interference with function at the somatic *neuromuscular junction.* Such drugs are routinely given at many types of surgical operation to promote muscle relaxation. Because respiratory muscles are affected, artificial ventilation is required.

## Investigations

Though the diagnosis of lower motor neurone damage can usually be made on clinical examination, additional information is sometimes sought from the following investigative procedures.

### Studies of Electrical Excitability

When the lower motor neurone is severed, changes occur in the response of the muscle to electrical stimulation. The analysis of these changes is complex but they indicate primarily a decrease in the ease with which a muscle can be made to contract by electrical stimulation in the patient. When a normal muscle is stimulated electrically through the skin, the contraction results *indirectly* from the excitation of nerve fibres which are more excitable than the muscle fibres. When the nerve fibres degenerate after a lower motor neurone lesion, a stronger stimulus is required to excite the muscle because the muscle fibres must be stimulated *directly*.

A normal muscle is most readily stimulated when the electrode is over the "motor point", the point where the motor nerve enters the muscle. After the motor nerve degenerates, this is no longer the case.

## Electromyography

When an electrical recording is made from a normal muscle at rest, action potentials are usually not evident. During voluntary contraction of the muscle, a burst of action potentials due to electrical activity in discrete motor units (motor unit potentials) is recorded in the electromyogram. In a denervated muscle, the spontaneous twitching of isolated muscle fibres (fibrillation) is associated with characteristic electromyographic changes. Fibrillation potentials are seen in the resting muscle occurring at irregular intervals as monophasic or diphasic spikes and have a short duration of about 1.0 millisecond (ms). In normal muscle, the action potentials are much longer (5-12 ms) and have a bigger amplitude, presumably because they represent activity in a whole muscle unit rather than a single muscle fibre.

## Nerve Conduction Studies

Damage to the motor neurones may reduce the rate of impulse conduction along them. For this reason, the velocity of impulse travel in nerves is sometimes measured. To do this, a recording electrode (R) is inserted into or applied over a muscle (Fig. 78), and the nerve supplying that muscle is stimulated by electrodes at two points, one near the muscle ($S_2$) and one some distance away ($S_1$). The latency between the time of stimulation and the time of appearance of the muscle action potential is measured in both instances. The difference between these times is the time for the impulse to travel the distance from $S_1$ to $S_2$. From these measurements the velocity of nerve conduction can be calculated. Normal values range from about 40 to about 70 metres/second in these large myelinated axons.

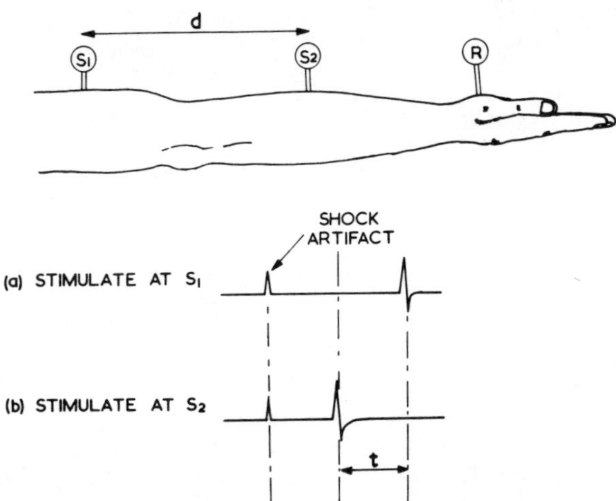


FIG. 78.—Measurement of conduction velocity in nerve. The nerve is stimulated at two points ($S_1$ and $S_2$) and an action potential is recorded (R) in a muscle supplied by the nerve. The difference in the latency between the shock artifacts and the appearances of the muscle action potentials (t) is the time taken for the impulse to travel the distance from $S_1$ to $S_2$ (d). Rate of conduction is d/t.

### Muscle Biopsy

Histological examination of a small sample taken from an affected muscle will give unequivocal evidence of muscle atrophy in severe cases. In the early stages, it is difficult to tell the difference between a normal and a denervated fibre on histological grounds.

## TREATMENT

To a large extent treatment depends on the cause of the neurone damage but there are several general points. Firstly, steps should be taken to *prevent overstretching* of the weakened muscles and deformities developing due to permanent shortening (*contracture*) of the paralysed muscles. This can be achieved by appropriate splinting. Physiotherapy including massage and passive movements of the affected muscles may also prevent the development of deformities. *Electrically induced contractions* of the paralysed muscles do not completely prevent atrophy of denervated muscles but reduce its degree so that more muscle fibres persist intact until recovery of innervation takes place.

A number of surgical procedures may be used in treatment. The severed ends of a cut nerve may be sewn together to facilitate regeneration of the nerve. The insertions of strong muscles may be transplanted so that they can effect the movements normally made by the paralysed muscles. Operations which fix joints in one position are sometimes used to stabilize limbs made unstable by muscle paralysis.

If the motor neurones responsible for ventilation are inactivated, as may happen in poliomyelitis, artificial ventilation of the lungs by an appropriate machine is required.

# 42. DISTURBANCE OF PYRAMIDAL SYSTEM FUNCTION

## Definition

THE PYRAMIDAL tract carries nerve fibres which are concerned in the execution of voluntary movements. These nerves have their cell bodies in the motor area of the cerebral cortex (Fig. 79). Their axons descend through the internal capsule, cross to the opposite side in the brain stem and travel down the spinal cord in the corticospinal (pyramidal) tracts. Their nerve endings synapse either directly or via intermediate neurones with the cell bodies of lower motor neurones in the anterior horn of spinal cord grey matter. The pyramidal tract also contains fibres of nerves which do not originate in the motor area of the cortex but the function of these fibres is less certain.

Activity in the pyramidal tract may be altered by disease. Occasionally the disease irritates a part of the tract giving rise to an increase in the traffic of impulses and uncontrolled muscular contractions and spasms. This condition is known as epilepsy and is dealt with later (Hyperactivity in the Nervous System). More commonly, disease destroys a part of the pyramidal pathway so that impulses cannot be conducted from the motor cortex to the anterior horn cells. The features associated with a pyramidal system lesion are described in this section.

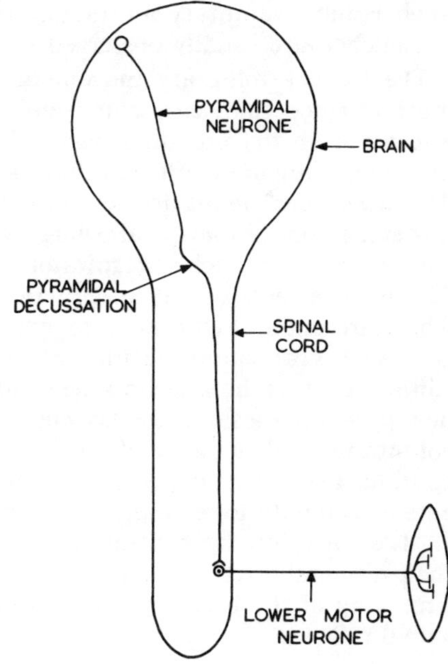


FIG. 79.—Nerves in the pyramidal tract have their cell bodies in the motor area of the cerebral cortex. The axons descend through the internal capsule and cross to the opposite side of the brain in the medulla oblongata (pyramidal decussation). They then descend in the lateral column of spinal cord white matter to synapse with lower motor neurones in the anterior horn of spinal cord grey matter.

A "*pyramidal system lesion*" is sometimes referred to as an "*upper motor neurone lesion*". The former term is preferred by the authors because there are two varieties of upper motor neurone, those responsible for the execution of voluntary movement and those responsible for the background of tone and posture on which the voluntary movements are superimposed. To avoid confusion, these systems are dealt with separately using "pyramidal system" as a functional description of the former and "extrapyramidal system" as a functional description of the latter.

## EFFECTS

Immediately after interruption of the pyramidal pathway there is often a stage which has many of the characteristics of spinal shock. There is paralysis of voluntary movement with loss of reflex activity and muscle tone. After some weeks, however, reflex activity and muscular tone begin to return and the characteristic features of the condition become manifest.

There are two main groups of effects, the negative effects due to absence of normal pyramidal function and the positive effects due to "release" of phenomena normally suppressed by the pyramidal system.

### Negative Effects

(i) **Loss or weakness of voluntary movement.**—Voluntary movements on one side of the body are weak (*hemiparesis,* Gr. *paresis* = relaxation) or lost (*hemiplegia,* Gr. *plege* = stroke). This differs from the paralysis in lower motor neurone lesions in that only voluntary movements are lost, not the power of individual muscles to contract. Involuntary movements such as tremors, reflex movements such as flexion reflexes and tendon jerks and synergistic movements such as the involuntary contraction of the wrist extensor muscles when the fist is clenched are usually preserved.

The loss of voluntary movement is not uniform. Skilled movements are more affected than coarse movements; the hands and feet tend to be more affected than the proximal parts of the limbs. In the upper limbs, extensor movements are more affected than flexor; the reverse is true in the lower limbs. The lower face tends to be more affected than the upper face and vocal movements more than swallowing movements. Movements which involve both sides of the body, such as respiratory and spinal movements, are usually spared. This may be because they have bilateral representation in the brain. Movements which are parts of emotional expression are also spared. Thus, a patient may contract his orbicularis occuli muscles when he smiles with happiness though he cannot contract the same muscle voluntarily to command. This again illustrates how pyramidal lesions are associated with loss of the ability to move muscles voluntarily and not loss of the ability of muscles to contract. The muscles continue to contract in involuntary movements and do not waste (atrophy) to the extent seen with lower motor neurone damage.

Loss of voluntary control of micturition and defaecation occurs only when the pyramidal tracts on both sides are interrupted as in spinal cord transection. Emptying of the bladder and rectum is then accomplished automatically by spinal reflexes.

(ii) **Loss of superficial reflexes.**—When the abdominal skin is lightly stroked, the underlying abdominal muscle contracts. When the skin on the inside of the male thigh is lightly stroked, the cremaster muscle contracts and raises the testis. These are examples of superficial reflexes. The exact pathway of the reflex arcs responsible for these reflexes is not known but the pyramidal tracts make up part of the path; in pyramidal tract damage, the superficial reflexes are lost.

**Positive Effects**

(i) **Increase in stretch reflex sensitivity and activity.**—Supraspinal centres normally facilitate reflex activity in the spinal cord. When this facilitation is removed, there is initially loss of reflex activity as in spinal shock. However, with the passage of time, the reflex centres develop the ability to act autonomously and the reflexes become abnormally exaggerated. When pyramidal influence is removed, reflex muscular contraction mediated through the spinal stretch reflex is eventually increased and altered in distribution to produce characteristic effects.

Tendon reflexes are exaggerated. Thus when the patellar tendon is tapped, there is an exaggerated contraction of the quadriceps muscle (knee jerk). Other tendon reflexes on the affected side are similarly exaggerated. The incrase in sensitivity of the muscle spindles is thought to be due to incrased impulse traffic in gamma efferent fibres which innervate the muscle in the spindles.

When the gastrocnemius muscle is passively stretched by dorsiflexing the foot of a patient with pyramidal tract damage, the muscle may undergo a series of rhythmic contractions. This is referred to as *ankle clonus* (Gr. *klonos* = a tumult) and is just another manifestation of the exaggerated stretch reflexes (Fig. 80).

*Muscle tone is increased.*—Following pyramidal tract damage, the resistance felt when the patient's limbs are passively flexed and extended is increased. This increase in tone is referred to as *spasticity* or *rigidity*. It is sometimes called *clasp knife spasticity* because the chief resistance is experienced at the

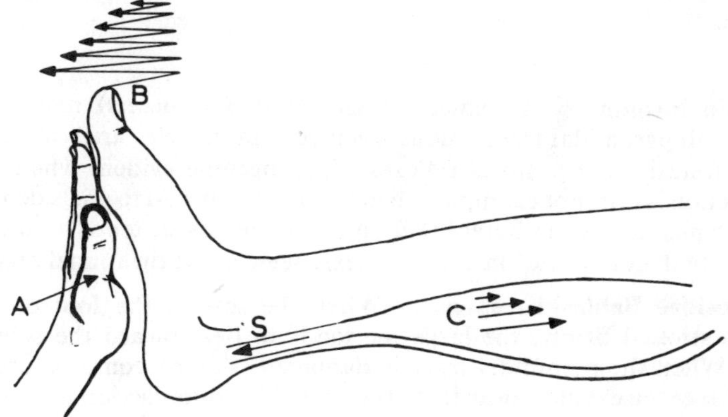


FIG. 80.—Ankle clonus. Pressure applied by the hand (A) to the foot stretches the Achilles tendon (S) and causes a rhythmic oscillation of the foot (B) by inducing a series of stretch reflex contractions in the gastrocnemius muscle (C).

beginning of the passive movement. After the initial resistance has been overcome, as with a clasp knife blade, the remainder of the movement is easy. The increase in tone is predominantly in the flexors of the upper limb and the extensors of the lower limb. It results from an increase in stretch reflex sensitivity caused by increased gamma motor neurone activity in the spinal cord.

As the increase in muscle tone is not uniformly distributed in all the muscles it leads to *abnormal posture* of the limbs (Fig. 81). The upper limbs tend to be held in a flexed position whereas the leg is adducted at the hip, and held in an extended position. The plantar flexion of the ankle makes the toes stick down and they tend to catch on obstacles when the patient walks.

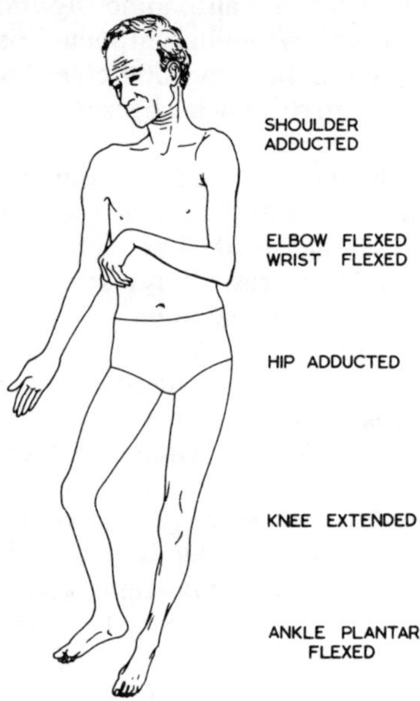


FIG. 81.—Abnormal posture of the left arm and leg with damage to the pyramidal system supplying that side.

Certain involuntary movements (*associated movements*) may be seen in patients with pyramidal tract lesions when certain muscles are stretched. These are manifestations of postural reflexes which become evident when pyramidal tract function is lost. For example, when the head is turned to one side the arm on that side tends to extend whereas that on the opposite side tends to flex (Fig. 82). These are analogous to the tonic neck reflexes seen in certain animal preparations.

(ii) **Positive Babinski response.**—When the sole of the foot of a normal subject is stroked firmly, the large toe tends to flex toward the sole (plantar flexion). When the pyramidal tract is damaged, this response is altered (Fig. 83). The large toe extends away from the sole. This dorsiflexion is referred to as a *positive Babinski response*. It is seen normally in babies in the first year of life before the pyramidal tracts become myelinated. As with the superficial reflexes the exact reflex pathway is not known.

The positive Babinski response is thought to be an exaggerated fraction of the spinal cord withdrawal reflex which becomes evident when pyramidal tract function is lost. In some cases a more full-blown withdrawal response occurs when the sole is stroked. The ankle, knee and hip may flex and the opposite leg extend. Painful *flexion spasms* may be a troublesome feature of a pyramidal neurone lesion.

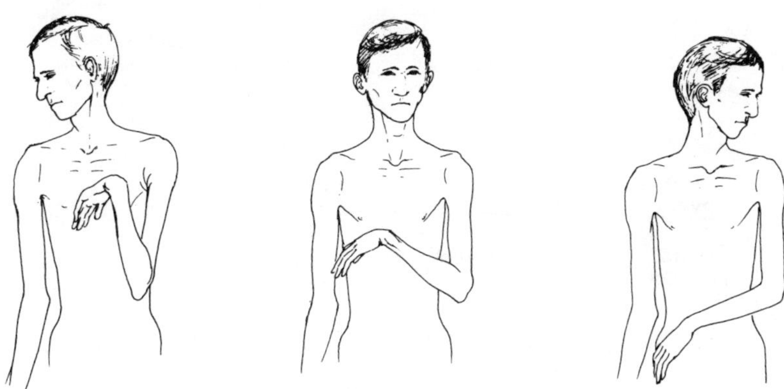

FIG. 82.—Associated movements seen with pyramidal system lesions affecting the left arm. Turning the head away from the affected side increases the flexion and adduction of the affected arm. Turning the head towards the affected side has the reverse effect. These movements are postural reflexes elicited by stretching certain neck muscles. The reflexes are normally suppressed by the pyramidal system.

## Causes

### Inadequate Cerebral Blood Flow

A *cerebrovascular accident* is the most common cause of pyramidal tract damage. The blood vessels (branches of the middle cerebral artery) supplying the pyramidal tract as it passes through the internal capsule are especially prone to pathological change. They may rupture to give cerebral haemorrhage, be occluded by clot formation to give cerebral thrombosis or may be blocked by a circulating embolus. However, anything which interferes with cerebral blood flow may cause damage. In hypertensive patients, the vessels may be occluded by a reversible spasm of the vessel wall. In all these cases the onset of signs and symptoms is quite rapid and the condition is described as a *stroke* until a more precise diagnosis can be made. *In many cases, the cause of a cerebrovascular accident lies outside the brain.* The cerebral vessels arise largely from the anastomotic *arterial circle* (of Willis) at the base of the brain. This circle is fed by the two common carotid arteries and, via the basilar artery, by the two vertebral arteries. Interference with flow in these feeder vessels for any reason can cause a cerebrovascular accident by depriving smaller distal vessels of the head of pressure necessary for adequate flow. If the smaller vessels in one part of the brain are already narrowed, then it is this part of the brain which will bear the brunt of the reduction in blood flow. The reserve of function in the carotid and

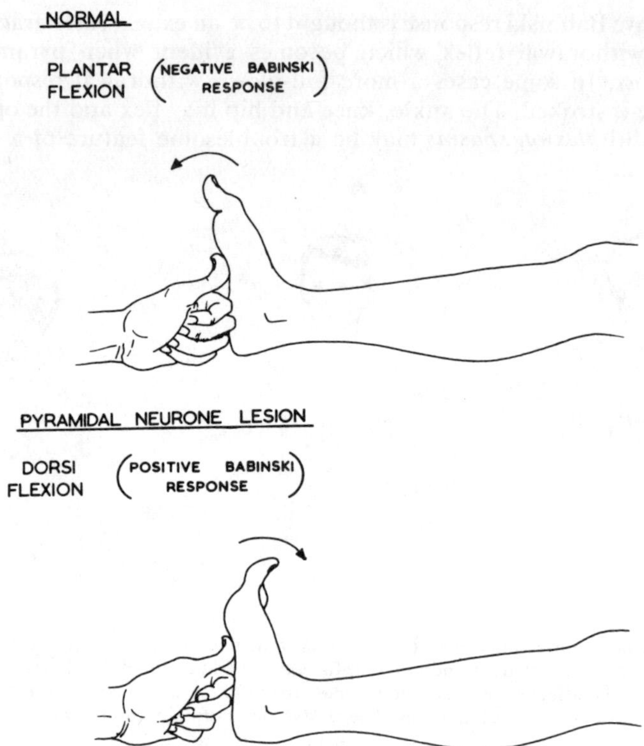


FIG. 83.—The Babinski response. With a pyramidal system lesion the response to stroking the sole of the foot is dorsiflexion and fanning of the toes.

vertebral arteries is large and in practice the supply of blood to the arterial circle is not usually seriously compromised unless flow in more than one of the feeding vessels is markedly reduced.

The anastomotic circle is shown in diagrammatic form in Fig. 84.

**Demyelinating Diseases**

Certain diseases such as *multiple* (or disseminated) *sclerosis* result in patchy areas of demyelination of axons here and there throughout the central nervous system. These areas may involve the pyramidal tract. Though some clinical recovery occurs after each attack, the repeated scarring of the central nervous system often eventually results in some permanent disability from pyramidal tract damage.

**Birth Injury**

Occasionally injury to the baby's head during labour damages the pyramidal system on both sides of the brain. These children show bilateral spasticity and are sometimes referred to as *spastics*. They often also show signs of damage in the extrapyramidal system and may be mentally retarded.

## Spinal Cord Injury

Injury to the spinal cord which causes complete transection, compression or interference with its blood supply may damage the pyramidal tracts. Spasticity is not so evident in these patients as it is when the damage is at the level of the internal capsule. However, the lack of inhibition of the withdrawal (flexion) reflex by the pyramidal tracts is sometimes associated with painful spasms of the flexor muscles of the legs, and if uncared for, may lead to flexion deformities.

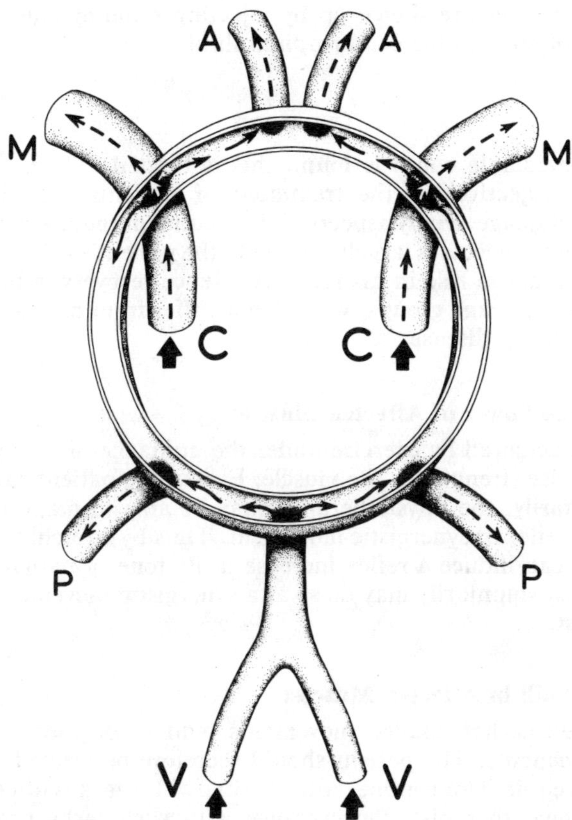


FIG. 84.—A diagrammatic representation of the main arteries supplying the brain. The aim is to show the functional significance of the anastomotic *arterial circle* at the base of the brain. This circle is fed by the paired carotid (C) and vertebral (V) arteries; blood flows from the circle to the paired anterior (A), middle (M) and posterior (P) cerebral arteries.

## Miscellaneous

Many diseases such as subacute combined degeneration of the spinal cord due to vitamin $B_{12}$ deficiency, cerebral tumour and head injury may involve the pyramidal tracts.

## Investigations

There are no special investigations which are particularly relevant in the *diagnosis* of pyramidal tract damage. The diagnosis and localization of the lesion can usually be made on clinical examination. However, a variety of investigations may be used in appropriate cases to identify the *cause* of the damage. These include X-ray examination of the skull, of the blood vessels supplying the brain, and of the cerebral ventricles, examination of cerebrospinal fluid, electroencephalography etc. When studying the arterial supply of the brain, radio-opaque material is injected into the carotid or vertebral arteries; similarly the ventricles are shown up by injecting radio-opaque material or a small quantity of air into the cerebrospinal fluid.

## Treatment

Where a remediable cause is found this is treated. There are, however, certain general objectives in the treatment of patients with long-standing pyramidal tract damage. Many aspects of the treatment require active co-operation by the patient. Success is largely related to the patient's will and intellectual ability to play an active role in his recovery. Hence recovery is best in people, usually in the younger age groups, who do not suffer from mental deterioration due to the underlying disease.

### Increasing the Power of Affected Muscles

This can be achieved by exercise under the guidance of a physiotherapist. Systematic exercise strengthens the muscle. Even if the patient cannot contract a muscle voluntarily, the physiotherapist may be able to make it contract by involving it in a reflex or synergistic movement. Thus by stretching a muscle the physiotherapist can induce a reflex increase in its tone; a patient who cannot dorsiflex his wrist voluntarily may do so as a synergistic movement when asked to clench his fist.

### Developing Skill in Affected Muscles

As mentioned earlier, skilled movements tend to be more affected than non-skilled movements. The patient should therefore be trained to regain the muscular skills required for routine daily activities. This task is often undertaken by an occupational therapist. Perseverance with such tasks presumably encourages the development of compensatory mechanisms.

### Reducing Spasticity

This may become necessary because spasticity interferes excessively with voluntary movement. Certain drugs are reputed to reduce muscle tone but they are not very effective. Passive flexion and extension of affected limbs by physiotherapists is probably more helpful than the use of drugs. Exercises, both active and passive, are very useful in preventing the development of abnormal postures.

### Abolishing Painful Flexor Spasms

For these spasms it may be necessary to inject a corrosive fluid around the nerves in the spinal canal. This works by interfering with the reflex pathways responsible for spasticity but may disturb the micturition reflex and cannot therefore be undertaken lightly.

# 43. DISTURBANCE OF EXTRAPYRAMIDAL SYSTEM FUNCTION

## Definition

THE EXTRAPYRAMIDAL motor system contains the upper motor neurones which travel to lower motor neurones by paths other than the pyramidal tracts. Whereas the pyramidal system is involved in the execution of voluntary movements, the extrapyramidal system is responsible for providing a suitable background of muscle tone and posture for those movements. It also seems to be concerned in the initiation of voluntary movements and the suppression of certain involuntary movements. Its normal physiological role is very poorly understood but when it is damaged by disease, characteristic clinical features occur.

The fibres in the extrapyramidal tracts arise from a wide variety of nuclei in the cerebral hemispheres and the brain stem. The basal ganglia, the substantia nigra, the reticular formation, the olivary and vestibular nuclei are among the most important of these. Extrapyramidal fibres exert their effect on skeletal muscle either by acting directly on alpha motor neurones, which innervate extrafusal muscle fibres, or by increasing the activity of gamma motor neurones which increase the activity of stretch receptors in the muscle spindles (Fig. 85).

Damage to the extrapyramidal system may cause different clinical pictures depending on which part of the system has been destroyed. However, the clinical syndrome usually associated with extrapyramidal damage is called *Parkinsonism* after Dr. James Parkinson who first described the condition in his *Essay on the Shaking Palsy* in 1817.

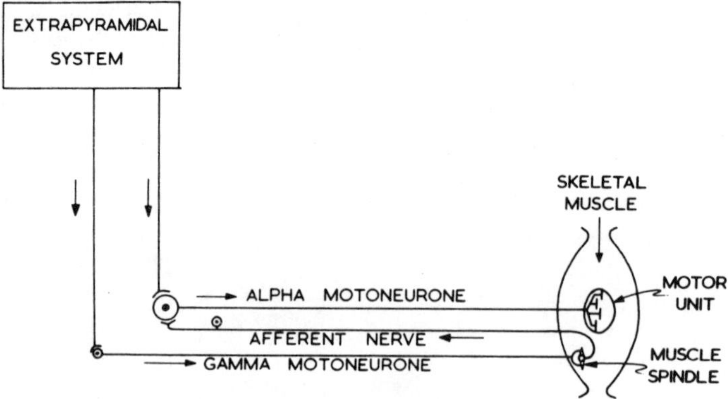


FIG. 85.—The extrapyramidal system can influence skeletal muscle either by a direct action on the lower motor neurones or indirectly by influencing the activity of gamma motoneurones and hence the afferent discharge from the muscle spindles.

## Effects

Though extrapyramidal damage causes many disturbances, many nervous functions remain relatively normal. There is no true paralysis indicating that the pyramidal and lower motor neurones still function. Stretch reflexes such as the knee jerk are little disturbed. Sensation is normal and the patient's mental powers are not impaired.

When a part of the central nervous system ceases to function, there are negative effects, i.e. the absence of phenomena for which the part is normally responsible, and positive effects, i.e. the appearance of phenomena normally suppressed by the part. With extrapyramidal damage, the negative effects include poverty and slowness of voluntary movement. The positive effects are more numerous and include an increase in muscle tone, disorders of posture and involuntary movements.

### Disorders of Voluntary Movements

Though there is no true paralysis or loss of voluntary movements, voluntary movements tend to be fewer, slower to start and weaker and slower to execute than normal. There is a pronounced poverty of voluntary movements. The face remains wooden and impassive and the eyes unblinking. The patient does not scratch his ear, cross his legs, swing his arms or make the numerous unnecessary movements normal people make. Hand writing may be small and illegible. With emotion, changes in facial expression can occur but the changes are very slow in onset and slow to develop. When the patient tries to walk he has difficulty in starting, takes small shuffling steps and tends to fall forward. The poverty of voluntary movement is sometimes called *akinesia* (Gr. *kinesis* = motion) and the slowness of execution of voluntary movement *bradykinesia*. The physiological mechanisms by which the normal extrapyramidal system prevents these disturbances are not understood.

### Disorders of Muscle Tone

Muscle tone tends to be increased to give rigidity of the limbs. It differs from the spasticity of pyramidal system damage in that the resistance to muscle extension is felt throughout the full excursion of the movement. It is referred to as *cog wheel rigidity* when the resistance is intermittent throughout the movement or *lead pipe rigidity* when it is constant. Both flexor and extensor muscles are equally affected and the proximal muscles of the limbs tend to be affected more than the distal muscles; this is the reverse of the spasticity seen in pyramidal system lesions.

The rigidity is thought to be a release phenomenon. It has been produced experimentally in animals by destruction of parts of the basal ganglia. Here it is due to facilitation of alpha motor neurones; this would imply that a normal physiological function of some parts of the extrapyramidal system is to inhibit the alpha motor neurones. The rigidity in experimental Parkinsonism does not disappear if the posterior roots are cut and does not depend therefore on facilitation of the gamma motor neurones. This would explain why the stretch reflexes are not exaggerated in this type of extrapyramidal disorder.

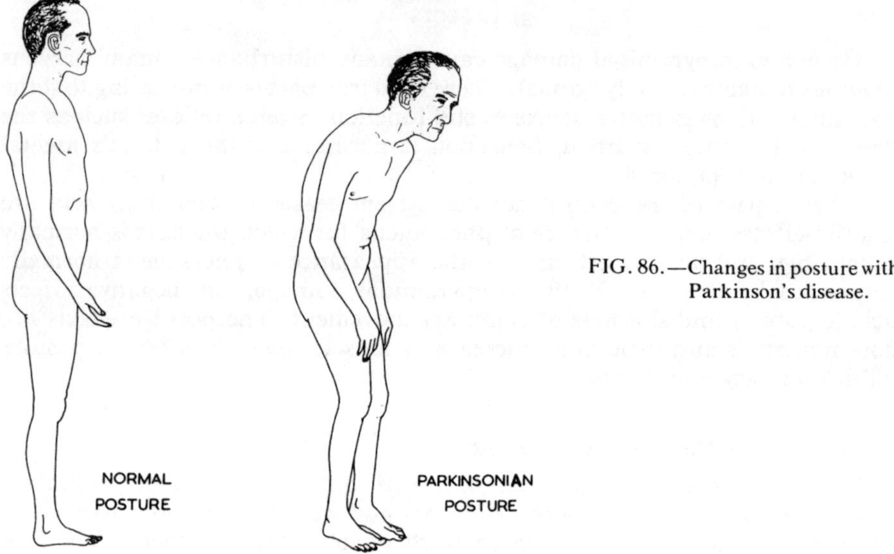


FIG. 86.—Changes in posture with Parkinson's disease.

**Disorders of Posture**

The abnormal muscle tone leads to a characteristic posture (Fig. 86). The body tends to be stooped forward and the head bowed so that the centre of gravity is shifted to the front in a somewhat simian fashion. The joints in the arms and legs all tend to be flexed with the exception of the interphalangeal joints.

**Involuntary Movements**

The most common form of involuntary movement is *tremor.* This consists of a rhythmic alternate contraction of opposing muscle groups. It is best seen in peripheral parts of the body such as the hands, feet, jaw and tongue. The frequency of the tremor is about 4-8 Hz. It disappears when the patient goes to sleep and usually when he makes a purposeful voluntary movement. This is unlike the tremor seen in cerebellar disease. Parkinsonian tremor becomes more obvious when the limbs are at rest and may be a real source of embarrassment. It is also made worse by emotional upset and stress.

The physiological basis of the tremor is not well understood. It is mediated through the pyramidal tract and pyramidal tract damage will abolish it. The ventrolateral nucleus of the thalamus also is essential for tremor. Recordings from cells in this nucleus have shown activity with the same frequency as the tremor. Destruction of this nucleus in patients with Parkinsonism may abolish the tremor. The ascending reticular formation may also play a part because the tremor disappears during sleep.

More rarely other forms of abnormal movements are seen in disorders of the extrapyramidal system. *Choreiform movements* (Gr. *choreia* = a dance) are irregular, jerky, non-repeating movements of the limbs and face. *Athetosis* (Gr. *athetos* = without fixed position) is the name given to slow writhing movements of the face and limbs. *Torsion dystonia* describes movements similar but slower than those of athetosis which spread to the proximal parts of the limbs and trunk. These torsion spasms may hold the patients for some time in twisted abnormal postures.

## CAUSES

**Parkinson's Disease (Paralysis Agitans)**

This accounts for about 80 per cent of all cases of Parkinsonism. It is a disease of unknown cause which comes on in middle-aged patients and gradually progresses. It is characterized biochemically by a reduction in the dopamine content of the basal ganglia. Dopamine, which is synthesized from tyrosine via dopa in the body, is thought to be a neurotransmitter substance in the brain. Nearly all the dopamine in the normal brain is found in the basal ganglia and substantia nigra. It is not known however whether the dopamine depletion in these patients is the result or the cause of the disease.

The cause of Parkinson's disease is unknown, but a number of known agents which damage the brain can cause all the signs and symptons of the disease. A Parkinson's disease-like state is known as "Parkinsonism". It may be seen in elderly people who have degenerative arterial disease which causes death of basal ganglia cells by restricting their blood supply. Some signs of Parkinsonism are seen in nearly all very old people. In younger people, brain injury, certain types of virus infections of the brain, carbon monoxide poisoning, and other agents may cause the syndrome.

**Huntington's Chorea**

In this hereditary disease, degenerative changes in the brain manifest themselves when the patient is about 30 years of age by causing bizarre choreiform movements and later mental deterioration. The most consistent lesion in Huntington's chorea is a loss of GABA-containing neurons, with a normal or increased density of dopaminergic nerve terminals. The choreiform movements are in some respects an exaggeration of the normal associated movements and gestures which are notably absent in Parkinsonism. Huntington's chorea and Parkinsonism are in some respects mirror-image disorders and, in fact, over-enthusiastic treatment of Parkinsonism with levodopa can lead to excessive movements similar to those in Huntington's chorea.

Choreiform movements also occur in *Sydenham's chorea* (*St. Vitus dance*), a hypersensitivity manifestation following streptococcal throat infections and in *Wilson's disease* (*hepato-lenticular degeneration*), a hereditary disease where copper is absorbed to an excessive degree and deposited in and around the lentiform nuclei of the basal ganglia and in the liver.

## TREATMENT

**Parkinsonism**

(i) **Levodopa.**—The finding that the dopamine content of the brain was reduced in patients with Parkinson's disease suggested that administration of levodopa, a precursor of dopamine, might help these patients by making good the deficiency. When used therapeutically in Parkinson's disease, levodopa reduces rigidity and aids in the initiation and execution of voluntary movements. It usually has little effect on the tremor. Because of its side-effects, its use requires constant supervision.

(ii) **Anticholinergic drugs.**—Drugs such as atropine are reputed to reduce rigidity in Parkinson's disease though not the disturbances of voluntary movement and tremor. Anti-histamine drugs have had a similar vogue. There is no good pharmacological reason why these drugs should have any effect on muscular activity; atropine does not block the effect of acetylcholine on skeletal muscle.

(iii) **Surgery.**—This may help patients who do not respond to drug treatment. Minute destructive lesions in the globus pallidus or the ventrolateral nucleus of the thalamus usually decrease rigidity and tremor but do not affect the disturbances of voluntary movement. If the disturbances of voluntary movement are the most incapacitating part of the illness, surgery should not be used. The surgeon usually applies local cooling to the area he contemplates destroying to see if it alters the condition. Destruction is carried out only when a positive result is obtained. The operation is usually carried out with local anaesthesia so that the patient may remain conscious and help the surgeon in evaluating the effects of the procedures.

(iv) **Sedatives.**—These help tremor, partly by depressing the central nervous system and partly by relieving anxiety and emotional stress which aggravate tremor.

(v) **Physiotherapy.**—Active and passive movements of the muscles are useful in retaining muscular skill and reducing rigidity.

### Other Extrapyramidal Disorders

A variety of drugs have been tried in these varied conditions; as with Parkinsonism treatment cannot be directed at the fundamental cause and is of limited value. In Wilson's disease where excessive copper deposition is a problem, attempts may be made to increase the excretion of copper from the body.

# 44. DISTURBANCE OF CEREBELLAR SYSTEM FUNCTION

## Definition

The cerebellar system consists of the cerebellum and its afferent and efferent connections. It is essential for the execution of smooth co-ordinated movements. It does not exert its effect directly on the muscles. The efferent connections of the cerebellum are to the neurones of the pyramidal and extrapyramidal motor systems (Fig. 87). Its effects are mediated through these systems. The main afferent connections of the cerebellum are from the motor cortex and from the muscles.

Information passing down the pyramidal tracts to the muscles is sent via collaterals to the cerebellum. The cerebellum thus receives a "copy" of the *instructions* for voluntary movement sent to the muscles. The cerebellum also receives information about the *actual performance* of the muscles. This information is picked up by proprioceptors in the muscles and elsewhere and conveyed to the cerebellum via the spinocerebellar tracts. The cerebellum can thus compare "instructions" with "performance" for any voluntary movement. Any discrepancy between the two is corrected by the cerebellum via its efferent pathways to the motor systems. The chief characteristic of damage to the cerebellar system is a lack of precision in voluntary movement.

The cerebellum also plays a part in maintaining equilibrium. It receives

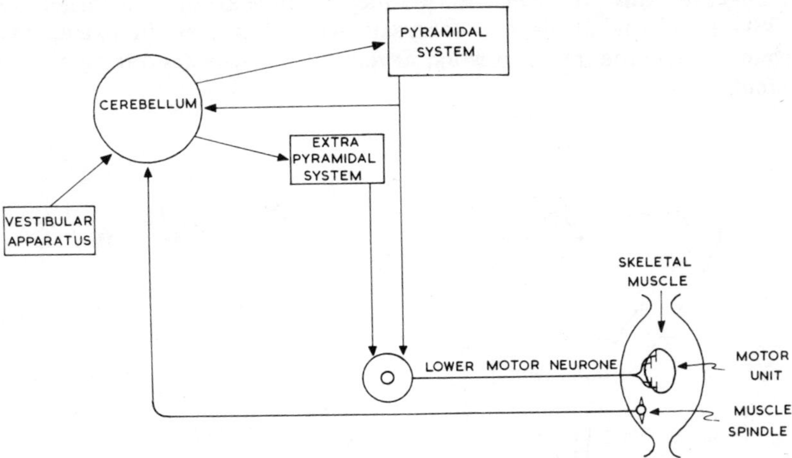


FIG. 87.—Afferent and efferent pathways of the cerebellum. The cerebellum receives information about instructions for voluntary movements via collaterals from the pyramidal system. It also receives information about the execution of these instructions via afferents from muscle spindles and other proprioceptors. The cerebellum compares the instructions with the performance. If there are differences, these are corrected via efferent connections with the pyramidal and extrapyramidal systems.

afferent information from the vestibular apparatus and co-ordinates many of the postural reflexes in response to the stresses of gravity and angular and linear acceleration on the body.

### Effects

Unlike damage to most parts of the brain, damage to the cerebellum causes disturbance of function on the same side as the damage.

**Disturbance of Voluntary Movement**

The general term to describe this is *cerebellar ataxia* (Gr. *ataxia* = loss of order), i.e. inco-ordination or clumsiness. Movements tend to be badly "damped"; they are slow to start and slow to finish. Sequential movements become disjointed. Unlike the ataxia seen with loss of sensory information from the limbs (sensory ataxia), it is not made worse by closing the eyes. Clinical manifestations of cerebellar ataxia or clumsiness include:

*Dysdiadochokinesia* (Gr. *diadochos* = repeated).—This is the inability to make smooth and rapid alternating movements such as supination and pronation of the forearm as is required when using a screwdriver.

*Dysmetria*.—This is the inability to execute the correct range of a voluntary movement with precision; the force of the movement is inappropriate. A patient with cerebellar disease when asked to pick up an egg may well break it. When asked to touch the examiner's finger with his own, his pointing finger tends to pass by the examiner's finger (*past-pointing*).

*Intention tremor*.—This is a coarse irregular tremor which develops in a limb when it approaches its target in the course of a voluntary movement. The movements are badly "damped" so that they oscillate around the target position (Fig. 88). This type of tremor contrasts with that seen in extrapyramidal disturbance where the tremor is worst at rest and disappears during a voluntary movement.

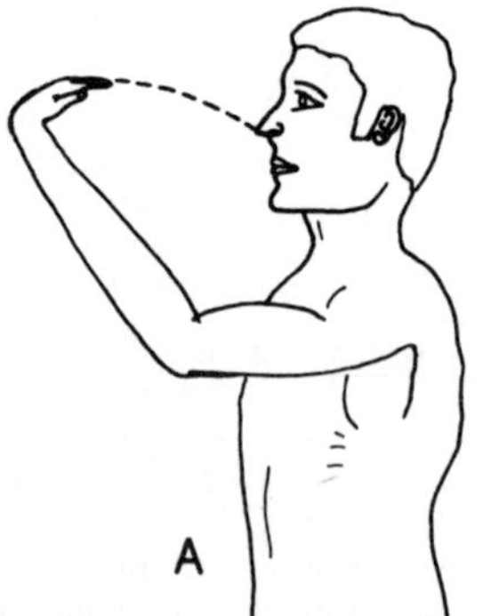


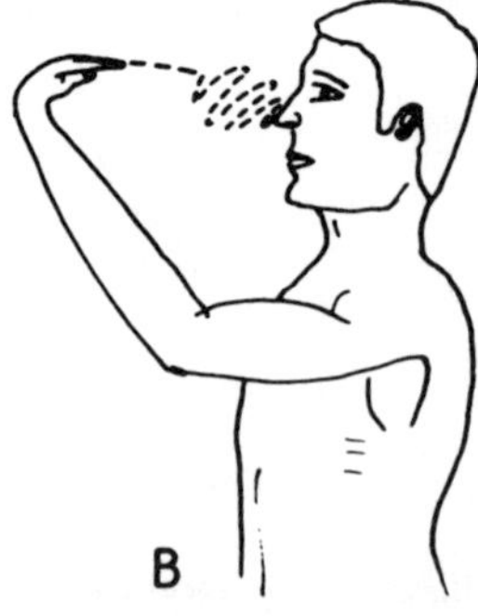


FIG. 88.—When asked to touch his nose with his finger, a normal person (A) can execute a clean arc of movement. With cerebellar damage (B), the finger oscillates clumsily as it approaches the target position (intention tremor).

*Rebound.*—When the patient flexes a limb against resistance and the resistance is suddenly removed, the limb continues to flex in an uncontrolled fashion. This is called rebound. A normal subject is able to stop flexion almost immediately the resistance is removed.

*Speech disorders.*—Lack of precision and inappropriate muscular force used for voice production results in speech having an irregular, jerky quality (staccato speech). It may be too loud and each syllable may not follow the preceding syllable in a smooth fashion. Slurring of speech can also occur.

*Nystagmus* (Gr. *nystazein* = to nod).—When the patient looks to one side, the eye movements are poorly damped so that they tend to oscillate around the point of fixation. It is usually most marked in the horizontal plane.

*Cerebellar gait.*—The gait is clumsy and the patient tends to stagger as he walks. When the damage is on one side, the patient deviates to that side.

### Disturbance of Muscle Tone

The normal activity of the cerebellum seems to facilitate muscle tone. When the cerebellum is damaged, tone is low and the muscles are flaccid. Despite this, the tendon (stretch) reflexes are still present. In fact, on tapping the patellar tendon the quadriceps contracts and the foot tends to oscillate in a pendular fashion for a longer time than normal. *"Pendular knee jerks"* are a sign of cerebellar disease (Fig. 89). There is no simple physiological explanation of these disturbances.

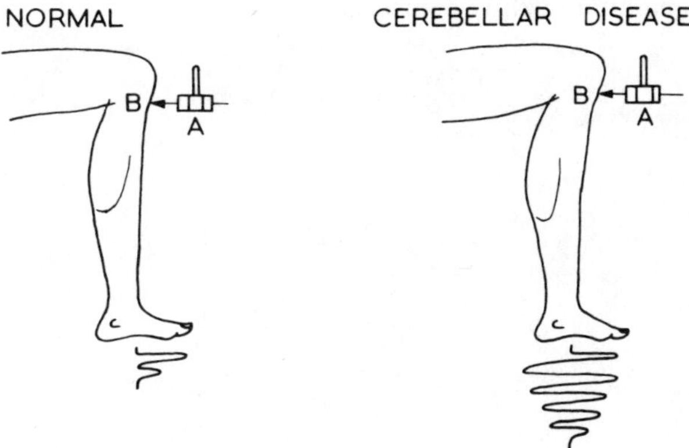


FIG. 89.—The pattern of reflex movement of the foot produced by tapping the patellar tendon (B) with a patella hammer (A) in a normal person and a patient with cerebellar disease. In the normal person the movements are rapidly damped after about two oscillations. In cerebellar disease the foot oscillates for longer, like a pendulum.

### Disturbance of Equilibrium

This is characteristically seen when the posterior part of the cerebellum, the flocculo-nodular lobe, is destroyed. This is the part which co-ordinates the reflexes originating in the vestibular apparatus. The patient may have an abnormal posture and lean toward the side of the lesion. He has difficulty in maintaining the upright posture and may fall towards the side of the lesion. He has difficulty in withstanding changes in gravitational stress. Children with tumours in the flocculo-nodular lobe are often brought to see a doctor because of their tendency to fall.

### Causes

Damage to the cerebellar system may be caused by interference with its blood supply, diseases such as multiple sclerosis which cause patchy demyelination, developmental abnormalities in the brain and a great variety of less common conditions.

### Investigations

Diagnosis of cerebellar dysfunction is made on clinical evidence. Malformation and tumours can be identified radiologically.

### Treatment

If possible the cause should be treated. However, in most cases the diseases are progressive and the cause is not amenable to cure. For such patients, physiotherapy is the main help. The patients should be encouraged to practice balancing exercises and skilled movements because they seem to reduce the degree of disability. Presumably these efforts enhance compensating mechanisms using remaining healthy neurones.

# 45. DISTURBANCE OF SENSORY SYSTEM FUNCTION

## Definition

Sensory information travels over a sequence of three sensory neurones (Fig. 90). The peripheral end of the primary sensory neurone is adapted to form a sensory receptor. Each receptor is especially sensitive to one variety of energy (its "adequate stimulus") which it transduces into electrical impulses. The primary neurones travel into the spinal cord via the posterior roots to synapse with secondary sensory neurones. In the peripheral nerves, fibres conveying all types of sensory information travel together. Inside the spinal cord they separate. Those serving conscious muscle-joint sense and half touch travel up the posterior column of the same side to the brain stem. They then cross to the opposite side and travel up in the medial lemniscus to the thalamus. Those serving pain, temperature and the rest of touch cross and travel up the spinothalamic tracts on the opposite side. In the brain stem they rejoin the fibres serving conscious muscle-joint sense and proceed to the thalamus. Most fibres

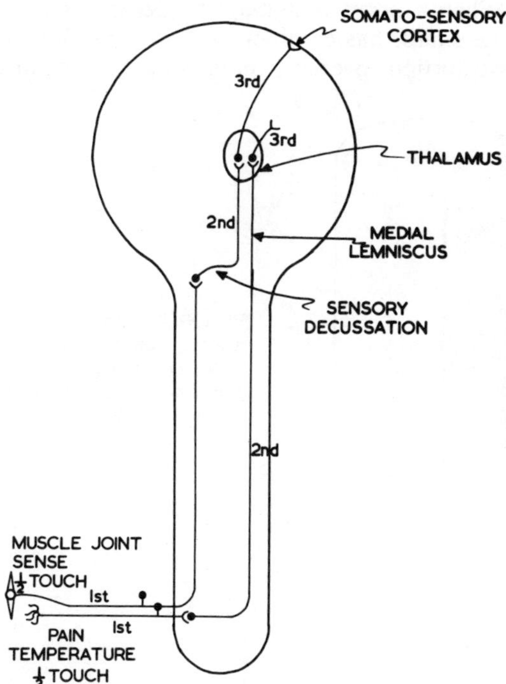


FIG. 90.—Sensory pathways in the central nervous system. It is thought that there are primary, secondary and tertiary neurones serving all modalities of sensation, but the destination of the tertiary neurones serving pain and temperature is not known.

are thought to synapse in the thalamus with tertiary sensory neurones which travel to the sensory part of the cerebral cortex. Sensory fibres serving unconscious proprioception travel up the lateral columns of the same side in the spinocerebellar tracts and then enter the cerebellum.

The effects of damage to the sensory system depend on what part of the system is damaged, but it is possible to make some generalizations.

## EFFECTS

Damage to the sensory system can result in *pain, abnormal sensation* (paraesthesia, Gr. *Para* = beyond, *aisthesis* = perception) and *loss of sensation.* The degree of each depends on the level of the damage.

### Damage to the Peripheral Nerves

(a) **Pain** is common with peripheral nerve damage. It is called *neuralgia* (Gr. *neuron* = nerve; *algos* = pain), and is felt in the distribution of the nerve. When it has a burning quality it is called *causalgia* (Gr. *kausos* = heat).

(b) **Paraesthesiae,** such as tingling, itching, numbness and hot flushing are also common in the distribution of the nerve.

(c) **Loss of sensation** usually involves all varieties (modalities). The loss may be partial or complete. Most skin areas are served by more than one cutaneous nerve (Fig. 91). When a nerve is damaged, the area served by it exclusively, which may be quite small, has complete loss of sensation. In the areas where adjacent nerve distributions overlap, there is partial loss of sensation.

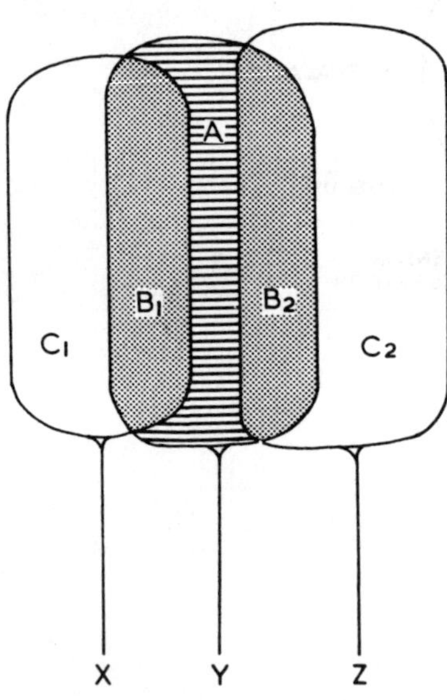


FIG. 91.—When cutaneous nerve Y is damaged by disease there is complete loss of sensation in the area (A) supplied exclusively by it. In the areas partly supplied by nerve Y ($B_1$ and $B_2$) sensation will be partially lost.

*Pain sensation* may be tested by pricking the skin with a pin. Loss of pain sensation is called *analgesia.* When its protective function is lost, signs of *damage* to the skin are often present. *Touch sensation* may be tested by applying a wisp of cotton wool to the skin. Its absence is called anaesthesia (Gr. *aesthesia* = sense of feeling). *Temperature sensation* is tested by applying test tubes containing hot and cold water to the skin.

*Muscle-joint sensation* (proprioception) may be tested in a number of ways. The patient, with his eyes closed, may be asked the direction in which his limbs or digits are passively moved. With his eyes covered, the patient may be asked to stand still with his feet together. If muscle-joint sense is impaired, his stance will be unsteady and excessive muscular action can be seen in his legs. This is known as a *positive Romberg's sign.* Loss of muscle-joint sense is associated with loss of *vibration sense.* The vibrations of a tuning fork applied to a bony prominence cannot be felt in this condition. A knowledge of the position of the limbs is essential for the precise execution of a voluntary movement. When muscle-joint sense is lost, movements are clumsy and are even more clumsy if the eyes are closed. With the eyes closed the patient is not able to compensate visually for the loss of muscle-joint sense. The clumsiness in this situation is known as *sensory ataxia.*

### Damage to Nerve Roots

The effects resemble those due to peripheral nerve damage, except that the abnormality of sensation tends to follow a segmental distribution.

### Damage to the Spinal Cord

When there is damage to sensory pathways in the spinal cord, pain and paraesthesiae are less common than when the peripheral nerves are involved. The loss of sensation does not usually affect all modalities to the same degree. This is because the different modalities of sensation take different routes in the cord and one lesion is unlikely to affect them all. If there is damage to the posterior columns, muscle-joint sense is lost to give sensory ataxia and light touch is impaired. If the anterior and lateral columns are involved, pain and temperature sensation are lost. This occurs in syringomyelia where the degeneration around the central canal impinges on the pain and temperature fibres as they cross to the opposite side of the cord. Light touch, however, is preserved and this is known as *dissociated anaesthesia.*

### Damage to the Brain Stem

As all the sensory fibres come together again in the brain stem, damage to the sensory tracts at this level tends to cause complete loss of sensation on the opposite side of the body. In the high brain stem, damage may cause loss of sensation in the face on the opposite side as well. Pain is uncommon with damage in the brain stem.

### Damage to the Thalamus

Damage to the tracts at this level gives a characteristic picture, the *thalamic syndrome.* There is diminution or loss of all sensation on the opposite side of the body as in brain stem injury, but spontaneous pain is common and may be excruciating. The threshold for pain may be raised on the affected side but when it is elicited it may become almost unbearable. It is accompanied by an excessive emotional reaction which makes it especially unpleasant.

**Damage to the Sensory Cortex**

This never leads to complete loss of sensation. All sensations can be experienced but sensory judgement and analysis are impaired. It would seem that this part of the brain evaluates the sensations which enter consciousness somewhere else. Sensory judgement can be tested as follows. If the patient is given an object to feel, he will be able to feel it but not identify it by its tactile characteristics. This defect is called *astereognosis* (Gr. *stereoz* = solid; *gnosis* = knowledge). The patient also loses the ability to realize that he is being touched at *two* points close together on the finger tips with a pair of dividers. This is loss of *two point discrimination.*

Disease of the parietal lobe can also lead to the phenomenon of *sensory inattention.* Visual fields, hearing and general sensation are normal when each side is tested separately. However, when stimuli are presented to both sides simultaneously, the stimulus on the affected side is persistently neglected.

Pain is not a feature of sensory cortex damage, nor is loss of pain sensation, suggesting that pain is not represented at this level. However, paraesthesiae such as tingling do occur.

## Causes

The factors which damage motor nerves also damage the sensory fibres. Thus peripheral nerves are commonly damaged by peripheral neuritis, the spinal cord tracts by demyelinating and other diseases, and the tracts in the brain by vascular accidents.

## Treatment

The treatment, if possible, is that of the cause. When the sensation of pain is lost, special precautions have to be taken to prevent tissue damage.

# 46. PAIN

## Definition

Pain is a sensation generated by neurones somewhere in the brain when impulses are carried to them in afferent fibres from pain receptors. Though the sensation is generated in the brain, it is projected to some part of the body. The projected sensation may be *well localized* to the area of stimulation, *badly localized and diffuse* in the area of stimulation or *referred* to a site remote from the area of stimulation.

Pain is difficult to describe except in terms of itself, e.g. a sore pain or a stabbing pain. It is also difficult to measure, because it is a subjective sensation and can be gauged only by the person experiencing the pain. It is not possible to have an impartial observer corroborate the severity of sensation. The subject may attempt to assess pain by grading it as mild, moderate or severe or according to some other grading system. The observer may attempt to assess pain by the reaction of the subject to the painful stimulus. Neither system is very reliable. People vary greatly in the way they react to similar painful stimuli. The degree of reaction is modified by early conditioning. Some cultures and ethnic groups tolerate painful stimuli such as childbirth with apparent impassivity. Others react to it with gesticulation and apparent anguish. It is the reaction rather than the stimulus that differs.

It is important to be aware of the two components in "pain", the actual *sensation* component and the individual *reaction* component. It is the reaction component which makes patients unhappy and seek aid from their doctors. The reaction to pain is greatly enhanced by anxiety and alleviated by comfort and sedation. Thus the "pain" arising from a suspected cancer may be much more distressing than experimental "pain" of similar "intensity" produced in a laboratory for a study of pain sensitivity. The study of pain is made very difficult by the sensation and reaction which are entangled. A "pain" becomes less distressing if the patient stops worrying about it.

Most sensations have some emotional overlay. Warmth is usually a pleasant sensation whereas cold is usually unpleasant. Touch on the skin in the form of a gentle caress is nice; touch due to an earwig in the external auditory meatus is nasty. The emotional overlay with pain is particularly strong and unpleasant so that it engenders a strong emotional drive to withdraw from, or inactivate the source of pain.

Pains arising from stimulation of receptors in different areas have somewhat different characteristics. For this reason it is convenient to classify pain as follows.

(a) **Cutaneous pain.**—This is due to stimulation of receptors in skin. It is a sharp, well-localized pain and not associated with much systemic disturbance.

(b) **Deep pain.**—This is due to stimulation of receptors deep to skin in the body wall. The receptors lie in the muscles, tendons, joint capsules, periosteum and pariental pleura and peritoneum. The pain has a dull aching quality and is felt diffusely in the area of stimulation. It is associated with reflex contraction of the local muscles and some systemic disturbance.

(c) **Visceral pain.**—This is due to stimulation of pain receptors in certain viscera such as the gut and heart. The pain is poorly localized and may be referred to a site remote from the area of stimulation. It is associated with considerable systemic disturbance mediated by the autonomic nervous system.

(d) **Headache.**—This is a convenient generic term to describe a wide variety of pains which have in common a projection to some part of the head.

## Effects

### Protection of Tissues

As pain sensation gives rise to a strong emotional drive to withdraw from injurious situations, to forego activities which cause tissue injury and to remove stimuli which are damaging to the tissues, it plays a vital role in protecting the tissues from damage. It supplements the withdrawal reflex, a spinal reflex in which a limb is withdrawn from a stimulus which causes tissue damage.

Pain also tends to reduce activity and movement in injured or threatened tissues. Because movement usually increases the pain in an injured limb, the patient is encouraged to rest the limb quietly. This permits faster and more effective healing and repair to take place. If the metabolic demands of active muscle, including heart muscle, exceed the capacity of the circulation to deliver blood, the resulting pain makes the patient reduce muscular activity to a level that his circulation can support.

If pain sensation is lost, damage to the tissues occurs quickly. In the skin, small cuts and abrasions, which are normally protected from further damage by pain, develop into ulcers which are slow to heal. Skin burns are readily sustained if pain sensation goes. Joints are normally protected against excessive wear and tear by the pain that occurs if the joints are harshly treated. With loss of pain sensation, the articular cartilage breaks down, effusion of fluid into the joint cavity occurs and the functional capacity of the joint is lost. Bone fractures are also a feature of loss of pain sensation.

Of all sensations, that of pain is probably the most beneficent and the most important for survival. Yet people tend to think of pain as a troublesome affliction which should be relieved. This is only valid in the sense that the cause of the pain should be removed or cured. Relieving pain for pain's sake is bad medical practice and should not be attempted before the cause of the pain is known. In the absence of pain, insufficient care is taken to protect the body and activity is not sufficiently limited to promote healing.

### Autonomic Effects

If pain is sufficiently severe, it is usually associated with increased activity of the autonomic nervous system. This is especially true for visceral pain. Nearly any pattern of autonomic activity may occur. Changes in heart rate and blood pressure are common and may result in fainting. Pallor, dilatation of the pupils, sweating and vomiting may also occur. Increased activity in intestinal muscle may in severe cases cause diarrhoea.

The function served by these changes is difficult to see but they increase the unpleasantness of the pain sensation.

### Reflex Increase in Muscle Tone

This is a marked feature of deep pain and its function seems to be the immobilization of the injured tissue. Thus pain arising in a joint leads to a reflex contraction of muscles which limit movement of the joint. Pain arising in the parietal peritoneum causes a reflex contraction of abdominal muscles which limits movement of the abdominal contents.

### Alteration in Mental State

Pain has an alerting effect on the individual which is associated with anxiety. This may be useful in the sense that it puts the subject on guard against further stimuli which might cause tissue damage. However, it impairs the subject's ability to think, concentrate and work.

### Loss of Sleep

The alerting effect of pain makes sleep difficult and sleeplessness is a troublesome feature of diseases which cause prolonged severe pain.

## Causes

### Cutaneous Pain

**Receptors.**—It is thought that painful stimuli act on specific pain receptors rather than by causing over-stimulation of receptors serving other modalities of sensation. Evidence for this idea includes the finding that pain sensation is occasionally lost leaving other types of sensation intact. It has also been noticed that warmth receptors stop generating impulses when the temperature applied to them rises into the painful range.

Unmyelinated nerve plexuses lying superficially in the skin are thought to be specific pain receptors. This is based on the finding that this type of nerve ending is found to be the only one present in areas such as the cornea, tympanic membrane and dental pulp, which can generate only pain sensation. In addition, small samples cut from skin under a point that responds to painful stimuli are found to contain bare nerve plexuses.

**Stimulation of receptors.**—Though there is little direct evidence, it is believed that painful stimuli are those which cause tissue damage and release of chemical agents which depolarize the bare nerve plexuses. A number of chemical constituents from tissues are known to cause pain when applied to the base of a skin blister after the covering epidermis has been cut away. These include histamine, 5-hydroxytryptamine, acetylcholine, potassium ion and certain polypeptides such as bradykinin, that are readily formed in tissue fluid. The idea is that tissue damage leads to release of some of these agents from cells and raises their concentration in extracellular fluid.

This may explain why the effect of a painful stimulus tends to be prolonged compared with that of say touch. With pain, there is a prolonged after discharge which is in keeping with the release of a chemical which is slowly cleared away, rather than with a direct effect of the pain stimulus on the receptor. It would also explain *hyperalgesia,* a common feature of many varieties of pain sensation. In damaged or inflamed tissue, the pain threshold is so lowered that even gentle stimulation of the tissue may cause pain. The increased sensitivity to pain (hyperalgesia) could be due to the depolarizing effects of products of tissue damage released in the affected area. This would bring the pain receptors closer to their threshold for firing so that very modest stimuli might excite them.

The skin forms the first line of defence against agents which damage the tissues and has in general a much lower threshold for pain than the deeper tissues. During surgery it is found that if the anaesthesia is sufficiently deep to allow the skin to be opened, it is deep enough to cover exploration of all the deeper tissues.

The threshold for cutaneous pain varies in different parts of the body. In the cornea it is especially low, i.e. pain can be elicited with a very weak stimulus. It is higher in abdominal skin, and higher still in forearm skin. It is very high in the skin of the sole of the foot and is highest of all in the skin of the finger tips, i.e. a very strong stimulus is required to elicit pain in these parts. This is surprising to many people in that they feel intuitively that their hands and feet are very sensitive to sensory stimuli. Though this is true for tactile stimuli, it is clearly less painful to progress across a gravel path on the soles and palms, than on elbows and knees, not to mention on the stomach.

The threshold for pain rises if the subject's attention is distracted by an interesting task or situation, such as taking part in a game of rugby football, a boxing match or sexual intercourse. It is lowered by agents such as sunburn and previous injury which cause tissue damage. Though a friendly slap on the back may hardly be noticed with normal skin, it may cause excruciating pain if the skin has been made hyperalgesic by sun-bathing to excess.

Compared with most cutaneous receptors, pain receptors show *poor adaptation.* This enhances the protective function of pain sensation because pain does not disappear until the stimulus has been removed.

**Nerve pathways.**—The route taken by cutaneous pain impulses is illustrated in Fig. 92. Two types of fibre act as primary sensory neurones and carry impulses back to the spinal cord, A$\gamma$ fibres which conduct at about 30 metres/second and C fibres which conduct at about 1 metre/second. When pain receptors in the skin are stimulated with a sharp pin, the first pain felt is a bright sharp pain and a little later a more diffuse burning second pain is experienced. A$\gamma$ and C fibres are thought to be responsible for the first and second pains respectively because the two types of pain can be blocked independently. Local anaesthetic agents such as cocaine which first block thin unmyelinated nerves can block the second pain and leave the first. Local obstruction of blood flow which causes hypoxia and blocks larger myelinated fibres first, can block the first and leave the second pain.

The primary sensory neurones synapse in the posterior horn with secondary neurones which cross to the opposite side of the spinal cord and ascend to the thalamus in the spinothalamic tracts. Pain may enter consciousness at thalamic level; the role of the sensory cortex in the appreciation of pain seems to be small. Stimulation of the parietal cortex in man does not cause pain and disease or removal of the parietal cortex does not abolish pain. If there is a tertiary sensory neurone serving pain sensation, its destination is unknown.

**Projection.**—The mechanism whereby impulses are translated into pain sensation by the brain is not understood, but the sensation is projected with great accuracy to the point of stimulation.

This is thought to be a learned phenomenon. The brain learns to associate impulses arising from one particular nerve ending with pain sensation in one particular area. Thus if a flap of chest skin is grafted with its nerve supply to a site on the arm, a painful stimulus applied to the grafted skin on the arm will be misinterpreted as arising in the chest.

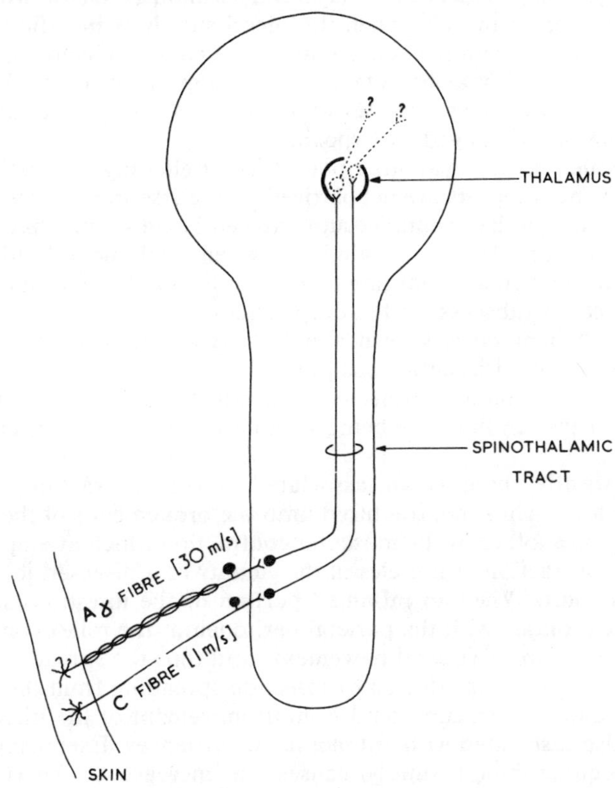


FIG. 92.—Nerve pathways for cutaneous pain sensation. Note that the fibres cross to the other side of the spinal cord soon after entering it.

Pain is sometimes referred by a patient to a limb which has been amputated (*phantom limb pain*). It may be due to irritation of nerve fibres from the limb at the site of amputation, the sensation being projected to the original location of the pain endings.

The pain sensation arising from the skin has a sharp clean quality compared with the dull ache that is characteristic of pain in deeper structures. It is not usually a "sickening" pain nor is it associated with pronounced autonomic disturbances. Nevertheless, the low threshold for skin pain receptors is very effective in maintaining the integrity of the skin.

### Deep Pain

**Receptors.**—Not much is known about these; they are thought to be bare nerve plexuses in the muscle fascia, tendons, joint capsules, periosteum, arterial walls, parietal pleura and parietal peritoneum.

**Stimulation of receptors.**—As with cutaneous pain, chemical products released from damaged tissue cells are thought to depolarize the nerve endings. Injections of potassium salts into muscle give rise to pain. Hyperalgesia is seen when structures in the body wall are damaged.

Accumulation of products of metabolism is another factor which causes deep pain. It is seen in muscle, when the blood supply is insufficient to clear metabolites effectively from the active tissue. The pain-producing metabolite is not known for certain; it may be potassium or histamine or some other product or products. This mechanism ensures that activity is not pursued to a rate at which the tissues are damaged by hypoxia.

A rise in tissue pressure is another potent factor eliciting deep pain. It is not known whether the receptors are depolarized by the rise in pressure directly or by the consequences of tissue damage and reduced blood supply that result from a rise in tissue pressure. However, pain is felt when a volume of fluid is injected intramuscularly or when an enlarging focus of infection leads to the formation of a pus-filled cavity (abscess) in the deep tissues.

**Pathways.**—Not much is known about these but they are thought to be similar to those involved in cutaneous pain.

**Projection.**—Deep pain is projected as a dull diffuse ache to the general area where the nerve endings are being stimulated but may be referred to the skin elsewhere.

Characteristically there is an associated reflex contraction of skeletal muscles in the area. Thus in a fractured limb the broken ends of the bone may be pulled over one another by the muscular contraction which attempts to splint the limb. The contraction of muscles in the vicinity of a diseased joint tends to immobilize the joint. When an inflamed portion of the intestines such as the appendix makes contact with the parietal peritoneum, the reflex contraction of the abdominal wall limits visceral movements and thus the spread of infection. When the parietal pleura is affected by infection spreading from the lung, pain and reflex muscular contraction tend to limit movement of the diseased lung. Deep pain is also associated with autonomic disturbances. Exercising a muscle with an inadequate blood supply causes an increase in heart rate and vasoconstriction and results in a rise in arterial blood pressure.

### Visceral Pain

**Receptors.**—The receptors responsible for visceral pain are again thought to be bare nerve plexuses which are scantily distributed in the walls of the intestines and elsewhere.

**Stimulation of receptors.**—The viscera do not normally give rise to localized sensation and people are normally unaware of visceral existence apart from such general sensations as those associated with micturition and defaecation.

A loop of gut delivered through a locally anaesthetized wound in the abdominal wall does not give rise to pain if it is nipped, cut, burned or ground up. However, there are several ways in which pain may be elicited from the intestines. These included distension of the gut by for example, the inflation of a balloon in its lumen, traction on the mesentery, and stimulation of intestinal wall made hyperalgesic by congestion, irritation or inflammation.

Distension is probably responsible for the pain and discomfort that follow the ingestion of a large meal after part or all of the stomach has been removed. It would also explain the pain that occurs when the bladder or ureters become distended with urine when urinary outflow is obstructed. Traction on the mesentery is probably not a cause of abdominal pain except in the artificial conditions of abdominal surgery under local anaesthesia. It may result from direct stimulation of the sensory nerves as they converge in high density to enter the posterior abdominal wall.

Lowering of the pain threshold by hyperalgesia is probably responsible for most of the pain arising from the intestines. In people with direct openings (fistulae) from the abdominal surface into the stomach, it has been found that healthy gastric mucosa may be pinched or given an electric shock without eliciting pain. However, if the mucosa is irritated and inflamed, these stimuli do elicit pain. Thus when the bowel is inflamed, stimuli such as the movement of food and peristalsis, of which the person is ordinarily unaware, give rise to pain. Very vigorous peristaltic movements, especially of inflamed or congested viscera give rise to very severe spasms of pain known as *colic*.

In a healthy person, the abdominal contents can usually be pressed or squeezed by abdominal palpation without causing an unpleasant sensation. When the viscera are congested or inflamed, such pressures give rise to pain. Pain on pressure is called *tenderness* and is important in diagnosing mischief in the viscera.

Some viscera such as the liver, spleen, kidneys and lung tissue do not give rise to much pain when damaged by disease. Others such as the heart, bladder, ureters and intestines can be the origin of excruciating pain.

**Pathways.**—Visceral pain impulses are carried back to the spinal cord with autonomic nerves. Most of the viscera are served by fibres which travel with the sympathetic system; fibres which travel with the parasympathetic nerves are responsible for pain only in the oesophagus and upwards and in the sigmoid colon downwards together with the base and exit from the bladder.

Thus cutting the splanchnic (abdominal sympathetic) nerves on both sides abolishes pain sensation from most of the intestines whereas bilateral cutting of the vagus nerves does not. Patients with spinal cord lesions in the lower thoracic segments can still experience abdominal pain because the fibres responsible for abdominal pain pass with sympathetic trunks into the thoracic segments of cord.

The pathways for visceral pain in the central nervous system are not well documented but are thought to be similar to those for cutaneous pain.

**Projection.**—Visceral pain is felt vaguely and diffusely in the midline and/or referred to an area remote from the site of stimulation. Thus pain in appendicitis before the parietal peritoneum becomes involved is felt as diffuse spasms of pain (colic) somewhere in the area of the umbilicus. Pain arising from the liver and gall-bladder may be felt in the upper abdomen or over the right scapula. Pain from the kidney may be felt on the affected side or groin. Pain from the heart may be felt in the chest or in the left shoulder and arm.

There are many theories to explain the projection of referred pain in visceral diseases. Though there is no agreement on the exact mechanism, they all depend on the fact that referred pain is felt in the skin area whose nerve supply is derived from the segments of spinal cord which supply the offending viscus. It has been suggested that the input of visceral pain impulses into a spinal cord segment increases the local excitatory state so that activity "spills over" to cause firing of neurones normally excited by inputs from other sensory nerves. Local anaesthetic injected into the skin area to which pain is referred may relieve the pain. This suggests that nerves in the area of referral are important in the genesis of pain.

Visceral pain is usually accompanied by strong autonomic activity and in that sense is a "sickening" pain. Pallor, sweating, nausea and vomiting are common. The blood pressure may fall. Visceral pain does not usually elicit the reflex muscle contraction which is such a feature of deep pain. Thus in disease of the abdominal viscera, the abdominal muscles are usually relaxed until the parietal peritoneum becomes involved.

### Headache

**Receptors.**—Pain receptors are found inside and outside the cranium. Inside the cranium they occur in the walls of arteries and the meningeal coats. The brain substance itself is devoid of pain sensation. Outside the cranium pain receptors are found in most tissues as elsewhere in the body wall. They are thought to be bare nerve plexuses.

**Stimulation of receptors.**—Intracranial pain receptors are stimulated by stimuli which cause *traction* on the blood vessels or meninges. Thus when cerebrospinal fluid is lost, the sagging brain pulls on these structures and causes headache. Similarly, a tumour growing within the cranial cavity may apply traction to the structures.

*Dilatation of intracranial vessels* stimulates the pain endings in their walls. This accounts for the headache experienced when histamine is injected. The headache in hypertension and meningitis has also been attributed in part to this mechanism. This headache has a throbbing quality synchronous with the heart beat and is relieved by factors which reduce pulsatile distension of the arteries, e.g. reducing arterial pressure or raising intracranial pressure.

*Raised intracranial pressure* may cause a headache which is worse on lying down and coughing and relieved by sitting up. The way in which the raised pressure stimulates pain endings is not clear. Experimental raising of cerebrospinal fluid pressure in man does not cause headache.

*Dilatation of extracranial vessels* has been cited as the cause of headache in patients with *migraine.* Migraine is a distinct disease entity characterized by periodic severe headaches which may be preceded by an aura of visual and autonomic disturbances and relieved by vasoconstrictor drugs. The mechanism underlying this type of "sick headache" appears to be of great complexity.

The very common headache which occurs in people who are worried and tense, subject to eye strain, or have arthritis of joints in the cervical vertebrae, has been attributed to *spasm of muscles* at the back of the neck. It may be that prolonged contraction of these muscles interferes with local blood flow and results in the local accumulation of pain-producing metabolites. Another common extracranial cause of headache is *sinus disease.*

**Pathways.**—Pain impulses arising from intracranial structures above the tentorium in the cranial cavity are carried to the brain with the first (ophthalmic) division of the Vth (trigeminal) cranial nerve. Those arising from structures below the tentorium travel to the cervical spinal cord with the upper cervical nerves. The subsequent route of these impulses in the brain is not clear but is probably analagous to those serving other varieties of pain sensation.

**Projection.**—The pain associated with stimulation of receptors above the tentorium is referred to the frontal region, the area supplied by the ophthalmic division of the trigeminal nerve (Fig. 93). Pain from sub-tentorial intracranial structures is referred to the occipital region of the scalp supplied by the cervical nerves. The pain may be throbbing or aching and is usually made worse by sudden movements of the head.

Extracranial pain is usually felt in the area where the pain nerve endings are being stimulated.

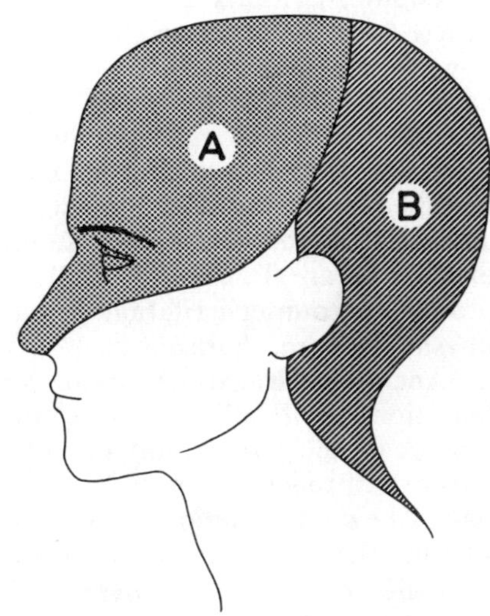


FIG. 93.—Pain arising in intracranial structures above the tentorium is referred to the area A, i.e. that supplied by the ophthalmic division of the trigeminal (Vth cranial) nerve. Pain arising in structures below the tentorium is referred to area B, i.e. that supplied by branches of the upper cervical nerves.

## TREATMENT

It is important to remember that pain is essentially a protective device and that it should not be relieved unless the cause of the pain is known and the decision has been made that the pain is serving no further useful function. Relief of pain in an abdominal disorder of unknown origin may result in disaster because it may mask the progression of a serious condition.

However, depression or temporary elimination of pain sensation is essential for most forms of surgery and has a place in the obstetric management of labour and in dental treatment. It also has a place in cases where there is pain in a progressive illness for which there is no known cure e.g. when secondary deposits of cancer affect the spine and compress nerve roots or in rheumatoid arthritis where progressive inflammatory changes in the joints can cause much pain and misery.

The approaches to the elimination or depression of pain sensation include the use of:

**Limitation of movement.**—A prominent feature of pain sensation is hyperalgesia, the increased sensitivity to pain after injury. In hyperalgesia, trivial stimuli may excite the pain endings. Thus procedures which limit the movement of damaged tissues reduce pain sensation. Splinting of broken limbs, or strapping with adhesive plaster of sprained ankles brings considerable relief. Bed rest is another method of limiting movement. Measures taken to relieve vomiting or coughing may also be helpful.

**Reduction of tissue pressure.**—A rise in tissue pressure can cause severe deep pain. This may occur when infection leads to inflammation and pus formation in spaces where expansion is limited by tight skin, fascial sheaths or bone. Evacuation of pus from such areas reduces tissue pressure and may cause a dramatic reduction in pain sensation.

**Counter-irritants and rubefacients.**—Agents which cause irritation of the skin (*counter-irritants*) such as mustard plasters or a local increase in skin blood flow (rubefacients, L. *rubor* = redness, *facere* = to make) such as a hot pad, have been used for the relief of pain since very early times. Yet there is no satisfactory physiological explanation of their effect. It has been suggested that when pain impulses arise from a viscus, their transmission to the brain may be modified by tne input of impulses from skin served by the same cord segment. There is little valid evidence to support this idea. The ancient Chinese procedure of *acupuncture* (L. *acus* = needle, *puctum* = prick) may relieve pain partly by producing counter-irritation; it cannot be explained on the basis of known physiological mechanisms.

**Analgesic drugs.**—A great variety of drugs are used to depress pain sensation specifically. The spectrum ranges from the weak and relatively safe aspirin to the strong and potentially dangerous morphine. They are more effective in relieving continuous moderate pain than sharp sudden and severe pain. The mode of action of these drugs is not understood and it is even uncertain whether they act in the brain or peripherally at the pain receptors.

**Relief of fear and anxiety.**—Fear and anxiety associated with, or in anticipation of pain, are often harder to tolerate than the pain itself. If pain can be divested of its emotional overlay and looked at more objectively by the patient as just another sensation, it loses much of its terror.

Sympathy is of importance in the relief of pain, and its effectiveness may be greatly enhanced by a strong belief by the patient in the ability of the sympathizer to relieve the pain. The mother's kiss is often a rapid and complete cure for the young child's minor injuries. Even in the adult, strong suggestion may at least temporarily relieve pain. This explains why virtually any procedure which the patient believes will relieve his pain may, in fact, do so for a time.

Tranquillizing and sedative drugs are often used in conditions where pain is expected or is present. Morphine is particularly useful in the relief of pain because it acts not only to reduce pain sensation but also to calm and sedate the patient. It has been found that drugs which relieve pain sensation only and do not relieve the anxiety associated with pain are much less useful in clinical practice. Drugs which induce a state of forgetfulness (amnesic drugs) are also used. Though they do not relieve pain, they allow the patient to forget its occurrence once the pain is over. It seems that pain which is not remembered is of relatively little importance.

**Local anaesthetic agents.**—These drugs relieve or prevent pain when infiltrated into the tissues by blocking conduction in nerve trunks for a limited time. Cocaine and its analogues and derivatives are effective for this purpose. They may be used for extracting teeth, suturing skin wounds or delivering babies without pain if the nerves supplying the appropriate parts are suitably infiltrated.

**Nerve section.**—This is used to produce permanent loss of pain sensation. The other modalities of sensation carried in the nerve trunk that is sectioned are also lost. There is a condition, *tic douloureux,* in which sharp, stabbing pains on one side of the face can make life miserable. The cause is not known and it is sometimes treated by sectioning the sensory root of the trigeminal nerve. Freedom from the tic pain is bought at the cost of loss of all feeling (anaesthesia) on one side of the face.

In cases of incurable cancer where severe pain results from involvement of bones or nerves in the spine and trunk, the spinothalamic tracts are cut as they pass up anterolaterally in the spinal cord. This causes permanent loss of pain and temperature sensation below the level of the section on the opposite side of the body.

**General anaesthetic agents.**—General anaesthetics interfere with pain sensation by causing loss of consciousness. They depress the central nervous system and, in suitable doses, produce unconsciousness by affecting higher centres without arresting activity in the respiratory and other vital centres in the brain stem. They are used for major surgery and can be inhaled e.g. ether, injected intravenously e.g. quick-acting barbiturates or drunk e.g. alcohol.

# 47. SPINAL CORD TRANSECTION

## Definition

THE SPINAL cord has three functions. It carries ascending sensory tracts on their way to supraspinal nuclei. It carries descending motor tracts which synapse with lower motor neurones. It is the integrating centre for a large number of spinal reflexes. When the spinal cord is transected, these functions are disturbed.

## Effects

### Stage of Spinal Shock

Spinal shock is a condition characterized by loss — below the level of the transection — of conscious sensation, voluntary movement and reflex activity. It does not refer to the fall in blood pressure or "shock" in the cardiovascular sense that may or may not accompany spinal injury.

The loss of conscious sensation is due to interruption of the sensory tracts, the loss of voluntary movement is due to interruption of motor tracts, and the loss of reflex activity is due to the loss of facilitation from supraspinal centres which is necessary for normal reflex activity.

The loss of reflex activity causes loss of muscle tone and stretch reflexes. Loss of the defaecation and micturition reflexes leads to constipation and urinary retention with dribbling overflow. If the lesion is above the region ($T_1$-$L_2$) where the sympathetic nerves leave the spinal cord, the loss of sympathetic tone causes a fall in blood pressure.

### Stage of Returning Reflex Activity

Spinal shock lasts for several weeks and reflex activity starts to reappear after this time. Reflex activity can usually be demonstrated first in the flexor muscles and strong flexor reflexes can eventually be elicited. When the skin of the foot is stimulated, the leg flexes to withdraw from the stimulus. The opposite leg may extend. With the loss of supraspinal influences, the reflex activity in the flexor reflex can spread (radiate) to involve many areas of the cord and lead to emptying of the bladder and rectum (mass reflex). Reflex activity later returns to the extensor muscles but is less pronounced than that in the flexor muscles. For this reason, flexion deformities of the lower limbs are likely to occur.

Reflex activity in the autonomic system also returns. Thus the bladder and rectum will empty reflexly without voluntary control when they are appropriately filled. Erection of the penis and ejaculation are possible in the absence of conscious sensation. Blood pressure later returns to more normal values. This return of vascular tone is probably due at least partly to relatively crude spinal cord reflexes which are normally overshadowed by reflexes operating via the vasomotor centre in the medulla oblongata. Spinal reflexes may also lead to unusual and troublesome episodes of sweating.

## Causes

The spinal cord may be transected by injuries which cause fractures or dislocations of the vertebrae. Sometimes local restriction of blood flow to a segment of cord causes death of the segment and a functional transection. This may occur if protrusion of an intervertebral disc or a tumour compresses the cord. Other important causes include pressure from an abscess on the dura mater (epidural abscess) and inflammation in the cord itself (transverse myelitis).

## Investigations

These are directed towards discovering the underlying cause if this is not apparent. In particular, it is important to discover a condition, such as a tumour or epidural abscess, which may be relieved surgically but which may, if not relieved, produce serious permanent damage. Radio-opaque material injected into the spinal cerebrospinal fluid may reveal such conditions on X-ray (*myelogram*).

## Treatment

If impaired spinal cord function is due to compression of the cord causing restriction of blood flow, surgical procedures to relieve the compression are indicated. Many of the effects are reversible with early treatment. Otherwise, because repair of nervous tissue cannot take place within the central nervous system, the treatment of spinal cord transection is mainly to teach the patient how to live with his disability.

In the stage of spinal shock, it may be necessary to drain the bladder with a tube (catheter) inserted through the urethra and provide artificial respiration if diaphragmatic muscle is affected. Good nursing is necessary to prevent the development of pressure sores.

As the reflexes return, physiotherapy is useful to prevent the development of deformities due to abnormalities of muscle tone. Over the years, techniques have been developed which allow handicapped people in wheel-chairs to lead relatively independent lives.

# 48. DEPRESSION OF THE NERVOUS SYSTEM

## Definition

WHEN THE ability of the central nervous system to function is impaired by drugs, lack of blood supply, metabolic disturbances, disease or injury, the effects vary from slight impairment of mental ability at one end of the spectrum to death from respiratory paralysis at the other.

The effects of depression of the nervous system are relatively independent of the cause. Thus raised intracranial pressure, inflammation of the brain (encephalitis), or its meninges (meningitis), alcohol, a fall or rise in body temperature, carbon monoxide poisoning, renal failure, severe diabetes mellitus, head injuries etc. can all give rise to a similar clinical picture. It is therefore useful to be familiar with the general features and it is these features which are discussed in the present chapter.

"Depression" in the psychological sense is a mood characterized by lowering of the spirits and melancholia; voluntary activity is reduced, slow and lethargic. *Psychological depression is not related to the physiological depression discussed in this section;* its origin is not understood but it is not due to general depression or inhibition of neuronal activity.

## Effects

When the nervous system is depressed, all functions are not equally affected. The higher mental skills such as learning ability and memory, which reside in the cerebral cortex, are most sensitive to disturbance and are therefore the first to go. With greater depression the brain stem structures, such as the reticular activating system, become involved and consciousness is first clouded and then lost. Finally, as the lower brain stem and spinal cord are affected, reflex activity gradually disappears and the activity ceases in vital parts, such as the respiratory centre.

For convenience, the effects of depressed neuronal activity are considered in three arbitrary stages despite the fact they really are a continuum and any division into stages is somewhat artificial.

The arbitrary steps are:

1. Stage of impaired mental ability.
2. Stage of depression of consciousness.
3. Stage of depression of reflex activity.

### Stage of Impaired Mental Ability

This occurs when the cerebral cortex responsible for higher intellectual functions is affected. Inability to concentrate, slowness of thought, forgetfulness, lack of initiative and emotional instability are among the first signs of the depression. The cerebral cortex is normally concerned in the suppression of the manifestations of the more primitive emotions such as anger and fear. When the cerebral cortex is depressed, there is an apparent inability to hide these emotions. If the depression comes on slowly, the changes may not be noticed by the patient or his doctor. However, the patient's ability to cope with the world and carry out his daily work is apt to deteriorate, especially if his work has a considerable intellectual content. This is often noticed first by his colleagues at work. The slowing of the higher mental functions is usually noticed also by his family. The affected person adapts to this impairment of function by avoiding tasks which require considerable intellectual effort.

With further depression of cerebral function, the patient may lose the ability to understand and interpret the significance of his sensations. This is called *agnosia* (Gr. *gnosis* = knowledge) and may appear when the sensory association areas of the cortex become involved. It causes peculiarities of voluntary behaviour. If the patient takes up a bar of soap and misinterprets it as a bar of chocolate, he may eat it, instead of using it for washing his hands. *Sensory* or *receptive aphasia* (Gr. *phasis* = speech) is an example of agnosia as it affects speech function. With sensory aphasia, the patient cannot understand the significance of written or spoken speech. For this reason he finds what other people say to be unintelligible. His own speech may be nonsense since he cannot understand and therefore regulate the meaning of his sentences. He may have difficulty with reading if he cannot understand the meaning of the symbols he sees. It is as if the pages were written in a language the patient did not understand.

Sometimes the patient may understand the evidence of his sensations yet be unable to recall the appropriate learned motor act to perform when faced with a particular situation. This condition is called *apraxia* (Gr. *praxia* = ability to perform regulated movements) and appears when the motor association areas of the cortex are affected. Here he understands the nature of familiar objects but cannot remember how to use them. His memory store for the patterns of learned motor responses has been damaged. If such a patient is given a tie, he recognizes the object to be a tie and knows that it is an item of dress. However, he is unable to put it on and knot it even though he is not paralysed in any way. His fumbling attempts to put it on are like those of a child who has seen his father knot a tie but has not yet acquired the skill to do it himself. *Motor or expressive aphasia* is an example of apraxia as it affects speech function. The patient can understand written and spoken speech but he cannot express himself since he cannot recall the patterns of voluntary movements necessary to speak words. In mild cases, this shows itself as a tendency to forget occasional words. In severe cases the patient may be speechless. When shown an object such as a key, they recognize it as a key and become irritated that they cannot remember the word "key" even though it is "on the tip of their tongue". They get around the difficulty by paraphrasing the object, saying that it is "for opening doors". Most people develop some degree of motor aphasia as they grow older. In motor aphasia, intellectual speech is affected before emotional speech. Thus a patient who appears speechless may emit a suitable expletive if a brick is dropped on his toe.

### Stage of Depression of Consciousness

This is the stage when brain stem structures begin to be involved. Consciousness is an awareness of the world and oneself. It depends on the brain, and the parts of the brain which are particularly concerned are the reticular activating system in the brain stem and the cerebral cortex. Damage to the reticular activating system in experimental animals can cause perpetual unconsciousness.

Consciousness is not an all or none affair; it may be gradually lost in a succession of ill-defined stages. The recognition of such depression and the assessment of whether the level of consciousness is rising or falling are important aspects of medical diagnosis.

Impairment of consciousness covers the spectrum from slight clouding of consciousness to deep unconsciousness or coma (Gr. *coma* = deep sleep). The level of consciousness falls in a series of steps or stages which can be recognized clinically. As the level of consciousness rises, these stages are seen in a reverse sequence. In many ways the stages run into one another. Their separation is a rather arbitrary exercise and there is some disagreement in nomenclature. Nevertheless, it is useful to have some mental framework on which to base assessment. One possible classification of the sequence of stages in decreasing consciousness is given below. Practice in the use of this classification may be gained by studying the behaviour of guests at parties where there is considerable consumption of alcoholic beverage. To get the maximum benefit, the observer should confine himself to non-alcoholic drinks.

*Alert.*—This is the description of normal consciousness. The subject is aware of his surroundings, is attentive, remembers familiar facts and complies with instructions in a normal and rational fashion.

*Stage of automatism.*—The patient is aware of his surroundings but may show abnormality of mood, in being unusually elated or irritable. He may show defects of memory and judgement and be unable to remember his actions later.

*Stage of confusion.*—At this stage the patient loses the ability to speak and think in a logical and coherent fashion. He responds to simple orders but may be disoriented in time and space.

*Stage of delirium.*—Restlessness and violent activity are characteristics of this stage. The patient is not capable of rational thought. He may be very troublesome and unco-operative and not comply with simple instructions.

*Stage of stupor.*—In stupor, the patient becomes quiet and uncommunicative. He is conscious but sits or lies with a glazed look and does not respond to orders, either because he cannot, or does not, understand the order. Bladder and rectal incontinence occur in this stage. To the unitiated the relative peace of this stage may seem an improvement on the previous wild stage.

*Stage of semi-coma.*—This is a twilight stage in which the patient passes fitfully into unconsciousness but may be aroused to a stuporose condition by vigorous stimulation.

*Stage of coma.*—This describes the state where the patient is deeply unconscious and does not "wake up" even with vigorous stimuli such as face slapping, nipping or verbal assault.

### Stage of Depression of Reflex Activity

This is the stage when depression of activity involves the neurones in the lower brain stem and spinal cord. These neurones are more resistant to disturbance than other neurones in the nervous system and are therefore the last to be affected.

As the depression deepens, muscle tone gradually diminishes and is finally absent. Eyeball movements disappear and the eyeballs become fixed. The pupils gradually dilate. Superficial and deep reflexes become harder to elicit and eventually disappear. One of the last to go is the corneal reflex, a reflex closure of the eye when the cornea is touched with cotton wool. Body temperature tends to fall because of the absence of muscular activity and the reflex thermoregulatory mechanisms. Respiration is gradually depressed, intercostal breathing usually disappearing before diaphragmatic breathing.

As death approaches, all reflex activity disappears. Deep or superficial reflexes cannot be elicited and the muscles are completely flaccid. The eyes are fixed in a staring position, the pupils being widely dilated and not constricting to retinal stimulation by light. The bladder and rectal sphincters relax and these organs may empty spontaneously. As the vasomotor centre succumbs, the high peripheral resistance normally maintained by vasoconstrictor impulses from the centre is greatly reduced. This results in a large reduction in arterial pressure. Without the normal reflexes, the body cannot compensate for this and circulatory failure develops. The skin becomes cold and pale as thermoregulatory reflexes are lost and the circulation fails. Breathing becomes periodic, i.e. periods of no breathing alternate with periods of breathing. In the final stages breathing is irregular and gasping before it finally ceases.

The heart, which can function independently of the brain, continues to beat for some minutes after breathing stops. However, it eventually stops from lack of oxygen. It is important to remember that if breathing has stopped for more than about 5-10 minutes, the brain is irreparably damaged.

## Causes

To maintain normal function, neurones require a very stable local environment. This is especially true for the neurones responsible for higher cerebral activities. *Any* condition which alters the local environment of these cells will tend to depress their function. Thus depression of the nervous system is seen when there is a fall in body temperature, serum potassium, pH, $Po_2$, cerebral blood flow, blood glucose or sodium concentration. Depression of function is also seen when there is a rise in body temperature, serum potassium, pH, or the pressure of carbon dioxide or carbon monoxide in the blood. Renal failure and liver failure are associated with depression because of the inability in these conditions to excrete or detoxicate all the toxic products of protein metabolism. In severe untreated diabetes mellitus, the water and electrolyte disturbances lead to a progressive depression of brain function.

Head injury, brain damage resulting from a vascular accident and oedema of the brain can also give a similar picture.

Many drugs depress the central nervous system. Alcohol and barbiturates are perhaps the ones most commonly used but all general anaesthetic and sedative drugs have this action.

Ageing tends to produce similar changes on a very much expanded time-scale.

## Investigations

### Electroencephalography

Investigations are primarily directed at identifying the cause of the depression of nervous system function. The electroencephalogram can be used to study the level of consciousness regardless of the cause of depression of function. It is not, however, used routinely (clinical observations serially recorded are used) but may provide an indication of "brain death" when respiration and circulation are being maintained artificially.

When electrodes are applied to the scalp, small potential changes can be recorded which reflect electrical activity in the underlying brain. Recordings are usually made from the frontal, parietal, occipital and temporal regions. Though the patterns of activity vary from place to place on the same side of the scalp, they are usually nearly identical at comparable points on opposite sides. A different pattern at two such points suggests that disturbance of brain function under one electrode is responsible for the asymmetry.

The "brain waves" vary in frequency and amplitude and are classified in terms of their dominant frequency (Fig. 94).

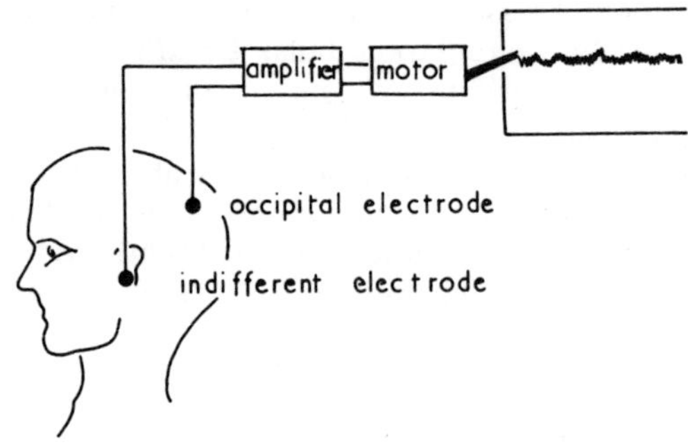


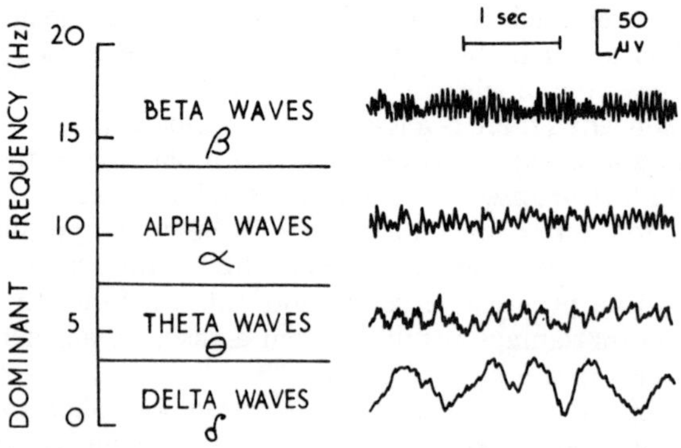


FIG. 94.—Electroencephalography. The potential between an electrode placed on the scalp and an indifferent electrode is suitably amplified to drive a pen motor and make a trace on moving paper. The potential waves are irregular and vary in frequency and amplitude. They are classified in terms of their dominant frequency.

*Alpha waves* are seen in most normal people when they are awake and in a state of mental rest. They are most prominent over the occipital region when the subject's eyes are closed. With active thought or an interesting visual stimulus, the alpha rhythm is "desynchronized" or "blocked" and replaced with a very fast, irregular, low amplitude rhythm.

*Beta waves* are a feature of tension and excitement. They are best recorded over the frontal and parietal regions.

*Theta waves*—These are frequently seen in children especially in the parietal and temporal regions. In adults they may appear for short periods during emotional stress, but persistent theta rhythms suggest a brain abnormality.

*Delta waves* are seen in deep sleep and occasionally in young children but their presence in an awake adult is suggestive of serious brain disease such as a brain tumour underneath the electrode.

The electroencephalogram (EEG) can be used to assess and follow the degree of depression of the nervous system (Fig. 95). When the patient is alert, an alpha rhythm is seen. When he becomes delirious and excited it is converted to a faster beta rhythm. With loss of consciousness, slow delta waves of large amplitude appear. When reflex activity is seriously depressed, the delta waves are reduced in amplitude and periods of activity alternate with periods of suppression (burst suppression). Eventually only short bursts of delta activity occur sporadically in an otherwise flat record. As death approaches, the EEG becomes quite flat except for an occasional ripple.

A completely flat EEG record suggests that recovery of useful brain function is very unlikely, but in certain circumstances, e.g. barbiturate poisoning, such recovery cannot be completely ruled out.

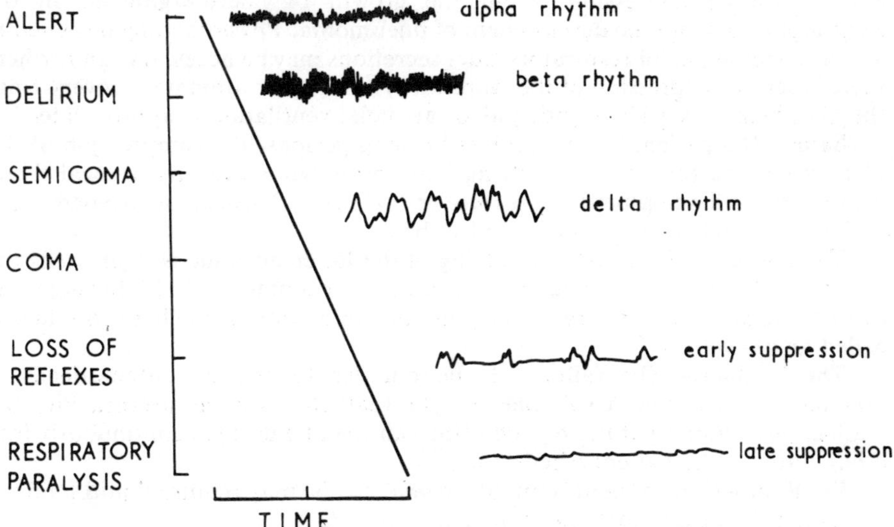


FIG. 95.—Changes in the electroencephalogram with gradual depression of the nervous system. The alpha rhythm of the alert state changes to a beta rhythm in the stage of delirium. A delta rhythm appears when consciousness is lost and is gradually suppressed until death occurs from respiratory paralysis.

## Treatment

The aims of treatment are to remove the cause, stimulate the nervous system where appropriate and provide the special nursing and management that is required by a person who is unable to look after himself.

The nervous system may be stimulated by natural methods or by drugs. All sensory tracts carrying sensory impulses to the thalamus send collaterals to the reticular activating system in the brain stem. Thus all sensory inputs result in impulses being sent to the reticular activating system with a resulting increase in its activity. This in turn leads to increased neuronal activity in the cerebral cortex. Thus a patient with a severely depressed nervous system may be roused if his face is slapped or if he is manhandled or made to walk up and down. A strong sensory input may rouse the patient so that he may breathe or cough more effectively.

Drugs are also used to combat depression of the nervous system. Coffee, because of the caffeine it contains, is a nervous system stimulant. Other more powerful stimulants can be injected, but their value is limited because the stimulation they cause is usually followed by a period of depressed neuronal activity. If there is an antagonist drug to the drug causing depression, it can be injected. For example, the drug nalorphine antagonizes the effects of morphine and may prevent death in morphine poisoning.

When the patient is unconscious, the problems in looking after him are considerable. Attention must be given to the following areas.

**Respiratory system.**—The patient should lie in such a position that his tongue does not fall back and block his airway. The position adopted should also prevent aspiration of vomit into the respiratory tract. The lack of movement and respiratory depression favour the growth of micro-organisms in the respiratory tract and the development of pneumonia. Physiotherapy designed to aid the coughing up of respiratory tract secretions may be necessary, and, where there is serious depression of the respiratory centre but ultimate recovery from the underlying disorder is anticipated, artificial ventilation is appropriate.

**Skin.**—If a patient lies motionless for long periods, the compression of the blood vessels in the tissues overlying bony prominences results in local tissue death and the development of *pressure* or *bed sores*. This can be avoided if the patient is regularly turned or moved in bed.

**The bowels.**—If the reflex emptying of the lower intestine is depressed, the rectum and lower colon can be evacuated by running in fluid through the rectum and allowing it to drain, bringing the faeces with it. Such a procedure is called an *enema.*

**The bladder.**—The failure of the bladder to operate automatically is overcome by inserting a soft plastic tube (catheter) via the urethra into the bladder and retaining it there. The urine can then be drained continuously into a bag hung at the patient's bedside.

**Feeding.**—If the patient is unable to swallow, he may require liquid food via a plastic tube passed into the stomach.

# 49. HYPERACTIVITY IN THE NERVOUS SYSTEM

## Definition

THE TERM "hyperactivity" is used here to describe states in which abnormal electrical activity in neurones disturbs the normal function of the nervous system. These states are usually characterized by involuntary muscular contractions and disturbances of consciousness. An *epileptic fit* (Gr. *epilepsia* = falling sickness) is a well-known variety of such hyperactivity.

The term is not used to describe hyperactivity in the behavioural sense. Thus conditions of obsessively energetic behaviour and activity are excluded as is the excited, overactive behaviour of people with the psychological disorder of *mania* (Gr. *mania* = fury).

## Effects

Abnormal activity in cerebral neurones can cause motor or sensory disturbances as well as disturbances of consciousness. The contribution of each of these to the total clinical picture depends on the area of brain involved. Though they usually occur together, they will be considered separately.

### Motor Disturbances

These may be localized to certain muscle groups or generalized throughout the body. They consist of vigorous involuntary muscle spasms which may be sustained (*tonic*) or intermittent (*clonic*). When generalized, such spasms are called *convulsions.* In attacks or *"fits"* of hyperactivity, the contractions are usually tonic at first so that the body is contorted into a fixed position. Later, as the neurones responsible for the contraction become fatigued, the contractions become clonic to cause violent irregular involuntary movements. Convulsive muscular activity rarely persists for more than a few minutes and this also has been attributed to neuronal fatigue.

When localized to the neurones responsible for one limb, the activity may result in a sustained spasm of the affected muscles or involuntary jerky contractions (*myoclonic jerks*) such as the familiar (normal) *"night starts"* which may awaken a person who is falling asleep. When the electrical disturbance in the brain starts in one area and spreads to involve other areas, a spreading pattern of muscular involvement can be seen. These episodes are termed *focal motor seizures.*

### Sensory Disturbances

These occur when sensory areas of the brain are involved. However, the patient is usually unaware of them because his consciousness is usually depressed during an attack. They are a more prominent feature when the focus of hyperactivity does not involve the reticular activating system and cause loss of consciousness. Peculiar smells, tastes, sounds, sights and other sensations may be experienced and may be a warning sign (aura) of a more severe attack to follow.

**Disturbances of Consciousness**

Consciousness depends on activation of the cerebral cortex by impulses passing up to it through the reticular activating system. Abnormal foci of activity involving the reticular formation interrupt this stream and therefore lead to depression or loss of consciousness. "*Automatism*" may be a feature of the depression of consciousness and the patient may be unable to remember actions he performed during an attack.

## Causes

Hyperactivity in the nervous system must involve an increase in the excitability of the neurones. Even though the change is a local one, the increase in activity generated by an active focus may activate adjacent areas and so the change may spread widely over the brain.

The interplay of excitatory and inhibitory influences on neurones in the central nervous system is little understood. However, any agent which interferes with inhibitory transmitter substances would swing the balance towards an excitatory state. The poison strychnine is thought to work in this way. By reducing inhibition, it facilitates synaptic transmission. Initially the reflexes are exaggerated and more easily elicited. Visual and auditory stimuli are experienced with increased intensity. Eventually the sensory input causes widespread reflex muscular activity which progresses into convulsions.

The excitability of neurones is a function of their transmembrane potential which in turn depends on the state of the membrane and the distribution of ions on either side of it. Agents which depolarize neurones increase their excitability by bringing their membrane potentials closer to the level for firing.

Another postulated mechanism for hyperactivity is, paradoxically, depression of the cerebral cortex. It is known that one of the most highly developed functions of the higher centres is their ability to suppress or inhibit lower centres. With depression of the higher centres, it has been suggested that the unrestrained activity of the reticular activation system causes hyperexcitability of the neurones responsible for the fit.

Whatever the cause of the hyperexcitability, one of the most impressive electrophysiological features is the general synchrony of neuronal discharge during a fit. Impulses from a pacemaking focus are able to spread much more easily than normal to cause repeated widespread synchronous discharges of cerebral neurones.

Hyperexcitability of this sort occurs in a great variety of conditions. Most epileptic fits (about 95 per cent) occur as isolated incidents in the course of another disease, e.g. the convulsions in a child with a high fever or in an hysterical patient who hyperventilates. In these cases, there is an obvious cause, either a disease process in the brain or a change in the composition of the fluids bathing the neurones. The generic term including these cases is *secondary* epilepsy, but they are also referred to as *organic* or *symptomatic* epilepsy. The main problems of epilepsy are encountered in the 5 per cent of cases where there is no obvious abnormality in the structure or composition of the brain. These cases have been called *functional* epilepsy but the terms *essential* or *idiopathic* epilepsy are also used. All three terms indicate that the cause of the epilepsy is not known.

**Secondary Epilepsy**

Convulsions (grand mal seizures) may occur in a great variety of situations which include:

*Lack of oxygen* (hypoxia).—A severe reduction in the delivery of oxygen to the brain is a common cause of convulsions. A convulsive fit may follow sudden arrest of the heart, the fall in blood pressure associated with fainting, vascular obstruction in the brain or breathing an atmosphere with a very low partial pressure of oxygen. It has been suggested that hypoxia causes convulsions by depressing the higher centres which normally inhibit the activating brain stem systems.

*Reduction in blood* $Pco_2$ (hypocapnia).—Overbreathing which lowers the carbon dioxide pressure in blood may result in convulsions. Two factors at least are involved. The low $Pco_2$ leads to constriction of cerebral blood vessels which reduces oxygen delivery to brain cells. Secondly the increased alkalinity of the blood allows the plasma proteins to bind more calcium ions. The fall in the concentration of free calcium ions increases the excitability of muscle and nerve.

*Reduction in blood glucose concentration* (hypoglycaemia).—Glucose is used in brain neurone metabolism and lack of glucose is as incapacitating as lack of oxygen. Convulsions may therefore be seen in diabetic patients who take an overdose of insulin or less commonly in patients who have a tumour of their islet cells which produces excessive amounts of insulin.

*Water and electrolyte disturbances.*—The distribution of electrolytes across the cell wall is responsible for the transmembrane potential and thus the excitability of neurones. It is not surprising therefore that electrolyte disorders may precipitate convulsions. Low serum calcium, magnesium or hydrogen ion concentrations are well-known causes of increased neuronal excitability, tetany and convulsions. Convulsions are also sometimes seen with the overhydration of cells in water intoxication, the salt and water retention that occurs in pregnancy (*eclamptic fits*) and the salt and water depletion that occurs in adrenal failure.

*Increase in body temperature* (hyperthermia).—In conditions such as heat stroke, some forms of malarial infection and some childhood infections, where the body temperature rises rapidly, convulsions may occur. The mechanism is obscure but the rise in temperature may, like hypoxia, inhibit cortical suppressor areas and result in release of the reticular activating system.

*Pyridoxine deficiency.*—Pyridoxine (Vit. $B_6$) forms a co-enzyme required for the metabolism of amino-acids. It is also involved in the synthesis of gamma amino butyric acid (GABA), an inhibitory neurotransmitter substance. Convulsions are seen with pyridoxine deficiency in infants and may result from a lack of GABA causing an excitatory state in the central nervous system.

*Local lesions in the brain.*—A tumour which grows within the brain or presses upon the brain may so irritate it that convulsions occur. Brain injuries sustained at birth or later life may leave scars which act as irritating foci. Infections within the brain (encephalitis) or of the meninges surrounding the brain can act similarly, as may cerebrovascular accidents on occasion.

It is interesting to note that many of the conditions which cause depression of the nervous system can also cause hyperexcitability and fits. Both depression and hyperexcitability are thus ways in which an unfavourable internal environment interferes with the normal functioning of the nervous system.

### Functional Epilepsy

Here the cause of the hyperexcitability is quite unknown. Anatomically, the brain appears to be normal. Physiologically, it shows a predisposition to hyperactivity which often has a hereditary basis. It manifests itself in two forms, *grand mal* and *petit mal*. In both cases the hyperactivity causes "fits" or "seizures" which occur as spontaneous isolated episodes.

In a typical episode of *grand mal*, there is a sequence of phases. The first phase is the *aura*, characterized by the experience of bizarre sensations such as peculiar smells or sounds. Secondly, there is the *tonic phase* when the muscles go into sustained tonic contraction. The patient falls unconscious to the ground with a cry as air is expelled by the contracting muscles past the larynx. The muscular contraction prevents breathing and the patient may become cyanosed. The third phase is the *clonic phase* characterized by alternate contraction and relaxation of the muscles. The vigorous thrashing movements which can last from a few seconds to a few minutes may cause injury. Froth may appear at the mouth. The fourth phase is a stage of *coma* in which the muscles are completely relaxed. The coma usually extends into a period of prolonged *sleep*.

In *petit mal*, the attacks consist of brief periods of loss of consciousness but there are no generalized convulsions. The patient is usually unaware that he has had an attack.

#### Investigations

By far the most important basis for diagnosing epilepsy is an accurate account by a competent observer of the fit. Electroencephalography may assist in certain cases. However, the problem is that the electroencephalogram (EEG) in normal people varies very widely and there is no specific EEG pattern present in epileptics between attacks.

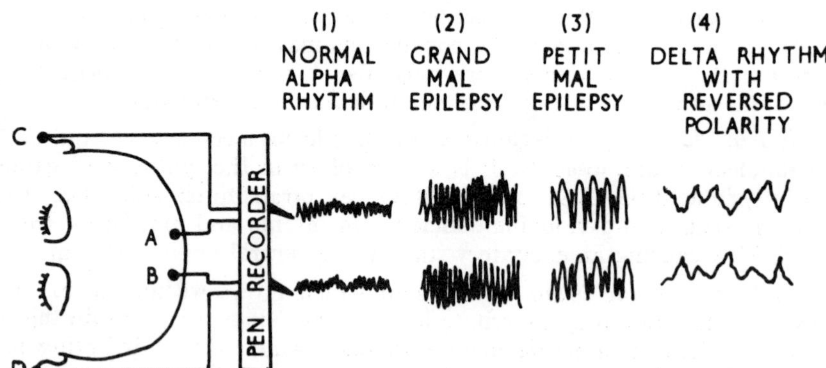


FIG. 96.—EEG records obtained from two sites (A and B) on the scalp. C and D are the indifferent electrodes. (1) Normal alpha rhythm. (2) Synchronous high amplitude, high frequency spikes of grand mal epilepsy. (3) Synchronous spike and dome pattern of petit mal epilepsy. (4) Delta rhythm with reversed polarity suggestive of a brain tumour lying between the electrodes A and B.

EEG examination can help in two ways in the diagnosis of epilepsy. In functional epilepsy, characteristic changes are seen in the EEG if the recording is made *during* an attack. In between times the patterns may be quite normal. In an attack of grand mal, high frequency, large amplitude spikes are recorded all over the scalp (Fig. 96). The synchronous activity all over the cortex suggests that the activity may be involving the reticular activating system in the brain stem, so that all parts of the cortex are "alerted" and rendered synchronously active. In attacks of petit mal, a characteristic spike and dome pattern is seen in the EEG synchronously all over the brain (Fig. 96).

The EEG may also provide evidence of a localized irritating lesion such as a tumour in the brain. Recordings at comparable points on both sides of the brain normally have the same polarity. If there is reversal of polarity between two points it suggests that activity is originating from a focus lying between the electrodes (Fig. 96).

## Treatment

The cause of the hyperexcitability should be identified and, if possible, treated. Otherwise, it is necessary to administer drugs such as barbiturates continuously to depress the nervous system. In this way, the hyperexcitability may be held in check. It is also necessary to modify the life style so that the patient is protected from situations in which he would endanger his or other people's lives if he had an attack. Swimming, car driving and tending heavy machinery may have to be disallowed.

# 50. DISORDERS AFFECTING CEREBROSPINAL FLUID

## Definition

Cerebrospinal fluid (CSF) is found in the ventricles within the brain and the subarachnoid space within the cranium and spinal canal (Fig. 97). The central canal of the cord is normally only a vestigial remnant in postnatal life. CSF has a volume of about 130 ml. It is formed by choroid plexuses in the ventricles of the brain at a rate which has been estimated to be about 0.2 ml/min. If true, this means that the CSF is replaced several times a day. However, the rate of formation may well vary depending on, e.g. postural changes, and changes in arterial and venous pressure. The pressure head set up by the secretory process drives the fluid down through the ventricular system and out into the subarachnoid space. There it travels up over the brain to be reabsorbed by arachnoid granulations into the venous system.

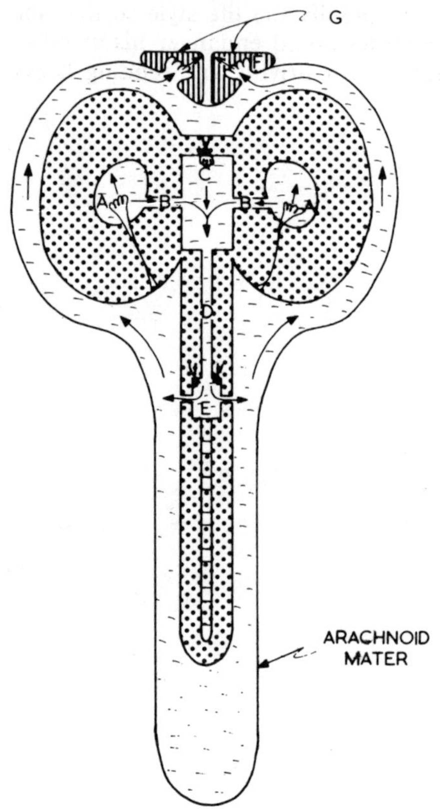


FIG. 97.—Cerebrospinal fluid is formed by the choroid plexuses in all four ventricles. It circulates, as indicated by the arrows, from the lateral ventricles (A) through the foramina of Munro (B), the third ventricle (C), the aqueduct of Sylvius (D) to the fourth ventricle (E). Here it passes out into the subarachnoid space and travels over the surface of the brain and spinal cord to be reabsorbed into venous sinuses (G) via arachnoid granulations (F).

It is a clear fluid and its formation involves ultrafiltration and active secretion by the cells of the choroid plexus. In composition it differs from plasma by having less protein, glucose, potassium, calcium, urea, uric acid, phosphate, sulphate, bilirubin etc., and more sodium, chloride, hydrogen and magnesium. Furthermore, the blood-CSF barrier tends to prevent substances in plasma such as certain drugs and ions, immunoglobulins etc., from entering the CSF. With the subject lying on his side, the CSF pressure recorded by inserting a needle into the subarachnoid space in the lumbar region is usually from 5-15 cm $H_2O$.

The functions of CSF are not concerned with the nutrition of the brain; the rate of formation is much too low. Nutrition is dealt with adequately by the blood vessels which supply the brain. The function of the CSF is to provide *mechanical protection*. It can compensate for changes in the volume of the brain. Shrinkage of the brain is associated with a rise in CSF volume and vice versa. It also protects the brain against the effects of gravity. If fresh brain is set on a table, the deformation due to gravity may rupture its surface. When it is suspended in fluid it experiences an upthrust (equal to the weight of water displaced) which reduces its apparent weight and therefore the effect of gravity upon it. In air, the brain weighs about 1.5 kg whereas it weighs only about 50 g when suspended in water.

This effect also protects the brain against the effects of sudden acceleration and deceleration of the head. The inertia of the brain is reduced by the apparent loss of weight, and the danger of cerebral damage correspondingly decreased. CSF also protects the cerebral blood vessels against the effects of gravity and acceleration. When a person lies down, his cerebral vessels are subjected to an increase in internal pressure. The tendency to distension is countered by an equivalent rise in the CSF pressure.

Various problems can arise in connection with CSF. Firstly, there may be a reduction in CSF volume as may occur following loss of the fluid after a skull injury. Secondly, there may be an increase in CSF volume (hydrocephalus) which may occur following blockage of the flow of CSF through the ventricular system or of absorption by the arachnoid granulations. Thirdly, the CSF may become contaminated by bacterial invasion, blood or other foreign material. Each of these problems is considered separately.

## REDUCTION IN CSF VOLUME

### Effects

Reduction in the volume of CSF gives rise to a *throbbing headache*. This is probably due to stimulation of pain sensitive nerve endings in the vascular walls by distension. When the CSF pressure is lowered, it does not provide the same counter pressure on the vessels to resist distension. The headache is therefore worse with each systole when the distending pressure is greatest.

Accelerations and decelerations of the head also give rise to headache in this situation. This is especially true when some of the CSF is replaced with air. As the fluid is drawn off, the brain does not receive the same support. Its apparent weight rises. In any movement of the skull therefore, its inertia, momentum and impact are greater. The pain probably arises from pulling on the meninges which have a pain-sensitive innervation.

In many cases the headache is associated with nausea and vomiting. This may be a direct consequence of the pain; pain is often associated with autonomic reactions.

### Causes

The volume of CSF falls when fluid is removed at lumbar puncture for examination. To sharpen contrast for skull X-ray pictures, CSF is sometimes removed by lumbar puncture and replaced by air. The loss of CSF in these situations may be greater than intended if CSF continues to leak from the wound in the arachnoid mater. Fractures, especially those involving the base of the skull, may tear the arachnoid mater and let CSF escape into the nasal cavity (cerebrospinal rhinorrhoea; Gr. *rhis* = nose, *rhoia* = flow) or elsewhere.

### Treatment

After lumbar puncture, the patient is made to lie flat in bed to ensure that CSF pressure at the puncture site is low enough to prevent excessive leakage and to maintain an adequate CSF pressure within the skull. To relieve headache after lumbar puncture, the patient is advised to drink fluids. The resulting dilution of the plasma proteins and therefore the colloid osmotic pressure of the blood, reduces reabsorption at the arachnoid granulations and favours the production of CSF at the choroid plexuses.

## INCREASE IN CSF VOLUME (HYDROCEPHALUS)

There are three main varieties of hydrocephalus. *Compensatory hydrocephalus* occurs when the CSF volume is expanded to compensate for a reduction in the size of the brain. This is seen in cerebral atrophy when wasting of the brain substance occurs with death of cerebral tissue. There is no rise in CSF pressure and the increased volume of CSF is beneficial and does not cause problems. *Obstructive hydrocephalus* results when the circulation of CSF is mechanically obstructed so that it cannot reach the arachnoid granulations. Reabsorption of CSF is thus impaired. The obstruction usually occurs at one of the openings or ducts in the ventricular system. *Communicating hydrocephalus* is also due to failure of reabsorption of CSF but is due to loss of function of the arachnoid granulations, e.g. due to inflammation of the meninges; there is free communication of fluid throughout the subarachnoid and ventricular systems. Obstructive and communicatiing hydrocephalus have very similar effects. This is because, when pressure rises in any part of the cranial cavity, the rise is transmitted to all parts of the cavity. From the point of view of intracranial pressure, it does not matter whether the hydrocephalus is communicating or not.

### Effects

The effects depend to a large extent on the age of the patient. In young children whose skull sutures have not closed, hydrocephalus leads to enlargement of the cranial cavity and an increase in the circumference of the head.

Most of the problems with this variety of hydrocephalus are due to deformation to the cranial contents and surrounding structures by the enlargement. The intracranial pressure is only moderately raised and few effects are due to raised intracranial pressure. When hydrocephalus occurs later in life, the reverse is true. After the sutures have closed, the skull cannot expand and failure to reabsorb CSF causes CSF pressure to rise. Most of the problems in the late form are caused by the raised intracranial pressure; deformation of intracranial structures is of less importance.

### Effects Due to Enlargement of the Cranial Cavity

*Skull deformity.*—The skull tends to balloon above the eyes. The bone becomes very thin and on tapping the scalp with the finger, a fluid vibration may be felt passing over the vault. The eye orbits are affected so that the eyes are pushed forwards and downwards. As the child lies with the back of his head on a pillow, the skull tends to be flattened so that it is broader from side to side than from front to back.

*Brain damage.*—The thinning of the brain with enlargement of the ventricles may damage the pyramidal system on both sides to cause a spastic paralysis of both legs and arms. Intelligence may be impaired, though it is frequently within the normal range despite gross thinning of the cortex.

*Cranial nerve damage.*—The deformation of the skull may affect the olfactory nerves to cause loss of smell (*anosmia*), the optic nerves to cause *blindness*, the oculomotor nerves to cause *squint* and the facial nerves to cause paralysis of the muscles of the face.

*Pituitary damage.*—Pressure on the pituitary, distortion of the bone which surrounds it and damage to the hypothalamus may cause a wide variety of endocrine disturbances. Impaired pituitary function may reduce activity in the thyroid, adrenal, and sex glands. Obesity may result from damage to the hypothalamic centres which regulate appetite.

### Effects Due to Raised Intracranial Pressure

*Headache.*—The headache with raised intracranial pressure is often worse during the night when the patient is lying down; it is improved when he sits up. Actions such as coughing or sneezing which raise intracranial pressure exacerbate it. The way in which a raised CSF pressure stimulates pain receptors is not clear; it may involve deformation of the meninges and their blood vessels but raising the CSF pressure in man experimentally does not cause headache.

*Papilloedema.*—The meninges envelop the optic nerve and a rise in CSF pressure is thus applied to these nerves. The consequent compression of the central vein causes oedema of the optic disc which bulges forward into the eye (Fig. 98). If the condition persists for some time, the retina in the region of the optic disc is damaged and causes enlargement of the blind spot.

*Vomiting.*—Traditionally this vomiting is not associated with much nausea and is projectile in character. The cause is thought to be stimulation of the vomiting centre by compression of the medulla oblongata.

*Slow pulse.*—A decrease in blood flow due to compression of the medulla oblongata by the raised pressure causes a reflex increase in arterial blood pressure. The object of this reflex is to maintain medullary blood flow. The rise in arterial pressure causes reflex slowing of the heart through the baroreceptor reflexes.

*Decreased pituitary activity.*—In long-standing cases, the pressure on the hypothalamus and the pituitary fossa may decrease pituitary activity to cause underactivity of the thyroid, adrenal and sex glands.

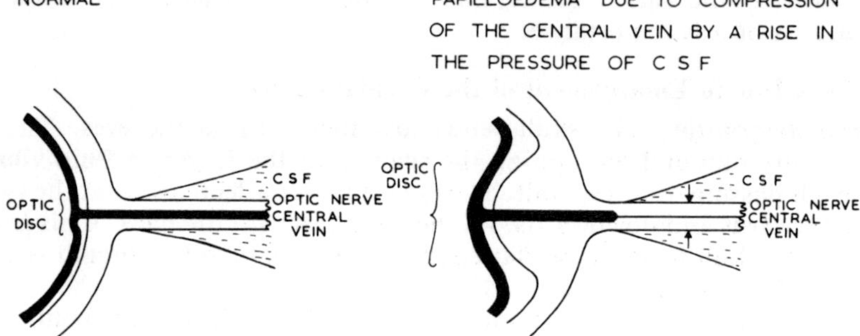


FIG. 98.—Papilloedema due to a rise in cerebrospinal fluid pressure. The subarachnoid space extends for some distance along the optic nerve towards the eye. A rise in CSF pressure thus compresses the vein in the optic nerve. The obstruction to venous drainage causes the veins to distend and oedema fluid to form in the region of the optic disc (papilla).

*Cerebral depression.*—The compression of the brain by raised intracranial pressure may result in a reduction in its blood supply. This leads progressively to impairment of mental function, clouding of consciousness, stupor and coma. Breathing is eventually depressed as the respiratory centre becomes affected. Alternate periods of no breathing (apnoea) may be interspersed with periods of breathing. This is called *periodic breathing* or Cheyne-Stokes breathing after the doctors who first described it. Death is usually due to respiratory failure.

The rise in pressure may occasionally lead to an increased excitability of brain neurones and unco-ordinated outbursts of muscular activity (convulsions).

## Causes

Obstructive hydrocephalus is usually due to a tumour, cyst or congenital abnormality in the brain substance obstructing the free flow of CSF through and out of the ventricular system. Infection of the meninges may also block the foramina opening from the fourth ventricle. Communicating hydrocephalus usually results from infection or haemorrhage in the subarachnoid space. Inflammatory changes and their consequences impair the ability of the arachnoid granulations to reabsorb CSF.

## Investigations

### Radiography

Thinning of the skull by raised intracranial pressure gives it a "beaten copper" appearance on X-ray. The bony cavity in which the pituitary gland lies (sella turcica) may be enlarged. Enlargement of the ventricles can be visualized by X-ray examination if some of the CSF in the ventricles is replaced with air to increase the X-ray contrast. The air can be introduced directly into the ventricle or via a lumbar puncture.

**Phenolsuphonylpthalein Test**

This test provides evidence whether hydrocephalus is of the obstructive or the communicating type.

The dye is injected into a lateral ventricle via a small hole drilled in the skull or fontanelle in a child. In a person whose CSF system is normal, it travels down the ventricular system, out into the subarachnoid space and appears in CSF at lumbar puncture within 2-3 minutes. It is reabsorbed by the arachnoid granulations and appears in the urine in 12-14 minutes. In obstructive hydrocephalus it is retained in the ventricular system and, if it appears at all in lumbar puncture CSF or urine, its appearance is greatly delayed. In communicating hydrocephalus the dye appears in the lumbar puncture CSF quite rapidly but its appearance in urine is greatly delayed.

### Treatment

When the hydrocephalus is communicating, removal of CSF by lumbar puncture is permissible to relieve the effects of raised intracranial pressure. However, *it is very dangerous to do this in obstructive hydrocephalus.* Lowering the pressure in the spinal CSF in the presence of a high intracranial pressure may cause the medulla to be pushed down into the foramen at the base of the skull (foramen magnum). This is known as *coning*. The compression of the medulla may stop blood flow to the respiratory centre and cause death by respiratory failure.

Various surgical techniques have been used with varying success to provide a permanent drainage system for CSF in patients with hydrocephalus. In obstructive hydrocephalus a tube with a one-way valve has been introduced into the brain connecting a cerebral ventricle with a large vein. The valve permits CSF to escape into the vein when CSF pressure rises, but does not allow reflux of blood. In communicating hydrocephalus, attempts have been made to establish an anastomotic connection between the subarachnoid space and the peritoneal cavity at the level of the lumbar spine. As CSF pressure rises, CSF drains off into the peritoneal cavity.

## CONTAMINATION OF CSF

The CSF is normally protected in its composition by the blood-CSF barrier. However, bacterial invasion or leakage of blood from a ruptured blood vessel may contaminate it and give rise to a characteristic clinical picture.

### Effects

**Headache**—This is usually the earliest symptom. It probably results from stimulation of pain nerve endings by inflammatory changes in the meninges. Release of vasoactive material such as histamine with cellular damage may cause dilatation of the cerebral blood vessels. This is associated with a headache which throbs with each heart beat and has a "bursting" character. It may radiate down the back of the neck and spine.

**Head retraction.**—The meningeal irritation causes reflex contraction of the muscles overlying the irritated areas. Thus there is contraction of the muscles at the back of the neck. This is noticed first as *neck stiffness*; attempts to flex the neck passively cause pain due to stretching of the meninges. Later the head becomes firmly retracted and extension of the spine may occur. Extension of the knee may become impossible when the hip is flexed because of the pain it causes by applying stretch to the meninges. This is known as *Kernig's sign*.

**Fever.**—This usually accompanies cellular contamination of the CSF. It probably results from the release of temperature raising substances (pyrogens) from disintegrating cells.

**Cranial nerve damage.**—The cranial nerves have to pass through the subarachnoid space and infection in this area may interfere with their function. Thus there may be difficulty in swallowing due to damage to the glossopharyngeal nerves, squint and unequal pupils due to damage to the oculomotor nerves, facial paralysis due to damage to the facial nerves, deafness due to damage to the auditory nerves or blindess due to damage to the optic nerves.

**Hypersensitivity to sensory stimuli.**—The patient may find bright light unpleasant (photophobia) or be unable to tolerate loud noises, but the reason for this is not clear.

**Depression of consciousness.**—This probably results from depression of brain function by toxins produced by the invading bacteria.

## Causes

Bacteria may gain access to the CSF from the blood, or from a brain abscess, or from an infected cranial sinus. Blood may enter CSF from a blood vessel which ruptures spontaneously at a weak part in its wall or as the result of a head injury.

## Investigations

Examination of a sample of CSF obtained by inserting a needle into the subarachnoid space in the lumbar region (lumbar puncture) is usually required.

*Cells.*—The CSF normally contains less than 5 leucocytes/$mm^3$. In infection it may rise to about 1000/$mm^3$. This is part of the inflammatory response.

*Colour.*—The normal CSF is crystal clear. In infection it becomes cloudy due to an increase in the number of leucocytes and protein it contains. Following haemorrhage, the CSF may be stained red with blood. Some time after the bleeding, the CSF turns yellowish due to bilirubin released from the breakdown of red cells.

*Protein.*—Normal CSF contains virtually no protein, about 0.02 g/100 ml. In infection, the increased vascular permeability allows the protein to escape from the blood. A value of about 0.2 g/100 ml may be seen in infection. CSF protein is also increased with some brain cancers; it may be that the capillaries supplying the cancerous tissue are more permeable to protein than ordinary brain capillaries. Certainly the tissues surrounding a brain cancer tend to be oedematous.

*Glucose.*—The normal value of about 60 mg/100 ml is greatly reduced in many infections affecting the CSF. This is because the invading organisms use the glucose for their own metabolism. It is useful to measure the blood glucose level at the same time as the CSF glucose. The CSF glucose is normally related to the blood glucose level. Hence a value of 60 mg/100 ml in the CSF would be abnormally low if the blood glucose level were markedly raised.

### Treatment

In the case of infection, the first step is to determine the offending organism and find its sensitivity to various antibiotics. The antibiotic to which it is sensitive is then given. It may be necessary to give the antibiotic into the subarachnoid space through a lumbar puncture needle if it does not pass the blood-CSF barrier readily.

## RECENT LEADING ARTICLES

Anguish unremembered. *Lancet,* 1974, **1,** 968.
Benign essential tremor. *Lancet,* 1972, **2,** 471.
Boxing and the brain. *Brit. med. J.,* 1973, **4,** 439.
Causalgia. *Lancet,* 1972, **1,** 1170.
Control of cerebral blood flow. *Lancet,* 1970, **2,** 918.
Control of spasticity. *Brit. med. J.,* 1973, **4,** 751.
Diabetic autonomic neuropathy. *Brit. med. J.,* 1974, **3,** 2.
Dopamine and G.A.B.A. in Huntington's chorea. *Lancet,* 1974, **2,** 1122.
Drug treatment of spasticity. *Lancet,* 1970, **2,** 918.
Insensitivity to pain. *Brit. med. J.,* 1973, **4,** 187.
Internal carotid stenosis. *Brit. med. J.,* 1974, **1,** 258.
Is your pain really necessary? *Brit. med. J.,* 1973, **2,** 323.
Levo dopa. *Lancet,* 1971, **2,** 533.
Management of hydrocephalus. *Lancet,* 1971, **1,** 587.
Measurement of intracranial pressure. *Lancet,* 1972, **2,** 749.
Mute of malady. *Brit. med. J.,* 1973, **1,** 755.
New studies of strokes. *Brit. med. J.,* 1971, **3,** 723.
Organic psychosis. *Brit. med. J.,* 1974, **3,** 214.
Presymptomatic detection of Huntington's Chorea. *Brit. med. J.,* 1972, **3,** 540.
Ptosis. *Brit. med. J.,* 1973, **2,** 679.
Speech on both sides. *Brit. med. J.,* 1972, **2,** 245.
Tests of acupuncture. *Brit. med. J.,* 1973, **2,** 502.
Understanding pain. *Lancet,* 1971, **1,** 1254.

brain capillaries. Certainly, the tissues surrounding a brain cancer tend to be oedematous.

*Glucose.* The normal value of about 60 mg/100 ml is greatly reduced in many infections affecting the CSF. This is because the invading organisms use the glucose for their own metabolism. It is useful to measure the blood glucose level at the same time as the CSF glucose as the CSF glucose is normally related to the blood glucose level. In fact a value of 60 mg/100 ml in the CSF would be abnormally low if the blood glucose level were markedly raised.

*Treatment*

In the case of infections the first step is to determine the infecting organism and its sensitivity to various antibiotics. The [illegible]

*Section VIII*

# DISORDERS OF THE SPECIAL SENSES

# 51. DISTURBANCES OF HEARING

## DEAFNESS

### Definition

Deafness is complete or partial loss of the sense of hearing. The condition is unlikely to cause much disability unless both ears are affected.

### Effects

Despite the fact that deafness is not generally considered nearly such a disabling condition as blindness, it is in some respects more unpleasant. Its main effect is to isolate the individual from the rest of society. This is because most communication at the personal level is by speech. The situation is aggravated by the fact that, while blind people are generally treated with patience and sympathy, deaf people tend to be treated with impatience, and condescension. They tend to become embarrassed and tired of constantly asking people to repeat themselves or write messages down for them and hence to a certain extent lose contact with their family and friends.

As with impairment or loss of any of the senses, there is an increased danger of injury due to failure to detect a warning signal, e.g. in the case of deafness, the sound of an approaching vehicle or its horn.

### Causes

Deafness may be subdivided into *conductive deafness* and *sensorineural deafness*. In conductive deafness, the sound wave vibrations are hindered in their passage to the organ of hearing (the organ of Corti) in the inner ear. In sensorineural deafness either the vibrations fail to be translated into nerve impulses in the sense organ of hearing, or the nervous pathway from the cochlea fails to conduct the impulses. Thus conductive deafness is due to disease of the external or middle ear; sensorineural deafness is due to disease of the inner ear or nervous pathways therefrom (Fig. 99).

**Conductive deafness** is commonly caused by excessive wax in the external auditory canal.

A more serious cause of conductive deafness is disease of the tympanic membrane and auditory ossicles. This delicate mechanism ensures the efficient *transmission* of sound waves from the air of the external auditory canal to the fluid of the cochlea. It also prevents the round window from receiving sound waves at exactly the same time as the oval window (*round window protection*). Such an effect would greatly dampen vibrations in the inner ear.

The extent of the deafness caused by disease of the tympanic membrane and ossicles depends on the type and severity of the damage. A small perforation of the eardrum has little effect on hearing ($<$5 decibels loss). Complete loss of eardrum and ossicles causes severe deafness in that ear (50 decibels loss, the difference between a quiet whisper and loud conversation). Stiffness of the

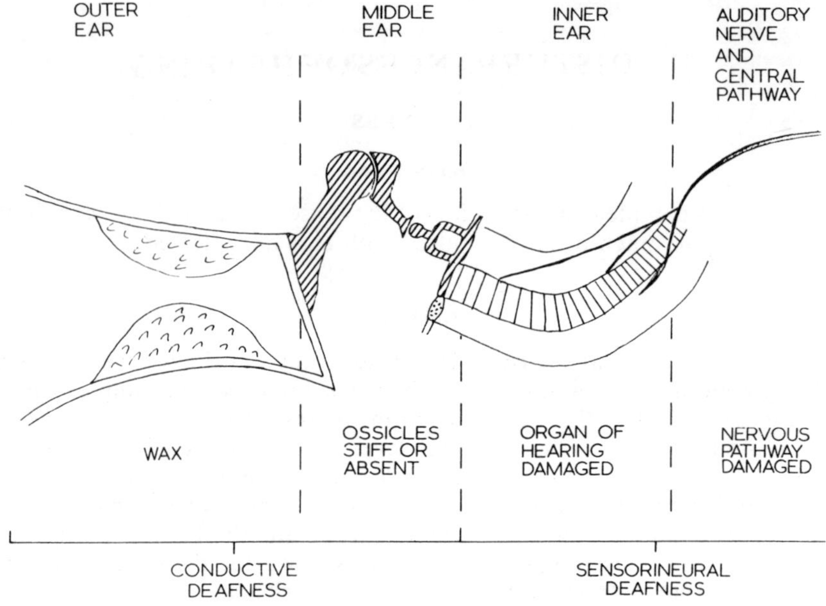


FIG. 99.—A diagrammatic representation of the main causes of deafness.

ossicles can, if severe, also cause severe deafness. One cause of such stiffness is a disease (*otosclerosis*; Gr. *otos* = ear, *sklerosis* = hardening) in which the footplate of the stapes becomes united to the bone around the oval window.

Temporary loss of hearing may be experienced when the pharyngotympanic tube is blocked due to inflammation of its mucosa, usually due to spread of infection from the throat. Air pressure in the middle ear can no longer be equalized with atmospheric pressure during swallowing. The ear drum tends to bulge inwards (as middle ear air is absorbed) and its vibration is reduced. Fluid collects to fill the vacuum.

**Sensorineural deafness** may be due to damage by a variety of diseases to the delicate structures in the organ of hearing. One cause is excessive vibration due to prolonged exposure to loud noise (*traumatic deafness* — Gr. *trauma* = injury). The noise has to be of an intensity of around 85 decibels or more to produce damage. This level is not uncommon in heavy traffic, around airports, near heavy machinery and with music played at high amplification. Typically a person who has been subjected to these noises is somewhat "deafened" (*temporary threshold shift*) just after exposure and hearing then gradually improves. Repeated exposure leads to a permanent loss of hearing.

According to the travelling wave theory, sound waves of high frequency cause only the first part of the basilar membrane to vibrate (at the same frequency as the sound). The lower the pitch the farther up the cochlea does the vibration travel. In traumatic deafness, damage is most marked in the more basal region of the cochlea and hearing loss is most marked for high-frequency sounds around 4,000 Hz.

The deafness of old age (*presbycusis*; Gr. *presbys* = old, *akousis* = hearing) is also associated with wear and tear changes in the cochlea together with nerve fibre degeneration, and again hearing is usually most severely impaired for the higher frequency sounds. The inability to hear high frequency sounds makes it difficult to distinguish between consonants and therefore to understand normal conversation. There is also a blurring of sounds due to impaired discriminatory ability in the ear. Misunderstandings arise when an elderly person with such deafness mistakes key words in sentences. Such people may be even more sensitive to an increase in sound intensity than normal people. They do not like to be shouted at nor do they get much help from a hearing aid, which works by amplifying sound intensity only. They are helped by a clearer and more deliberate enunciation of consonants by the people who speak to them.

Deafness can, of course, be produced by damage to any part of the auditory nerve pathway leading from the cochlea. *Central deafness* is the name of the rare condition in which the ear and its nervous pathways are intact, but damage has occurred in the brain so that sounds, if heard, have no significance for the patient (psychic deafness, receptive dysphasia or sensory aphasia).

## Investigations

An approximate assessment of deafness may be made by comparing the patient's hearing with that of the examiner under the same conditions (assuming the examiner's hearing is normal). Each ear must be tested separately. The test stimulus can conveniently take the form of whispered and then spoken words, making sure that the patient does not lip-read.

Having established that deafness is present, the next step is to decide whether it is of the conductive or sensorineural variety. Two simple tuning fork tests help in the decision—the Rinne and the Weber tests (Fig. 100). They are based on the fact that bone conduction is not impaired in conductive deafness.

In the *Rinne test,* the tuning fork is held beside the ear to test air conduction and then its end is placed on the mastoid process to test bone conduction. Normally, and also in the presence of partial sensorineural deafness, the sound is louder and persists longer in the case of air conduction. Conductive deafness interferes with the efficient middle ear transmission of sound and the sound is heard better by bone conduction.

The *Weber test* is used when one ear is normal and the other abnormal. The tuning fork is placed on the centre of the patient's forehead. A normal person would hear the noise equally in both ears. In a person with conductive deafness, external masking noises are excluded and the sound is localized to the "bad" ear. In the case of sensorineural deafness, the sound is best heard in the "good" ear.

Precise measurement of hearing requires the use of an *audiometer.* This instrument produces sounds of a known frequency and intensity. In order to avoid masking by other noise, the test should ideally be carried out in a sound-proof room. A record (audiogram) is kept of the lowest intensity (in decibels) at which the patient can hear notes of differing frequencies. The patient should not be able to see when the instrument key is depressed to produce the test sounds so that he must rely on hearing alone to tell the operator

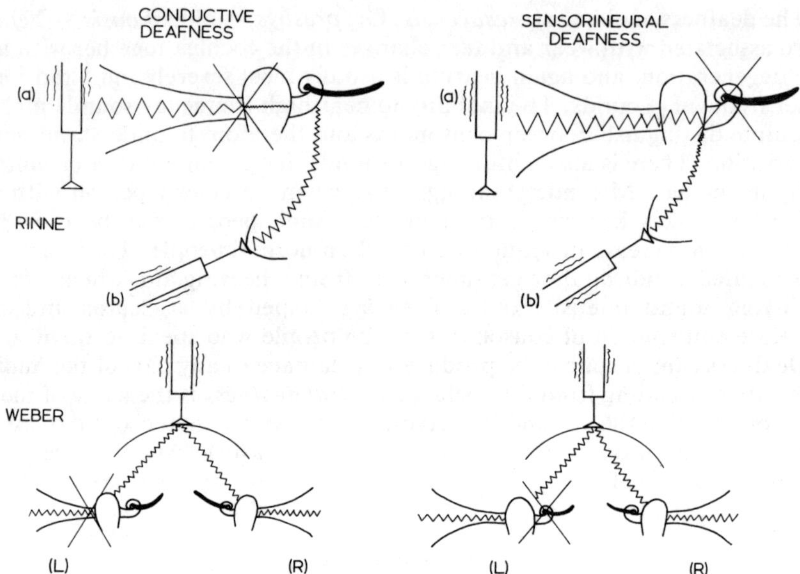


FIG. 100.—Two tests for distinguishing conductive and sensorineural deafness. In conductive deafness, bone conduction (b) is superior to air conduction (a)—Rinne test. In sensorineural deafness, the sound, if heard at all, is heard better by the air conduction route as in normal people. The Weber test is used when one ear is abnormal. In conductive deafness, the sound is best heard in the affected ear (L); the reverse is true in sensorineural deafness.

when the note begins and ends. For convenience, the hearing *loss* is often plotted. A person with normal hearing will have approximately zero loss at all frequencies. If at a certain frequency a person requires a sound of 40 decibels above the normal threshold before he can hear it, he has a hearing loss of 40 decibels.

Some typical *audiogram* patterns are shown in Fig. 101. In conductive deafness, air conduction is impaired relative to bone conduction; in sensorineural deafness bone and air conduction are equally impaired. In traumatic deafness, the hearing loss for high-pitched sounds is very marked, reaching a maximum loss around 4,000 Hz. An audiogram showing this pattern is important evidence in the case of a person claiming that his deafness is due to working at a very noisy occupation.

In addition to testing the hearing of single sounds, it is useful to test the ability to recognize the component parts of actual speech. Monosyllables are played to each ear at varying degrees of loudness. In some cases a person may be able to hear noise at a certain level but may not be able to identify the syllable.

## Treatment

Removal of excess wax in the external auditory canal is a common procedure which may relieve deafness when hearing is otherwise satisfactory.

Damage to the ear drum and ossicles may be corrected by reconstructive surgery. The tympanic membrane is replaced by fascia from the patient's nearby temporalis muscle (*tympanoplasty* — Gr. *plassein* = to form). Ossicles which have been damaged, whose joints have become stiffened or which have become fixed to the oval window (*otosclerosis*) may also be replaced. These delicate operations are carried out with the surgeon viewing the operation field through a microscope.

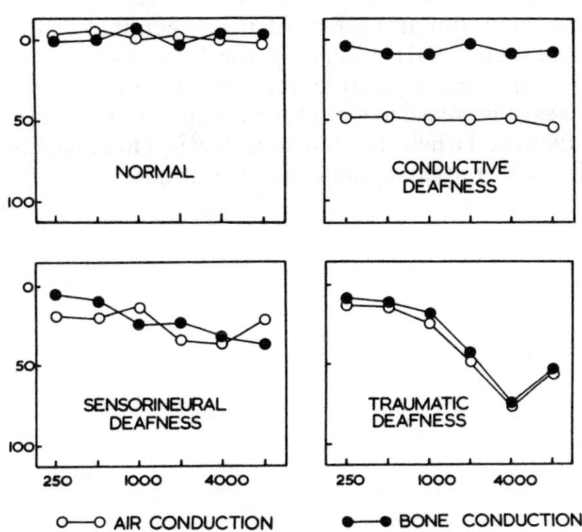


FIG. 101.—Some typical audiogram patterns. The vertical scale shows hearing loss in decibels. The horizontal scale shows frequency/second(Hz).

When the mechanisms of hearing cannot be improved, a hearing aid may help. Hearing aids tend to improve conductive deafness more than sensorineural deafness. Where poor sensory discrimination is the problem, so that sounds are blurred, a hearing aid may be of no benefit.

## DISTORTION OF HEARING

The abnormal hearing patterns to be considered are "ringing in the ears" (tinnitus), and distortion of the sensation of loudness (hyperacusis and loudness recruitment).

*Tinnitus* (L. = *tinkling*) or "ringing in the ears" is a noise (buzzing, hissing, roaring, humming, etc.) which the patient hears independently of external sound. It may be caused by some disease which irritates and hence stimulates the organ of hearing, the auditory nerve or its central connections. Not uncommonly, the cause is obscure, e.g. it may persist after division of the auditory nerve. A form of tinnitus can be produced by obstruction of the external ear. This can be experienced by obstructing the ear with a finger.

Occasionally tinnitus is due to a real sound in the patient's head due to turbulent blood flow in abnormal vessels. Such sounds tend to occur in time with the arterial pulse and may be heard by the examiner using a stethoscope.

Tinnitus should be distinguished from auditory hallucinations. These sounds are more elaborate and organized than tinnitus and may occur in epilepsy affecting the temporal lobe.

*Hyperacusis* is a condition in which sounds appear unpleasantly loud. It is caused by paralysis of one or both of the small muscles (tensor tympani, stapedius) which reflexly pull on the auditory ossicles and damp down their vibrations, especially in the presence of loud noise. The tensor tympani is supplied from the mandibular branch of the trigeminal (Vth) nerve and the stapedius from the facial (VIIth) nerve in the facial canal.

In *loudness recruitment,* a slight increase in sound intensity causes a sudden severe and unpleasant increase in sound sensation. In this condition, the patient may be severely disturbed when shouted at in an effort to make him hear better. It tends to occur in older patients; the cause is unknown.

# 52. DISTURBANCES OF VESTIBULAR FUNCTION

## Definition

The *vestibular apparatus* consists of the three semicircular canals, the utricle and the saccule. The term "labyrinth" includes both the vestibular apparatus and the cochlea. The three semicircular canals are stimulated by rotatory acceleration in the three dimensional planes, the utricle by linear acceleration and by gravity. The function of the saccule is obscure; it may be sensitive to vibrations. Thus the vestibular apparatus provides information on the orientation of the head relative to the pull of gravity and on linear and rotatory accelerations of the head. This information is used to maintain normal body posture by means of the various postural reflexes. These reflexes are particularly important when a person slips or stumbles and must regain his balance to prevent a fall. Visual and general proprioceptive information can partly, but only partly, compensate for deficient vestibular information. Information from the vestibular system is used also in the reflex control of eye movements, whereby vision is fixed on an object despite movements of the head.

Disturbances of vestibular function can be divided into

1. loss of function
2. excessive or unbalanced function.

The second type of disturbance produces the more dramatic effects.

## LOSS OF VESTIBULAR FUNCTION

### Effects

Loss of vestibular function impairs the ability to maintain normal body posture and balance. Visual and general proprioceptive information can compensate in many situations, but in the dark and following a slip or slight stumble, a person lacking vestibular function is much more likely to fall. Such a person who has his vision obstructed when swimming under water, may not be able to tell in which direction the surface lies and may consequently drown.

### Causes

Vestibular system dysfunction is often caused by a disease which also affects the cochlea e.g. Ménière's disease. This is because these structures adjoin each other and are very small (the whole labyrinth is only about 2 cm across), and because the fluids contained in (endolymph) and surrounding (perilymph) the membranous labyrinth in the cochlea are in free communication with endolymph and perilymph in the vestibular apparatus.

Vestibular function is lost also if the vestibular nerve and its central connections are damaged. Such damage may occur when the blood supply to these central pathways is impaired (*vertebro-basilar insufficiency*). Intermittent symptoms of this condition may be produced when flow through the vertebral arteries is interfered with by certain neck movements.

### Investigations

Function in the vestibular system may be investigated by the *caloric test*. In this test, water is run into the outer ear at a temperature just above or below body temperature (e.g. 30°C, 44°C). This warms or cools the nearby vestibular system which, like the cochlea, adjoins the inner wall of the middle ear. The changing temperature sets up convection currents in the endolymph within the membranous labyrinth and these convection currents stimulate hair cells in the vestibular system. Stimulation of this kind provides the brain with false information about head movements and leads to two main consequences. Firstly, nystagmus is produced (Gr. *nystazein* = to nod). This consists of repetitive jerky eye movements due to derangement of the normal reflex which ensures that the eyes remain focused on the desired object, despite changes in head position. Secondly, particularly if the temperature of the water in the ear is much different from body temperature, the stimulation may be so strong that a sensation of rotation, with nausea and vomiting, is produced.

If vestibular function is impaired by disease, the response to caloric stimulation is reduced. This may be quantitated by observing the duration of nystagmus after a known thermal stimulus. For example, a patient may have nystagmus for 2 minutes after caloric stimulation on his normal side, whereas the nystagmus may persist for only one minute on the side where the vestibular apparatus has been seriously damaged. An advantage of the caloric test is that each side can be tested separately. Before carrying out the test, it is necessary to check that the ear drum is not perforated lest the water runs directly into the middle ear; this could provide a very strong nauseating stimulus to the vestibular system and might lead to middle ear infection. Warm or cool air may be used in such a case.

The semicircular canals may also be stimulated by rotating the patient and then abruptly stopping the rotation. When rotation stops, the momentum of the endolymph causes it to swirl round the semicircular canals and vigorously stimulate the hair cells. Again the duration of nystagmus is noted. The canals on both sides are stimulated together in this test, but by positioning the head appropriately, it is possible to stimulate one pair of canals, e.g. the horizontal, at a time.

Ability to maintain balance can also be tested objectively, e.g. by noting the response to body tilting.

### Treatment

If a tumour is impairing vestibular function, improvement may follow the appropriate surgical treatment. However, in many cases where vestibular function is impaired, the cause is unknown and no treatment is known which will restore function.

## EXCESSIVE OR UNBALANCED VESTIBULAR FUNCTION

### Effects

Excessive or unbalanced activity in the vestibular apparatus causes severe symptoms. Such activity may be caused by external stimuli (as in travel sickness), or by imbalance in the impulses received from the vestibular apparatus or its nervous pathways on the two sides. The effects are vertigo, loss of balance, nystagmus, nausea and vomiting. *Vertigo* is a sensation of the external world revolving around the patient (L. *vertex* = whirl) and is the sensation experienced by a normal person who spins round several times and suddenly stops. It is quite different from dizziness or lightheadedness which are vague sensations of unsteadiness.

As well as nausea and vomiting, excessive vestibular stimulation may cause other autonomic disturbances including pallor and sweating, a fall in blood pressure and heart rate, and diarrhoea. These autonomic disturbances are probably mediated by reflex centres in the brain stem. Similar effects can be produced by excessive pain or psychological stimulation.

When the vestibular upset is abrupt and severe, the patient may be quite incapable of standing and falls to the ground, vomiting and suffering from extremely unpleasant waves of nausea and vertigo.

The effects of excessive or unbalanced vestibular function are summarized in Fig. 102.

The symptoms due to abrupt loss of function in one labyrinth decrease gradually as the brain adjusts to the new situation.

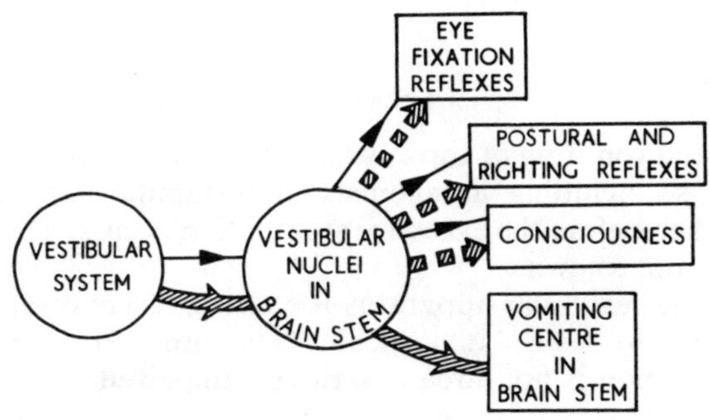


FIG. 102.—A diagram of the vestibular connections. In health (narrow lines) vestibular information is used in eye fixation, postural and righting reflexes.

In disease leading to excessive or unbalanced vestibular function (wide lines), inappropriate information is passed to these regions and the vomiting centre may be stimulated.

## Causes

A common, but fortunately temporary, cause of excessive vestibular stimulation is travel-sickness, particularly sea-sickness. Irregular pitching and tossing is effective in producing this condition.

Disease of the vestibular system and its nervous pathways produces vertigo, nausea, etc. when there is an *imbalance* in the information coming to the brain from the two sides. This imbalance may be due in some cases to increased stimulation from the affected side due to an irritative process but decreased stimulation from the affected side due to decreased sensitivity produced by disease is more common.

Severe disturbance of vestibular function is one of the main features of *Ménière's disease*; a disease of unknown cause, in which there appears to be disturbance of the endolymph within the membranous labyrinth. Involvement of the cochlea causes tinnitus and increasing deafness; simultaneous involvement of the vestibular apparatus causes vertigo, nausea, etc. and increasing loss of vestibular function. The disturbance occurs intermittently and leads to episodes of severe vertigo and nausea. As the condition progresses and the labyrinth is destroyed, attacks of sudden nausea and vertigo tend to be less troublesome; at the same time loss of hearing becomes more marked.

A similar condition can be caused by invading micro-organisms infecting the labyrinth. Various other diseases, such as tumours, may have similar effects.

Vertebro-basilar insufficiency, as referred to in the previous section, is a common cause of temporary or permanent distortion of vestibular function.

## Investigations

When there is disease of the labyrinth (e.g. Ménière's disease), caloric tests show impaired vestibular sensitivity on the affected side and there is usually a parallel impairment of hearing (sensorineural variety). Caloric tests are normal if the disease is due to interference with central pathways (e.g. vertebro-basilar insufficiency).

## Treatment

The nausea and vomiting of travel-sickness and labyrinthine disease may be relieved by drugs, including antagonists of histamine and of acetylcholine, particularly if taken before the excessive stimulation occurs. The mode of action of these drugs is not known.

If disease of the vestibular apparatus is causing severe disabling symptoms, it may be desirable to destroy it surgically. The aim is to spare the cochlea if possible, if its function is not already seriously impaired.

# 53. DISTURBANCES OF VISION

## LOSS OF VISION

### Definition

Loss of vision may be partial or complete. The term blindness implies complete or almost complete loss of vision. From a practical point of view, a person is generally considered blind if his vision, with or without glasses, is so poor that he is unable to do any work requiring vision.

### Effects

The general effects of total loss of vision (*blindness*; *amblyopia* — Gr. *amblys* = dull, *ops* = eye; *amaurosis* — Gr., darkening) are too well known to require description. Information from the eyes is normally used in the reflex regulation of body posture and equilibrium. Blind people are more likely to stumble and fall than normal people. However, they learn to make better use of their other senses such as touch, hearing and muscle joint sense to evaluate their environment. It is questionable, however, whether this amounts to an increase in sensitivity of the other senses; it is more likely to be a learned adaptation to sensory deprivation.

Possibly the most dangerous situation for a patient with impaired vision is when impairment is only partial, particularly if there is marked loss of peripheral vision of which the patient is unaware, because central vision may remain satisfactory. This state of affairs greatly increases the risk of accidents, particularly when driving, and is an argument in favour of accurate testing of the fields of vision and visual acuity in someone who intends to drive.

Loss of vision in one eye presents certain difficulties. It reduces the field of vision and impairs the ability to perceive depth and assess distance (stereoscopic vision). These restrictions make a person with one eye vision more prone to accidents and less able to carry out skilled work involving vision.

### Causes

As with deafness, blindness has two main causes — failure of light to reach the sensory cells (the rods and cones of the retina) and disease of the sensory cells and their neural connections. Obstruction to the passage of light is often remediable by surgery; blindness due to retinal or nerve disease can only very rarely be relieved.

*Obstruction to the passage of light to the rods and cones* can be caused by disease of any of the media through which the light rays pass — the cornea, aqueous humour, lens, vitreous humour, and the superficial layers of the retina through which the light must pass to reach the sensory cells (Fig. 103). Obstruction to the passage of light through the cornea is caused by inflammation due to infection or injury in which opaque scar tissue is laid down. An

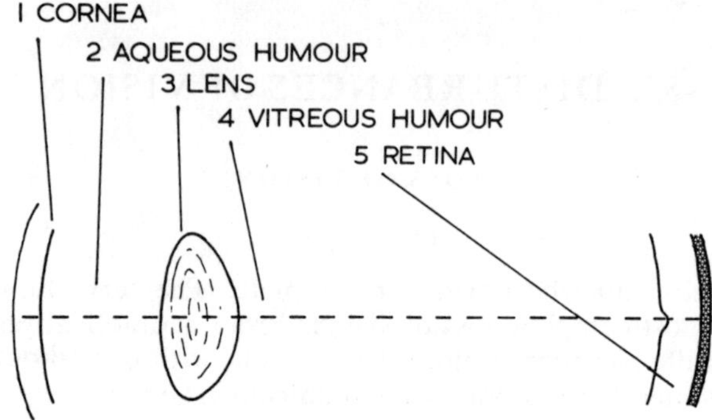


FIG. 103.—Sites at which the passage of light to the rods and cones (hatched in diagram) may be obstructed.

example of this is the scarring produced by the *trachoma* (Gr. = roughness) virus which causes blindness on a very large scale in hot, dry regions where hygiene is poor.

Vitamin A is necessary to maintain normal conjunctival translucency. In its absence, the conjunctiva becomes increasingly keratinized and the normal lachrymal secretions are no longer produced (*xerophthalmia*; Gr. *xeros* = dry). Eventually the cornea becomes dry, wrinkled and opaque. Failure of lachrymal secretions for any reason leads to drying of the cornea and the risk of its becoming opaque.

Anything opaque, e.g. blood, pus, in the aqueous humour must interfere with vision. Much commoner, however, is opacity of the lens (*cataract*). This condition becomes increasingly common with increasing age and is typical of the tissue changes associated with ageing. There is loss of elasticity so that the ability to vary lens convexity decreases. Opacity is due to degeneration of the lens proteins and, in some cases, calcification.

Vitreous humour opacity may be produced by bleeding and by infections. It has also been found that infants treated with high oxygen levels are subject to a condition referred to as *retrolental fibroplasia* (growth of fibrous tissue behind the lens — L. *retro* = behind). Blood vessels and fibrous tissue grow from the retina into the vitreous and may cause complete blindness. This condition occurs in premature infants who have been exposed to high concentrations of oxygen while being nursed in an incubator after birth. The incompletely vascularized retina is sensitive to a raised oxygen tension and its blood vessels constrict. When the oxygen tension is subsequently lowered, the vessels "over-react" by growing rapidly and invading the vitreous, accompanied by fibrous tissue. When oxygen therapy is required for an infant, the concentration is kept below about 40 per cent or oxygen is given intermittently because of the risk of retrolental fibroplasia.

*Detachment of the retina* from the underlying choroid layer may occur in retrolental fibroplasia and in a number of other conditions. The detached retina tends to curl over into the vitreous, obstructing the passage of light to normal regions of retina.

*Disease of the sensory cells and their neural connections* corresponds functionally to sensorineural deafness. Gradual destruction of the retina occurs in a variety of diseases, including hypertension and diabetes mellitus. In both diseases, abnormalities of the arterioles and small haemorrhages can be seen using the ophthalmoscope. Retinal damage is produced by interference with blood supply and the small scars which follow the damage are also characteristic of these conditions. More dramatically, thrombosis of a major retinal artery or vein can lead to sudden loss of vision in the affected eye.

A particular form of curable impairment of retinal function is *night blindness* due to vitamin A deficiency. Vitamin A is required for the synthesis of rhodopsin, the visual pigment of the rods which are responsible for vision in poor light. Rhodopsin is formed from retinene (vitamin A aldehyde) and opsin.

Just as deafness can be caused by exposure to loud noise, so retinal damage can be produced by exposure to intense light e.g. by looking directly at the sun. Such damage is often referred to as an *"eclipse burn"* because it is not uncommonly obtained by observing an eclipse of the sun without a protective dark screening material in front of the eyes. The lens of the eye focuses the incoming light on a relatively small area of the retina.

Increases in pressure on either side of the optic nerve as it leaves the eye at the optic disc can lead to destruction of nerve fibres. When the pressure is from behind the eye (e.g. due to a general increase in intracranial pressure) or within the nerve (inflammation), the condition leads to bulging forwards of the optic disc — *papilloedema* (oedema of the papilla, or optic disc).

Increased pressure within the the eye causes the reverse condition — depression of the optic disc away from the eye — the appearance as seen through the ophthalmoscope is described as *cupping of the disc*. The condition of raised intraocular pressure is referred to as *glaucoma* (Gr. *glaukos* = silvery — the eye in advanced glaucoma has a greyish metallic appearance). Aqueous humour pressure may increase four-fold or more. The increased pressure is due to an imbalance between the rate of active secretion by the ciliary processes behind the iris and the rate of absorption of the aqueous. It seems likely that the disease is usually due to impaired passage of aqueous into the circular canal of Schlemm situated at the junction of cornea and sclera. Glaucoma may be due to closure by the root of the iris of the angle between cornea and iris where the aqueous humour enters the canal of Schlemm (angle closure glaucoma). In other cases the angle is normal (open angle glaucoma) and there is probably local obstruction of the passage of aqueous between the anterior chamber and the canal. Glaucoma may arise suddenly, causing severe pain in the eye, or so gradually that no symptoms are produced, despite the fact that vision is steadily deteriorating.

In glaucoma, peripheral vision is particularly affected, so that eventually only central (macular) vision may remain (*"tunnel vision"*). Preservation of macular vision may be related to the relatively large number of nerve fibres which mediate macular vision as opposed to peripheral vision. Another feature of glaucoma is that rings of rainbow tints ("haloes") may be seen around lights. This is related to oedema of the cornea causing chromatic aberration (light of different wavelengths is focused on different regions of the retina). The oedematous cornea also tends to cause blurring of vision, so that both conduction and reception of light signals are impaired in glaucoma.

*Damage to the visual pathway* at various points causes a variety of typical changes in the fields of vision. These changes can be predicted fairly readily if four points are borne in mind: (i) damage to the optic nerve can affect vision only in the corresponding eye, (ii) damage to the optic chiasma affects the crossing fibres, which come from the nasal side of both retinae and mediate temporal vision on both sides, (iii) damage between the optic tracts and the occipital cortex impairs vision in both eyes on the opposite side to the damage. Thus damage to the optic pathway on the left damages fibres coming from the left side of both retinae and the patient cannot see objects to his right, (iv) cortical representation of the macula is distinct from cortical representation of the rest of the visual field; therefore in cortical damage, macular vision may be selectively spared or affected.

*Sympathetic ophthalmia* is a condition which can destroy vision in an initially normal eye following a perforating injury to the other eye. The initial injury can be fairly slight. Damage, typically in the ciliary region, is followed in about 10—14 days (occasionally much longer) by inflammation in the ciliary body, iris and choroid of the other eye. This condition is referred to as *uveitis*. It involves the pigmented region of the anterior eye which resembles in some respects clusters of miniature black grapes (L. *uva* = a grape). This inflammation may completely abolish vision in the "sympathetic" eye; the originally injured eye is also affected by the inflammation.

The condition may be due to an antigen-antibody reaction because, (i) the injury would tend to release into the bloodsteam antigens from parts of the eye not normally in contact with the circulation, (ii) the tissues affected in the "sympathetic" eye correspond at least partly to those damage in the injured eye, (iii) the latent period corresponds to the latent period of antibody production and (iv) glucocorticoids tends to suppress the condition. The condition can usually be prevented by removing the damaged eye at an early stage before sympathetic ophthalmia has become apparent. It is too late to remove the damaged eye after sympathetic ophthalmia has appeared. If the original damage is only slight, the decision as to whether to remove the eye may be a very difficult one.

## INVESTIGATIONS

*Visual acuity*, like auditory acuity, is assessed by comparing a patient's acuity with an arbitrary standard of normality. Each eye is tested separately, the other eye being covered. Letters of a standard design (Snellen's types) are often used. Each size of letter is designated a number, which is the distance in metres, feet, etc. at which the average normal person can read the type. If, on the metre scale, a person reads size 6 letters at 6 metres, his vision is described as 6/6 — average normal. If he can read size 4, his vision is 6/4 — better than average. If he cannot read letters smaller than size 12 at 6 metres, his vision is 6/12 — below normal. Acuity is thus expressed as patient/average normal, where the figures used in the ratio are the distances (in any units) at which each can just read the letters in question. The ratio can be expressed as a fraction or decimal; it exceeds unity for better than average vision and is less than 1.0 for impaired vision.

Children and illiterate people can be tested by asking them to orientate a card with a symbol printed on it in the same position as the test symbol (which is presented to them in varing sizes). A person is regarded as blind if his vision is below a certain value, e.g. less than 3/60, i.e. he cannot identify at 3 metres when an average person can identify at 60 metres.

*Perimetry* is the precise measurement of the visual fields. With each eye covered in turn, the patient fixes his gaze on a white spot at a fixed distance from the eye. A second spot is moved along an arc so that it remains the same distance from the eye but gradually moves out of or into the field of vision. A chart is thus built up of the area within which the test spot can be seen. The patient's field of vision is compared with the standard visual field for that eye.

*Tonometry* is the measurement of the pressure within the eye (intra-ocular pressure). It is an indirect method in which the footplate of the tonometer is placed on the cornea (previously anaesthetized with local anaesthetic) and slight constant pressure applied. The footplate moves forward slightly, displacing the cornea. The higher the intra-ocular pressure, the less movement there will be. The actual movement can be converted into the intra-ocular pressure in mm Hg using an appropriate calibration. If the pressure is maintained on the cornea for a few minutes, the footplate moves in slightly further. The extent of this secondary movement gives some indication of the freedom with which the aqueous can drain from the eye.

*Slit-lamp microscopy* enables the conjunctiva, cornea, anterior chamber, iris, posterior chamber and lens to be examined in great detail. The structures are viewed through a microscope while illuminated by a narrow beam of intense light which can be moved gradually across the eye.

## Treatment

The restoration of sight where it has been lost due to corneal or lens opacity is one of the most dramatic cures in the practice of medicine. In the case of both corneal grafting and cataract extraction, however, vision cannot improve if there is also serious retinal damage. It may not be possible to tell in advance whether the retina is healthy.

**Corneal grafting** differs from many other examples of organ transplant in that rejection of the tissue by the recipient's immune system occurs much less often. This may be related to the scarcity of cells in the cornea, so that it is a weak antigenic stimulus. Also, as the nutritional requirements of the cornea are met by the aqueous humour rather than by circulating blood, antigens may be prevented from leaving the region; similarly antibodies may not readily reach it. The cornea for grafting is taken shortly after death from a donor who has previously indicated that he or she wished this to be done. The tissue may be preserved at a low temperature for a day or two until required.

**Lens extraction** is usually carried out only when the lens has become moderately opaque and hardened (mature). Otherwise it may be difficult to remove it completely. Removal of the lens leaves the eye without the ability to accommodate for near vision. This can be compensated for by the use of a convex lens of appropriate power.

**Fixation of a detached retina** by a process of "tacking", using a coagulating current or a laser beam, is a delicate operation which, if successful, may markedly improve vision.

**Glaucoma** may be treated by drugs and by surgery. The early treatment of glaucoma is very important in the prevention of blindness. Drug treatment can modify both the formation and the drainage of aqueous. The normal rate of aqueous formation is quite low — probably around 5—10 ml/day. As secretion of bicarbonate ions appears to play an essential role in the formation of aqueous humour, the use of a carbonic anhydrase inhibitor reduces aqueous formation by slowing the reaction $CO_2 + H_2O \rightleftharpoons H_2CO_3 \rightleftharpoons H^+ + HCO_3^-$ at its first stage where carbonic anhydrase acts as a catalyst.

A drug which constricts the pupil tends to decrease the tendency of the periphery of the iris to overlie the entrances to the canal of Schlemm. An anti-cholinesterase is generally used. Conversely, an atropine-like drug which blocks the action of acetylcholine (often used to dilate the pupil for a better view of the retina) should not be given to a patient who is likely to suffer from glaucoma, as it can make matters very much worse.

If drug treatment is inadequate, aqueous drainage may be improved by (a) removing a segment of iris (usually towards the top of the eye where it is partly hidden from view by the upper eyelid) in the case of angle closure glaucoma, or (b) making a small hole between the anterior chamber and the subconjunctival space near the cornea-scleral junction. Some of the principles discussed in connection with glaucoma are illustrated in Fig. 104.

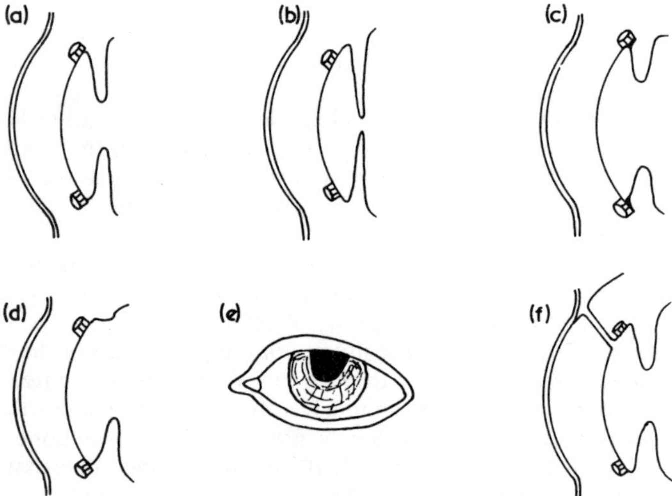


FIG. 104.—A diagrammatic representation of mechanisms in the treatment of glaucoma. When the peripheral iris partly blocks the entrances to the canal of Schlemm (a), constriction of the pupil (b) may relieve the obstruction and dilatation (c) may make it worse. Glaucoma may be treated surgically by removal of part of the iris (d, e) and by draining the aqueous humour into the subconjunctival space (f).

## REFRACTIVE ERRORS

### Definition

A refractive error is present when parallel light rays (i.e. from a distant object) are not brought to a focus on the retina when the ciliary muscle is at rest (i.e. the lens is at its least convex).

If the parallel rays are brought to a focus in front of the retina, in the resting eye, the condition is *myopia* (short-sightedness). If parallel rays are focused behind the retina, it is *hypermetropia* (long-sightedness).

### Effects

When the *normal eye* is fully relaxed, so that its lens is at its least convex (Fig. 105) parallel rays from a distant object ("at infinity") are brought to a focus

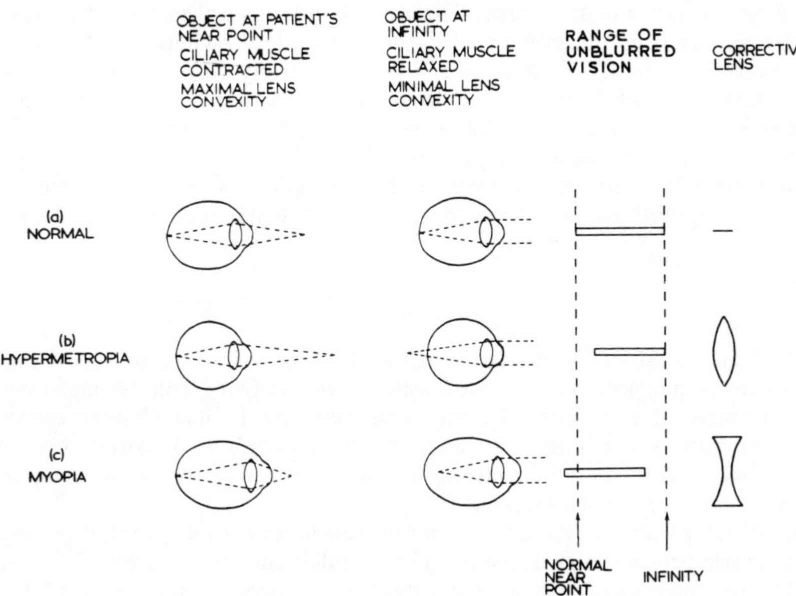


FIG. 105.—How hypermetropia and myopia affect the range of unblurred vision. In hypermetropia the near point recedes. In myopia the near point approaches the eye but distant objects are blurred.

precisely on the retina. When the lens convexity is maximal, an object at, e.g. 15 cm from the eye is focused on the retina. In this case, 15 cm is the 'near point' and objects closer to the eye cannot be seen clearly. Thus in the normal eye it is possible to focus clearly on objects between the normal near point and infinity. The near point tends to recede with age, e.g. it should be around 10 cm at age 20, but has increased on average to 80 cm at age 60.

In the resting *hypermetropic* eye, parallel rays are focused behind the retina and lens convexity must be increased to bring distant objects into sharp focus.

Thus part of the focusing capacity of the eye is used up in bringing distant objects into focus. There is therefore an impaired ability to focus on nearby objects; the near point is further than usual from the eye, and nearer objects appear blurred. The range of unblurred vision reaches to infinity as usual, but does not come as close to the eye as usual.

When the *myopic eye* is at rest, distant objects are focused in front of the retina. Nothing can be done to decrease resting lens convexity, so distant objects are always blurred. However, at maximal lens convexity, the myopic eye can focus on objects closer to the eye than the normal near point. The range of unblurred vision extends from closer to the eye than usual, but does not include distant objects, e.g. on the opposite side of the street.

Refractive errors constitute one of the causes of *squinting* by interfering with the precise visual feed-back necessary for the development of the finely balanced eye movements required for simultaneous fixation with both eyes.

*Ocular headaches* are associated with the efforts required to compensate for a refractive or other ocular error. Their mechanism is unknown. The headache may be experienced mainly around the eyes, or elsewhere in the head. In hypermetropia, the ciliary muscle is used excessively when focusing on fairly near objects, e.g. while reading. If there is a tendency to squint, the muscles which move the eye may be worked excessively to try to overcome the squinting tendency. Ocular headaches are particularly related to prolonged use of the eyes for close vision (reading, clerical work). They should be relieved by correction of the refractive error. If not, then the cause of the headaches must be sought elsewhere.

### Causes

The basic cause of a refractive error is that the refractive power of the lens plus cornea is inappropriate for the length of the eye from front to back (see Fig. 105). Theoretically, this could be due to an abnormality in the lens or cornea, or an abnormality of eye length. In fact, in young people it is usually eye length that is abnormal for genetic reasons, being greater than usual in myopia and less than usual in hypermetropia.

With increasing age, the condition referred to as *presbyopia* (Gr. *presbys* = old; compare presbycusis) develops. There is decreasing elasticity of the lens so that the maximal convexity it can attain is reduced. This accounts for the recession of the near point with age. The condition resembles hypermetropia in that nearby objects tend to be blurred.

Certain *drugs* can paralyse accommodation by interfering with the action of the acetylcholine released by parasympathetic nerves supplying the ciliary muscle. The convexity of the lens is then at its minimum, so that nearby objects are blurred. This effect is produced by atropine and related drugs and by autonomic ganglion-blocking drugs.

### Investigations

Investigations are aimed at determining the type of refractive error present (myopia or hypermetropia) and measuring its severity. The methods available can be classified as *objective* (in which the patient plays no active part in the assessment) and *subjective*, in which the patient states whether the use of various lenses improves his vision.

Objective determination of a refractive error is made by means of a *retinoscope* (Fig. 106). The patient gazes at a distant object (or has accommodation paralysed by an atropine-like drug). Light from a lamp beside the patient's head is reflected from a mirror into the patient's eye, and back again, so that in certain positions of the mirror the examiner can see the reflected light through a hole in the mirror. Light rays emanating from the retina of a normal relaxed eye are parallel and do not form an image. As shown in Fig. 106 light rays emanating from the retina of a myopic eye converge, to form a "real" image in front of the eye. Light rays from the retina of a hypermetropic eye (Fig. 106) diverge, so that a "virtual" image is formed behind the lens of the eye.

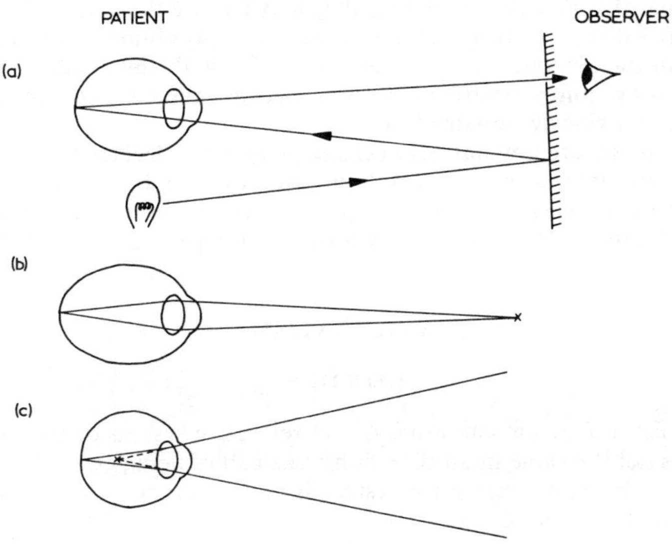


FIG. 106.—Some features of retinoscopy. In (a), the path of the light is shown, from beside the patient's head to the mirror; to the patient's retina and back to the mirror and the examiner's eye. When light emerges from the long myopic eye (b) the rays converge, to form a "real" image in front of the eye. Light from the short hypermetropic eye (c) diverges and a "virtual" image is formed behind the lens of the eye.

With experience it is thus possible to tell at once if the eye is myopic or hypermetropic. If a refractive error is present, compensating lenses are placed in a frame just in front of the patient's eye. The lens which converts the optics of the patient's eye to normal indicates the refractive error and the necessary corrective lens the patient should wear in his spectacles.

The advantage of the method is that, provided the patient's ciliary muscle is relaxed (and this can be secured by drugs) and that he will keep his eye still, no communication is required between patient and examiner and there can be no question of the patient feigning a refractive error.

In most cases, however, the patient is asked in addition to read test letters using the corrective lens, and report whether vision is indeed improved.

### Treatment

Treatment consists of the use of a lens which makes the refractive system of the eye appropriate to eye length. As shown in Fig.105, when the refractive power of the eye is too small for eye length (hypermetropia), a convex lens is required. In myopia (Fig. 105) a concave lens is used to decrease refractive power.

In most cases, spectacles are the standard treatment. Where the refractive error is very large, *contact lenses* (worn on the cornea and held there by surface tension forces) may provide a more perfect correction, because the lens directly adjoins the refractive system of the eye, rather than being 1—2 cm away. A disadvantage of contact lenses is that they may irritate the cornea. This may be at least partly due to the fact that the cornea normally obtains part of its oxygen supply from the air and a contact lens interferes with this. Clearly the use of contact lenses requires expert supervision, because the development of corneal opacity could seriously impair vision.

Contact lenses are popular with certain people who feel that wearing glasses would diminish their attractiveness. Some adapt to the difficulties of inserting, wearing and not losing these small expensive lenses better than others. In all cases, great care must be taken that corneal damage does not result.

## ASTIGMATISM

### Definition

In this condition, the curvature of the refractive system of the eye (cornea plus lens) is not the same in all directions. Instead of being equivalent to a slice cut from a sphere, the refractive system is equivalent to a slice cut from an ovoid, parallel to its long axis.

### Effects

Since the refractive power of the eye varies in different planes, part of the visual field tends to be blurred. Rays in one plane may be accurately focused on the retina while rays in another plane are focused in front of or behind the retina. When a person with astigmatism looks at two similar black lines at right angles to each other, the line in one plane appears darker than the other.

### Causes

Theoretically, either the cornea, or the lens, or even the curvature of the retina, may be faulty. In practice, it is usually the cornea which lacks symmetry. The cause of the asymmetry is not known.

### Investigations

Astigmatism can be assessed *objectively* using the retinoscope as a measure of refractive error. Light reflected from the patient's eye is viewed in different planes and the refractive power of the eye noted in those planes. This shows in which plane the eye is abnormal and measures the extent of the abnormality.

*Subjective* examination consists of asking the patient to look at a diagram consisting of groups of parallel bars radiating like spokes of a wheel. The patient reports which groups of bars are blurred and these indicate the plane of astigmatism. The effect of correcting lenses can be checked using this diagram and also letters.

### Treatment

A cylindrical lens is required, that is, a lens which is not symmetrical in all directions so that its refractive power in the plane of the patient's astigmatism compensates for the abnormality in the patient's eye. If someone with normal vision were to wear spectacles containing such a lens, he would himself suffer from astigmatism.

As with a severe refractive error, contact lenses if tolerated by the patient's cornea may in some cases provide rather better correction than spectacles, because again the correction is carried out in the immediate vicinity of the refractive system of the eye.

## SQUINTING (STRABISMUS)

### Definition

Squinting is present when the visual axis of only one of the two eyes is directed at the object of gaze (Fig. 107). This eye is the *fixing eye*. If the other eye (the *squinting eye*) is "turned in", the squint is *convergent;* if "turned out", the squint is *divergent*.

### Effects

The effects of squinting depend on the age of the patient at the time a squint develops. Squinting is not uncommon in young children and tends then to result in *suppression of vision* and eventual blindness in the squinting eye. The suppression of vision in one eye is a compensatory effect which protects the child from double vision which would otherwise occur if the visual axes diverged; however it deprives the child of the benefits of binocular vision and of a "reserve" eye should vision in the other be lost.

*Suppression of vision* appears to be a central nervous system phenomenon, whereby signals from the eye concerned are increasingly "ignored" by the brain. Visual acuity in the affected eye falls progressively, and eventually the eye becomes completely blind, despite the fact that the organ itself may be perfectly normal. Although partial visual impairment can be reversed by appropriate treatment, if the eye is allowed to become completely blind, vision cannot be restored. This type of blindness in one eye is sometimes referred to as *amblyopia ex anopsia* (blindness due to disuse of vision).

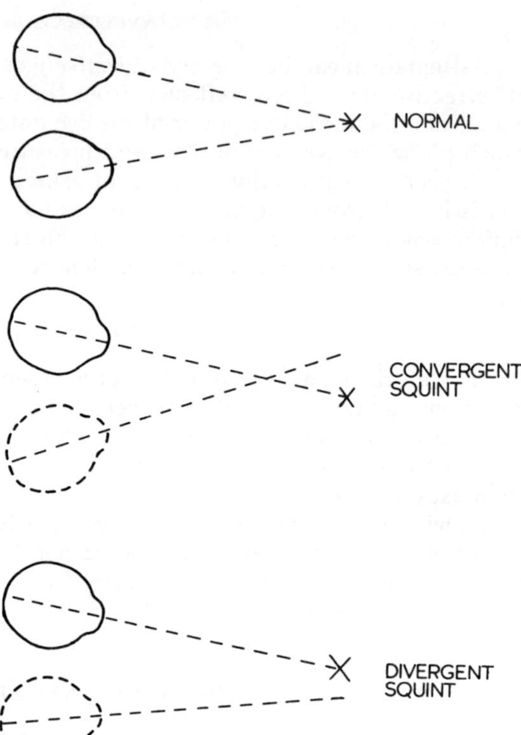


FIG. 107.—The visual axes (direction of light rays focused on the macula) in convergent and divergent squint. The fixing eye is drawn in continuous line; the squinting eye in interrupted line.

If a squint develops in adult life, the image from the squinting eye cannot be suppressed and troublesome *double vision* (*diplopia*) results. The patient may suffer from headaches which have been ascribed to his increased oculomotor efforts to correct the direction of his visual axes.

## CAUSES

Squinting generally arises in early childhood due to some disability which makes it harder than usual for the child to learn to use his two eyes together. Normally, vision of nearby objects involves simultaneous (a) constriction of the pupil plus increased convexity of the lens — accommodation and (b) convergence of the visual axes. This *accommodation/convergence reflex* is gradually acquired in the early years of life just as co-ordination of the legs in walking is acquired. Just as some children are slower to learn to walk for a great variety of reasons, so the causes of squinting, which is a disorder of eye fixation, are varied, and not always apparent in the individual case. Causes are genetic in many cases and will be considered under five headings.

The first three causes produce squinting in which the two visual axes bear a fairly constant relationship to each other. Such a squint is described as *concomitant* (L. *cum* = together, *comes* = comparison)—the two eyes move together. In contrast, the fourth cause—eye muscle paralysis—leads to a varying angle between the visual axes, which is the hallmark of the *paralytic* squint.

**High degree of refractive error.**—This is a common cause of squinting. The refractive error tends to make objects blurred and disturbs development of the accommodation/convergence reflex. The child's efforts to compensate for the refractive error are added to the effort required to avoid squinting. At best, the result is a somewhat blurred picture on account of the refractive error. In these circumstances, the child tends to give up the struggle, uses one eye for fixing, and suppresses vision in the other eye.

**Poor vision in one eye.**—If function in one eye is impaired for any reason, its contribution to vision is relatively small and may tend to be ignored so that the eye concerned no longer fixes on the object of vision.

**Extraocular muscle imbalance.**—In some cases, there seems to be a fault in the muscles which move the eye, such that the resting position of the eyes is abnormally convergent or divergent. The muscles which correct this convergence or divergence are at a mechanical disadvantage. As with a refractive error, excessive effort is required to maintain fixation with both eyes and, again, the effort is subconsciously given up.

**Eye muscle paralysis.**—If, at any age, one or more of the eye muscles are paralysed, squinting and consequent double vision must occur in certain directions of gaze. For example, if the right lateral rectus muscle, which swings the right eye laterally, is paralysed, gaze to the left is normal but when the patient attempts to look to the right, the right eye will not move sufficiently laterally and will show a convergent squint, associated with double vision.

**Stress.**—Tiredness, ill health or mental stress may unmask a latent squinting tendency if, for any reason, normal fixation requires an effort of which the child is only just capable. This is analogous to the situation where a child who has just gained nocturnal bladder control is subjected to stress and starts to wet the bed again.

## INVESTIGATIONS

The first step is to confirm in doubtful cases that a squint is indeed present. Sometimes a child with rather widely separated eyes may appear to have a convergent squint when none is, in fact, present. The *cover test* generally confirms or excludes the presence of a squint and shows which eye is squinting; or whether there is any alternating squint with the eyes taking it in turn to act as the fixing eye. The test consists in covering each eye in turn while the child looks steadily at some object. When the fixing eye is covered, the squinting eye moves, so as to bring its visual axis into line with the object; the fixing eye may also move, maintaining the angle of squint. When the fixing eye is uncovered, the eyes may move again to take up their original position. In an alternating squint, there may be no further movement at this stage. In the absence of squinting, eye movements do not take place during the cover test. Figure 107 may be of use in working out the effects of the cover test.

Having confirmed the presence of a squint, it is necessary to test eye movements so that paralysis of eye muscles may be confirmed or excluded as a cause of squinting.

*Assessment of vision* in both eyes is very important, so that suppression of vision can be detected as early as possible. Visual acuity is assessed, and, in addition *binocular vision*. In assessing binocular vision, the aim is to find whether the child is *using* the visual impressions received by both eyes. Instruments have been developed in which separate pictures can be presented to the two eyes simultaneously. For example, the left eye may be shown a picture of a bird cage, while the right eye is shown a picture of a bird. If binocular vision is present, the impression will be of a bird in a cage. If visual images from one eye are being suppressed, the child may report seeing only a cage, or only a bird.

## Treatment

Treatment can be considered under four headings: (i) optical correction, (ii) occlusion, (iii) orthoptic training and (iv) surgery.

**Optical correction.**—This will usually consist of supplying spectacles which correct the refractive error and also astigmatism if present. As children do not usually enjoy wearing spectacles, those who have to do so will require considerable support and encouragement from parents and teachers.

**Occlusion.**—If the vision in one eye is being seriously suppressed, then the other ("good") eye must be covered up (occluded) so that the "bad" eye will be used and its vision improved. This procedure may be even more unpleasant for the child than wearing spectacles, but the struggle is worthwhile, because the prize to be gained is more or less normal vision in the affected eye, rather than complete loss of vision in that eye. Should injury to the good eye subsequently destroy its vision, the preserved vision in the other eye may prevent complete blindness.

**Orthoptic training** (Gr, *orthos* = straight, normal).—This consists in presenting separate images to each eye (as in the assessment of binocular vision) and encouraging the child to make the effort to fuse the two images into a single picture. This training helps the child to use both eyes optimally and make use of information derived from the two simultaneously (binocular vision). It is a matter of co-ordinating the movements of the eye muscles appropriately; such co-ordination is encouraged when the brain can make use of the extra information obtained when images are focused on the two maculae simultaneously.

**Surgery.**—Surgery is urgently required if vision in one eye is being suppressed, despite the application, where appropriate, of optical correction, occlusion and orthoptic training. The aim is to alter muscle balance so that the eyes are in the optimal relationship to one another in the resting position. If one eye is being turned in towards the midline, then the action of the lateral rectus must be strengthened relative to that of the medial rectus. This can be done by separating the lateral rectus from the eyeball and reattaching it more anteriorly so that it swings the eye more laterally. Alternatively, or additionally, the attachment of the medial rectus can be moved back, thus decreasing its effectiveness. This is to a certain extent a matter of trial and error and in some cases the operation may need to be repeated.

Even though vision will not be affected by an operation (e.g. in an older person), the psychological benefits of a more normal appearance generally make it worthwhile to correct the squint.

## COLOUR BLINDNESS

### Definition

Normal colour vision depends on the presence in the retina of three types of cone which are sensitive to light of three distinct wavelength bands. Colour blindness is present when the function of one or more of the three cone types is absent. Taking the seven traditional colours of the rainbow, the wavelength bands to which the three types of cone are sensitive cover respectively (i) red-orange-yellow, (ii) green, (iii) blue-indigo-violet light. The corresponding cones may be termed the red, green and blue cones respectively. Absence of these cone types is sometimes referred to as *protanopia, deuteranopia* and *tritanopia* respectively, from the Greek, *protos, deuteros, tritos* = first, second, third; and *anopia* (or anopsia) meaning defect of vision.

### Effects

Absence of one or more types of cones reduces the range of colours which an individual can distinguish. Thus, if either the red cones or the green cones are absent, red and green cannot be distinguished from each other, but will appear different from blue. This "red-green" colour blindness is by far the commonest form of the condition. Varying degrees of severity of colour blindness exist. Colour blindness is of little importance to the affected person in most cases but it may debar him from occupations where precise identification of colours is important.

### Causes

Abnormal colour vision is usually due to an inherited defect. The cause appears to be an abnormality of the X chromosome because colour blindness behaves as a sex-linked recessive trait like haemophilia. It is common in men (about 8 per cent have some degree of abnormality) but rare in women (0.4 per cent). A woman is unlikely to be colour blind unless she has the abnormal gene on both X chromosomes; if she has the gene on one chromosome only, she is a carrier and, on average, half of her sons will have defective colour vision. All the daughters of a colour blind man will carry the defective gene.

### Investigations

These consist in colour matching tests and charts of the Ishihara type. In the Ishihara charts, numerals are made up of coloured dots against a background of different coloured dots. Only the person who can distinguish the particular colours used can read the figure.

### Treatment

No treatment is available.

## FALSE VISUAL IMPRESSIONS

These are analagous to tinnitus, in that the patient receives the impression of stimuli (in this case visual stimuli) in the absence of any external source. The impressions can vary from small specks which float before the eyes (*muscae volitantes*, L. = flitting flies) to a completely false visual picture as may occur in temporal lobe epilepsy. The cause of the muscae volitantes is unknown and, like tinnitus, they tend to be less troublesome if ignored.

False visual impressions are a recognized feature of migraine. The visual phenomena, if present, usually occur in the hour preceding the onset of the headache. They may take the form of defects in the visual fields, or lights or patterns of various colours. The cause is thought to be local impairment of the retinal or cerebral blood flow by the vascular spasm which precedes the vasodilatation responsible for the headache.

# 54. DISTURBANCES OF SMELL

## LOSS OF THE SENSE OF SMELL

### Effects

Loss of the sense of smell (*anosmia*) is unlikely to be noticed by the patient unless both sides are affected. Severe loss of smell causes food to lose most of its flavour, because the flavour of food is largely determined by volatile aromatic materials which enter the nasal cavity when food is chewed in the mouth. Most "tasty" food also has an appetising smell. The primary tastes—sweet, salt, sour, bitter—are, of course still present when smell is lost, but cannot convey fully the subtleties of flavour.

As with loss of hearing and vision, an impaired ability to detect smells can cause an individual to miss certain warning signals, notably the smell of fumes which may be associated with carbon monoxide poisoning.

### Causes

The sense of smell may be affected peripherally by the common cold viruses when they affect the olfactory mucosa. A similar temporary loss of smell can be produced by holding the nose when taking unpleasant medicine. With increasing age, possibly due to the onslaughts of repeated head colds, the sense of smell, like the other senses, deteriorates. This could be referred to as "presbyosmia". Injuries, tumours, etc. in the frontal lobe region can interfere with the central pathways of smell—usually on one side only. For genetic reasons, some people are unable to appreciate certain types of odour; these deficiencies are analagous to the varieties of colour blindness.

### Investigations

In order to detect unilateral loss of smell, each nostril must be occluded in turn. Some fairly strong aroma is then presented to the other nostril. It is necessary to avoid the use of substances such as ammonia which could be detected by their irritant effect rather than by stimulation of smell receptors.

### Treatment

No specific treatment is available. Loss of smell due to inflammation of the olfactory mucosa is generally only temporary.

## DISTORTION OF THE SENSE OF SMELL

Patients who suffer from epilepsy arising in the uncus (uncinate fits) not uncommonly experience olfactory hallucinations. Occasionally persistent unpleasant olfactory sensations may occur following head injury (*parosmia*).

# 55. DISTURBANCES OF TASTE

Loss of the primary tastes (sweet, salt, sour, bitter), as opposed to loss of the flavour of food, is a very rare complaint. Localized loss due to nerve or central tract damage must be specially tested for, because the tongue is supplied by two cranial nerves on each side as far as taste is concerned (chorda tympani branch of VII for anterior two-thirds; IX for the rest of the tongue).

Distortions of taste sense can occur as with smell in epilepsy due to an irritative condition in the uncus.

## RECENT LEADING ARTICLES


A human pheromone? *Lancet,* 1971, **1,** 279.
Auditory perception. *Brit. med. J.,* 1973, **2,** 728.
Better aids for the deaf. *Brit. med. J.,* 1973, **2,** 569.
Giddiness. *Brit. med. J.,* 1972, **4,** 743.
Glaucoma. *Lancet,* 1972, **2,** 73.
Noise in hospital. *Brit. med. J.,* 1973, **4,** 625.
Squint. *Brit. med. J.,* 1974, **3,** 430.
Surgery to the stapes. *Brit. med. J.,* 1974, **3,** 298.
Symposium on glaucoma. *Brit. med. J.,* 1972, **2,** 66.
Taste and smell. *Brit. med. J.,* 1971, **4,** 508.
Tests of hearing in school. *Brit. med. J.,* 1974, **2,** 3.
The importance of being short-sighted. *Lancet,* 1974, **1,** 393.
The schoolchild with a squint. *Lancet,* 1971, **1,** 1342.
Tissue typing in corneal grafting. *Brit. med. J.,* 1973, **2,** 500.
Vision on the road. *Brit. med. J.,* 1973, **4,** 628.
Vitreous surgery. *Brit. med. J.,* 1973, **2,** 258.

## *Section IX*

# DISORDERS OF THE LOCOMOTOR SYSTEM

The ability to make free voluntary movements, i.e. normal locomotion, is dependent on adequate health in (a) the joints, (b) the bones and (c) the neuromuscular system. Factors which impair function in joints, bones or muscle impair locomotion. It will be obvious that the list of factors which may affect these tissues is virtually endless. Nevertheless, there are some general features which characterize loss of function in each type of tissue and these will be described in the following chapters.

# 56. DISTURBANCE OF JOINT FUNCTION

## Definition

Joints are the articulations where the different parts of the skeleton meet. The amount of movement that can occur at a joint varies greatly throughout the body; it may be very free as in the shoulder joint or it may be very limited as in the symphysis pubis.

Disturbance of function may affect the amount of movement at a joint (generally by restricting it, occasionally by allowing it to become excessive) or may lead to pain at rest or during movement.

Joint function is often disturbed when mechanical stresses cause structural damage. However, joints can also be damaged by infections with micro-organisms and by a number of diseases whose mechanism is obscure. When joint structures are damaged the condition is referred to as *arthritis,* even though in some cases the primary cause of the disease is degeneration of joint surfaces rather than inflammation (as suggested by termination "-itis").

## Effects

The general structure of a joint is shown in Fig. 108. The articulating bone surfaces are covered in cartilage. This is a very resilient tissue which can stand a lot of abuse. It has no blood supply and receives nourishment via the underlying bone, synovial fluid and the subsynovial vessels at the periphery of the cartilage surface. The joint capsule consists of an inner layer of secretory synovial membrane and an outer tough fibrous coat which is attached to the periosteum of the articulating bones.

The joint cavity contains synovial fluid which lubricates the surfaces and nourishes the cartilage. It is rich in mucoprotein. Normal joint fluid is about 95 per cent water but because of its high content of mucoprotein (about 1 per cent) its viscosity is usually more than 100 times that of water (blood viscosity is only about 4 times that of water). Uric acid and glucose concentrations are similar to those in blood. Total protein content is usually less than 2 per cent and the albumin:globulin ratio is about 20:1. The mucoprotein content and hence the viscosity is reduced in most types of arthritis. This may be due to dilution of the mucoprotein in the excessive effusion of fluid or destruction of mucoprotein by enzymes produced by invading organisms.

When joint tissues are damaged either by injury or some disease process, certain characteristic effects are produced:

**Swelling.**—Joints usually become swollen because of increased production of synovial fluid, but oedema of the surrounding tissues may contribute. The normal mechanism of formation of synovial fluid is not understood and it is not known whether damage to the joint increases the rate of formation or decreases the rate of reabsorption. In many joint diseases, the concentration of protein in synovial fluid is increased. This would suggest that the permeability of the synovial vessels was increased. A rise in the protein concentration in synovial

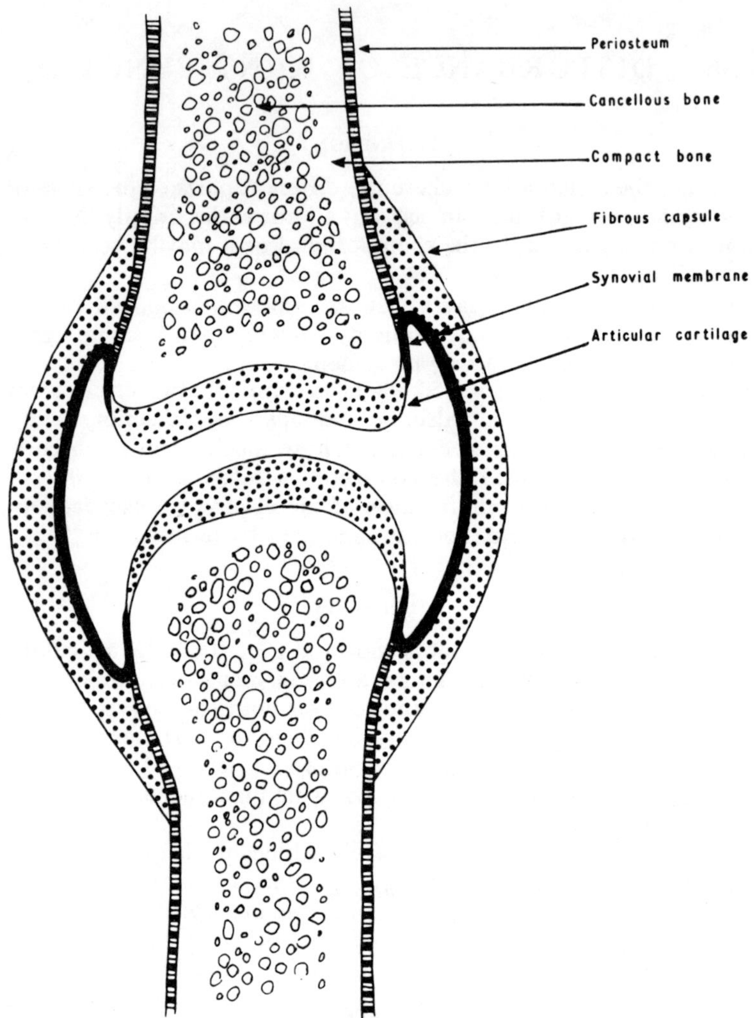


FIG. 108.—The general structure of a joint.

fluid would reduce the colloid osmotic pressure gradient tending to return fluid to the blood and so favour the persistence of fluid in the joint capsule. In some joint infections, the glucose content of the synovial fluid is reduced; this is probably due to glucose utilization by the invading organisms for their own metabolism.

**Pain.**—The tissues surrounding joints have a rich innervation with pain-sensitive nerves. Stretching or tearing of these tissues causes pain which can be extremely incapacitating. Pain is usually least when the joint is immobilized in the position of maximum capacity of the joint cavity. It is increased when movement accentuates the stretching of the periarticular tissues.

**Abnormal posture.**—A diseased joint tends to be held stationary in the position which provides maximum comfort. This is usually one of slight flexion in which the capacity of the joint cavity is maximal. In this position, the increased synovial fluid can be accommodated with the least tension and stretch of the tissues surrounding the joint cavity.

**Limitation of movement.**—A diseased joint cannot usually execute its full range of normal movement. Some of this is due to the pain which increases progressively as movement reduces the capacity of the joint cavity. Swelling and oedema of the periarticular tissues may also limit movement. A further limiting factor is the reflex contraction of the surrounding muscles which help to stabilize the joint. In some cases where the disease process is prolonged, scarring of the joint capsule and other periarticular tissues freezes the joints into near rigid structures. This is called *fibrous ankylosis* (Gr. *ankyloz* = crooked).

When articular cartilage is damaged or removed in the adult it does not regenerate but is replaced by loose connective tissue. This results in an irregular joint surface which hastens degenerative changes by affecting the smooth operation and efficiency of the joint. The integrity of articular cartilage also seems to depend on the constant mechanical influences associated with joint action. When two joint surfaces are separated (dislocated) the articular cartilage shows degenerative changes. Damage to articular cartilage and the reduction in joint movement associated with damage tend to start a vicious circle which progressively reduces the usefulness of the joint.

If the articular cartilage is completely destroyed by the disease process, the underlying bone of the two articular surfaces can undergo bony fusion. This is known as *bony ankylosis*. Though the joint is now useless, there is the consolation that it is a stable, pain-free structure.

**Reflex muscle contraction.**—In the early stages after joint damage, the muscles which act on the joint are held in a state of reflex contraction. The afferent limb of the reflex probably arises from the pain receptors in the joint capsule. By limiting movement of the joint, the reflex contraction diminishes pain and reduces the danger of further mechanical damage to the joint surfaces.

**Muscle wasting.**—In long-standing cases of joint disease, the muscles associated with the movements of the affected joints tend to waste. This is probably due in part to lack of use of the affected muscles (disuse atrophy). It contributes to the limitation of movement at diseased joints.

**Local changes in bone consistency.**—In long-standing cases of joint disease, the bone tissue adjacent to affected joints tends to become abnormal. In some cases, the bone decreases in density (*osteolytic change,* Gr. *lysis* = dissolution); in other cases, the reaction to the joint disease leads to a zone of increased density near the joint (*osteosclerotic change,* Gr. *sklerosis* = hardening—*cf.* arteriosclerosis). The cause of these changes is obscure.

## Causes

The commonest cause of joint damage is *mechanical injury.* Excessive movements of joints may tear the joint capsules or even disarticulate the joint surfaces. This is known as *dislocation.* Such tears may also cause *instability* of joints; this interferes with the maintenance of normal posture. Fractures of the bone may involve their articular surfaces.

*Infection by micro-organisms* may also cause joint damage. Organisms can reach the joint by the blood stream or by spread from neighbouring bone.

There are a large number of diseases whose natures are little understood which produce chronic and progressive damage to joints. One of these is *rheumatoid arthritis* (Gr. *rheuma* = a humour). This disease which may be a manifestation of auto-immunity, causes chronic inflammatory changes in many joints, especially the peripheral ones. The disease, which is associated with general systemic disturbance, shows exacerbations and remissions but may progress gradually to produce severe disablement.

*Anklyosing spondylitis* (Gr. *spondelos* = vertebra) is a rare condition where chronic inflammatory changes occur in the joints of the spine and pelvis which eventually make the spine into a rigid pillar. Though it resembles rheumatoid arthritis in many ways, and its cause is unknown, it differs from that disease by affecting central rather than peripheral joints and men rather than women.

*Osteoarthritis* is another type of chronic arthritis. It appears to be a premature appearance of the degenerative changes that come to joints with age and is not associated with much systemic disturbance. Some authors prefer the term *osteoarthrosis,* "-osis" implying an abnormality, "-itis" an inflammation. It may develop in a single joint which sustained injury in youth or it may be more generalized, affecting joints in the fingers and spine. Characteristically, the margins of the articular surfaces show small bony outgrowns (*osteophytes,* Gr. *phyton* = a plant). Unlike most other types of joint changes, osteoarthritis causes the bone in the region of the joints to become more dense.

Inflammatory changes in joints are also seen in *gout.* The name is derived from the french word *goutte* meaning drop. In medieval times gout was thought to be due to drops of morbid material passing to the joints with the blood. This is not far from the truth. Gout is a disorder of purine metabolism in which the blood uric acid rises and sodium monourate crystals are laid down adjacent to cartilage in the joints and ear lobes. In the joints the crystals cause inflammatory changes in the synovial capsule and the characteristic effects of arthritis described above.

Some conditions of uncertain cause can produce severe joint pain which may persist for months and then clear completely. The shoulder is not uncommonly affected, the condition being appropriately referred to as "*frozen shoulder*".

**Loss of pain sensation.**—It should be noted that the well-being of joints depends to a large extent on their sensitivity to pain. In diseases where pain sensation is lost, the patient does not protect his joints from excessive loads and rough treatment. When this happens, the articular cartilage and underlying bone eventually break down and fragment. Loose bodies in the joint cavity exacerbate the joint injury and lead to further accumulation of synovial fluid. The joint becomes quite unstable and the articulating surfaces may be dislocated.

### INVESTIGATIONS

X-ray examination of the affected joints often reveals characteristic changes which enable a diagnosis to be made. X-ray examination may reveal changes in the joint outline and increased or decreased density (sclerosis or lysis) of the bones adjoining the joint.

There are a number of blood tests which can be used in diagnosing the variety of arthritis which is present in a patient. Thus a raised blood uric acid level would suggest gout. The presence of certain immune globulins (rheumatoid factor) in the plasma would suggest rheumatoid arthritis.

A sample of joint fluid may be obtained when local infection is suspected in a joint (*septic arthritis*). The white cell count in synovial fluid rises above the normal value of about $400/mm^3$ in most types of arthritis but the greatest increases are seen where the arthritis is due to invasion by pus-forming organisms. Organisms may be cultured and determination of their identity and sensitivity to antibiotics is very useful in making the diagnosis and choosing the best antibiotic to use in treatment. Urate crystals may be found in synovial fluid if the patient suffers from gouty arthritis.

## Treatment

There are few general rules for treatment of joint disorders. The cause should be dealt with where possible. Thus micro-organisms are destroyed with appropriate antibiotics. In patients with gout, blood uric acid may be lowered by restricting purine-rich foods like liver and kidney in the diet. It also helps to give drugs which increase uric acid output in the urine or inhibit the enzyme xanthine oxidase which is essential for the formation of uric acid in the body.

In most cases of arthritis, however, the cause is not known with certainty. In these cases, the treatment has to be empiric, e.g. the treatment of rheumatoid arthritis with colloidal gold, or symptomatic, e.g. the treatment of osteoarthritis with short wave diathermy to relieve pain. Glucocorticoids have been used locally and systemically in the treatment of many types of chronic inflammatory arthritis. Their effect may be due to a depression of the inflammatory response and scar formation in the joints.

There are some general lines of attack in the treatment of arthritis which are listed below.

**Relief of pain.**—Pain is one of the most troublesome features of arthritis. Pain-relieving drugs of appropriate strength are extensively used in treatment. Liniments, rubs and local heat are used as counter-irritants in the relief of pain. In very severe cases, the injection of a local anaesthetic such as procaine into the tissues around the joint may be tried. If a large single joint has been so damaged by disease that it causes severe and persistent pain and discomfort, the formation of a bony union between the articulating bones by surgery may be tried. This destroys the joint but makes the part stable and alleviates the pain of movement. The procedure is known as *arthrodesis.*

**Rest and exercise.**—In a sense, these two aspects of treatment are in conflict. Rest provides favourable conditions for repair processes to restore the joint to normal and reduces the possibility of further damage to the diseased joint. Exercise, however, is useful in maintaining mobility of the parts and preventing the development of deformity. Each situation has to be judged on its merits. In the early acute stages of arthritis, rest is normally given priority whereas exercise is given priority in the later recovery period.

**Physio- and occupational therapy.**—There is a place for the massage, exercises and the other ploys used by physiotherapists to encourage the re-use of joints after attacks of arthritis. This can help prevent permanent disability and deformity. There is also a place for the occupational therapist in encouraging the patient to develop skills such as basket work and weaving which are consistent with the degree of disability. This not only provides useful exercise but has the psychological advantage of focusing interest away from the arthritis. Prevention of deformity can be achieved to some extent by the judicious use of exercises, manipulations and splints.

**Surgical provision of artificial joints.**—In cases where a large joint such as the hip joint has been destroyed by say osteoarthritis, good results have been obtained by replacing the diseased joint by a metal or plastic substitute. This is of value if other joints are little affected and the general health of the patient is good.

# 57. DISTURBANCE OF BONE FUNCTION

## Definition

Bone forms the skeleton on which the body is built. The failure of bone to disintegrate after death as rapidly as other tissues gives the incorrect impression that living bone is an inactive tissue. In fact, living bone is a very dynamic tissue. It is continually being broken down by osteoclasts with the release of calcium salts. Simultaneously bone tissue is being built up by osteoblasts and calcified. This equilibrium may be disturbed by factors which affect the amount of calcium in the body or the formation of the protein matrix in which the calcium salts are deposited. Bone may also be damaged by mechanical stress, infection with micro-organisms and factors which reduce its blood supply.

## Effects

### Pain

Many disorders of bone give rise to pain which may be very severe. The main site of pain-sensitive nerve endings is thought to be the periosteum. Tumours or infections in bone cause pain probably by the tension they cause in the overlying periosteum. Weakening of the bone by excessive demineralization also results in pain which may result from excessive deformation of the bone and hence the overlying periosteum.

### Reflex Muscular Contraction

Severe damage to a bone, such as a fracture, results in a reflex contraction of the muscles in that area. Afferent impulses from pain-sensitive receptors may be responsible for this. The contraction results in the immobilization of the part, which limits further movement of and damage to the bone and adjacent structures. In the case of a complete fracture of a long bone, the contraction results in shortening of the limb as the ends of the fractured bone are made to override one another. For this reason, traction of the limb is an essential feature in the setting of a limb fracture.

Even when the bone is not actually fractured, the local muscle contraction associated with bone damage may interfere with normal locomotion.

### Spontaneous Fracture

Though severe injury may cause a healthy bone to break, trivial stresses may cause fractures in bones affected by disorders which cause local or general bone weakness. As can be imagined, severe generalized bone weakness makes life miserable because the patient can not withstand the mild buffeting, which is a normal part of everyday life, without sustaining multiple fractures.

### Deformity

As the general form of the body is determined by the bony skeleton disorders which affect the shape of bones lead to deformity, e.g. a badly set fracture can lead to a shortening of a limb.

Softening of bone may cause deformity. In the child with vitamin D deficiency (*rickets*), this results in bowing of legs and the development of a "pigeon chest". In the adult, increased curvature of the spine (*kyphosis,* Gr. *kyphos* = bent) and loss of height can result as the vertebrae are compressed. The pelvis may be distorted in such a way that the inlet to the pelvis is narrowed.

Bone infections may also cause deformity. Infections of vertebrae by tubercle bacilli may result in partial collapse so that the spine becomes hunched and twisted. This is known as *kyphoscoliosis* (Gr. *kyphos* = bent; *scolios* = twisted).

Besides the unsightly appearance of deformity, there are other disadvantages. Muscles may not be able to work to the same mechanical advantage on deformed bones. This can result in loss of power and efficiency. Thus the work of breathing is much greater in a person with kyphoscoliosis than a normal person. The respiratory reserve is reduced and coughing impaired. Thus, there is an increased risk of serious chronic lung disease and respiratory failure.

The child whose legs are bowed with rickets is unlikely to be able to run quickly. Deformity of one part may throw undue stresses on other parts of the body. For example if, as the result of a previous fracture, one leg is shorter than the other, abnormal stresses are applied to the hip joints during walking. This often results in the appearance of premature degenerative changes (osteoarthritis) in those joints.

Bony deformity may also cause problems by exerting pressure on neighbouring structures. Deformity in the spine may exert pressure on the nerves leaving the spine causing pain, paralysis and loss of sensation in the areas of the body served by those nerves. Deformation of the skull may cause blindness, deafness and loss of smell sensation due to cranial nerve damage. Finally, bony deformity may make normal labour impossible if it affects the pelvis so that the promontory of the sacrum is pushed forward and the side walls of the bony pelvis are pushed together. This causes severe narrowing of the inlet to the pelvis and may prevent the passage of the baby's head.

### Formation of Stones in the Urinary Tract

Conditions associated with demineralization of bone such as hyperparathyroidism or prolonged recumbency in bed sometimes feature the formation of stones or calculi (L. *calculus* = a pebble) in the kidney or ureter because the blood calcium level is raised and the urinary output of calcium is increased. The stones are composed of calcium phosphate and oxalate and may damage the substance of the kidney or block the flow of urine. When this happens, the output of urine falls and bacteria proliferate in the stationary urine. This may lead to the bacterial infection of the pelvis of the ureter and the kidney called *pyelonephritis* (Gr. *pyelos* = vessel, in this case referring to the ureter; *nephros* = kidney).

### Anaemia

Blood cells are formed in red bone marrow, especially that of the flat bones. Though most disorders in bone are not sufficiently widespread to affect blood formation, those conditions which tend to obliterate or replace the marrow may lead to anaemia. This may be seen with cancers originating in the marrow, such as multiple myeloma. It may also occur with cancers starting elsewhere and spreading to the marrow to cause widespread marrow destruction. Very rarely, the marrow is replaced by dense bone; this condition causes "osteosclerotic anaemia".

## CAUSES

One of the commonest causes of bone disorder is *accidental trauma* resulting in fracture.

*Infection by bacteria* can also occur to cause an osteomyelitis (Gr. *osteon* = bone, *myelos* = marrow). The organisms may spread to the marrow from the blood stream or from an adjacent site such as an infected tooth socket.

A variety of *tumours* can arise in bone, some of which are compatible with life (*benign*) and some of which progress, grow and spread to cause death of the individual (*malignant*).

There are many causes of *loss of bone density*. The normal structure of bone is dependent on the stresses to which it is subjected. If these stresses are reduced, the bone atrophies, an example of *disuse atrophy*. Thus demineralization of bones is seen in the patient who is confined to bed for long periods, the astronaut who is liberated from gravitational stresses and the limb which is immobilized in a plaster cast. Normal structure is also dependent on the adequate formation of collagen to build the bony matrix. Factors such as cortisone or protein starvation which interfere with collagen synthesis also lead to loss of bone substance. Local breakdown of bone is seen in bone adjacent to a joint immobilized by arthritis, a bone compressed by an external mass or a bone with an impaired blood supply. *Excessive production of parathormone* stimulates osteoclast activity and may cause generalised rarefaction and cysts to develop in bones. In these patients there is considerable osteoblast activity. The osteoblasts release alkaline phosphatase and the blood alkaline phosphatase level is high. *Inadequate deposition of calcium salts* in bone occurs in children who have inadequate dietary vitamin D and calcium and little exposure to sunlight. A similar situation occurs in adult women with inadequate dietary calcium and frequent pregnancies.

A condition which produces characteristic bone changes is *osteitis deformans* (*Paget's disease of bone*). In this condition, the bones become both abnormally thick and abnormally weak and liable to distortion. It appears that there is increased osteoclastic activity and that the poorly-calcified thickened bones are undergoing a diffuse process rather like the healing of a fracture, where local thickening occurs due to the laying down of bony material which is at first poorly calcified. Calcitonin improves the situation in osteitis deformans, probably by limiting the excessive bone resorption.

Two terms—*osteoporosis* and *osteomalacia*—are widely used in the description of bone diseases involving loss of bone substance. These terms tend to cause more confusion than enlightenment to students and will be considered in some detail in view of their common usage.

The terms are basically descriptions of bone changes as seen by the pathologist. They carry implications as to the cause of the loss of bone density but cannot be distinguished by radiology alone.

*Osteoporosis* (Gr. *poros* = passage) refers to an abnormal porousness or rarefaction of bone due to enlargement of the bone canals or to abnormal cavities which develop diffusely. There is thus a patchy loss of both the collagen matrix and the calcium salts of bone. It is common in old age (senile osteoporosis) and with immobilization of bone.

In contrast, *osteomalacia* (Gr. *malakia* = softness) is characterized by preservation of the collagen matrix with diffuse loss of calcium salts. This condition is associated with shortage of calcium (e.g. due to dietary deficiency of vitamin D) or with excessive release of calcium from bone (e.g. due to a raised parathormone level).

It is, as stated, impossible to distinguish these conditions radiologically. They may be differentiated only by features such as distribution of bone abnormalities, the clinical picture and biochemical tests. Rather than use the two terms in describing X-rays, it is preferable to describe the condition as *loss of bone density* or *rarefaction,* possibly in addition hazarding an intelligent guess as to the causative disease.

## Investigations

Because bones contain such a high concentration of salts which are X-ray dense, bone disorders are readily visualized by X-ray examination.

Measurement of the blood concentration of bone constituents such as calcium, phosphate, etc., and of bone enzymes such as alkaline phosphatase may also be useful.

Histological examination of a small specimen (biopsy) of bone excised from a superficial bone may also be helpful in establishing the diagnosis.

## Treatment

The treatment of bone disorders are those specific to each condition. There are no general principles of treatment.

# 58. NEUROMUSCULAR DISORDERS

## Definition

Nerves and muscles normally work harmoniously together to move the skeleton in locomotion. Locomotion may be disturbed by disorders affecting the muscles, the nerves which control them or the central connections which integrate their activity.

## Effects

Though countless disorders may cause impairment of neuromuscular function, and hence locomotion, impairment of function manifests itself in a limited number of ways as described below.

**Pain and stiffness.**—Most people are familiar with the pain and stiffness which occur in muscle subjected to severe and unaccustomed exercise. Many factors can cause trouble of this sort in various groups of muscles and result in severe incapacity. This is responsible for a large proportion of the man-hours lost through illness in industry. The causes of pain and stiffness in muscle are given later.

**Weakness and paralysis.**—Incapacity by weakness and paralysis is less common than incapacity by pain and stiffness. Nevertheless it may be caused by a wide variety of disturbances. In milder forms, it may interfere with the capacity of the patient to carry out manual work but in severe generalized forms it endangers life by interfering with respiration and swallowing.

**Involuntary movements.**—Various diseases of the nervous system may cause muscle tremors or more violent involuntary movements which interfere with the execution of normal locomotion.

**Muscular inco-ordination.**—The nervous system is responsible for programming muscle action so that voluntary movements are carried out in a smooth precise fashion. Damage to the nervous system may affect this control system so that voluntary movement becomes inco-ordinated and clumsy.

## Causes

### Causes of Pain and Stiffness

*Muscle strain.*—This is one of the commonest causes of muscle pain and stiffness. The pain in this case is protective because it reduces movement of the muscles while they heal. Over-enthusiastic pushing of a car which is reluctant to start or the man-handling of young relatives can cause quite incapacitating discomfort in the next few days. Similarly, sudden muscle stretching due to slipping or stumbling may lead to pain in the affected limb. As the pain tends to follow the injury after an interval of some hours, the precipitating cause may not be recognized. Occasionally pain in chest muscles is mistaken for a heart attack, pain in the abdominal muscles for an abdominal disorder, etc.

*Fibrositis* (non-articular rheumatism).—This refers to a very common variety of conditions which cause aching pain, stiffness and limitation of movement in muscle without demonstrable changes in the joints. Though the cause is not known, it is thought to involve inflammatory changes in the fibrous connective tissue associated with muscles and tendons. Attacks often come on suddenly to give a painful stiff neck, shoulder or back, but the arms and legs can also be affected. The pain and stiffness cause limitation of movement. Difficulty may be encountered in putting on clothes.

Attacks may be precipitated by exposure to draughts and cold or by fatigue. Muscular aches and pain frequently accompany the onset of infectious illnesses. In fact, the cause in the majority of cases is completely obscure. Damp conditions have also been incriminated and there is extensive folklore on the subject.

*Tenosynovitis.*—The synovial sheaths which surround tendons in certain areas may shown inflammatory changes when they become infected by an invading organism or they are subjected to excessive wear and tear. The pain and scarring found in this condition can make normal movement very difficult.

*Intervertebral disc lesions.*—If part of an intervertebral disc becomes extruded from between the vertebrae it may press on one of the spinal nerves and cause pain in the distribution of that nerve. It is usually accompanied by stiffness because of the reflex contraction of local muscles. This type of trouble may be responsible for pain radiating down the back of the leg (*sciatica*), pain in the back muscles (*lumbago*), pain radiating down the arm (*brachial neuritis*) and pain in the neck (*cervical neuritis*).

### Causes of Weakness and Paralysis

*Pyramidal system damage.*—Damage to the pyramidal tract causes paralysis of the appropriate muscles on the opposite side of the body. Reflex movements persist in the muscles and account for the increase in muscle tone. Because of this, muscle wasting is slight.

*Lower motor neurone lesions.*—These cause flaccid paralysis and severe wasting of the affected muscles. All voluntary and reflex movements are abolished.

*Myasthenia gravis.*—This is a condition where the muscles are readily fatigued. Though muscle action is initially strong, it rapidly tires and becomes weak. The weakness is less obvious in the morning, but gets greater as the day continues. The trouble is a fault in neuromuscular transmission. Either inadequate acetylcholine is liberated from the nerve endings or the receptors at the muscle end-plate do not respond normally to the liberated acetylcholine. The condition is alleviated by drugs which inactivate cholinesterase and prolong the action of acetylcholine.

The disorder causing myasthenia is not understood. There is some relationship to the thymus gland and the condition may be improved by removing the thymus. This would suggest that there was some fault in the immune system and myasthenia gravis sometimes occurs in patients who show other signs of autoimmune disease. One possibility is that antibodies are formed which react with antigens in the area of the neuromuscular junction. Histological changes in the muscle fibres and nerve terminals have been described.

*Muscular dystrophy.*—This term include a variety of hereditary conditions in which progressive degeneration and weakness occurs in some or all of the muscles of the body. There is no obvious defect in their innervation but there is some evidence that the nerves may be responsible for the muscular changes. The muscle fibres initially swell and fat may be deposited in them but wasting occurs later and the muscles are replaced by fibrous tissue. Muscular dystrophy usually appears in early life and results in premature death from respiratory failure. There is no effective treatment.

### Causes of Involuntary Movements

Tremors and involuntary movements are characteristic of damage to the extrapyramidal system. Tremor may also be seen in cerebellar disease. *Tics* or *habit spasms* are involuntary, co-ordinated movements which are made repeatedly by some people in the absence of any damage to the central nervous system. For example, the person concerned may repeatedly contract the muscles on one side of his face or blink or sniff. The movements cease during sleep and are made worse by excitement. The underlying cause is not known. Tics have most of the features of a bad habit like nose picking. Despite good intentions, it is very hard to stop.

### Causes of Inco-ordination

Clumsiness of movement may result from cerebellar disease (cerebellar ataxia) or from loss of muscle-joint sense from the affected muscles (sensory ataxia).

## Treatment

Treatment of neuromuscular disorder should be directed at the underlying process where possible. However, physiotherapy and occupational therapy can do much to help patients make better use of their impaired facilities.

## RECENT LEADING ARTICLES


Articular cartilage. *Lancet,* 1971, **2,** 81.
Calcitonin and metabolic bone disease. *Lancet,* 1971, **1,** 1168.
Cracking joints. *Lancet,* 1971, **2,** 649.
Gamekeeper's thumb on the ski slopes. *Brit. med. J.,* 1974, **1,** 213.
Growing pains. *Brit. med. J.,* 1972, **3,** 365.
Immobilization and bed rest in rheumatoid arthritis. *Lancet,* 1971, **1,** 1281.
Injury after small falls. *Lancet,* 1972, **1,** 945.
Management of myasthenia gravis. *Lancet,* 1972, **1,** 780.
Mechanism of gouty inflammation. *Brit. med. J.,* 1973, **4,** 125.
Muscle, nerve, or what? *Lancet,* 1974, **1,** 1025.
New knees for old. *Brit. med. J.,* 1972, **2,** 363.
Osteoporosis. *Lancet,* 1970, **1,** 180.
Pathogenesis of myasthenia gravis. *Brit. med. J.,* 1971, **2,** 1.
Pathogenesis of osteoarthrosis. *Lancet,* 1973, **2,** 1131.
Rheumatoid cervical luxation. *Brit. med. J.,* 1973, **1,** 759.
Scoliosis. *Brit. med. J.,* 1973, **2,** 192.
Tennis elbow. *Lancet,* 1973, **2,** 1426.
Thymectomy for myasthenia gravis. *Brit. med. J.,* 1972, **3,** 543.
Total replacement of the hip. *Brit. med. J.,* 1972, **2,** 177.
Voluntary dislocation of the shoulder. *Brit. med. J.,* 1973, **4,** 505.

*Section X*

# DISORDERS OF THE BLOOD

# 59. DEFICIENCY OF HAEMOGLOBIN (ANAEMIA)

## Definition

ANAEMIA is a condition characterized by a deficiency of circulating haemoglobin. The red cells may be deficient in number or in haemoglobin content or both. There is generally a fall in all three primary red cell indices—the red cell count, haematocrit (packed cell volume) and blood haemoglobin level.

For reasons of convenience, the haemoglobin level is most often used to diagnose or exclude anaemia in patients. The normal value is about 15 g/100 ml—slightly higher in men than in women. There is quite a wide normal variation, but values below 12 g/100 ml in women and 13 g/100 ml in men are suggestive of anaemia.

It should be noted that the haemoglobin level depends not only on the amount of circulating haemoglobin, but also on the plasma volume. Factors such as pregnancy, which increase plasma volume, reduce the haemoglobin level and the reverse is true in sodium depletion when plasma volume falls. Judgements about the total amount of circulating haemoglobin should take into account possible abnormalities of the plasma volume.

## Effects

For a given decrease in oxygen pressure ($Po_2$) anaemic blood releases less oxygen to the tissues than does normal blood (Fig. 109). The effects of anaemia (Fig. 110) stem from (*a*) the direct effects of lack of oxygen on the tissues and (*b*) the compensatory changes brought into play by the tissue hypoxia. It must however, be emphasized that even when anaemia is severe, the patient may complain of few, if any, symptoms. This is because of the adequacy of the compensatory changes and because the condition often develops so slowly that the patient adjusts unconsciously to the impaired bodily functions.

The fall in the circulating haemoglobin level accounts for the pallor so characteristic of anaemia. As with cyanosis, the pallor of anaemia can be diagnosed definitely only if superficial capillaries are well filled with blood. Thus, pallor of mucous membranes—conjunctival and inside the mouth—is generally a reliable indicator of anaemia, whereas marked skin pallor due to vasoconstriction may coexist with a normal haemoglobin level.

### Tissue Hypoxia

Tissue hypoxia tends to be slight in mild anaemia because the effects are mitigated by the compensatory changes. It increases markedly when the anaemia is so severe that its effects cannot be countered by the compensatory mechanisms. Function of skeletal muscle is impaired. With a low $Po_2$, anaerobic metabolism occurs in muscle during activity. This would account for the weakness and ready fatigue during exercise. With anaerobic metabolism

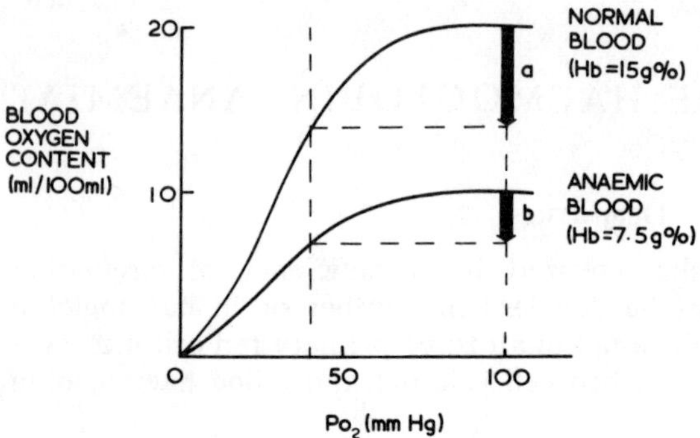


FIG. 109.—Oxygen dissociation curves for normal and anaemic blood. For a given reduction in $Po_2$ (e.g. from 100 to 40 mm Hg as normally occurs in blood traversing the tissue capillaries) less oxygen is released from anaemic blood (B) than normal blood (A).

there is an increase in the production of lactic acid. The resulting metabolic acidosis stimulates breathing and may account in part for breathlessness on exertion seen in anaemic patients.

In severe anaemia, the heart muscle may be diffusely damaged by the lack of oxygen giving rise to pain on exertion and eventual heart failure. Temporary heart failure and pulmonary congestion on exertion is a second possible mechanism of the breathlessness on effort in anaemia.

The reduced arterial oxygen content renders the brain more sensitive than usual to any reduction in blood flow, making the patient prone to light-headedness and fainting attacks. It may also account for the lethargy and headache which are common in anaemia. Hypoxia in the peripheral nerves may be responsible for the tingling sensations; similar sensations may be felt when the blood flow to peripheral nerves is temporarily arrested, e.g. by a sphygmomanometer cuff or by a posture which interferes with local circulation. Tests of liver and kidney function may reveal impairment of function. Alimentary tract dysfunction may be manifested by loss of appetite and abdominal discomfort.

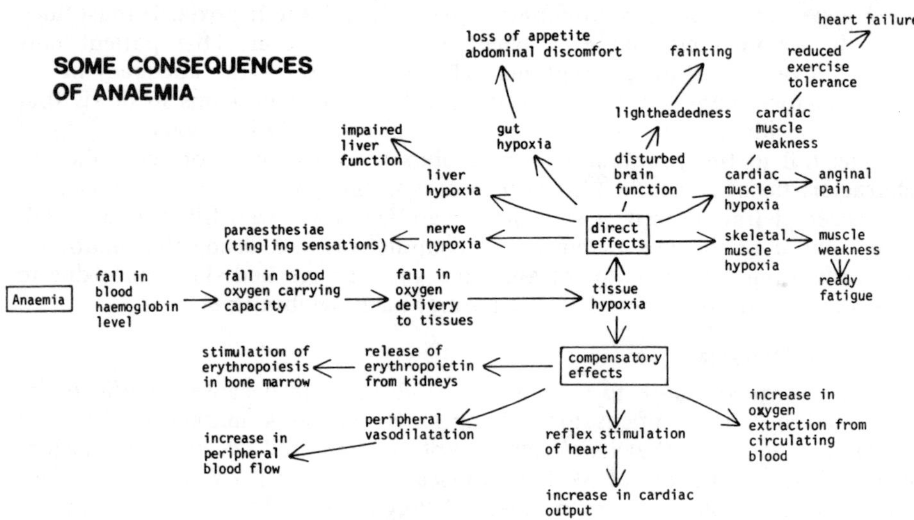


FIG. 110.—Mechanisms of symptoms and signs associated with anaemia.

## Compensatory Mechanisms

In anaemia the oxygen carrying capacity of the blood is reduced. All other things being equal, this would result in a fall in the delivery of oxygen to the tissues. The body compensates for this in several ways which increase oxygen delivery to the tissues. Firstly the peripheral blood vessels dilate. Secondly there is a reflex increase in cardiac output. Thirdly there is an increased oxygen extraction from the blood perfusing the tissues.

Hypoxia in the tissues causes a generalized vasodilatation of resistance vessels in the systemic circuit. The resulting reduction in peripheral resistance, arterial blood pressure and stimulation of arterial baroreceptors causes a reflex stimulation of the heart and increase in cardiac output. There is an increase in both heart rate and force of contraction. The forceful contractions of the heart which keep up systolic pressure together with the low peripheral resistance which lowers diastolic pressure result in a large pulse pressure. The combination of a rapid heart rate with an increased pulse pressure is referred to as a *hyperdynamic circulation.* Patients are often unpleasantly aware of the forceful heart beat, describing the sensations as *palpitations*. The cardiac output and sensations experienced at rest may resemble those in healthy people immediately after moderately severe exercise.

The circulatory changes in anaemia favour turbulence in the blood and hence an increased chance of hearing abnormal sounds over the heart (a systolic murmur) and major arteries, especially the carotids (a systolic bruit; Fr. *bruit* = noise). The increased turbulence is favoured by two factors—increased velocity of blood flow due to the increase in cardiac output and decreased viscosity of blood due to the fall in haematocrit (Fig. 111). Blood density falls slightly, which tends to reduce turbulence. However, this effect is small because

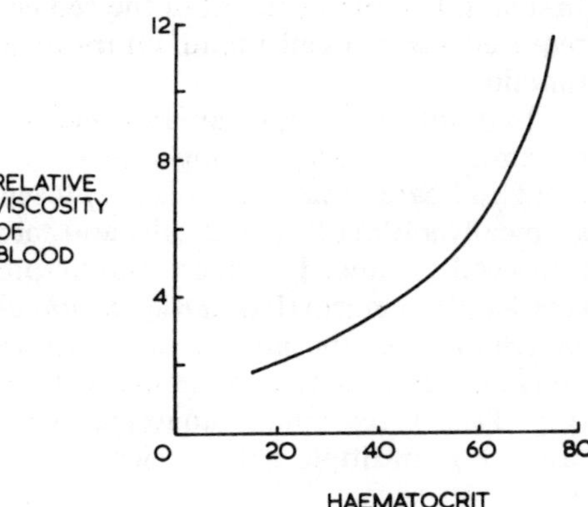


FIG. 111.—The effect of haematocrit on the viscosity of blood relative to that of water.

red cell density is only slightly greater than that of plasma. In the anaemic patient, murmurs and bruits should not be taken as evidence of valvular or arterial abnormality unless they persist after the anaemia has been treated.

Like hyperthyroidism and pregnancy, which also produce a hyperdynamic circulation, anaemia may "unmask" heart disease. If circulation to the heart in anaemia is impaired by coronary artery narrowing, severe hypoxia of the hyperactive myocardium may occur during exertion or excitement and cause chest pain (angina pectoris). In extreme cases, damage to heart muscle (myocardial infarction) may result. When there is a tendency to develop heart failure (e.g. due to previous myocardial damage, valvular disease or hypertension) anaemia may precipitate typical features of heart failure at rest.

As arterial oxygen *pressure,* as opposed to *content,* is normal in anaemia, the chemoreceptors are not stimulated and ventilation *at rest* is not usually increased. However, as mentioned earlier, shortness of breath *on exertion* is a common complaint. In addition, all the abnormalities caused by anaemia become exaggerated during exercise, reducing exercise tolerance and causing excessive tiredness and loss of energy.

Any stress, such as an infective illness, produces a more severe than usual effect in the anaemic patient because the body generally is working at a disadvantage due to the hypoxia. Sudden blood loss, e.g. during childbirth, is particularly dangerous when anaemia is present.

### Causes

Failure to maintain an adequate pool of circulating red cells has two fundamental causes—an excessive rate of removal of cells from the pool and an inadequate rate of replenishments of the pool. Excessive removal may be due to loss of blood from the body or an increased rate of breakdown of red cells within the body. However, the body is capable, in response to a raised erythropoietin level, of manufacturing red cells at 5-10 times the normal rate, *provided* that bone marrow function is normal and there is no deficiency of raw materials. Therefore, to cause persistent anaemia, blood loss must be persistent or associated with raw material deficiency; it tends to cause iron deficiency due to loss of the iron in the red cells lost from the body. Similarly, haemolysis must be severe before it overtakes the ability of the body to replace the lost cells. Inadequate replenishment of the red cell pool is due to deficiency of substances required for red cell manufacture or a disease which depresses bone marrow function.

In practice, it is not generally necessary to consider all the possible causes of anaemia when investigating a given patient. Once the haemoglobin measurement indicates that anaemia is present, the next step is to examine microscopically a film of the red cells and measure the secondary indices, such as the mean cell colume. It is then usually possible to conclude that the red cells are smaller than normal (*microcytic anaemia*), bigger than normal (*macrocytic anaemia*) or of the normal size (*normocytic anaemia*). The causes, investigations and treatment of anaemia will be subdivided under these three headings (Fig. 112). Occasionally anaemia in a given patient may have more than one cause, e.g. multiple deficiencies due to intestinal malabsorpton.

## MICROCYTIC ANAEMIA

### Causes

Microcytic anaemia is due to iron deficiency and is by far the commonest type of anaemia. Iron deficiency anaemia is one of the commonest diseases in all countries of the world. The main causes are dietary iron deficiency, malabsorption of iron and blood loss.

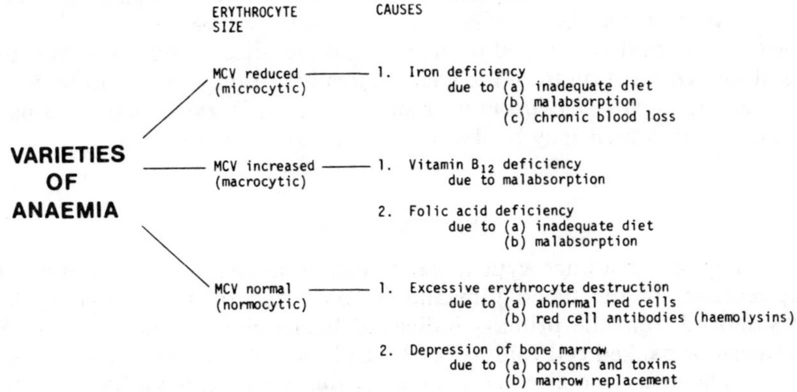


FIG. 112.—A classification of the commoner causes of anaemia based on red cell size as judged from blood films and indices. Iron deficiency anaemia is far commoner than all other anaemias combined.

The ease with which iron deficiency can develop is related to the fact that the body's intake and loss of iron are finely balanced. A normal diet contains 10-20 mg of iron, of which about 10 per cent (1-2 mg) is normally absorbed (in severe iron deficiency the proportion may exceed 50 per cent. Daily loss of iron amounts to approximately 1 mg. During menstruation, 50 ml blood (25 mg iron) may be lost. Hence any persistent loss of blood (e.g. from an ulcer in the alimentary tract) can easily upset the balance, especially in the female.

**Dietary deficiency of iron** is extremely common in countries where general malnutrition is a problem. In the more highly developed countries where food shortage is uncommon, dietary deficiency is not often the primary cause of anaemia but may aggravate iron deficiency due to excessive blood loss. It is most likely to occur in infancy, in women during the reproductive period, and in old people. The normal infant has at birth a store of iron which is used during the early months of life, when a largely milk diet contains little iron. The stores may be inadequate if the baby is born prematurely or if the mother suffered from iron deficiency during her pregnancy. Women during the reproductive period have a steady loss of iron due to menstruation and may have a much greater loss of iron due to pregnancy and childbirth. Any deficiency of iron in the diet is likely to cause or aggravate anaemia in these circumstances. Elderly people who live alone tend to eat food which is cheap and easily prepared. Such a diet tends to be high in carbohydrate and low in iron content.

**Malabsorption of iron** is relatively uncommon. It is part of the picture in diseases causing general malabsorption. It also occurs following surgery to the stomach and duodenum. Iron is absorbed mainly in the duodenum, possibly because of the relatively low pH there. When part of the stomach is removed, and particularly if the duodenum is bypassed, iron absorption is often impaired.

**Loss of blood** involves loss of iron, protein and porphyrin, but the loss of iron is by far the most serious, and iron deficiency due to prolonged relatively mild blood loss cannot be made good by normal dietary intake, even though iron deficiency modifies iron absorption so that a greater proportion of the dietary iron is taken into the body. The loss of blood may be fairly obvious, e.g. increased menstrual loss, blood in the urine, or vomited blood. However, blood loss at a steady, low rate from the gastro-intestinal tract is not readily detected by the patient (*occult bleeding*) and requires special investigation. It is usually due to an ulcer, which may be due to a malignant tumour.

## Investigations

The diagnosis of a microcytic anaemia can be made, with experience, from the appearance of the blood film and is confirmed by the secondary blood indices derived from the primary indices of haemoglobin concentration (Hb), haematocrit or packed cell volume (PCV) and red cell count (RCC). *Mean cell volume* can be measured directly or it may be derived from the volume of cells in 1 $mm^3$ divided by their number. If the PCV is 45 per cent, the volume of cells in 1 $mm^3$ is 0.45 $mm^3$. The number of cells in 1 $mm^3$ is the RCC, say $5 \times 10^6$. Hence mean cell volume (MCV) $= \dfrac{0.45}{5 \times 10^6} = 9 \times 10^{-8}$ $mm^3$ or 90 $\mu^3$. MCV falls as iron deficiency anaemia progresses.

*Mean corpuscular haemoglobin concentration* (MCHC) is estimated by dividing Hb (g/100 ml) by PCV. Thus if 100 ml blood contain 15 g of haemoglobin and the PCV is 45 per cent, 45 ml of red cells contain 15 g of haemoglobin, a concentration of 33 per cent. The MCHC as calculated in this way also falls in iron deficiency anaemia. It has been suggested recently that this is a spurious result due to a falsely high PCV in iron deficiency anaemia because the cells do not pack so closely. Direct measurements of MCHC have usually given normal results in iron deficiency anaemia.

The finding of a low serum iron level confirms that anaemia is due to iron deficiency. Particularly in the case of a man or a post-menopausal woman, iron deficiency anaemia with no obvious cause may be an early sign of a gastro-intestinal tumour. If this is the case, blood can usually be detected chemically in the faeces. Radiological studies of the alimentary tract are then required to supplement the history and clinical examination in the search for the cause of the bleeding.

The bone marrow shows characteristic (hyperactive) changes in iron deficiency states. However, most patients with microcytic anaemia can be diagnosed and treated successfully without marrow puncture.

## Treatment

Replacement of iron cures the anaemia. Oral administration of ferrous salts is satisfactory in most cases. Provided that malabsorption is not present, the intestinal absorption of iron increases markedly when iron deficiency is present.

Injection of iron may be required when there is severe malabsorption and when the patient cannot or will not tolerate oral iron.

If severe anaemia is discovered late in pregnancy, the only treatment acting sufficiently rapidly is blood transfusion. When loss of blood is the cause of the anaemia, the underlying condition requires attention.

## MACROCYTIC ANAEMIA

### CAUSES

Macrocytic anaemia is due in most cases to deficiency of vitamin $B_{12}$ or folic acid.

Vitamin $B_{12}$ and folic acid play an essential role in the synthesis of DNA. Deficiency of either will affect rapidly dividing cells in general. Cells formed in the bone marrow are particularly affected. Red cell formation is depressed, and deficient polymorph and platelet formation can often be demonstrated in addition. Epithelial renewal, particularly on the surface of the tongue, also tends to be inadequate when there is a deficiency of $B_{12}$ or folic acid.

Deficiency of vitamin $B_{12}$ is almost always due to failure of absorption, because most diets contain sufficient amounts. The commonest cause is absence of "intrinsic factor" which normally combines with $B_{12}$ so that it can be absorbed in the terminal ileum. This may be due to atrophic gastritis in which the gastric mucosa degenerates—in some cases at least due to auto-antibodies against gastric mucosa—and loses its secretory functions, including the secretion of intrinsic factor. Auto-antibodies can also be produced against intrinsic factor itself. As a result, vitamin $B_{12}$ cannot be absorbed adequately and anaemia results. The condition is sometimes referred to as *pernicious anaemia,* a name it acquired when there was no known treatment. Occasionally, as a result of surgery, a patient acquires a blind loop of intestine which may harbour organisms which consume $B_{12}$, depriving the patient of it. If disease should affect the terminal portion of ileum, where $B_{12}$ is usually absorbed, the vitamin may not be absorbed adequately, even though there is no shortage of intrinsic factor.

As well as causing anaemia, $B_{12}$ deficiency also affects the nervous system by interfering with the metabolism of myelin sheaths. In the spinal cord, degeneration of the posterior columns leads to impaired vibration and muscle-joint sense; the motor tracts may also be damaged. Peripheral nerves and brain function may also be adversely affected.

Deficiency of folic acid may be dietary in origin, or due to malabsorption. Dietary deficiency is particularly liable to occur during pregnancy when there is an increased demand for folic acid from the rapidly dividing cells of the fetus. Malabsorption of folic acid is caused by the diseases which lead to generalized malabsorption from the intestine. Certain drugs, including some used in the treatment of epilepsy, interfere with the metabolism of folic acid and hence produce a macrocytic anaemia.

### INVESTIGATIONS

A macrocytic anaemia is diagnosed from the appearance of the blood film and from the finding of an increased mean cell volume. There is often also a reduction in the numbers of circulating polymorphs and platelets. The diagnosis may be confirmed by examination of a sample of bone marrow, usually obtained from the sternum or iliac crest. The erythrocyte precursors are also larger than normal, giving rise to the term, megaloblastic (Gr. *megale* = large; *blastos* = germ).

Having diagnosed a macrocytic anaemia, the next step is to determine whether it is due to $B_{12}$ deficiency or folic acid deficiency. Measurement of the blood levels of these substances helps to do this. Analysis of gastric contents may show absence of the normal acid secretions, thereby suggesting the presence of atrophic gastritis, and the associated inability to produce the intrinsic factor essential for $B_{12}$ absorption.

A more direct test of the patient's ability to secrete intrinsic factor is known as the *Schilling test.* A large dose of $B_{12}$ is injected, enough to meet the patient's requirements. Radioactive $B_{12}$ is then given by mouth. If intrinsic factor is present, the $B_{12}$ is absorbed and, as it is not required, is excreted in the urine where it may be detected. Should excretion be below normal, the procedure is repeated with the addition of oral intrinsic factor. Increased $B_{12}$ excretion in this second part of the test confirms that the patient cannot secrete adequate intrinsic factor. If excretion is still below normal, the $B_{12}$-intrinsic factor complex is not being absorbed; this suggests disease of the terminal ileum, the site of $B_{12}$ absorption. The two parts of the test may be combined, using two forms of radio-active $B_{12}$ each containing a different radio-active isomer. One form of $B_{12}$ is bound to intrinsic factor and one is not. Both are administered and assayed in the urine. If both forms of $B_{12}$ are excreted normally, then $B_{12}$ absorption is normal; if only the bound $B_{12}$ is excreted, there is a deficiency of intrinsic factor; if neither form of $B_{12}$ is excreted in normal amounts, disease of the terminal ileum is likely.

Folic acid deficiency can be demonstrated by the *Figlu test.* The amino-acid histidine is metabolized in the body to formiminoglutamic acid (figlu for short). Figlu is then further metabolized, folic acid being essential for this further metabolism:

$$\text{HISTIDINE} \longrightarrow \text{FIGLU} \xrightarrow[\text{acid}]{\text{folic}} \text{OTHER PRODUCTS}$$

In folic acid deficiency, figlu accumulates. In the figlu test, a histidine load is given, and urinary figlu subsequently measured. A rise in urinary figlu excretion above the normal level indicates folic acid deficiency. The value of this test as an indicator of folic acid deficiency is reduced by the fact that histidine metabolism may be disturbed in liver, thyroid and other diseases.

Where generalized malabsorption is suspected, the appropriate tests are carried out.

The final test, carried out after treatment has been commenced, is measurement of the *reticulocyte response.* A raised peripheral reticulocyte count indicates that the bone marrow is working at well above the normal rate. If specific treatment for a deficiency anaemia has just been commenced, the raised reticulocyte count suggests that the treatment is correct.

### Treatment

The deficiency of folic acid or $B_{12}$ must be made good. In the absence of malabsorption, folic acid may be given orally. As vitamin $B_{12}$ deficiency is almost always due to impaired absorption, replacement must be by injection. The absorption of even large oral doses tends to be unreliable.

If there is doubt as to which of the two treatments, $B_{12}$ or folic acid, is required, $B_{12}$ is given first, to avoid the risk of further nervous system damage.

## NORMOCYTIC ANAEMIA

### Causes

The main causes of a normocytic anaemia are excessive breakdown (haemolysis) of red cells, and bone marrow depression. A normocytic anaemia also results from the sudden loss of a considerable quantity of blood; the remaining blood is diluted as the blood volume is restored by fluid drawn from the tissue spaces. However, the management of this condition is concerned mainly with the loss of blood and its cause, and will not be considered further here.

**Excessive haemolysis.**—As mentioned earlier, haemolysis of red cells must exceed 5-10 times the normal rate before it overtakes the body's ability to replace the loss. Most of the iron from the broken-down cells is retained by the body so that iron deficiency does not result from the haemolysis.

The excessive breakdown leads to excessive release of bilirubin. The liver, provided it is normal, excretes increased amounts of bilirubin and in some cases this crystallizes in the gall-bladder to form bile pigment gallstones. When haemolysis is marked, bilirubin is added to the blood at a greater rate than it is removed and conjugated by the liver. The patient becomes jaundiced, but does not excrete bile in the urine. This is because unconjugated bilirubin is protein-bound and cannot pass through the glomerular filter. The condition is referred to as *acholuric jaundice*—jaundice without bile in the urine (Gr. *chole* = bile).

There are two main causes of excessive haemolysis—red cell abnormality and the presence of substances which break down normal cells. Red cell abnormalities are generally genetic in origin and the cells are more susceptible than usual to the normal body mechanisms which break down the older degenerating red cells. In some cases, e.g. hereditary spherocytosis, they have an increased tendency to lyse in hypotonic solutions. This is referred to as increased *osmotic fragility*. Other abnormal red cells may show increased resistance to osmotic lysis, despite being readily broken down in the body.

Normal red cells may be broken down excessively by a great variety of toxic substances—various poisons, including lead and arsenic; drugs to which the patient is particularly sensitive; and products of infective agents, of which malaria is the most notable example. In addition, antibodies against red cells (haemolysins) may cause haemolysis in certain situations—haemolytic disease of the newborn; transfusion with incompatible red cells; and acquired haemolytic anaemia, in which the patient develops antibodies against his own red cells.

*Haemolytic disease of the newborn* is due to the passage across the placenta to the fetus of maternal antibodies against the fetal blood. As anti-Rhesus antibodies are smaller and pass the placenta more readily than ABO antibodies, they more often cause haemolytic disease in the fetus. In cases of Rhesus incompatibility, the fetus is Rhesus positive, i.e. its red cells have the Rhesus antigen, and the mother Rhesus negative, i.e. her cells do not have the Rhesus antigen and her immune system recognizes the antigen as foreign. Whereas a person whose red cells lack the A or B antigen has the corresponding antibody in his plasma, a person whose red cells lack the Rh antigen (Rhesus negative cells) does, in contrast, not have the corresponding antibody unless he has been sensitized by exposure to Rh-positive cells. Rhesus haemolytic disease of the newborn can develop only if the mother is Rhesus negative and then only if two conditions are fulfilled—(i) the mother is immunized against (sensitized to) the Rhesus factor, (ii) her fetus is Rh-positive (Fig. 113). Rhesus disease thus develops in two phases—*immunization* and *haemolysis*.

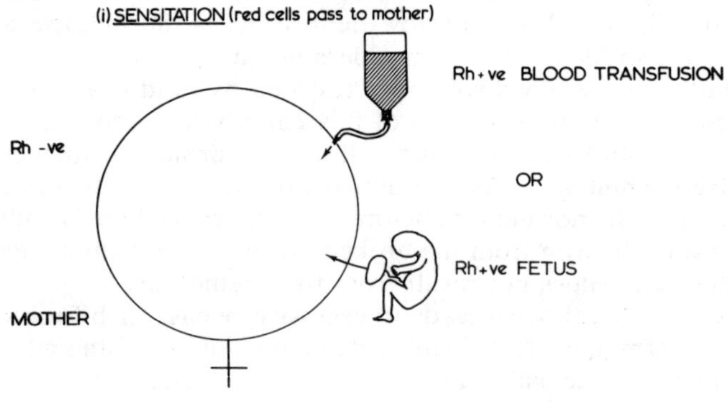

FIG. 113—Two phases in the development of haemolytic disease of the newborn due to anti-Rhesus antibodies.

*Immunization* may occur by the transfusion of Rh + ve blood to a Rh—ve woman. This is generally avoided by giving only Rh—ve blood to Rh—ve individuals and by using Rh—ve blood in the rare emergencies when there is not time to group the patient's blood before a transfusion is given. Immunization may also occur when Rh + ve red cells pass from a fetus to its mother. Small amounts of fetal blood leaking through the placenta do not usually provoke an appreciable immune response, but the larger amounts which pass to the mother at delivery (or at abortion) do. Immunization can usually be avoided by giving anti-Rh antibody to the mother at delivery. This neutralizes the antigens of the Rh + ve cells before they can provoke an immune response.

The second phase of haemolytic disease of the newborn—*haemolysis*—results from the passage of antibodies to the fetus. If a woman has been sensitized after one pregnancy, the small amounts of blood which leak across the placenta from the fetus during the next pregnancy may boost the antibody titre so that, towards the end of the pregnancy, antibody crosses to the fetus in large amounts. Fetal cells are broken down to cause increasing anaemia. The resultant excess bilirubin is dealt with by the mother's liver. However, after birth, the immature fetal liver cannot deal with the excessive load. Unconjugated bilirubin accumulates rapidly in the fetal circulation. This bilirubin is not prevented from entering the brain (as it is in the adult due to the blood-brain barrier) and brain damage may occur, the risk increasing with the bilirubin level.

Finally, it must be emphasized that because the reactivity of the immune system and the permeability of the placenta vary from person to person, only a small proportion of women who might be expected to have infants with haemolytic disease do in fact have affected babies.

*Transfusion with incompatible red cells* occasionally occurs despite strenuous effort to avoid it and leads to rapid and severe haemolysis which may be rapidly fatal due to circulatory failure associated with renal damage by the haemoglobin released from lysed red cells. Incompatible plasma is less likely to cause trouble because the infused antibodies are diluted by the recipient's plasma.

**Bone marrow depression** may be due to disease specifically affecting the bone marrow, or secondary to disease elsewhere in the body. The marrow may be directly affected by a variety of chemicals, including drugs; by exposure to X-rays and other forms of radiation; and by replacement of the normal red bone marrow by fibrous material (myelofibrosis) or by cancer cells, notably leukaemia cells. When red bone marrow is so replaced, formation of blood may take place in the liver and spleen (*extramedullary haemopoiesis*—normal in the fetus), thereby markedly enlarging those organs.

Marrow function may be depressed by any severe disease. In renal failure it is particularly often depressed, probably due to erythropoietin deficiency. As thyroid and glucocorticoid hormones are necessary for normal blood formation, absence of these hormones is associated with anaemia.

## INVESTIGATIONS

Features which suggest a *haemolytic anaemia* are a raised level of bilirubin (the excess being due to the unconjugated form) and an increased proportion of circulating reticulocytes. This *reticulocytosis* is due to the increased bone marrow activity. The increased bilirubin excreted into the intestine is converted into urobilinogen (same as stercobilinogen). Some of this is reabsorbed and excreted in the urine to give an increased level of urinary urobilinogen. An increased tendency for red cells to haemolyse in hypotonic saline (osmotic fragility) suggests that haemolysis is due to red cell abnormality; a positive *Coombs' test* suggests that it is due to antibodies. The Coombs' test is positive if the patient's circulating erythrocytes have antibodies attached to them. The test is carried out by mixing the patient's red cells with antibody to human globulin (the antibody is obtained by injecting human globulin into animals). If gamma globulin (haemolysing antibody) is attached to the cells (a stage in the process of lysis) the anti-globulin antibody leads to agglutination (Fig. 114). The red cells are "washed", to remove free circulating globulins, before the test is carried out.

Examination of a sample of bone marrow may help in the diagnosis of conditions such as myelofibrosis, the leukaemias and secondary bone marrow invasion by a tumour.

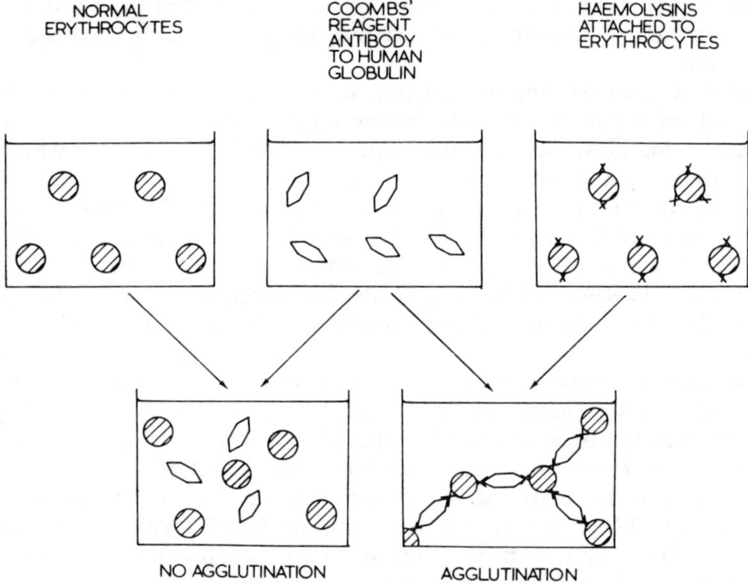


FIG. 114.—The Coombs' test for the presence of haemolysins as a cause of haemolytic anaemia. When normal erythrocytes are mixed with Coombs' reagent, there is no agglutination. When erythrocytes with haemolysins attached to them are mixed with the reagent, the Coombs' antibody unites with the haemolysins and agglutination results.

## Treatment

When *haemolysis* is due to a drug, poison or toxin, removal of the causative agent is required. When the red cells are abnormally fragile, removal of the spleen—the major site of red cell breakdown—usually improves the anaemia. Severe cases of haemolytic disease of the fetus may benefit from intra-uterine transfusion—the blood is passed into the peritoneal cavity of the fetus, from which it is absorbed by an unknown mechanism. More commonly, haemolytic disease of the newborn is treated shortly after birth by *exchange transfusion.*

Regular small volumes of blood are removed from the Rhesus-positive infant and replaced, not by Rhesus-positive blood, but by blood containing Rhesus-negative cells, which are not affected by the circulating anti-Rh antibody. This treatment raises the red cell count and washes out some of the excess bilirubin. A similar procedure may be used to treat a severe transfusion reaction in an adult.

*Haemolytic disease due to Rhesus incompatibility* can be *prevented* in most cases by administering anti-D antibody (D is the commonest of the Rhesus antigens, C, D, E) to the Rhesus-negative mother immediately after the birth of her baby or after an abortion. In this way, the Rhesus positive fetal cells in the mother's circulation are eliminated and cannot induce the antibodies which would cause harm in a subsequent pregnancy.

When haemolysins produced by the patient against his own red cells are a problem, glucocorticoids help by their suppression of the immune mechanisms. Severe haemolytic anaemia may require blood transfusion to raise the red cell count.

*Bone marrow depression* by general disease can be cured only if the general disease is relieved. For example, a successful kidney transplant cures the anaemia caused by renal failure. Otherwise repeated transfusions may be required. If very large numbers of transfusions are given, the patient tends to become sensitized to various foreign red cell antigens, so that eventually it may be very hard to obtain blood which will not be haemolysed by the patient's plasma.

# 60. INACTIVATION OF HAEMOGLOBIN

## Definition

In the conditions described under this heading, oxygen carriage is impaired, not by a shortage of haemoglobin, but by a chemical change in haemoglobin so that it can no longer combine with oxygen in a suitable fashion for oxygen carriage in the body. The common causes are conversion of haemoglobin to carboxyhaemoglobin (COHb), methaemoglobin or sulphaemoglobin.

## Effects

These are similar to anaemia. The altered haemoglobin, being unable to form a readily reversible compound with oxygen, is useless for oxygen carriage. Thus if 40 per cent of the circulating haemoglobin is inactivated, the condition is equivalent to an anaemia where the haemoglobin level is 60 per cent of normal. In addition, however, both COHb and methaemoglobin tend to shift the $O_2$ dissociation curve of the remaining normal haemoglobin to the left so that it gives up its oxygen to the tissues less readily. In the case of carbon monoxide (CO) poisoning the condition usually develops rapidly, e.g. in a quarter or half an hour, and the main effect is severe generalized hypoxia leading to unconsciousness due to depression of brain function. If treated, the condition may be completely cured, but in some cases there is permanent brain damage.

The abnormal pigments have their own characteristic colour and result in characteristic coloration of the skin, provided that the skin vessels are well filled. Patients with CO poisoning tend to look pink, those with methaemoglobinaemia, blue.

About 10 per cent of the circulating haemoglobin (Hb) is bound to CO in smokers and those exposed to heavy exhaust gas pollution, e.g. in traffic. When 20 per cent of the Hb is bound to CO, mild impairment of brain function may occur; 60 per cent COHb leads to unconsciousness; death usually occurs if the COHb level reaches 80 per cent. As $Po_2$ remains normal there is no dyspnoea, and the victim may be unaware that he is being poisoned.

## Causes

Carbon monoxide poisoning arises most commonly from the inhalation of coal gas, internal combustion engine exhausts and oil heater fumes in poorly ventilated spaces. As CO has about 300 times the affinity of oxygen for haemoglobin, it follows that if CO is breathed for long enough at 1/300 of the concentration of oxygen (i.e. about 0.1 per cent in air) half of the circulating haemoglobin will be rendered useless to the body by conversion to COHb. Larger concentrations can rapidly reduce the free haemoglobin concentration to a level at which brain function cannot continue and unconsciousness and death result. The exhaust from a petrol engine contains about 7 per cent CO.

Methaemoglobinaemia may be caused by a number of drugs and poisons. It may also result from a genetic abnormality of the erythrocytes. The iron in the haemoglobin is oxidised from $Fe^{2+}$ to $Fe^{3+}$. Sulphaemoglobinaemia has similar causes to methaemoglobinaemia and the two may be present together.

### Investigations

The diagnosis of CO poisoning is usually clear from the victim's environment. Its presence in the blood can be determined spectroscopically.

Two fairly simple tests confirm the presence of abnormal blood pigments such as methaemoglobin. If blood containing such pigments is shaken in air for a few minutes, it fails to turn the normal bright red colour. Spectroscopic examination shows a pattern distinct from those due to oxygenated and reduced haemoglobin.

### Treatment

In the case of CO poisoning, the aim is to displace the CO from the Hb by the administration of oxygen at a high partial pressure. Oxygen and carbon monoxide compete for the combining sites on the haemoglobin molecule. The first step is to remove the victim from the CO atmosphere *immediately,* to stop further absorption of CO. Administration of $O_2$ should be started as soon as possible. If the patient's respiration is depressed by the direct effect of hypoxia on the respiratory centre, artificial ventilation may be required. If available, oxygen should be administered at high pressure, e.g. 2-3 atmospheres (*hyperbaric oxygen*). The high $Po_2$ has two beneficial effects: (i) it enables $O_2$ to compete on more equal terms with CO for haemoglobin, (ii) the oxygen dissolved in arterial blood rises to amounts which are capable of providing most of the tissue requirements. If the $Po_2$ in arterial blood is raised from 100 to 1500 mm Hg, the dissolved $O_2$ rises from a negligible 0.3 ml to 4.5 ml/100 ml blood. Such a quantity is sufficient to meet the resting needs of the body for oxygen.

In the case of the abnormal blood pigments, removal of the cause is usually sufficient, but severe methaemoglobinaemia may be treated with a reducing agent such as methylene blue or ascorbic acid. These reduce the iron in the haemoglobin from the ferric to the ferrous form.

# 61. INCREASE IN HAEMOGLOBIN (POLYCYTHAEMIA)

## Definition

THIS IS A condition in which the total amount and often the concentration of circulating haemoglobin are above normal. It is in this respect the reverse of anaemia.

## Effects

The amount of oxygen carried by the blood at all oxygen tensions is increased in proportion to the degree of polycythaemia. More oxygen than normal is released with a given fall in $Po_2$ (Fig. 115). This is an important compensating mechanism in certain conditions where there is a tendency towards tissue hypoxia.

This beneficial effect of increased oxygen carriage is counteracted by an increased blood volume and viscosity. As the haematocrit rises above the normal value of around 45 per cent, the viscosity rises increasingly steeply, so that at a

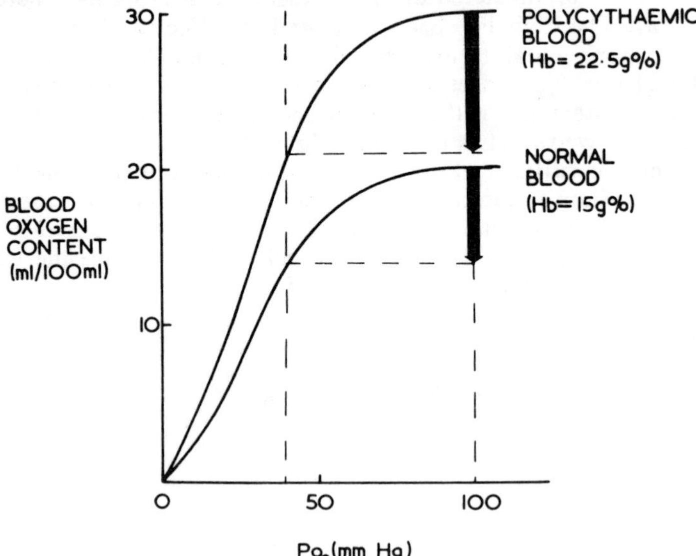


FIG. 115.—Oxygen dissociation curves for normal and polycythaemic blood. For a given fall in $Po_2$ from 100 to 40 mm Hg (as normally occurs in blood perfusing the tissues) more oxygen is released from polycythaemic (b) than from normal (a) blood.

haematocrit of 55 per cent, it is about one and a half times normal, and at 65 per cent it is twice normal (Fig. 111). The increased blood volume leads to an increase in cardiac output and the increased viscosity leads to an increased peripheral resistance. These effects tend to raise the blood pressure, and to increase the work of the heart and the risk of heart failure.

There is also an increased chance of the blood clotting within the circulation, e.g. in peripheral vessels and in those supplying the heart and brain. The cause is not known. It may be due to slowing of the blood in certain vessels because of the rise in viscosity, or an increase in the platelet count which occurs in some types of polycythaemia. Both of these factors would increase the liability to intravascular clotting.

The patient's appearance is the reverse of anaemia, with increased skin redness which may be accompanied by a bluish tinge due to an increased amount of reduced haemoglobin in the small superficial vessels.

## Causes

Polycythaemia has a number of causes:

**Hypoxia.**—Polycythaemia is a normal response to hypoxia and polycythaemia secondary to hypoxia is a physiological compensation rather than a blood disease. There is no increase in granulocyte and platelet formation.

Polycythaemia is part of the normal physiological response to living at altitudes above about 3,000 m where there is a marked reduction in the inspired $Po_2$. The hypoxia causes release from the kidney of erythropoietin which stimulates erythropoiesis in the bone marrow.

Diseases which cause the arterial $Po_2$ to fall cause polycythaemia in the same way. These include certain lung diseases, and certain heart abnormalities in which venous blood is shunted to the arterial side of the circulation, bypassing the lungs.

**Excessive circulating erythropoietin.**—Occasionally renal tumours cause polycythaemia by releasing inappropriate amounts of erythropoietin.

**Bone marrow abnormality.**—In this case (*primary polycythaemia*), the abnormality seems to be in the bone marrow which increases its formation not only of red cells, but also of neutrophils and platelets. Confirmation of the suggestion of a primary bone marrow abnormality lies in the fact that after some years leukaemia or general depression of bone marrow activity may develop. The erythropoietin level is normal or low in primary polycythaemia. This also is consistent with an organ which has become independent of its normal controlling mechanism.

A raised haematocrit occurs when fluids but not red cells are lost from the circulation. This is sometimes referred to as "relative polycythaemia", but it is probably better described as haemoconcentration. It is not a genuine polycythaemia because the total circulating red cell mass is not increased.

## Investigations

Although polycythaemia (increased red cell mass) may occur with a normal haematocrit if the blood volume increases in proportion to the red cell increase, for practical purposes it is diagnosed by finding an increase in one or more of the primary blood indices— the haemoglobin concentration, haematocrit or red cell count. The finding of a raised neutrophil and platelet count suggests that the polycythaemia is primary, rather than secondary to tissue hypoxia.

### Treatment

When the condition is a compensation for tissue hypoxia, the polycythaemia is not treated unless it becomes so marked that the risks of increased viscosity outweigh the benefits of increased oxygen carriage. Primary polycythaemia requires treatment to combat the steadily rising red cell mass and haematocrit. The ancient medical procedure of blood-letting or venesection lowers the blood volume, and, by subsequent haemodilution, the haematocrit. This treatment is usually accompanied by treatment with a drug which suppresses bone marrow activity.

# 62. DEFICIENCY OF GRANULOCYTES (AGRANULOCYTOSIS)

## Definition

In this condition, the number of circulating neutrophil polymorphonuclear granulocytes (neutrophils) is below normal. The term *neutropenia* tends to be used when there is a moderate fall below the normal level and *agranulocytosis* tends to be used when there are few, if any, circulating neutrophils.

When deciding whether there is a deficiency or an excess of a certain type of white cell, it is better to think in absolute numbers than in percentages. The number of circulating neutrophils is usually around 3,000-6,000/$mm^3$ and they normally comprise around 60 per cent of circulating white cells. If the total white cell count is raised to 50,000/$mm^3$ by an excess of lymphocytes and the neutrophils comprise 10 per cent of the total, then there are 5,000 neutrophils/$mm^3$. Although the *percentage* is much below normal, the actual number/$mm^3$ is normal and the condition should *not* be described as a neutropenia.

## Effects

As circulating neutrophils by their phagocytic action are an important part of the body's defence against bacterial invasion, the effects of their absence are the effects of uncontrolled multiplication in the body of the bacteria which are constantly invading it. The situation is analogous to a strike of a city police force when criminal tendencies normally suppressed find greatly increased expression. The severity of the condition depends on the extent of the reduction in the neutrophil count. If the count is virtually zero (agranulocytosis), overwhelming infection spreads throughout the body, bacteria are disseminated through the blood stream (*septicaemia*) and death may occur in a few days. If the neutropenia is relatively mild, the patient is unlikely to die but suffers from the weakening effects of inadequately controlled bacterial infection.

Most bacteria enter the body by inhalation and are deposited at the back of the throat. A typical early complaint in neutropenia is a sore throat. The infection spreads over the throat and mouth. If severe, destruction of the mucosa and epithelium is seen as ulceration. There is little sign of pus—pus consists of neutrophils which have done their work of ingesting bacteria and have been killed by toxic breakdown products from the dead bacteria. Products released from bacteria and damaged tissue lead to stimulation of the mechanisms which raise body temperature; in some cases, there may be intense shivering (rigors; L. *rigor* = stiffness).

Thus the effects of neutropenia are, in a mild case, sore throat, weakness and fever; in a severe case, death from overwhelming bacterial infection.

## Causes

A variety of drugs can depress formation of neutrophils. For reasons which are unknown, only a small proportion of patients who receive these drugs develop neutropenia as a result. In these patients, the drug may combine as a hapten with a protein in the membrane of the neutrophils. When antibodies develop, the resulting antigen-antibody reaction may result in destruction of neutrophils. Naturally, the greater the dose of the drug and the longer it is given, the greater the risk. Ionizing radiation may also cause agranulocytosis; it acts by interfering with the mitosis and multiplication of primitive white cells in the bone marrow.

Any disease which interferes with bone marrow function in general must cause a neutropenia. Such diseases are the leukaemias and fibrosis of the bone marrow. Suppressive treatment for a leukaemia or other cancer may itself cause a neutropenia. As might be expected, it is not always possible to tell why neutropenia has developed in a given patient.

### Investigations

With experience it is possible to diagnose a neutropenia by a brief examination of the blood film. The diagnosis is confirmed by calculating the neutrophil count/$mm^3$ (total white cell count multiplied by percentage of neutrophils). Examination of the bone marrow may show abnormality or absence of the neutrophil precursors (myeloblasts); if neutropenia is secondary to general bone marrow disease this can be verified.

If there is evidence of severe infection, strenuous attempts should be made to identify the bacteria responsible and determine to which antibiotic drugs they are sensitive. This involves culturing material from a site of infection, and culturing the blood, urine and possibly cerebrospinal fluid.

If it is essential to administer a drug which carries a high risk of causing neutropenia, the neutrophil count should be measured at intervals while the drug is given.

### Treatment

If a drug is suspected of causing neutropenia, its administration should be stopped at once and the neutrophil count checked. If neutropenia is due to a drug, neutrophil formation will usually recommence in a few days when the drug is stopped.

A severe but temporary shortage of neutrophils can be overcome by (i) giving appropriate antibiotic drugs to deal with infection if present, (ii) avoiding, as far as possible, fresh infection and (iii) transfusing blood as a source of neutrophils. Blood from a patient with a high neutrophil count due to an abscess or certain forms of leukaemia may be used as a rich source of neutrophils.

*Barrier nursing* is usually adopted to minimize the risks of infection. This procedure involves restricting to the minimum the number of people who come in contact with the patient. In particular, contact with other patients is avoided, usually by nursing the patient in a separate room. Those who do approach the patient wear clean, preferably sterile, gowns, masks and caps as for a surgical operation.

Finally, neutropenia can be prevented by withholding potentially dangerous drugs if they are not essential. If a patient has once suffered from neutropenia due to a specific drug, he must never again be given that drug.

# 63. ABNORMAL PRODUCTION OF WHITE CELLS (LEUKAEMIA)

## Definition

LEUKAEMIA is a disease in which there is excessive multiplication of abnormal white cells. There is usually, but not invariably, a raised white cell count in the peripheral circulation.

The condition is described as acute or chronic according to whether it progresses quickly or slowly and it is also named after the predominant cell type. Thus chronic lymphatic leukaemia is a slowly developing condition where the predominant cell is the lymphocyte. Acute myeloid leukaemia develops rapidly and the predominant cell is the neutrophil granulocyte. "Myeloid" here refers to the fact that these cells are derived from the bone marrow (Gr. *myelos* = marrow).

## Effects

The effects are in some respects similar to those of cancer in that large numbers of useless cells are produced, which drain body nutritional resources and interfere with the normal function of the organ (bone marrow in the case of leukaemia). Leukaemic cells spread throughout the bone marrow and are also found in the liver and spleen which may be regarded as reserve haemopoietic organs not normally required after birth. In lymphatic leukaemia the lymph nodes are also involved in the leukaemia process.

Interference with normal bone marrow functions has three main groups of effects—those of anaemia, neutropenia and thrombocytopenia. Even though the neutrophil count may be extremely high in myeloid leukaemia, the neutrophils tend to be abnormal and functionally useless. There is, in fact, a deficiency of normal cells. Thus the weakening effects of anaemia are combined with those of inadequate control of bacterial infection and of an increased risk of haemorrhage throughout the body.

The severity of leukaemia varies considerably. In the more acute cases, the disease may be fatal within a few weeks. In some of the more chronic cases there may be a few symptoms for years.

## Causes

In general, the cause is unknown. Excessive irradiation increases the risk of developing leukaemia.

## Investigations

There is usually, but not invariably, a raised white cell count, higher in the more chronic cases. The predominant cell type usually indicates the type of leukaemia. Precursor cells (primitive cells) usually confined to the bone marrow may be seen in the peripheral blood; the more there are and the more primitive they are, the more acute is the leukaemia. Bone marrow examination confirms the diagnosis by revealing excessive numbers of primitive cells.

### Treatment

There is no effective *cure* for leukaemia but various measures may prolong life and relieve symptoms. General treatment is directed at suppressing the leukaemia cells. Drugs which interfere with cellular multiplication are used; because leukaemia cells multiply at a very high rate, they are more sensitive to these drugs than other tissues. However, general ill-effects of the drugs limit their use. Specific treatment of anaemia, neutropenia and thrombocytopenia may be given as required.

# 64. ABNORMAL BLEEDING (HAEMORRHAGIC DISORDERS)

## DEFINITION

IN THE normal individual, *spontaneous bleeding* is prevented by (i) the integrity of the normal *vascular wall,* (ii) rapid *plugging* by platelets of leaks which develop in blood vessels and (iii) *coagulation* of the blood in and around damaged blood vessels. Excessive bleeding occurs when one or more of these main defence mechanisms breaks down.

## EFFECTS

Although individual haemorrhagic diseases tend to have distinguishing features which help in deciding the cause of the trouble in a particular patient, the effects overlap to a considerable degree. Bleeding may arise "spontaneously", i.e. in the absence of any injury which would be expected to cause bleeding in normal people. Secondly, bleeding following a definite injury tends to be more severe and prolonged than it would be in a normal person who had suffered the same injury.

Spontaneous bleeding is most easily detected when it occurs in the skin or the mucous membrane of the mouth. This excessive superficial bleeding is referred to as *purpura* (L. *purpura* = purple) because of the purple discoloration of the overlying skin. The appearance may be that of multiple large bruises. Sometimes there are large numbers of very tiny haemorrhages which give rise to purple spots 1-2 mm or less across—these are known as *petechiae* (Ital. *petecchia* = flea bites) and, unlike spots due to vascular dilatation, do not disappear with pressure on the skin. The colour of the bruises or spots gradually changes to shades of brown, yellow and green due to bile pigments produced from the breakdown of haemoglobin, as in an ordinary bruise.

Spontaneous bleeding may arise in any part of the body e.g., nose bleeding, coughing up of blood from the lungs, gastro-intestinal bleeding or bleeding from the urinary tract. Painful bleeding may occur into joints, in which case it is followed by the laying down of fibrous scar tissue as part of the healing process. If this happens repeatedly the joint is likely to become stiff and deformed. Bleeding from cerebral blood vessels may cause serious or fatal brain damage.

In conditions where bleeding is normally to be expected, patients with haemorrhagic disorders tend to lose much more blood than usual. Bleeding after such procedures as removal of tonsils, or extraction of teeth, may be fatal if not adequately treated.

The main serious effects of haemorrhagic disease are thus excessive blood loss and damage to tissues such as the brain by the local effects of the bleeding.

## CAUSES

A simple classification of the commoner haemorrhagic diseases is given in Fig. 116.

### Vascular Fragility (Vascular Purpura)

The ability of capillary walls to withstand distending pressure may be reduced by a wide variety of general diseases, including severe infections, liver and kidney failure and vitamin C deficiency. Antigen-antibody reactions sometimes lead to capillary damage (anaphylactoid purpura). Certain drugs and poisons also have a tendency to damage capillaries. Some people tend to bruise much more readily than others. This is presumably due to a slight vascular weakness, but is not to be regarded as a disease if the individual is otherwise healthy. An increased bruising tendency is not uncommon in elderly people, when it is referred to as *senile purpura.*

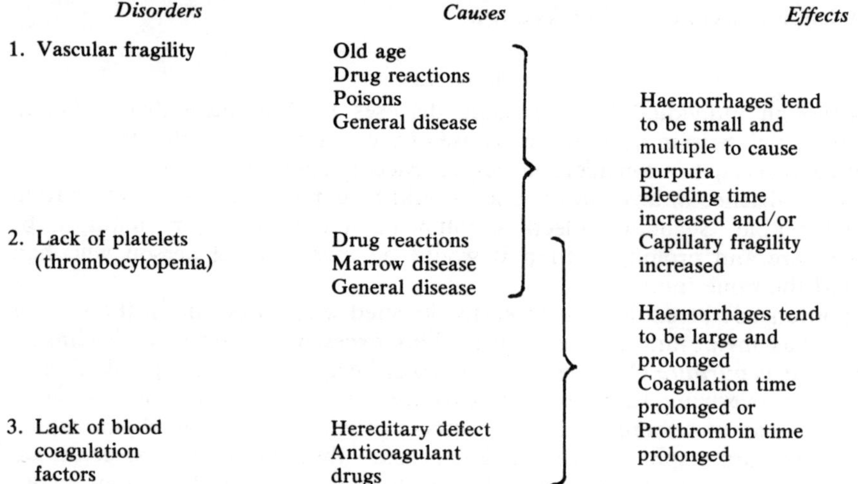


FIG. 116.—A classification of the haemorrhagic disorders, with typical clinical and laboratory findings.

Occasionally patients are found to have a number of small clusters of abnormal, dilated, weak-walled blood vessels (*telangiectasia;* L. *tela* = web, Gr. *angeion* = vessel, *ektasis* = dilatation). These may occur in the skin, lung, alimentary tract etc. and bleed profusely if damaged.

### Lack of Platelets (Thrombocytopenia)

As with anaemia and neutropenia, this may arise from depression of bone marrow function due to fibrosis, leukaemia, severe general disease or drugs. In some cases, antibodies are produced against the patient's own platelets. Not uncommonly, the cause of the thrombocytopenia cannot be discovered.

There is normally a large reserve of platelets. Although the lower limit of the normal count is around 150,000/mm$^3$, abnormal bleeding does not usually occur until the count falls below 40,000/mm$^3$.

### Lack of Blood Coagulation Factors

If any one of the factors required for coagulation is seriously deficient, the chain of reactions necessary for clotting is greatly slowed down. Again the normal amounts in the circulation are above the minimum required for normal clotting.

The greater the deficiency, the more readily is serious bleeding produced. For example, in the case of haemophilia (factor VIII deficiency), a factor VIII concentration of 1-2 per cent of normal is likely to be associated with serious spontaneous bleeding episodes. When the level is about 10-20 per cent, trivial injury is likely to cause serious and prolonged bleeding. Lesser deficiencies may be associated with excessive bleeding after serious injury.

The commoner coagulation defects (apart from platelet deficiency) can be classified as genetic or acquired. Genetic deficiencies e.g. haemophilia, tend to be single factor deficiencies. Acquired deficiencies tend to involve several factors with deficiency of one factor, commonly prothrombin or fibrinogen, predominating. Impaired coagulation is not caused *in patients* by calcium deficiency. Hypocalcaemia would kill the patient by causing severe tetany before the blood calcium level had fallen low enough to interfere with clotting.

*Haemophilia* is due to factor VIII deficiency. The disease has a genetic basis and is presumably carried on the X gene as a recessive character. Women with one abnormal gene do not suffer from the disease but will transmit the gene to, on average, half their children, so that half their sons will suffer from haemophilia and half their daughters will be carriers. Men with the abnormal gene suffer from the disease. Their sons will not suffer from haemophilia, but their daughters will all be carriers.

Prothrombin formation takes place in the liver, vitamin K playing an essential role. Most of the body's requirements of vitamin K are met by intestinal bacteria which synthesize the vitamin. Such bacteria are not present at birth, and this may explain why infants are particularly prone to *prothrombin deficiency* (haemorrhagic disease of the newborn). In adults, liver disease and malabsorption of vitamin K lead to prothrombin deficiency. As vitamin K is fat-soluble, anything which interferes with fat absorption, including obstruction of the common bile duct, is liable to cause vitamin K deficiency and a bleeding disorder. A common cause of deliberately induced prothrombin deficiency is oral anticoagulant therapy. The drugs used interfere with the synthesis of prothrombin in the liver. As with insulin therapy, the dose must be strictly regulated and one of the serious drawbacks of these drugs is the risk of severe haemorrhage should the prothrombin level in the blood fall unexpectedly. A number of other drugs, e.g. aspirin, can interfere with coagulation as a side-effect. The other common form of anticoagulant therapy, heparin by intravenous or subcutaneous administration, exerts its effects by antagonizing thrombin, the active form of prothrombin.

*Fibrinogen deficiency* may also be due to liver disease, but most often occurs in pregnancy. It is thought that the placenta may produce substances which lead to deposition of fibrin throughout the vascular system. This does not generally lead to serious vascular blockage, but uses up the fibrinogen so that it is not available when required.

## Investigations

Results are not always clear-cut, because more than one abnormality may be present and results of tests must be interpreted in the light of the clinical history and findings.

*Capillary resistance* (fragility) is assessed by inflating a sphygmomanometer cuff on the patient's arm to a pressure midway between systolic and diastolic for 5 minutes. This manoeuvre obstructs venous return and the capillaries become distended as the pressure within them rises to the cuff pressure. If the capillaries are abnormal, or if there is a platelet deficiency, large numbers of haemorrhagic spots (petechiae) appear. In normal people only a few tiny spots are produced.

*Bleeding time* is measured by making a small prick in the skin, e.g. on the ear lobe, and blotting off the blood at intervals until no more appears. The bleeding time is prolonged in platelet deficiency and in some cases of capillary abnormality.

The *platelet count* provides the diagnosis in thrombocytopenia.

The *coagulation time* is measured in small clean glass test tubes. Blood from a clean venepuncture is placed in the tubes which are tilted at one minute intervals to see whether clotting has occurred. The coagulation time is a relatively insensitive test. It is prolonged in severe haemophilia and when there is a shortage of fibrinogen. When there is prothrombin deficiency, the test may be within normal limits.

The *prothrombin time* is measured by mixing blood with a standard tissue thromboplastin. The more prothrombin that is present, the sooner coagulation occurs, unless there is a deficiency of fibrinogen for the thrombin to act upon.

If one of the factors, such as factor VIII, responsible for the reactions leading to prothrombin activation, is absent, this can be discovered by adding various factor preparations in turn to the blood. The one which restores coagulation to normal is the one which is missing in the patient.

## Treatment

Where the haemorrhagic disease is secondary to general disease, the appropriate treatment for this should be given. If the bleeding has led to severe anaemia, iron therapy or blood transfusion may be required.

Platelet deficiency may be temporarily relieved by the platelets contained in a transfusion of fresh blood (platelets degenerate rapidly when stored). Glucocorticoid therapy may relieve thrombocytopenia when this is due to auto-antibodies. Otherwise removal of the spleen may be required. This generally leads to a rise in the platelet count, because the spleen normally destroys platelets at a high rate (the platelet count rises also in normal people after removal of the spleen).

If prothrombin deficiency is due to malabsorption of vitamin K, injection of the vitamin is required. If excessive oral anticoagulants have been given, administration of vitamin K speeds up prothrombin synthesis. When concentrates of the deficient factors such as antihaemophilic globulin are available, injection of these in adequate amounts will restore coagulation to normal. Otherwise transfusion of fresh normal blood, which contains all necessary coagulation factors, will often stop serious bleeding by temporarily restoring normal coagulation.

If antihaemophilic globulin (AHG) were available in a form which could be injected regularly into the patient, the condition might be kept permanently under control. A number of serious problems stand in the way of this ideal solution. The material would probably have to be injected intravenously because, even with, say, twice daily intramuscular injections, the effect of the AHG would wear off between injections and haemorrhage would tend to occur at the injection site. As with frequent blood transfusions, there would be the problems of transmitted infections (e.g. hepatitis) and antigen-antibody reactions.

Thus, at present, treatment is given only when serious bleeding is taking place or when there is a risk of this, e.g. during tooth extraction or a surgical operation, major or minor. Where the injury is local and accessible, the application of foam impregnated with thrombin to the wound may cause coagulation in patients with haemophilia. This converts their fibrinogen to fibrin, bypassing their defective generation of active thromboplastin.

# 65. INTRAVASCULAR CLOT FORMATION (THROMBOSIS)

## Definition

THROMBOSIS is the clotting of blood within the circulation. It may occur in blood vessels of all sizes and in the heart.

## Effects

### Local Interference with Circulation

Effects differ, depending on whether or not the blocked region can be bypassed using alternative circulatory routes (collateral blood vessels) and on whether the vessels involved are arterial or venous.

The adequacy of a collateral circulation varies from region to region. In addition, if thrombosis follows a gradual obstruction of the circulation, e.g. by intimal thickening, there is a greater chance of an adequate collateral circulation having developed.

In general, blockage of the arterial side of the circulation tends to cause death of the tissues supplied. Blockage of the venous side of the circulation causes oedema due to a raised hydrostatic pressure at capillary level.

### Embolism

A clot (thrombus) in a large blood vessel or in the heart frequently releases fragments of various sizes which are carried along as emboli (Gr. *en* = in, *ballein* = to throw) by the circulating blood until they are trapped in a vessel too small for them to pass through. Factors such as clot retraction and the erosive action of the blood flow facilitate the dislodgement of fragments of clot. The effect of embolism is to halt abruptly blood flow through the vessel in which the embolus is trapped. In addition, the arrival of the embolus may cause local constriction of nearby vessels; chemicals, released from the thrombus (e.g. 5-hydroxytryptamine from platelets) may be responsible for the vasoconstriction. Embolism may affect the *pulmonary* or the *systemic* circulation.

*Pulmonary embolism.*—An embolus from a systemic vein (often in the leg or pelvis) travels to the right side of the heart. It is then expelled into the pulmonary circulation and trapped there. An embolus originating in the right side of the heart has a similar effect.

If the *pulmonary embolus* is very large, it may cause death because blood flow through the pulmonary circuit and hence the systemic circuit virtually ceases; there may also be a reflex abnormality of cardiac rhythm. Massive pulmonary embolism is a major cause of sudden death among patients who have recently had a surgical operation, among women who have recently given birth, and among patients who are confined to bed (immobitlity of the legs is the major cause of the leg vein thrombosis which leads to pulmonary embolism).

A smaller pulmonary embolus which fails to cause sudden death can produce severe effects. If more than two-thirds of the pulmonary vascular bed is blocked, the flow of blood through the lungs at rest is seriously impeded. Peripheral circulatory failure may result. The dammed-up blood dilates the right side of the heart and engorges the systemic veins. Output from the left side of the heart falls abruptly. The blood pressure may fall severely, despite compensatory vasoconstriction. Often, but not invariably, pulmonary tissue is damaged (*pulmonary infarction*) by impaired blood flow and bleeding occurs into the damaged region. The patient may then cough up blood. The effects of pulmonary embolism are summarized in Fig. 117. The main symptoms are discomfort or pain in the chest, shortness of breath and, if the damaged region of lung includes the visceral pleura, a sharp pain in the chest with each breath (pleuritic pain).

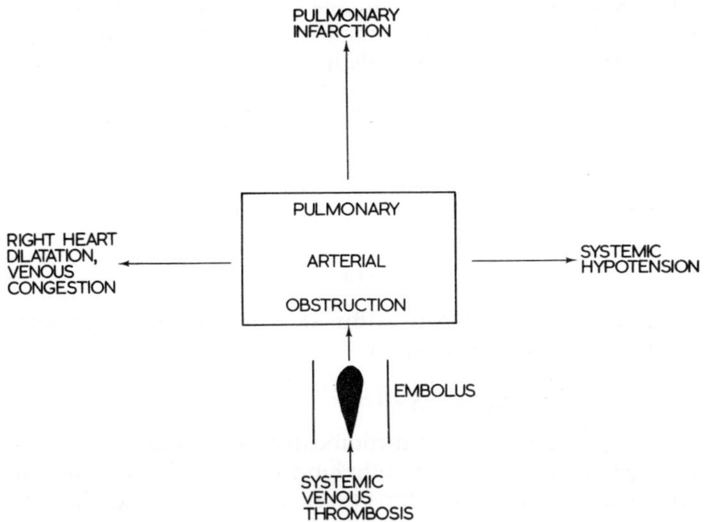


FIG. 117.—Cause and possible effects of a pulmonary embolus, large enough to obstruct more than two-thirds of the cross section of the pulmonary arterial vessels.

If the patient survives, his condition tends to improve, as pulmonary blood flow increases due to gradual shrinkage and erosion of the embolus. However, there is a high risk of a second, possibly larger embolus, unless appropriate treatment is given and is successful.

In some cases, multiple small emboli are thrown into the pulmonary circulation over a period of time. These small emboli may not produce any immediate effects, but eventually may cause a serious rise in pulmonary vascular resistance. This leads to pulmonary hypertension. The right ventricle hypertrophies and may then fail. Lung function deteriorates. The picture may closely resemble chronic obstructive lung disease, e.g. emphysema.

*Systemic embolism.*—Systemic embolism occurs when a fragment of clot separates from the wall of the left side of the heart or a major artery, e.g. the aorta. Effects are similar to those following thrombosis of the arterial vessel which is eventually blocked. Thus cerebral embolism due to a thrombus in the left side of the heart produces a "stroke"; embolism to a leg artery may cause severe pain and gangrene of the limb. In such cases, symptoms arise abruptly because blood flow is interrupted instantaneously.

### Clotting Factor Deficiency

The reserves of clotting factors can easily withstand the mild depletion caused by the clotting associated with most injuries. Rarely this is not the case. A deficiency of clotting factors can arise after extensive clotting in which widespread deposition of fibrin occurs within the body. This depletes reserves of clotting factors, especially fibrin. The condition is referred to as *disseminated intravascular coagulation.*

Disseminated intravascular coagulation and a consequent haemorrhagic tendency may be caused by tissue damage due, e.g. to a severe crush injury, premature separation of the placenta, or an incompatible blood transfusion. The damaged tissues (or lysed red cells in the case of the incompatible transfusion) release tissue clotting factors which lead to the widespread coagulation of blood, typically in small vessels. The condition may, in some cases, be alleviated by the use of heparin. This prevents the widespread clotting and conserves clotting factors. In this situation, heparin reverses its normal role by *preventing* haemorrhagic disease.

## CAUSES

The cause of clotting of blood within the circulation is not known. It is, however, favoured by three conditions.

### Local Damage to the Vessel Wall

By far the commonest cause of thrombosis is abnormality of the vessel wall. It is not clear how damage to the endothelium of a blood vessel initiates clotting which was previously prevented by the presence of normal endothelium. Thrombosis of coronary and cerebral vessels is a major cause of death. Almost invariably, there is disease of the vessels involved. The cause of this arterial disease, which consists of a degeneration and thickening of the intima, is one of the great unsolved mysteries of medicine. Various factors such as dietary patterns, hypertension and mental stress may be of some importance, but the fundamental mechanisms remain obscure.

Venous vessels are somewhat less commonly responsible for the initiating of thrombosis. The vessels of the legs and the pelvic veins (especially after surgery to the region) are the commonest site of disease (phlebitis—inflammation of a vein, Gr. *phlebos* =vein) and associated thrombosis.

### Stasis of Blood Flow

Slowing or stopping of the blood flow in a vessel (Gr. *stasis* = a standing still) is associated with an increased risk of thrombosis. Presumably there is an increased chance of the accumulation of a large thrombus when there is no swift blood flow to sweep away the small beginnings of such a thrombus. Stasis of blood flow probably plays a large part in the tendency towards leg vein thrombosis in patients who have undergone surgery or have been confined to bed or have had a limb immobilized following injury. It has been found that the increased blood flow associated with exercise of the leg muscles in such patients reduces the tendency to thrombosis.

Thrombosis also tends to occur in an atrium which is fibrillating. Again blood flow in the periphery of such a chamber is sluggish as compared with the normal situation.

### Excessive Clotting Tendency

Following injury, surgery and childbirth, there is an increased tendency of the blood to clot. This compensatory mechanism is partly responsible for the tendency to thrombosis in such patients. It is associated with a raised platelet count, an increased platelet adhesiveness and a decreased coagulation time.

As mentioned earlier, fibrin deposition throughout the vasculature, especially the small vessels (disseminated intravascular coagulation) may be caused by widespread tissue destruction. This results in release into the circulation of tissue clotting factors which can lead to activation of prothrombin and hence formation of fibrin.

## Investigations

Investigations in patients who have suffered a thrombosis are concerned largely with the effects of the thrombosis on local blood flow. Thus in a patient who has suffered thrombosis of a coronary artery, evidence for myocardial damage is sought in the electrocardiogram and in the level of circulating enzymes released from a damaged myocardium. Occasionally in coronary artery disease and in cerebrovascular disease, an attempt may be made to identify the region of blockage of the artery by X-ray studies after the injection into the circulation of an X-ray opaque material. X-ray studies of the pulmonary arteries (pulmonary angiography) may be made in order to identify a pulmonary embolus.

The presence of a *deep venous thrombosis* in the legs cannot be diagnosed reliably on clinical grounds. Investigations which may aid the diagnosis include *phlebography* (X-ray studies of the leg veins after the injection of an X-ray opaque material) and studies using *radio-active fibrinogen.* The fibrinogen is obtained from human blood and radio-active iodine is attached to it. It is then given to the patient and, if thrombosis is occurring, the radio-active fibrinogen is incorporated into the thrombus and can be detected in the leg by a scintillation counter.

Disturbances in clotting factors (e.g. deficiencies due to intravascular coagulation) are investigated as described in the section on haemorrhagic disorders.

## Treatment

The main treatment to be desired is prevention, because thrombosis, particularly arterial thrombosis, frequently causes permanent tissue damage, e.g. to the heart muscle or brain. In the absence of sufficient knowledge, little can be done to prevent arterial thrombosis and the condition is becoming commoner. The situation is more hopeful with respect to venous thrombosis. By reducing total bedrest to the absolute minimum and by encouraging regular exercise of the leg muscles, the muscle pump can be utilized to reduce venous stasis. Rhythmical electrical stimulation of the leg muscles of unconscious patients and intermittent compression of the lower leg in conscious patients by means of inflatable leggings also appear to improve venous flow and reduce the risk of venous thrombosis.

In certain cases, particularly following embolism, it may be possible to clear a vascular obstruction surgically.

Prevention or removal of thrombosis by anticoagulant or fibrinolytic therapy respectively appears a logical form of treatment for an excessive clotting tendency. Possibly because arterial thrombosis is due mainly to local arterial disease, such therapy has not been found effective in preventing arterial thrombosis. It appears to reduce the risk, however, of thrombosis in veins and in fibrillating atria.

**Anticoagulant therapy** is given in two main forms. (i) *injected heparin* has an immediate effect, mainly by antagonizing the action of thrombin. (ii) A number of *orally administered drugs* suppress the synthesis of prothrombin by the liver. These drugs are coumarin derivatives; they compete with and block the action of vitamin K in the synthesis of prothrombin and their effects can be reversed by the injection of vitamin K. They require several days to have their full effect because the circulating prothrombin present at the beginning of treatment takes that length of time to fall to a level which makes clotting less likely. Similarly, their effect takes some days to wear off.

Anticoagulant therapy carries the risk of serious haemorrhage, either because the patient may bleed from a damaged internal or external vessel or because an excessive dose is given and bleeding occurs spontaneously. As well as serious blood loss, internal damage, e.g. to the brain, may result from the haemorrhage. Other drugs taken incidentally can modify the clotting time. Thus care is needed in selecting suitable patients for anticoagulant therapy and in addition they *must* have regular checks of the prothrombin level in their blood.

## RECENT LEADING ARTICLES


Anticoagulants in mitral valve disease. *Brit. med. J.*, 1972, **1,** 641.
Better results in childhood leukaemia. *Brit. med. J.*, 1973, **2,** 624.
Complications with clotting factors. *Lancet,* 1972, **1,** 729.
Component transfusion therapy. *Brit. med. J.*, 1974, **3,** 544.
Control of oral anticoagulant treatment. *Brit. med. J.*, 1971, **4,** 317.
Detection of venous thrombosis. *Lancet,* 1971, **2,** 306.
Erythropoietin. *Brit. med. J.*, 1972, **1,** 263.
Haemoglobinometry in new units. *Brit. med. J.*, 1972, **3,** 129.
Haemopoiesis. *Lancet,* 1972, **1,** 1056.
Iron. *Lancet,* 1971, **2,** 475.
Management of haemophilia. *Brit. med. J.*, 1966, **2,** 1608.
Mechanism of eosinophilia. *Lancet,* 1971, **2,** 1187.
"Normal" leucocytosis. *Brit. med. J.*, 1972, **1,** 328.

Platelets for transfusion. *Lancet,* 1972, **1,** 25.
Platelet transfusions. *Brit. med. J.,* 1971, **1,** 2.
Prevention of post-operative deep venous thrombosis. *Lancet,* 1971, **1,** 1226.
Prevention of Rhesus haemolytic disease. *Brit. med. J.,* 1967, **4,** 3.
Prophylaxis in haemophilia. *Lancet,* 1971, **1,** 118.
The bleeding time. *Lancet,* 1971, **2,** 579.
The uses of plasma. *Lancet,* 1972, **2,** 1130.
Thromboembolism and oral contraceptives. *Brit. med. J.,* 1974, **1,** 213.

*Section XI*

# DISORDERS OF IMMUNITY

# 66. THE NORMAL IMMUNE RESPONSE

THE TERM "immunity" is derived from the Latin word *Immunis* meaning exempt. It was used to describe people who were exempted from military service and, in that sense, were protected from harm. As applied to man, immunity describes the factors which protect the body against invasion by micro-organisms, poisons or other foreign material. Some factors such as the phagocytes, skin, mucous membranes, and certain substances in the blood provide *non-specific* immunity in that they act against a wide variety of substances and organisms. Other factors such as free-circulating antibodies and sensitized lymphocytes provide *specific* immunity in that they react only with the substances which elicited their formation.

This section deals mainly with the specific forms of immunity and their disorders. Knowledge in this field is very incomplete so the subject is difficult to understand. Terminology is especially difficult because there is not general agreement on the meaning and usage of words; for this reason it is especially important that the student thinks carefully about the meaning of the terms he does use.

There is not as yet general agreement on the classification of disorders of immunity; while immune processes are often *suspected* as the cause of certain diseases, proof is often lacking and the exact mechanisms are obscure. However, this section begins with a short review of normal immune responses and then attempts to describe certain problems arising from immunity in terms of basic concepts.

The immune system normally protects the body against invasion by foreign antigens. *Antigens* are usually large complicated molecules such as proteins or carbohydrates. Small molecules which would not otherwise stimulate the immune system may become antigenic by joining on to body proteins. Such molecules are called *haptens* (Gr. *hapticos* = able to lay hold of). The proteins of the body do not normally elicit an immune response; this is because animals learn to tolerate all proteins they experience in fetal life and do not mount an immune response against them subsequently.

The way the immune system works is illustrated schematically in Fig. 118. When the antigen enters the body (A) it is taken up by the lymphatic system (B) and carried to a neighbouring lymph node. After this one of two courses may be followed, the antigen stimulates the production of either *humoral immunity* (C,E,F) or *cell-mediated immunity* (D,G).

**Humoral immunity.**—In this case the antigen provokes lymphocytes in the lymph nodes to develop into plasmablasts (C). These multiply rapidly and mature into *plasma cells* whose cytoplasm is highly adapted for protein synthesis. Protein manufactured by plasma cells is released as free *antibody* to circulate in the blood with the globulin fraction of the plasma proteins. The globulins with antibody activity are *immunoglobulins* (Ig). They may be classified in several ways but that most frequently used is based on their biochemical structure. The more important immunoglobulins in man include IgG, IgM and IgA.

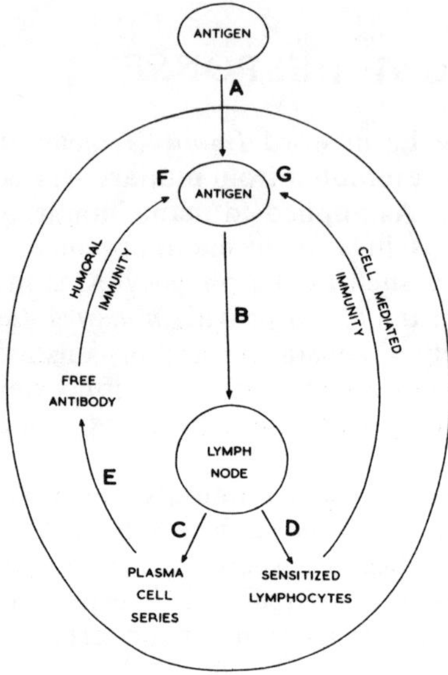


FIG. 118.—The genesis of immunity. Antigen which enters the body is carried to lymph nodes where it elicits the production of free antibody (*humoral immunity*) or sensitized lymphocytes (*cell-mediated immunity*). Free antibody and sensitized lymphocytes can react specifically with the antigen.

*IgG* is the immunoglobulin with the highest concentration in blood and is found in the gamma globulin fraction of the plasma proteins. It also has the lowest molecular weight; because of its small size it can cross the capillary wall into the tissue spaces and thus guard the body fluids generally. It can also cross the placenta from the mother to the fetus. It is responsible for the acquired immunity following bacterial and other invasion.

*IgM* is a large macroglobulin (Gr. *makros* = large). It has a molecular weight of about one million. It cannot cross the capillary wall or the placenta and is concerned with immune reactions in the blood. The "naturally occurring" antibodies of the ABO blood group system are of this type.

*IgA* is another large molecule which is found in blood and in the secretions from the upper respiratory tract, salivary glands, lachrymal glands and the breasts at the beginning of lactation. It is thought to be concerned in the protection of the pharynx and nasopharynx from infection.

Free antibody formed by plasma cells reacts specifically with the antigen which triggered its formation (F). It is thought that the structure of the antibody is complementary to that of the antigen ("key and lock") so that they interact by local physical forces. The *interaction of antibody with antigen* can have many effects, but in general the effect is to neutralize the antigen. When the antigen lies in the surface membrane of a cell the interaction may result in clumping of the cells containing antigen. This is referred to as *agglutination* and the antibodies are called *agglutinins*. In other cases the interaction results in breakdown or *lysis* of the antigen-containing cell and the antibodies are called *lysins*. When antibodies called *opsonins* react with antigens on the surface of bacteria, the bacteria become more liable to phagocytosis. When the antigen is a free-circulating protein, such as toxin, interaction of the antibody (*antitoxin*) with the antigen results in the formation of a non-toxic product. Antibodies called *precipitins* cause precipitation of circulating protein antigens.

In a number of these antigen-antibody reactions circulating plasma components, known as *complement,* are necessary for the occurrence of secondary manifestations of the reactions e.g. red cell lysis. Complement consists of several factors; its reactions occur in stages, require co-factors and rival in complexity blood coagulation reactions. In these reactions complement is consumed.

**Cell-mediated immunity** is seen when a foreign tissue is grafted to the body or when an extract of tubercle bacilli (tuberculin) is injected intradermally. In these cases the antigen "sensitizes" small lymphocytes (D) to produce antibody-like receptors on the cell membrane; antibody is not released free into the plasma. In cell-mediated immunity, these receptors are carried by the lymphocytes back to the antigen. In the case of grafted tissue, clusters of lymphocytes are seen surrounding the graft. The interaction between the sensitized lymphocytes and the antigen (G) results in release of factors from the lymphocyte which have a variety of effects on blood vessels, cell viability and local phagocytic activity. In the case of grafted tissue, it results in damage to the local blood vessels, causing tissue death and rejection of the graft.

The main importance of cell-mediated immunity may be the suppression of cancer cells in the body. The immune system is thought to maintain an "immunological surveillance" for cancer cells which differ antigenically from normal cells. In response to cancer cell antigens, sensitized lymphocytes are formed which in turn kill the cancer cells.

It is thought that two distinct varieties of lymphocyte are involved in the two types of immunity.

The lymphocytes responsible for *cell-mediated immunity* are thought to be derived from lymphoid tissue in the thymus gland and are called *T-cells.* Removal of the thymus gland in a young animal makes it unable to mount the cell-mediated immune response needed to reject a graft of foreign tissue in later life.

The lymphocytes responsible for *humoral immunity* are thought to be derived from lymphoid tissue in the bone marrow in man. They are called *B-cells* because in chickens they are derived from a distinct organ, the bursa of Fabricius. Removal of the bursa of Fabricius (a lymphoid appendage of the lower bowel) makes chickens unable to produce circulating immunoglobulins in later life. B-cells can be differentiated from T-cells by their high surface concentration of immunoglobulin.

As might be expected, most of the circulating lymphocytes are T-cells; the B-cells do not need to circulate to carry out their function.

Disease may occur if the body is unable to mount an adequate immune response against invading organisms (*deficient immunity*). It usually takes the form of an increased susceptibility to infections. Another problem is that certain people become unusually sensitive to one or more foreign antigens and the subsequent antigen-antibody reactions may cause disease (*hypersensitivity*). Occasionally, the immune system mounts an immune response against the tissues of the body (*autoimmunity*). *Abnormal production of immunoglobulins* may occur if the multiplication of plasma cells becomes completely unrestrained because of a cancerous change. These cells produce large quantities of immunoglobulin which is not related to any antigenic stimulation. Finally *blood transfusion* and *tissue transplantation* may cause disease through immune reactions if the tissues are not compatible.

# 67. DEFICIENT IMMUNITY

## DEFINITION

DEFICIENT immunity as discussed here is an impaired ability to mount a humoral or cell-mediated immune response to invasion by a foreign antigen.

## EFFECTS

**Susceptibility to infection.**—Patients with an inadequate immune reponse are subject to recurring infections, particularly of the upper respiratory tract and skin. They are much more likely to succumb to a bacterial or viral infection than a normal person.

**Failure to thrive.**—Young people with recurring infections have impaired growth. They are smaller and weaker than normal children. This is probably related to chronic tissue damage with the repeated infections. Animals whose immune system is damaged about the time of birth develop "runt disease"; they grow into small, feeble and unhealthy animals. Conversely, animals reared in germ-free conditions are larger and healthier than normal.

**Tolerance of grafted tissue and transfused blood.**—In patients with poor immune responses, grafted tissue survives for longer than in normal people. This is mainly seen with loss of cell-mediated immunity but it also occurs with loss of humoral immunity. Therapeutic depression of immunity is used to lessen the chances of rejection of grafted tissue.

A normal person with Group A blood has anti-B agglutinin circulating in his plasma. In states of deficient immunity, the titres of such agglutinins may be low or absent. Such people are therefore less likely to develop blood transfusion reactions when transfused with incompatible blood.

## CAUSES

A child does not develop his full ability to produce humoral antibodies until he is about six months old. Impaired ability to mount immune responses is normal before that age and the child has to rely on maternal antibodies for protection against infection.

Deficient immunity may occur after this age if there is a congenital abnormality which interferes with the development of the immune system or if the immune system is damaged by drugs, poisons or other agents. Humoral or cell-mediated immunity or both may be involved.

### Congenital Deficiency

(a) *Impaired humoral immunity* (*B-cell failure*).—Here there is a defect interfering with the development of B-cells and thus humoral immunity. There is an absence of plasma cells which produce immunoglobulin. The IgG component of plasma is reduced from the normal value of about 1 g per cent to less than 0.2 g per cent (*hypogammaglobulinaemia*). The thymus is normal. The condition is usually noticed about the age of 9 months because of complaints about recurring infections. These patients reject foreign grafts but do so more slowly than normal people. Surprisingly, such patients tolerate viral infections as well as normal children, though their ability to form anti-viral antibodies is reduced; this puts in question the role of antibodies in overcoming primary viral infections.

(b) *Impaired cell-mediated immunity* (*T-cell failure*).—This is a rare condition in which the structures derived from the 3rd and 4th branchial pouches in the embryo fail to develop. The thymus is one of these structures and its failure to develop results in the absence of T-cells and cell-mediated immunity. The patients do not reject grafts as rapidly as normal people. Lymphoid tissue is depleted throughout the body and the lymphocyte count is low. Though the IgG level is normal, these patients do not resist infections as well as normal people.

(c) *Impaired humoral and cell-mediated immunity* (*B-plus T-cell failure*).—This is a very severe form of impaired immunity in which both the B-cells and the T-cells fail to develop. There is thus a very low IgG level and lymphoid, including thymus tissue is scanty. Recurrent infection usually causes death before the age of 2.

### Acquired Deficiency

Many agents can interfere with the function of the immune system. Lymphocytes are very sensitive to ionizing irradiation and death following accidental exposure to high doses is often due to inability to resist infection. Cytotoxic drugs which interfere with the rapid multiplication of lymphocytes and plasma cells also reduce resistance. Drugs which interfere with protein synthesis such as adrenal glucocorticoids have a similar effect. Cancer affecting the haemopoietic tissues is another way in which the normal development of lymphoid tissue is depressed.

## INVESTIGATIONS

### Assessment of Humoral Immunity

The easiest way to do this is to look at the pattern of immunoglobulins in plasma. This is usually done by separating the proteins by electrophoresis into their various components in a plate of agar jelly. The different components, i.e. IgG, IgM, IgA etc. are identified using anti-human serum. The precipitin antibodies in the serum precipitate the immunoglobulins which can then be seen as white lines in the jelly.

Humoral immunity may also be tested by injecting a harmless antigen and monitoring the antibody response.

### Assessment of Cell-mediated Immunity

This can be tested by applying to the skin a chemical which is known to elicit a sensitized lymphocyte reaction. The degree of the reaction is a function of cell-mediated immunity. A low lymphocyte count is usually seen when cell-mediated immunity is deficient. *In vitro* tests for lymphocyte sensitization are also used.

## Treatment

When immunity is deficient, the patient should be kept away from risk of infection as far as possible. Some improvement has been found when patients are given weekly injections of gamma globulin from pooled samples of donor blood. Attempts have also been made to increase the patient's own production of gamma globulins by active immunization. The most valuable therapy is to give antibiotics, either continuously or for each infection. In the future, it may be possible to transplant immunologically competent cells from a normal donor to the patient. If the patient is unable to reject these cells, they should form humoral antibodies and sensitized lymphocytes. However, trials of this idea have not been very successful so far.

# 68. EXCESSIVE RESPONSE TO FOREIGN ANTIGENS (HYPERSENSITIVITY)

HYPERSENSITIVITY is the state where a person becomes unusually sensitive to an antigen or hapten compared with the population as a whole. Antibodies which are elicited by the antigens to which the subject is hypersensitive, react with them in a way which disturbs normal tissue or body function. There has been much controversy about the classification of the different varieties of hypersensitivity. However, it is currently popular to divide hypersensitivity into four types of reaction, the *anaphylactic reaction,* the *cytolytic reaction,* the *serum sickness reaction* and the *sensitized lymphocyte reaction* (Coombs and Gell, 1968). These reactions are described below.

*Allergy* is often used synonomously with hypersensitivity but this is not good usage. Allergy meaning "altered in reaction" (Gr. *allos* = altered; *ergon* = reaction) was defined as such by von Pirquet who first used it. When used in this sense, allergy should include both immunity and hypersensitivity. In both of these, the body's reaction to an antigen has been altered.

## TYPE I HYPERSENSITIVITY THE ANAPHYLACTIC REACTION

### DEFINITION

In this condition, antibodies known as *reagins* (*IgE*) which form in response to invasion by foreign antigens, become fixed to mast and other tissue cells. The antigen-antibody interaction which follows subsequent invasion by antigen causes the immediate stimulation of the mast cells with release of their vaso-active and smooth muscle stimulating contents by a process of degranulation (Fig. 119).

### EFFECTS

In all cases the response to exposure to antigen is immediate and this type of hypersensitivity is sometimes called *immediate hypersensitivity.* The effects may be localized to one tissue or generalized throughout the body and may vary greatly in severity. They include:

(a) **Asthma.**—In this case, the antigen is usually inhaled into the respiratory tract. Its reaction with the reagin-sensitized cells in the walls of the bronchi leads to their breakdown and the release of substances such as histamine, bradykinin and 5-hydroxytryptamine. These cause dilatation of arterioles, increased capillary permeability and contraction of bronchial smooth muscle. The effect of this in the bronchi is to reduce the calibre of the airways by smooth muscle contraction, oedema of the mucosa due to exudation of fluid from the capillaries and increased mucus secretion. The increased resistance to air flow makes ventilation of the lungs difficult.

(b) **Hay fever.**—This is a common condition where the interaction between inhaled antigen (usually grass pollen) and reagin-sensitized cells occurs in the mucous membrane of the nose area. The resulting vasodilatation and fluid exudation results in sneezing, nasal obstruction and streaming of fluid from the nose and eyes. It is usually most troublesome in the spring and summer when the amount of pollen in the air increases.

(c) **Urticaria** (L. *urtica* = a stinging nettle).—This is a condition where the skin looks as if it had been stung with nettles. It is sometimes called nettle rash. It may follow rapidly the ingestion, inhalation or injection of an antigen. The interaction with reagin-sensitized mast cells occurs in the skin and the release of histamine, bradykinin and other chemicals is responsible for the effects. The arteriolar dilatation is responsible for the increase in skin temperature, the capillary dilatation is responsible for the red colour and the increased capillary permeability is responsible for the local oedema (weal) of the overlying skin. The pain and itch in urticaria is also due to release of these substances; both histamine and bradykinin cause pain and itch when applied to the base of a blister on human skin.

(d) **Anaphylactic shock** (Gr. *an* = against, *phylaxis* = protection).—This is a generalized form of the anaphylactic reaction. The widespread degranulation of reagin-sensitized mast cells on exposure to antigen causes a condition similar to histamine poisoning. There is widespread nettle rash, constriction of the bronchi and dilatation of blood vessels. The drop in peripheral resistance by vasodilatation and the drop in blood volume due to fluid loss from capillaries may result in peripheral circulatory failure and underperfusion of the tissues with blood (shock).

## Causes

Anaphylactic reactions occur in people who have an inherited predisposition to develop them. The nature of this predisposition is not known. The antigens which are most commonly involved include grass pollens, house dust, insect stings, desquamated material from animal skin, shell fish, fruit such as bananas and drugs such as penicillin and aspirin. Invasion by these antigens or haptens results in the formation of IgE which circulates in the blood and becomes fixed to mast cells throughout the body. The antigen may enter the body by ingestion, inhalation or injection.

## Investigations

### Skin Tests

In Type I hypersensitivity, mast cells in the skin are sensitized with an antibody (reagin, IgE) which reacts specifically with the antigen which provoked its formation. This provides a method of identifying the offending antigen or antigens. If suspected antigens are scratched or injected into the skin, a local urticaria will occur at the site of injection if the antigen is responsible for the hypersensitivity.

However, these skin tests are tedious for the investigator and the patient.

Very often, they do not give clear-cut results and may result in severe anaphylactic reactions. Even if the skin test is positive for a given antigen, it does not necessarily mean that that antigen is responsible for the hypersensitivity being manifested in other tissues, e.g. the bronchi; the response in the skin may not parallel the response elsewhere. Because of this, skin tests are usually employed with circumspection and the clinical history is used to identify the origin of the antigen. If attacks only occur when the patient is exposed to a certain environmental situation, the antigen must be sought in that situation.

## Treatment

### Reduce Exposure to the Antigen

This may be easy if the antigen is known and not ubiquitous. Thus offending fish or fruit can be omitted from the diet and substitutes found for offending drugs. With antigens in house dust, pollens and animal scales, the problem is much more difficult. An attempt may be made to have bed coverings composed of or enclosed in material which is non-antigenic. Pets may have to be discarded and country walks eschewed.

### Suppress the Effects of Antigen-antibody Reactions

Some of the effects of Type I reactions are due to the release of histamine. Antihistamine drugs are thus a logical method of treatment where the antigen cannot be avoided. In more severe cases, including anaphylactic shock, adrenaline is useful because of its vasoconstrictor effect in skin etc. and its dilator effect on the bronchi.

Drugs, e.g. cromoglycate salts, which appear to block the degranulation process in mast cells which releases the vasoactive and smooth muscle stimulating substances, are very useful in the prevention of anaphylactic reactions.

### Suppress the Antigen-antibody Reactions

This may be attempted by drug therapy or by immunological methods. Glucocorticoids from the adrenal cortex antagonize antigen-antibody reactions in some way and the inflammatory response which may follow them. For this reason ACTH or cortisone may be used to prevent the more severe types of anaphylactic reaction.

*Desensitization* is the name given to the immunological procedure used to suppress Type I hypersensitivity. Graded doses of the offending antigen are given subcutaneously at regular intervals. The results of this treatment are not always satisfactory. It is thought to work by provoking the production of IgG or IgA antibodies rather than the IgE reagins responsible for Type I hypersensitivity. When the patient is subsequently exposed to antigen, the antigen is neutralized by the circulating IgG or IgA before it reaches the sensitized mast cells.

There is little point in trying to suppress the entire immune system in Type I hypersensitivity. Not only would it make the patient more prone to infection but incomplete suppression would not necessarily prevent hypersensitive reactions.

## TYPE II HYPERSENSITIVITY
## THE CYTOLYTIC REACTION

### Definition

In this variety of hypersensitivity, the invading antigen or hapten becomes incorporated in the membrane of cells such as those in blood. When antibody is formed, the antigen-antibody reaction results in the disintegration (lysis) of the cells (Fig. 119).

The cytolytic reactions are analogous in some respects to certain blood group reactions. The antigens are fixed to cell membranes and the antibodies circulate in the globulin fraction of the plasma proteins. In many cases a plasma component (complement) is necessary for lysis and is consumed in the reaction.

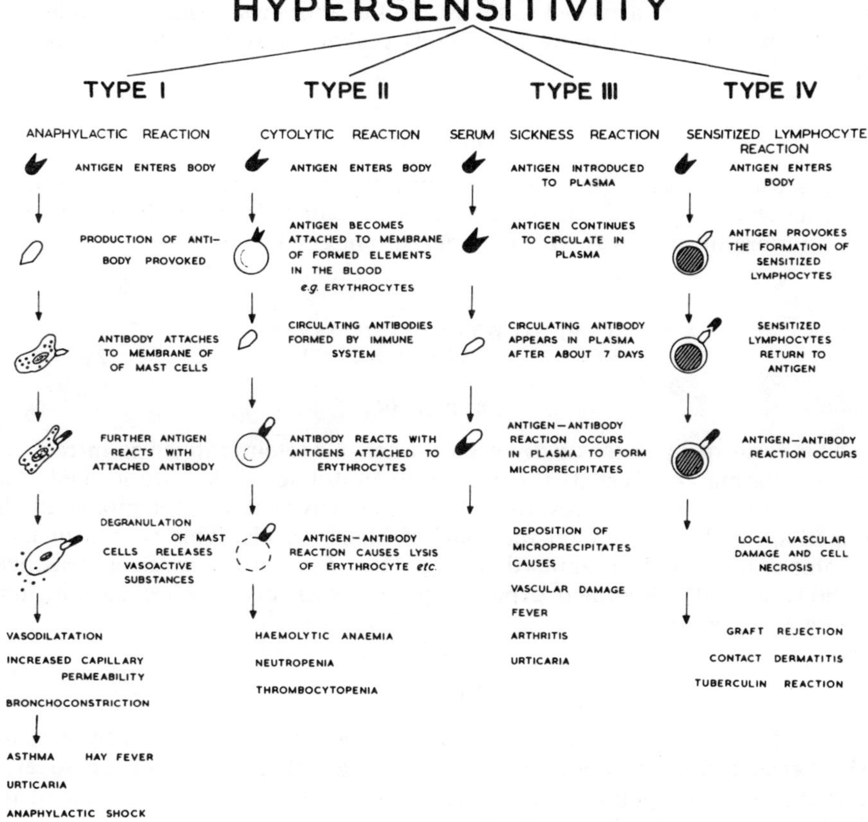


FIG. 119.—Varieties of hypersensitivity.

### Effects

The effects of Type II hypersensitivity depend on which cells are involved. If the antigen becomes attached to the membranes of erythrocytes, antigen-antibody reactions result in lysis of these cells (*haemolytic anaemia*). If granulocytes are affected, the granulocyte count is reduced (*agranulocytosis*) and there is increased susceptibility to infection. Reduction in the number of thrombocytes (*thrombocytopenia*) by this mechanism results in an increased bleeding tendency and small "flea bite" haemorrhages (petechiae) in the skin.

It has been suggested that this type of reaction may cause other diseases by affecting cells other than those in blood. The evidence is not conclusive.

### Causes

Drugs which act as haptens are perhaps the most common cause of Type II reactions but bacterial products can act similarly. Agranulocytosis and thrombocytopenia are quite common complications of drug therapy. In a sense, haemolytic transfusion reactions fit into this group. The antigens are introduced into the body already attached to the red cell membranes. When antibodies are formed, the subsequent antigen-antibody reaction results in lysis of the red cells.

### Treatment

This usually means stopping the administration of the offending antigen immediately. Glucocorticoids may be used to damp down the lytic response; they interfere in some way with antigen-antibody reactions.

## TYPE III HYPERSENSITIVITY SERUM SICKNESS REACTION

### Definition

This form of hypersensitivity occurs when a protein antigen is introduced into the plasma and persists there until precipitin antibodies are formed. The antigen-antibody reaction results in the deposition of microprecipitates of the antigen-antibody complex on the walls of blood vessels. This results in local vascular damage and inflammation (Fig. 119). Exposure to antigen sensitizes the body so that subsequent exposure produces effects which come on more rapidly and are more severe.

### Effects

Type III hypersensitivity is slow in onset compared with Type I hypersensitivity because time is required for antibody formation. A week or so after exposure to the antigen the patient develops a fever, urticaria, swelling of the lymph glands and pains in the joints.

The mechanism behind these changes is thought to be as follows. The antigen, a protein or a hapten attached to a protein continues to circulate in the protein fraction of the plasma until precipitating antibodies are formed. The subsequent precipitation of antigen by antibody results in microprecipitates of the antigen-antibody complex being deposited on the walls of blood vessels. By an unknown mechanism, which involves activation of complement and release of substances which attract leucocytes, these precipitates produce inflammatory changes in the vascular wall. Similar effects have been found following intravenous injection of such precipitates prepared *in vitro*. The inflammatory changes result in the release of histamine and other vasoactive products from cells. Platelets and leucocytes adhere to the damaged endothelium and clotting and obstruction can occur in the minute vessels. This makes the tissue damage and inflammatory changes worse.

A similar reaction can occur if antigen is injected into soft tissues such as muscle. After a week or so, circulating antibodies are formed. As the antigen diffuses towards the small blood vessels, it meets antibody diffusing out and microprecipitates of the antigen-antibody complex are deposited in and around the walls of the vessels. This causes local vascular damage, aggregation of platelets and leucocytes and intravascular clotting which may result in local tissue destruction. As in serum sickness, the antigen is in excess of the antibody. Such a reaction is known as the *Arthus phenomenon*.

### Causes

Serum sickness used to occur frequently when injections of sera prepared in horses and other animals were given to provide immunization. It is now quite uncommon because animal sera are seldom used for this purpose. Type III hypersensitivity is more commonly seen now as a complication of drug therapy. A drug like penicillin may combine as a hapten with a plasma protein and the subsequent formation of antibody results in the deposition of microprecipitates.

### Treatment

Drugs such as the adrenal glucocorticoids, which interfere with antigen-antibody interactions and suppress inflammation, are of more use in this condition than drugs which block the effects of histamine.

## TYPE IV HYPERSENSITIVITY
## SENSITIZED LYMPHOCYTE REACTION

### Definition

In this type of hypersensitivity, the cell-mediated type of immune response predominates. The antigen provokes the production of sensitized lymphocytes. The reaction of the sensitized lymphocytes with the antigen results in local vascular damage and cell destruction (Fig. 119).

### Effects

These depend on the nature and location of the antigen. If the antigen is a chemical, which combines with local protein as a hapten when applied to the skin, the sequence of events is as follows. About 24 hours after application of the chemical, the skin becomes red, swollen and hard. In severe cases, tissue death (necrosis) may occur to form an ulcer. Histologically, the affected tissue is heavily infiltrated with lymphocytes and other cells; most of the signs of inflammation are present.

### Causes

The reason why some antigens and not others should provoke this type of hypersensitivity is not known. However, Type IV hypersensitivity is seen with those chemicals, e.g. detergents, pot plants and some industrial chemicals, which cause *contact dermatitis* and certain bacterial products such as the protein from tubercle bacilli which causes the *tuberculin reaction*. As these agents are introduced locally into the skin, they produce a local response.

### Treatment

In contact dermatitis the antigen should be identified and avoided. This may be difficult if the antigen is an unavoidable ingredient of the patient's work. The identity of the antigen can be established by applying patches containing suspected antigens to the skin and seeing which provoke a reaction. Attacks of contact dermatitis are not helped by antihistamine drugs, but adrenal glucocorticoids may alleviate the condition.

## REFERENCE

COOMBS, R. R. A., and GELL, P. G. H., Eds. (1968). In *Clinical Aspects of Immunology*. Chap. 20, pp. 575-596. Oxford: Blackwell.

# 69. EXCESSIVE IMMUNE RESPONSE TO BODY TISSUES (AUTOIMMUNITY)

## Definition

This is the condition where the body produces antibodies against its own tissues (autoantibodies). Normally the immune system is not stimulated by body antigens which were present in fetal life.

There has been a tendency in the past to attribute many diseases of unknown origin to autoimmune processes because of the presence of autoantibodies. However, such antibodies may be present in normal people and in most cases it has not been possible to establish a causal relationship between the autoantibodies and the disease.

## Effects

The effects of autoimmunity depend on which tissue is affected by autoantibodies and can therefore present in a great variety of ways. Diseases which have been attributed to autoimmunity include:

*Certain haemolytic anaemias.*—In these patients, antibodies that can cause agglutination and haemolysis of erythrocytes can be identified in the plasma. The effects of haemolytic anaemia are described in the section on blood. In some cases the antibodies are of the IgG type and are most effective at 37°C. In others, the antibodies are of the IgM type and are most effective at about 4°C. These are known as *cold agglutinins* and cause trouble only when the peripheral parts of the body become excessively chilled.

*Thyroid disease.*—In this condition (Hashimoto's disease), the thyroid gland is enlarged, hard and shows chronic inflammatory changes. Thyroid function may be increased, normal or decreased. Antibodies which react with thyroglobulin and other thyroid proteins can be identified in the patient's plasma. However, similar antibodies have also been found in normal people and it is not clear whether they are the cause or the result of the disease process.

*Sympathetic ophthalmia.*—This is the condition where inflammatory changes occur in an uninjured eye some weeks or months following a perforating injury to the opposite eye. It has been postulated that antigens released from the injured eye provoke the production of antibodies which attack both the normal and the damaged eye. The condition is prevented by immediate removal of the perforated eye after injury and, if it develops, it is treated by adrenal cortical hormones which suppress antigen-antibody reactions.

*Pernicious anaemia.*—In about 80 per cent of patients with this condition, an antibody can be found in the plasma which is active against the parietal cells of the gastric mucosa. Some of the patients also have an antibody which reacts with intrinsic factor. It has been suggested that these antibodies are responsible for the atrophy of the gastric mucous membrane and the failure to absorb vitamin $B_{12}$ which occur in this condition.

*Rheumatoid arthritis and systemic lupus erythematosus.*—These are generalized diseases in which inflammatory changes occur in connective tissues throughout the body. Abnormal autoantibodies are found in both conditions and both conditions are helped by treatment with corticosteroids. However, there is not a good correlation between the amount of these antibodies and the severity of the diseases.

*Myasthenia gravis.*—In this condition, degenerative changes occur in the vicinity of skeletal neuromuscular junctions associated with ready fatigue and weakness of the muscles. Autoantibodies have been demonstrated and the condition is sometimes helped by removal of the thymus gland which is important for cell-mediated immunity. However, there is not general agreement that myasthenia is an autoimmune disease.

*Ulcerative colitis.*—In ulcerative colitis the mucous membrane of the colon becomes ulcerated, causing diarrhoea with the passage of blood and pus. The condition is subject to exacerbations and remissions and may be alleviated by adrenal cortical steroids. In some patients, plasma antibodies have been demonstrated which react with human fetal colon, supporting the idea that ulcerative colitis is at least partly an autoimmune disease. However, there is also evidence that psychológical stress is partly responsible for ulcerative colitis.

## Causes

The body is thought to become tolerant to its own proteins during fetal life so that it will not subsequently mount an immune response against them. A number of theories have been suggested to explain how this tolerance may be lost. (a) If tissue proteins undergo a chemical change because of a cellular disturbance, e.g. by mutation or virus infection, they may become unrecognizable or "foreign" and so stimulate the body's immune system. (b) If tissue proteins, which are sequestered in the tissues in early life, e.g. certain eye tissues, are later released to make contact with the immune system, they may appear "foreign" to and stimulate the immune system. (c) If the immune system undergoes a pathological change so that it loses its tolerance to body proteins, it may react abnormally against them. (d) If micro-organisms possess antigens closely related to body antigens, they may stimulate the production of antibodies which cross-react with the body tissues.

All these possibilities are hypothetical; the mechanism of the development of autoimmunity is not known with certainty.

## Treatment

Some autoimmune diseases can be helped by adrenal glucocorticoids which antagonize antigen-antibody reactions. Otherwise the treatment is that of the disturbance of function caused by the antigen-antibody reactions.

# 70. ABNORMAL PRODUCTION OF IMMUNOGLOBULINS

## DEFINITION

IN THIS condition, cancerous proliferation of plasma cells results in excessive production of immunoglobulins. It is called multiple myeloma (Gr. *myelos* = marrow; *oma,* thought to be derived from the Greek word *onkoma* = swelling) because the plasma cells spread by seeding to form multiple tumours scattered throughout the bone marrow.

## EFFECTS

**Increase in plasma immunoglobulin.**—The myeloma cells produce large quantities of immunoglobulin. As all the cells are the same, they produce the same immunoglobulin which has no particular value for immune purposes. Different myelomas can, of course, produce different immunoglobulins. The high plasma levels of globulin result in a high erythrocyte sedimentation rate and may increase blood viscosity.

**Loss of protein in urine.**—In about 50 per cent of patients with multiple myeloma, the plasma cells produce a protein of low molecular weight (22,000) which can traverse the glomerular filter and escape in the urine. This is called Bence-Jones protein and can be recognized in the urine by the fact that it forms a cloudy precipitate when the urine is heated to about 70°C and disappears again when the urine is heated further. The deposition of protein in the renal tubules may cause urinary obstruction and lead to renal failure.

**Bone erosion.**—The growing masses of cells result in bone erosion and the formation of cavities which show up readily on X-ray. Bone pain is common and spontaneous fracture may occur.

**Marrow replacement.**—The growing masses of myeloma cells replace normal marrow and result in a fall in the red cell, white cell and platelet counts in the blood.

## CAUSES

The cause of multiple myeloma, as of cancer, is unknown. It usually occurs in older people.

## INVESTIGATIONS

The combination of widespread bone erosion as seen in bone X-ray pictures, a high erythrocyte sedimentation rate, Bence-Jones protein in the urine and a raised level of abnormal globulin in plasma, provides good evidence of multiple myeloma.

## Treatment

The rapidly proliferating myeloma cells are very susceptible to ionizing radiations and cytotoxic drugs which interfere with mitosis. Thus radiotherapy or chemotherapy with cytotoxic drugs may be used to repress the cancer and prolong life.

# 71. TISSUE TRANSPLANTATION

## Definition

TISSUE transplanted to another site is known as a *graft*. If transplanted from one site to another on the same individual, it is called an *autograft*. If transplanted from another person it is called a *homograft*. If transplanted from another species it is called a xenograft (Gr. *xenos* = foreign). This section deals mainly with homografts.

## Effects of Homografts

When skin is transplanted from an individual who is not genetically identical (*allogenic homograft*), the grafted skin usually dies. After some days, an intense inflammatory reaction occurs around the grafted area. There is vascular engorgement, stasis and thrombosis. The skin turns black, forms a scab and is eventually sloughed off.

Antigens in the graft seem to elicit a sensitized lymphocyte reaction which leads to the rejection of the graft. During rejection the tissues around the graft are infiltrated with lymphocytes and plasma cells. Circulating antibodies seem to be of less importance than cell-mediated immunity in graft rejection.

Exposure to one graft sensitizes the body to the antigens in that graft. A second graft of the same material is rejected much more rapidly and violently than the first graft.

Similar reactions are seen when kidneys or hearts are transplanted from genetically dissimilar people. However, certain tissues are less liable to be rejected. Corneal transplants tend to survive if transplanted into the eye, perhaps because of their relative avascularity. This is useful for restoring vision to patients whose corneas have become scarred. Cartilage and bone are other tissues which survive transplantation better than most. Again it may have something to do with relative avascularity.

Another curious feature of transplantation is that tissues which are transplanted into the anterior chamber of the eye are less liable to be rejected; it may be that the immune system does not make contact with this area.

When tissue, such as a patch of skin, is transplanted from an individual who, as an identical twin, is genetically identical (*isogenic homograft*) there is no antigenic incompatibility. The skin becomes vascularized and "takes" in its new situation.

## Causes of Homograft Rejection

Grafts are rejected because some of the antigens they contain are different from those of the recipient. The recipient therefore mounts an immune response against these antigens and the subsequent antibody-antigen reaction causes vascular damage and rejection of the graft.

The antigens responsible for the graft reaction are known as the *transplanta-*

*tion antigens* and their presence or absence in the tissue is determined genetically. Thus, if two individuals are genetically identical, so are their transplantation antigens. The transplantation antigens are found in all nucleated cells including the leucocytes. Some tissues, such as skin, contain high concentrations of the antigens whereas others, such as skeletal muscle, contain low concentrations.

The immune response provoked by these antigens is not fully understood, but has a humoral and cellular component. Following rejection of a skin graft, immunoglobulins can be identified in the patient's plasma, which agglutinate the donor's blood and are toxic to the donor's tissues. However, it is difficult to transfer immunity to grafts from one individual to another by simply transferring plasma. This suggests that the material responsible for rejection is not always contained in plasma. Nevertheless, it is generally felt that circulating antibodies play some part in graft rejection.

The evidence that the immune response involved in graft rejection is partly cell-mediated is based on the finding that immunity to grafts can be transferred from one individual to another by the transfer of lymphoid tissue and that the histological features of graft rejection are similar to those seen in the sensitized lymphocyte reaction. Diseases which interfere with the development of cell-mediated immunity are associated with an increased tolerance of homografts.

It is likely that both immunoglobulins and sensitized lymphocytes are responsible for the antigen-antibody reactions which cause the vascular damage and local cellular necrosis seen in graft rejection.

## Investigations

The greater the similarity of the antigens in the tissues of the recipient and donor, the greater are the chances of a homograft taking successfully. Because of this, a number of investigations are employed before transplantation to compare the antigenic make-up of the recipient and the potential donor.

The most simple comparison is that of the blood group systems. The finding that the blood groups of the recipient and donor are identical is taken as evidence of genetic similarity. However, red cells do not contain the transplantation antigens and better tissue typing is possible if the various antigens in the lymphocytes of the recipient and donor are identified using various antisera. Lymphocytes do carry the transplantation antigens so, if the lymphocytes match, it is likely that the other tissues match. Unfortunately, tissue typing by identification of *lymphocyte antigens* is difficult and time-consuming. Because of this, tissue typing is usually confined to a few centres which can serve large populations. It is then possible to justify the effort and expense of maintaining the specialized typing sera that are needed. Computers can be used to match possible donors and recipients from information sent in from many countries.

## Prevention of Graft Rejection

### Matching the Donor and Recipient

The greater the genetic similarity between recipient and donor, the less the chances of graft rejection. Thus transplants between identical twins present no rejection problems. However, a non-identical twin, a brother, a sister or a parent is more likely to have a similar genetic make-up to the recipient than a random member of the population.

When a potential donor is found, his tissue antigens should be compared with those of the recipient by tissue typing. The best results are obtained with those patients that show the least incompatibility in their transplantation antigens. However, successful grafting can occur even when the match is not perfect.

**Immuno-suppression**

This is normally achieved by suppressing the immune response in the recipient so that he will not react to the foreign antigens. The difficulty of this is that the recipient is no longer able to mount an immune response against invading micro-organisms. To mitigate the effects of lowered resistance, the patient may have to be kept temporarily in a nearly germ-free environment and treated with antibiotics which act against a wide range of organisms.

The immune response can be suppressed by agents which lower the lymphocyte count. These include deep X-ray therapy, cytotoxic drugs, serum containing antibodies which destroy lymphocytes (*antilymphocyte serum*) and adrenal glucocorticoids.

**Production of Immunological Tolerance**

As mentioned above, the generalized depression of immunity causes serious problems and efforts are now being made to produce *selective* suppression of immunity. The idea here is to make the recipient tolerant to the antigens of the graft and yet retain his intolerance of other foreign antigens. An effective way of producing selective tolerance has not yet been devised.

# 72. BLOOD TRANSFUSION

THE TRANSFER of blood from one individual to another is by far the most widespread and most successful form of tissue transplantation. The reasons are: (i) the tissue can be obtained readily from living donors, who suffer no permanent effects, (ii) blood is relatively easy to group, so that donor and recipient can be matched for compatibility, (iii) the transfused blood is, in most cases, required only temporarily by the recipient. In cases where repeated transfusions are required because the patient cannot produce his own blood, incompatibility problems increase with time as the patient becomes sensitized to an increasing number of red cell antigens.

## BLOOD DONATION

### Selection of Donors

Before accepting blood from a donor, the transfusion organization should be satisfied that, as far as possible, the procedure will be free of risk for the donor and the recipient.

**The donor** is protected by a refusal to accept his blood if he is suffering from any obvious illness and by checking his haemoglobin level. The aim is to exclude anyone suffering from anaemia. The copper sulphate method is a very convenient screening test. The density of whole blood is determined mainly by its red cell content. A solution of copper sulphate is made up so that its density equals that of blood with a haemoglobin level of, say, 90 per cent Haldane. When a drop of blood whose Hb is less than 90 per cent is added, it floats and the donor is rejected. When the haemoglobin content is greater than 90 per cent, the drop sinks to the bottom and the donor is accepted.

**The recipient** is also protected by the rejection of anyone obviously suffering from a communicable disease. In addition, a person receiving treatment by drugs is not accepted, in case the unknown recipient should be sensitive to such drugs. The donor is excluded if he has suffered from any one of a variety of conditions which could possibly be passed on to the donor by organisms circulating in the blood or by some allergic mechanism.

One of the more serious risks of blood transfusion is the transmission of serum hepatitis, an infective liver disease probably caused by a virus. Anyone who has suffered from jaundice is taken as a possible carrier of the virus, which can persist in the bloodstream for years after the infection. In addition, blood for transfusion is screened for the presence of *Australia antigen*. This antigen was first found in an Australian aborigine but it has since been shown to be commonly present in patients suffering from, or carrying infectious hepatitis. However, because mild hepatitis without jaundice can occur, and because it is not possible to detect all carriers of the disease, infective hepatitis remains a risk of blood transfusion.

The strictness of donor screening varies from place to place and this will influence to a certain extent the safety of blood transfusion. It has been suggested that paid donors may be less likely than voluntary donors to admit to a condition which excludes them from donating blood.

### Effects of Blood Donation

Usually 5-10 per cent of the circulating blood volume is removed during blood donation. The loss of this amount of blood is compensated for, as with blood loss for any reason, by (a) sympathetic nervous system reflexes, (b) transfer of fluid from the extravascular to the vascular compartment, (c) retention of water and electrolytes of the kidneys and (d) increased synthesis of plasma proteins, erythrocytes, white cells and platelets.

The immediate effects of blood donation are the only obvious ones. The risk is that a donor may faint due to imperfect compensation by the sympathetic nervous system. To minimize this risk, the donor lies down for 10-15 minutes or so after the donation, and for longer if he feels faint on sitting up. During this time the blood pressure usually stabilizes due to (a) and to a certain extent, (b). Factors which antagonize the necessary rise in vasomotor tone increase the risk of fainting. The most important is cholinergic sympathetic vasodilatation activated by psychological factors—fainting is most common after a first donation. Vasodilatation is also induced by a hot drink and this should be avoided, particularly after a first donation.

Blood volume and electrolyte content are restored to normal in a day or so, and all deficits, including the red cell deficit, are made up within about six weeks. Provided the donor is healthy and has a satisfactory diet, no supplementary iron therapy is required.

## BLOOD TRANSFUSION

### Compatibility Testing

The first step is to select blood of the same ABO and Rhesus group as the patient. The next step is to test compatibility directly, by mixing the donor cells with recipient plasma or serum and looking for agglutination. The recipient cells may also be mixed with the donor plasma. The presence of antibodies to red cells not sufficient to cause agglutination may be investigated by the Coombs' test for human globulin. The compatibility testing procedure undertaken before transfusion must be more elaborate and prolonged when previous transfusions have been given and particularly when previous transfusion has led to a reaction.

In severe emergencies when there is no time for grouping or cross-matching, group O, Rhesus—ve blood is given. This contains none of the common antigens and is usually safe.

*Meticulous care over labelling* is essential if serious incompatibility is to be avoided. The correct label must always be applied to the correct blood container, and, in addition, the patient's identity must be established without any shadow of doubt, especially when there are two patients with the same name in a single ward unit.

**Risks of Blood Transfusion**

The main risks of blood transfusion are (i) a reaction due to incompatibility or some harmful material in the transfused blood, (ii) heart failure due to an excessive rise in blood volume and (iii) hepatitis.

**Transfusion reactions** often manifest themselves after only a small volume of blood has been given. For this reason, the transfusion should be started at a slow rate and under expert supervision. Incompatibility is suggested by the onset of shivering, a rise in temperature and a feeling of general discomfort. Later urinary output falls and haemoglobin may appear in the urine. In the unconscious patient the condition may be recognized only when signs of peripheral circulatory failure develop. Transfusion must be stopped if a reaction is suspected. A severe reaction may require exchange transfusion.

**Heart failure** is a risk in anyone suffering from heart disease. In such patients, packed red cells rather than whole blood are often given to minimize the rise in blood volume. If the jugular venous pressure tends to rise, the transfusion must be stopped.

**Hepatitis** is a long-term risk from either whole blood or plasma particularly in patients who receive frequent transfusions.

## RECENT LEADING ARTICLES

Antilymphocyte globulin. *Lancet,* 1973, **2,** 833.
Australia antigen—the pace quickens. *Lancet,* 1973, **1,** 85.
Control of humoral immunity by T cells. *Lancet,* 1972, **1,** 133.
Disodium cromoglycate. *Lancet,* 1972, **2,** 1299.
Fatal insect stings. *Brit. med. J.,* 1974, **2,** 345.
Function of thymus-dependent lymphocytes. *Brit. med. J.,* 1973, **1,** 4.
Hyperviscosity syndrome in multiple myeloma. *Lancet,* 1973, **1,** 359.
Immunosuppression and intracerebral lymphomas. *Lancet,* 1971, **2,** 143.
Is HL-A matching worthwhile in kidney transplantation? *Lancet,* 1974, **1,** 1150.
Kidneys from living donors. *Brit. med. J.,* 1974, **2,** 344.
Liver transplantation. *Lancet,* 1974, **2,** 29.
New light on inflammation. *Brit. med. J.,* 1971, **3,** 61.
Prevention of graft rejection. *Lancet,* 1971, **1,** 901.
Radiation, immunity and cancer. *Lancet,* 1972, **2,** 217.
Standards for immunoglobulins. *Lancet,* 1971, **2,** 82.
The lymphocyte. *Lancet,* 1973, **1,** 409.
Transfer factor. *Brit. med. J.,* 1974, **2,** 397.
Variable immunodeficiency. *Lancet,* 1971, **1,** 959.

*Section XII*

# DISORDERS OF THE BODY FLUIDS

# 73. INTRODUCTION TO WATER AND ELECTROLYTE DISTURBANCES

THE TOTAL quantities and concentrations of certain electrolytes such as sodium, potassium, magnesium, calcium and hydrogen in the body fluids are normally maintained with great precision. Large changes in these values occur only when the control systems which regulate them are overcome. The physiological disturbances caused by such changes are described in this and subsequent chapters. The normal distribution of electrolytes in intra- and extracellular fluid is shown in Fig. 120.

### Electrolyte Units

Chemical reactions between substances in solution depend mainly on the number of molecules or ionic charges present; the weights of the individual molecules are relatively unimportant. Therefore, when thinking about the behaviour of chemicals in solution, it is helpful to use units which indicate the number of molecules or ionic charges rather than their total weight. If one is interested primarily in the number of molecules present one should use the "gram molecular weight" or "mole". One mole of glucose (180 grams) will react with six moles of $O_2$ (192 grams). The fact that the weights of the glucose and oxygen molecules are different is unimportant from the point of view of the chemical reaction.

The concentration of substances in biological fluids is often expressed in millimoles per litre (mmol/l) rather than moles per litre because the amounts in solution are usually small. These units are the ones recommended for expressing concentrations using the Système International d'Unités (S.I. units) which is now widely accepted.

If one is primarily interested in the number of electric charges carried by the ions in solution one may express the concentration in "gram equivalent weights" or "equivalents" per unit volume. For example, one equivalent of hydrogen ions (1.0 gram) contains the same number of charges as, and will react with, one equivalent of chloride ions (35.5 grams). For the concentrations of ions normally seen in biological fluids, it is more convenient to use milli-equivalents than equivalents. For example, the concentration of potassium in human plasma is 0.004 equivalents per litre. Calling it 4 milli-equivalents per litre (mequiv./l) gets rid of the tedious decimals. Though concentrations of electrolytes are still frequently expressed as mequiv./l, mmol/l is the preferred term using S.I.Units.

The use of this terminology helps in understanding and predicting chemical events. To take a very loose analogy, suppose one intends to throw a large party. It would seem absurd to think of the guests in terms of their total weight. One would not invite so many kilograms of men and so many kilograms of women. The total number of reactive individuals is the important thing. From the party point of view, one male guest is equivalent to, and can react with, one female guest, regardless of any disparity in their respective weights. One can predict

## DISTRIBUTION AND COMPOSITION OF BODY FLUIDS

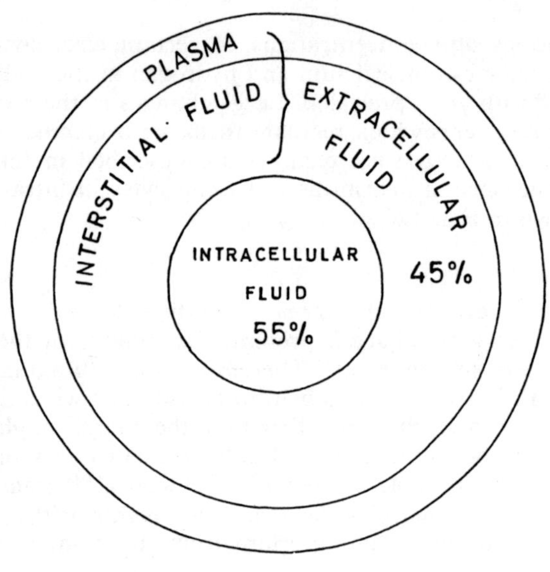


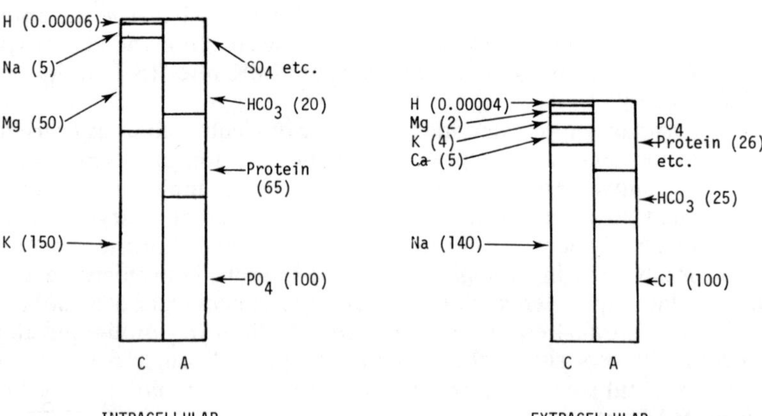


FIG. 120.—The normal distribution and composition of body fluids. In intracellular fluid which comprises about 55 per cent of the total, the chief salts are potassium phosphate and proteinate. In extracellular fluid which includes plasma and interstitial fluid and comprises about 45 per cent of the total, the main salt is sodium chloride. The figures in brackets show the approximate concentrations in mequiv./l.

the course of a party better by thinking in terms of the concentration of reactive units, rather than in terms of their total weight.

**Concentration or Total Amount**

Electrolyte in the body can be thought of either in terms of its concentration in extracellular fluid or in terms of its total amount in the body. The concentration of the common ions in extracellular fluid is easy to measure by examining samples of blood plasma with a flame photometer. The total amount of any ion in the body is so difficult to measure that such measurements are rarely made in clinical work. *The diagnosis of a change in the total amount of an ion in the body is generally made on clinical grounds rather than on biochemical tests.*

It is important to realize that these two values, concentration and total amount, are different things which can vary independently of one another. Thus a low value for total body potassium may exist with a low, normal or high value for serum potassium concentration. Unfortunately, by common usage, a term like potassium depletion is used to describe either a low total body potassium, or a low serum concentration of potassium or both. However, an attempt is made in this section to keep the reader aware of the variety of ion derangement being discussed.

**Varieties of Disturbance**

From the student's point of view it is helpful to look at each ion disturbance separately. From the clinical point of view, however, these conditions are not of equal importance because some are much more common than others. Though it is useful to know something of them all, it is advisable to concentrate on those which are frequently seen in clinical practice.

*Sodium ion depletion* is very common and arises whenever excessive extracellular fluid is lost, as for example, in severe vomiting or diarrhoea. This reduces the total amount of sodium in the body. Water is usually lost proportionally so that sodium concentration in extracellular fluid is little changed. *Sodium ion retention* is also common. As with depletion, water usually accompanies the sodium so that the chief effect is an expansion of extracellular fluid volume. Any condition which causes generalized oedema causes sodium ion retention.

*Disturbances of hydrogen ion concentration* are seen frequently and may be associated with disturbances of other ion systems. The loss of hydrogen ions in vomiting, and of bicarbonate ions in diarrhoea causes hydrogen ion, in addition to sodium ion, problems. Respiratory disorders which alter carbon dioxide output are another common cause of hydrogen ion disturbance.

*Disturbances of the amount of water in the body* without proportionate changes in sodium are rare events except for the water depletion which occurs when a person is unable to drink water. *Potassium and magnesium ion disturbances* are also quite rare as separate clinical entities; they are usually associated with other forms of electrolyte disturbance. Retention can occur in acute renal failure and depletion may result when extracellular fluid loss is replaced by saline solutions without regard to these ions. As potassium is mainly in the intracellular compartment, total body potassium can fall without much change in the serum concentration. However, it is the change in extracellular concentration rather than the change in total body potassium that is responsible for most of the physiological changes with potassium disturbances. Body calcium is mainly in bone. However it is the alteration in concentration of the tiny fraction of the total that is ionized in extracellular fluid that produces the physiological disturbances in calcium ion depletion and retention. The concentration of calcium ion is usually well maintained unless parathyroid gland function or hydrogen ion concentration is disturbed.

The syndromes associated with each electrolyte disturbance have features in common. This is not surprising because a stable internal environment is necessary for the proper function for most cellular activities. When the internal environment is disturbed by *any* change in its constituents, some of these common functions will be deranged.

# 74. DISTURBANCE OF SODIUM BALANCE

## SODIUM ION DEPLETION

### Definition

This is the condition where the total amount of sodium ion in the body is reduced. It occurs when losses of sodium from the body exceed sodium intake. As water tends to follow sodium chloride for osmotic reasons, sodium depletion is usually accompanied by some degree of water depletion.

### Effects

Sodium chloride is the main salt in extracellular fluid. When sodium is lost, chloride and water are usually lost as well. Thus the main effect of sodium depletion is a fall in the volume of extracellular fluid. If the normal volume of body water is maintained by water intake, the resulting hypotonicity of extracellular fluid results in a movement of water from the extracellular to the intracellular compartment. Thus, the volume of intracellular fluid expands and the volume of extracellular fluid shrinks further. This gives rise to widespread changes in the body (Fig. 121).

**Changes in appearance.**—The skin loses its elasticity and takes on a "pinched" look. The patient's eyes become "sunken" and he appears to have suddenly aged. The tongue is dry but there is not much sensation of thirst, presumably because the extracellular fluid does not become hypertonic.

**Circulatory effects.**—Because plasma is extracellular fluid, a fall in extracellular fluid volume is associated with a fall in the circulating blood volume. This tends to cause peripheral circulatory failure. The blood pressure falls and this elicits a reflex increase in heart rate together with constriction of the arterioles and veins via the baroreceptor reflexes. The loss of plasma volume without a corresponding loss of cells raises the haematocrit and makes the blood more viscous, as does the rise in the concentration of plasma proteins. The combination of the fall in perfusion pressure, increase in blood viscosity and the peripheral vasoconstriction may result in underperfusion of the tissues. This may result in death of certain sensitive tissues from hypoxia, e.g. the renal tubules.

However, some factors operate in sodium depletion to maintain the blood volume. The fall in capillary pressure due to reflex arteriolar vasoconstriction, and the rise in plasma osmotic pressure with the rise in protein concentration, favour the mobilization of interstitial fluid to restore the blood volume.

**Urinary effects.**—The kidneys excrete a very small volume of urine which is highly concentrated and low in sodium content. The fall in blood pressure and the consequent reduction in the glomerular filtration rate are partly responsible for the reduction in urinary volume. The fall in blood volume is thought to change the activity of "volume receptors" in the circulation to cause the reflex release of antidiuretic hormone. This tends also to reduce urinary output.

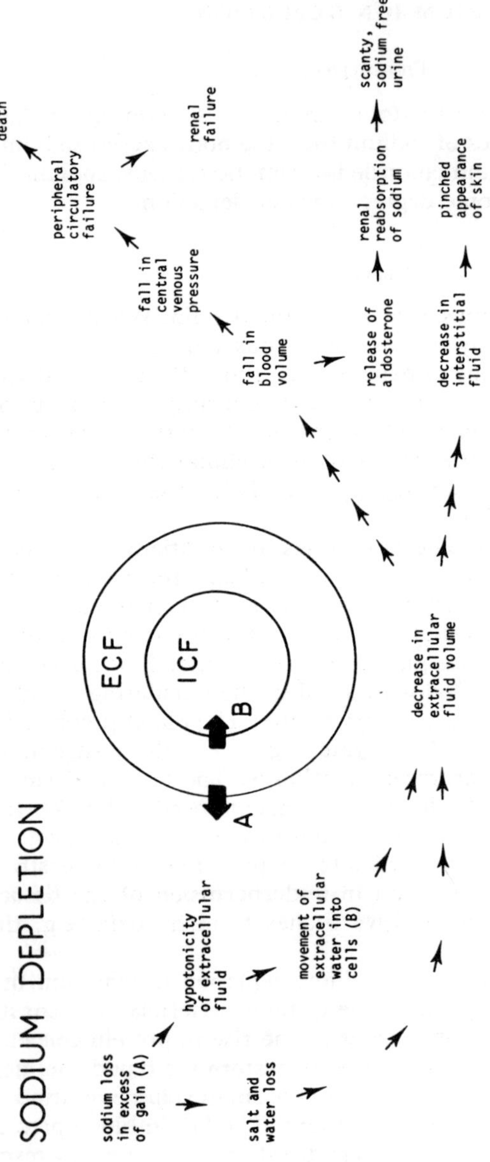


FIG. 121.—The consequences of sodium ion depletion. ECF, extracellular fluid, ICF, intracellular fluid.

Release of aldosterone by the adrenals through the renin-angiotensin mechanism may explain the almost complete reabsorption of sodium.

In severe cases, urinary output may cease, causing retention of urinary waste products in the blood. Loss of appetite, lassitude, vomiting and confusion may result from this type of "*pre-renal uraemia*".

## CAUSES

### Loss of Body Secretions or Fluids

Severe vomiting, diarrhoea, sweating, haemorrhage or burns cause excessive loss of salt and water from the body. Though these may cause pure sodium depletion if the water deficit is made good by drinking, there is usually depletion of both salt and water.

### Failure of Kidneys to Conserve Sodium

Each day about 1.2 kg of sodium chloride are filtered by the renal glomeruli. Normally about 99 per cent of this is reabsorbed back into the blood by an active transport mechanism in the renal tubules. Though most of this reabsorption occurs automatically, a proportion of it is dependent on mineralocorticoid secretion from the adrenal cortex.

Thus the body is depleted of salt via the kidneys in those conditions where glomerular filtration persists but tubular reabsorption is impaired. This is seen in the diuretic stage of *acute renal failure.* It also occurs when a big *osmotic load* is presented to the kidneys which limits reabsorption and causes an osmotic diuresis, e.g. the glucose load in severe diabetes mellitus. In such conditions, the retention of fluid within the tubules leads to a more rapid flow and less time for sodium reabsorption.

Certain *diuretics* which act by depressing sodium reabsorption may also lead to salt depletion. In *adrenal insufficiency,* the reduction in mineralocorticoid activity permits excessive urinary loss of sodium.

## INVESTIGATIONS

Measurements of plasma sodium concentration are easily made using a flame photometer. They are not of much value in the evaluation of sodium depletion because the values depend on the degree of hydration at the time the sample is taken.

Measurements of urinary sodium are useful in certain circumstances. When the sodium depletion is due to loss of body secretions, the urine will be virtually sodium free. However, in sodium depletion due to failure of the kidneys to conserve sodium, sodium is found in the urine.

Total body sodium can be measured by isotope dilution techniques but their complexity makes them unsuitable for routine clinical use.

A rise in the haematocrit and a rise in the plasma protein concentration as indicated by an increase in plasma specific gravity are useful indicators of the fall in extracellular fluid volume that accompanies sodium depletion. However, these values may be within normal limits if they were low before depletion.

### Treatment

Sodium depletion can be treated by giving salt and water by mouth. This is an unpleasant drink which is liable to cause vomiting, thus exacerbating the sodium depletion. Intravenous administration of isotonic saline solutions is the usual treatment and the only possible one when vomiting is the cause. It is difficult to know how much saline to give. In severe cases, up to ten litres may be necessary, but overloading with saline may lead to pulmonary oedema. Rather than base the amount to be given on any calculation of plasma sodium values, it is best to rely on factors such as the restoration of arterial pressure, heart rate and central venous pressure, as evidence that the depletion has been corrected. In elderly patients or where there is a risk of heart failure, central venous pressure should be monitored either clinically or directly using a catheter passed up through an antecubital vein to the right side of the heart.

## SODIUM ION RETENTION

### Definition

The sodium chloride in the diet is normally in excess of that required by the body and the excess is excreted by the kidneys. Should this mechanism fail, sodium retention occurs. In sodium ion retention there is an increase in the total amount of sodium in the body. As chloride and water are usually retained in proportion there is often no change in the concentration of sodium in extracellular fluid.

### Effects

Sodium is the predominant extracellular cation. Retention of sodium and the resulting retention of chloride and water leads to an expansion of the extracellular fluid compartment (Fig. 122). This expansion results in oedema which can be detected clinically when the volume rises by more than about 10 per cent. An increase in weight occurs also. The expansion of the blood volume results in a rise in venous pressure, cardiac output and arterial pressure. In severe cases, the rise in pressure in pulmonary blood vessels results in pulmonary oedema.

### Causes

Sodium retention occurs when the output of sodium in the urine is less than the sodium intake in the diet. Healthy kidneys have no difficulty in eliminating very large dietary intakes of sodium. However, if urinary output is reduced by *renal failure,* retention will tend to occur. Urinary output of sodium is also reduced in diseases where excessive secretion of *adrenal cortical hormones* results in excessive sodium reabsorption. *Heart failure* is another cause of renal failure to excrete sufficient sodium and it is also associated with sodium retention. Factors which cause net movement of fluid from the blood to the interstitial fluid also reduce the urinary output of sodium because the fluid sequestered in the tissues is not presented to the kidneys. Thus any condition which causes oedema results in sodium retention.

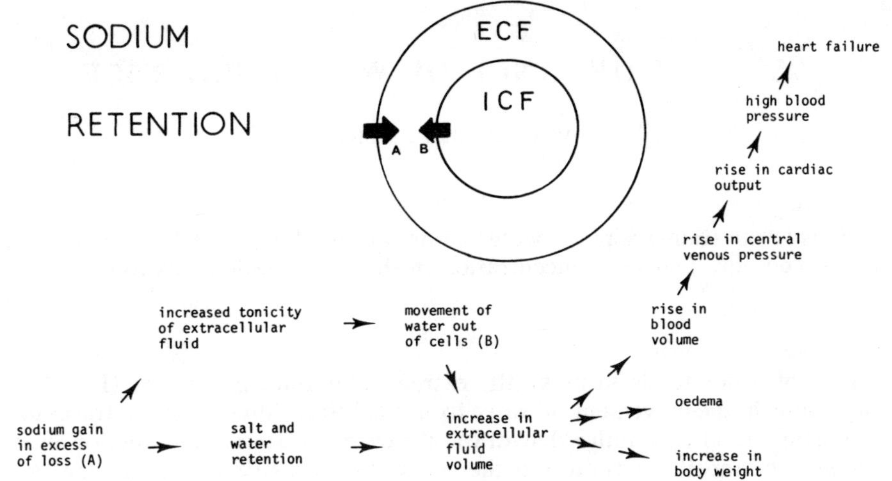


FIG. 122.—The consequences of sodium retention. ECF, extracellular fluid. ICF, intracellular fluid.

## Investigations

As in sodium depletion, measurements of plasma concentrations give little indication of the amount of sodium actually retained. Sodium concentration is usually normal because of the associated retention of water. The measurements of total body sodium or extracellular fluid volume can be made by indicator dilution techniques but are too complex for routine clinical use. Measurements of body weight are most useful for following progress in cases of sodium retention. Measurements of central venous pressure and arterial pressure may provide supplementary evidence.

## Treatment

Besides the treatment of the specific cause of the condition, the aim should be to reduce dietary intake, and increase urinary output of sodium. Thus salt in the diet may be restricted and diuretics which inhibit sodium reabsorption by the renal tubules may be used to increase sodium output. Such treatment needs to be supervised carefully because *overtreatment* may lead to sodium depletion.

# 75. DISTURBANCE OF WATER BALANCE

## WATER DEPLETION

### Definition

This is the state where the water content of the body is reduced relative to the salt content. The salt concentration in the body fluids tends to be high.

### Effects

Loss of water tends to make the extracellular fluid hypertonic (Fig. 123). This results in the movement of water from the intracellular to the extracellular compartment. Thus cellular dehydration prevents a dramatic reduction in the volume of extracellular fluid and the water loss is distributed in all the body fluid compartments. This is why the circulatory disturbances in water depletion are relatively less severe than those in sodium depletion.

Hypertonicity of the body fluids elicits the sensation of thirst and this may be intense in water depletion. It also affects the osmoreceptors in the hypothalamus so that the output of antidiuretic hormone from the posterior pituitary is increased. This results in increased water reabsorption by the kidney so that the urine is small in volume and highly concentrated. In severe cases, the patients become mentally confused, weak and irrational. Death occurs when the body water is reduced by about 25 per cent (10 litres). It is thought to be due to hypertonicity of intracellular fluid interfering with cellular function.

### Causes

**Insufficient Water Intake**

Difficulty in swallowing may result in some patients having inadequate intake to replace the inevitable water loss via the skin, respiratory tract and kidneys. This may occur if the pharynx or oesophagus become obstructed, during prolonged loss of consciousness or in conditions which interfere with the nerves and muscles involved in the swallowing reflex.

**Excessive Water Loss**

This may occur in patients with diabetes insipidus or other conditions where the ability of the kidneys to concentrate urine is diminished. Because sweat is hypotonic, excessive sweating can lead to water depletion. However, with free water intake, people who sweat heavily are more likely to develop a sodium depletion than a water depletion.

### Investigations

As water depletion is always associated with hypertonicity of the extracellular fluid, measurements of plasma sodium concentration are useful. So are

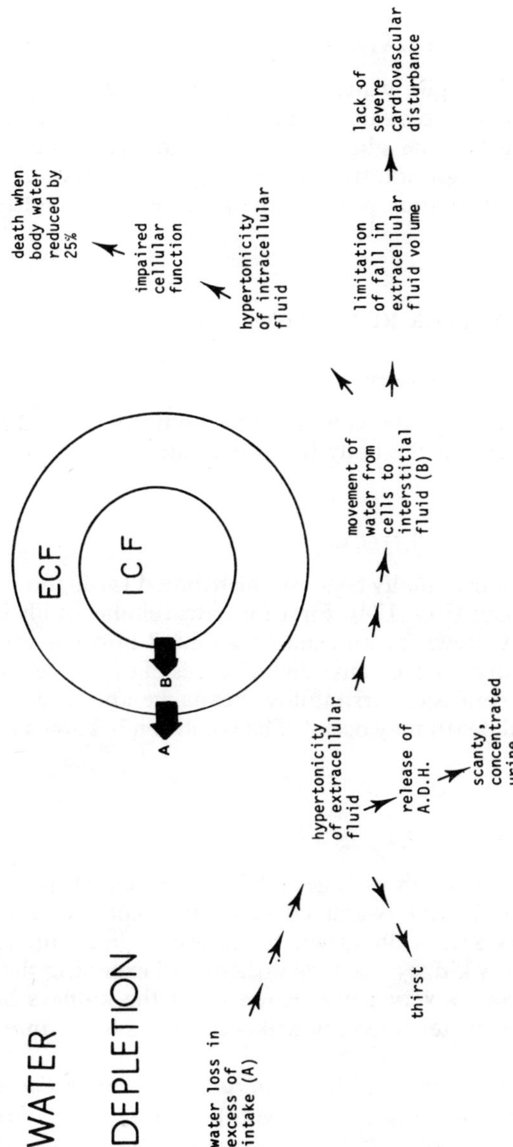


FIG. 123.—The consequences of water depletion. ECF, extracellular fluid. ICF, intracellular fluid.

measurements of plasma specific gravity. Though the total amount of water in the body can be measured by an indicator dilution technique using deuterium oxide, the technique is more suitable for research than clinical purposes.

### Treatment

The simplest treatment is to give water by mouth until thirst is quenched. However, if the patient is unable to swallow water, 5 per cent glucose in water should be given intravenously. The glucose is added to make the infusate isotonic with blood. Pure water given intravenously might cause lysis of the red blood cells. When the glucose solution is given, the glucose is metabolized and the patient is left with the water.

## WATER RETENTION

### Definition

This is the condition where the water content of the body is increased relative to its salt content. The tonicity of the body fluids is reduced.

### Effects

When water is retained it eventually becomes distributed throughout all the fluid compartments of the body (Fig. 124). First the extracellular fluids become hypotonic and then water is drawn by an osmotic gradient into the cells. The movement of water into brain cells can cause disturbances of the higher cerebral functions. Initially there is confusion, irritability, headache and dizziness, but later convulsions, coma and death may occur. The condition is known as *water intoxication.*

### Causes

Normally, the total water which is ingested is in excess of the volume required to make good the obligatory water losses via the skin and respiratory tract and the obligatory loss via the kidneys which is necessary to eliminate waste solutes from the body. Healthy kidneys have no difficulty in excreting the excess water even when water intake is very great. However, if the kidneys lose the ability to form a dilute urine, water retention will occur when large amounts of water are drunk.

Excessive water ingestion by patients with *renal failure* or congestive *cardiac failure* who are unable to excrete a dilute urine will thus cause water retention. In times of stress, such as post-operative convalescence, the release of *anti-diuretic hormone* (ADH) may prevent the excretion of a water load. Excessive drinking at this time may cause water intoxication. In some patients, water retention results from inappropriate secretion of ADH. The ADH can be released from some tumours and from tuberculous lung tissue.

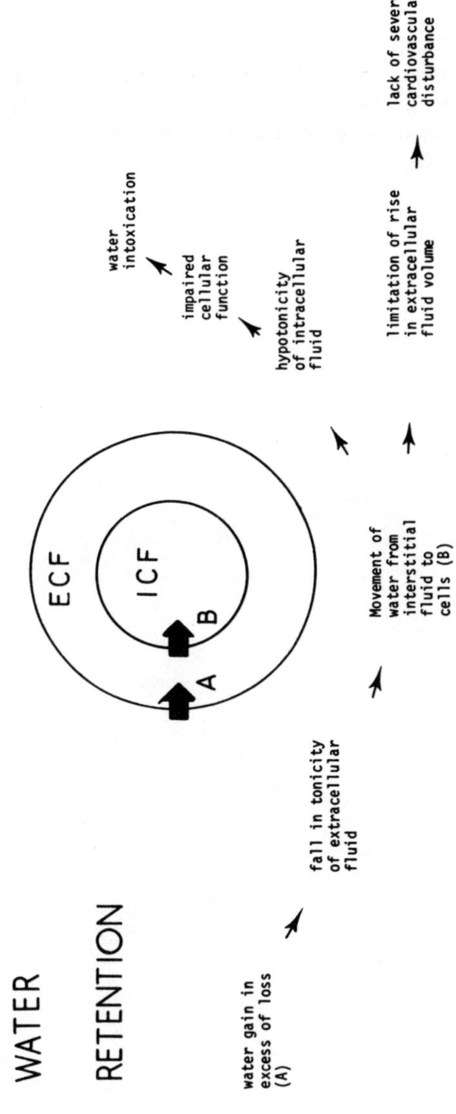


FIG. 124.—The consequences of water retention. ECF, extracellular fluid. ICF, intracellular fluid.

### Investigations

This is one condition where a measurement of the plasma sodium concentration is helpful. Normally it should be about 140 mEq/l. Values of less than 130 mEq/l suggest water retention.

### Treatment

Restriction of water is the safest treatment but in urgent cases hypertonic (5 per cent) saline may have to be given intravenously. This, by increasing the osmolarity of the extracellular fluid draws water from the cells and corrects the overhydration of the cerebral cells.

# 76. DISTURBANCE OF POTASSIUM BALANCE

## POTASSIUM ION DEPLETION

### Definition

THE TERM "potassium depletion" is used to describe a fall in the total body potassium stores or in the serum potassium concentration or in both. A low concentration of potassium in the serum is called *hypokalaemia*.

### Effects

Potassium depletion, if it is mild, may not be reflected in the level of plasma potassium. However, if it is severe, both the total body potassium and its extracellular concentration are reduced. Reduction in total body potassium without much change in extracellular concentration can cause the symptoms and signs of potassium depletion (Fig. 125). If the extracellular concentration is reduced as well as the total body potassium, the effects are similar in nature but more severe. The physiological basis of the effects which are seen in potassium depletion are not understood. They include:

**Mental changes.**—The most striking feature is drowsiness, from which the patient is roused only with difficulty. Apathy and confusion gradually develop into coma. After the episode is over the patient may not be able to remember it. This is called *amnesia* (Gr. *amnesia* = forgetfulness).

**Muscular weakness.**—All types of muscles are involved. The explanation of these changes is obscure. A low extracellular potassium would allow greater potassium efflux from excitable cells. Theoretically, loss of positively charged ions would cause hyperpolarization and decreased excitability. However, potassium retention also causes muscular weakness and eventually paralysis, so it would be foolish to oversimplify the explanations.

*Skeletal muscles* are the most obviously affected by potassium depletion but *smooth muscles* may also cause trouble. Paralysis of the gut smooth muscle (paralytic ileus) may cause intestinal obstruction. *Cardiac muscle* is also weakened. This may lead to dilatation of the heart, low blood pressure and heart failure. Changes in the excitability of heart muscle may cause abnormal heart rhythms. Characteristic changes in the ECG include a small T wave, a prolonged QT interval and depression of the ST segment (Fig. 126).

If death occurs, it may follow respiratory failure due to paralysis of the respiratory muscles or cardiac failure due to weakening of the myocardium.

**Disturbance of renal function.**—In potassium depletion, the kidney is unable to concentrate or dilute urine (*isosthenuria*). It does not respond to antidiuretic hormone. Large volumes of urine are therefore passed (*polyuria*) and the bladder needs to be emptied during the night (*nocturia*). This water loss gives rise to thirst. The reason for these changes is not certain. They may be due to interference with the normal mechanism for concentrating urine by producing a

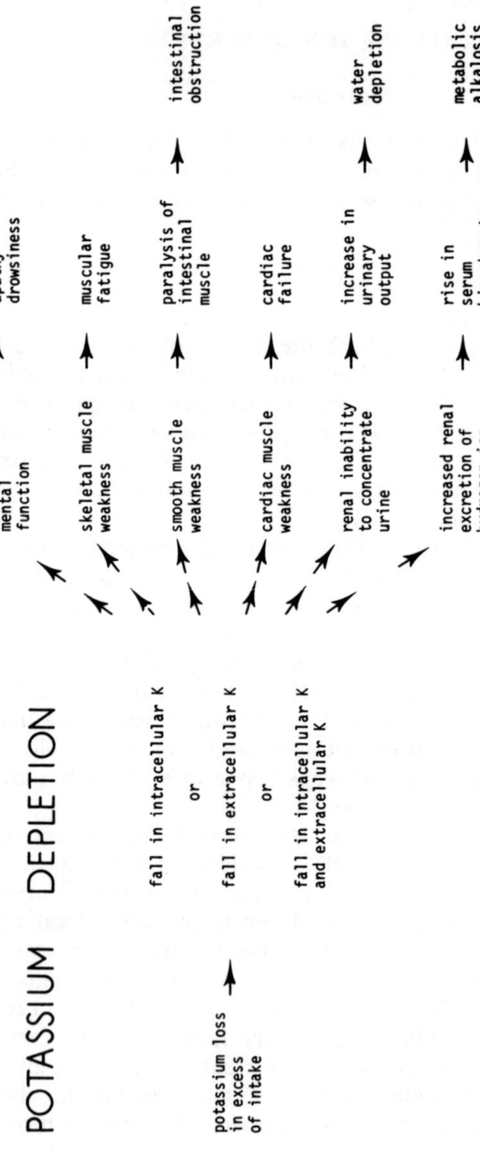


FIG. 125.—The consequences of potassium depletion.

high concentration of sodium chloride in the medullary interstitium. The pump which is responsible for the active movement of sodium normally exchanges a sodium for a potassium or hydrogen ion. A low potassium level in the body might interfere with this exchange.

Potassium normally competes with hydrogen for secretion in exchange for sodium in the distal convoluted tubules. Potassium depletion results in excessive hydrogen excretion through this mechanism. As a consequence, there is an increased manufacture of bicarbonate in the tubular cells. Its reabsorption raises the blood bicarbonate level and causes a *metabolic alkalosis*.

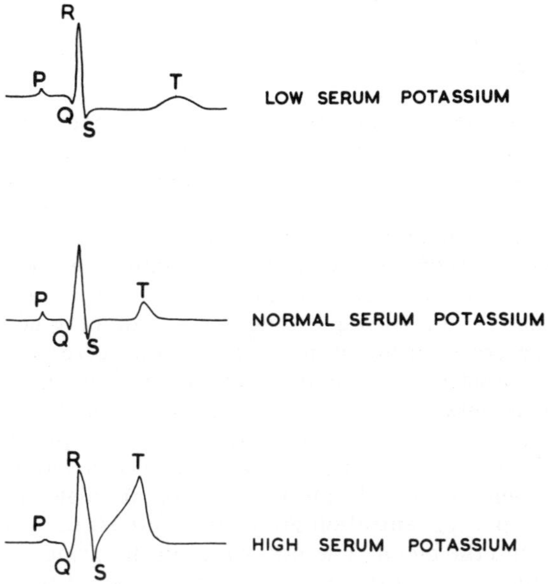

FIG. 126.—The effect of changes in serum potassium on the electrocardiogram. With a low serum potassium there is a prolonged QT interval, a depressed ST segment and a diminished T wave. With a high serum potassium, these changes are reversed, the T wave being large and "peaked".

Since $K^+$ and $H^+$ compete for secretion in the renal tubules in association with $Na^+$ reabsorption, depletion of *either* ion causes excessive loss of the other (Fig. 127).

## Causes

Most of the potassium in the body is in intracellular fluid; less than 5 per cent is in extracellular fluid where the concentration is maintained at about 4 mmol/l.

With a normal diet the daily intake of potassium is in the range of 50-100 mmol. The same quantity is eliminated daily via the urine and, to a lesser extent, the faeces. Potassium depletion occurs when the losses exceed the

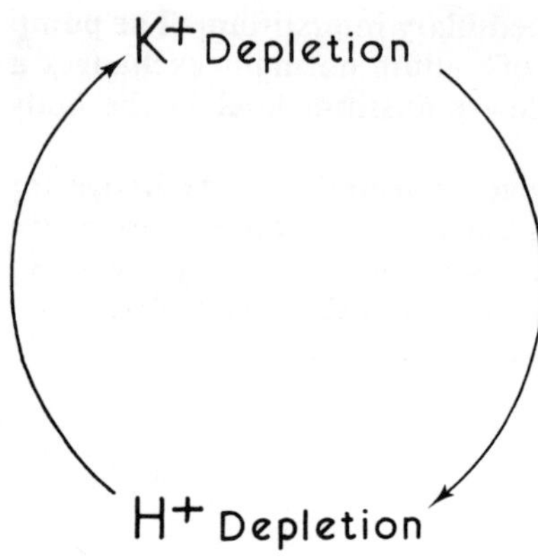


FIG. 127.—In view of competition for excretion of $K^+$ and $H^+$ by renal tubular cells, deficiency of one ion causes increased loss of the other ion which may lead to depletion.

intake. A fall in serum potassium without depletion of total body potassium can occur with agents which increase the uptake of extracellular potassium by cells.

**Loss via the alimentary tract.**—Severe vomiting or diarrhoea leads not only to loss of sodium and water but also the potassium content of the cellular debris derived from the mucosal epithelium. In the case of severe vomiting, the cause of potassium loss is rather more complex. If vomiting leads to a fall in the volume of extracellular fluid, the resulting secretion of aldosterone promotes potassium secretion in the distal tubules. The fall in blood hydrogen ion concentration in severe vomiting promotes potassium secretion, because hydrogen competes with potassium in sodium-potassium exchange in the tubules.

**Loss via the kidneys.**—Most of the potassium which is filtered by the glomeruli is reabsorbed in the proximal tubule. This is probably the passive consequence of the active reabsorption of sodium. When sodium is reabsorbed it is followed passively by chloride and water to maintain electrical and osmotic balance. This sets up a concentration gradient for potassium which moves from the tubules back into the blood. The potassium in the urine is that secreted in exchange for sodium in the distal convoluted tubule. Hydrogen competes with potassium at this site.

There are many causes of excessive urinary loss of potassium. They include:

*Increase in secretion of adrenal cortical hormones.*—As these hormones promote potassium secretion by tubular cells, potassium depletion may occur in Cushing's syndrome (excess glucocorticoid secretion), Conn's syndrome (excess mineralocorticoid secretion), salt and water depletion and following injury. In all these conditions there is increased adrenal cortical secretion.

*Increase in tissue breakdown.*—When tissues are broken down, as may occur in starvation, following surgery or after death of tissues deprived of their blood supply (infarction), the potassium content of the affected cells is excreted in the urine. About 3 mmol of potassium are released for each gram of protein broken down.

*Decrease in blood hydrogen ion concentration (alkalosis).*—As hydrogen ion competes with potassium in sodium-potassium exchange in the kidney, alkalosis leads to increased potassium loss in the urine.

*Decrease in renal reabsorption of sodium.*—Factors which interfere with sodium reabsorption such as diuretics or large osmotic loads, also limit the reabsorption of chloride, water and potassium in the proximal convoluted tubules. These factors favour potassium depletion.

**Loss into the cells.**—When insulin or glucose are given, uptake of extracellular potassium by the skeletal muscle cells is increased. It seems to help the deposition of glycogen. The explanation of this phenomenon is not clear. Presumably the activity of the sodium-potassium exchange pump is increased because potassium has to be moved against a high concentration gradient. Such movements can lead to quite low values of serum potassium. Therapeutically, insulin and glucose are given to treat hyperkalaemia.

There is a rare hereditary disorder known as *familial periodic paralysis* where the patients show episodic weakness of their skeletal muscles. The attacks may be precipitated by emotion or carbohydrate ingestion, insulin, etc. It is thought that they are due to the lowering of serum potassium due to cellular uptake of the ion.

**Fall in potassium intake in diet.**—This leads to potassium depletion only when food intake is restricted to near starvation levels.

## Investigations

Because potassium is mainly intracellular, estimations of plasma potassium may not always provide a clear indication of the potassium balance in the body. The total body potassium can be measured by an indicator dilution technique using radioactive potassium. Technically this measurement is too difficult for routine use. Muscle biopsy has been used to estimate the intracellular potassium concentration but corrections have to be made for the amount in extracellular fluid in the specimen. The potassium content of washed red cells gets over this difficulty but it may not reflect that in the body as a whole except in steady state conditions. The changes in potassium concentration in red cells tend to lag behind those in other parts of the body.

In clinical practice, the conclusion that intracellular potassium stores are depleted is usually reached by deduction from the clinical situation rather than by actual measurements. The electrocardiogram shows the changes illustrated in Fig. 126, and the serum bicarbonate level is usually raised, either as a cause or consequence of the hypokalaemia.

## Treatment

Potassium depletion must be treated by removing the casue of depletion and then making good the potassium deficit. Dealing with the cause may involve removing a tumour, stopping diuretic therapy, etc. To make good the deficit, potassium salts may be given by mouth or intravenously. Oral administration is safer because it is less likely to cause a fatal rise in serum potassium. However, it is irritant to the stomach and patients do not tolerate it well. For this reason it is sometimes given in tablets which are so coated that the potassium is not released until the tablets reach the small intestine (enteric coated tablets). Even then, the potassium may irritate the intestine at the site of release.

If it is not possible to use the oral route, carefully limited amounts may be given intravenously. Potassium depletion may be prevented from occurring in the treatment of sodium ion depletion if potassium salts are judiciously added to the intravenous saline infusions.

## POTASSIUM ION RETENTION

### Definition

The term "potassium retention" usually refers to a rise in the potassium *concentration* in the extracellular fluid. It may be called *hyperkalaemia.* The total body potassium is usually raised but the signs and symptoms of potassium retention can occur even if it is low.

### Effects

Curiously, the effects of potassium retention are almost indistinguishable from those of potassium depletion (Fig. 128). None of them is specific or can be explained in physiological terms. They include:

1. Mental changes. Lethargy, apathy, confusion and coma may be seen.
2. Skeletal muscle weakness. When extreme, this results in a flaccid paralysis.
3. Depression of cardiac muscle. As the serum potassium rises, cardiac muscle becomes weaker, the rate slows and may become irregular. Characteristic changes in the ECG include peaking of the T wave (Fig. 126). When the serum level rises above 7 mmol/l, death may occur from cardiac arrest.

### Causes

In a healthy individual the potassium level in the blood does not rise with a high potassium intake in the diet because the excess is readily excreted in the urine. A rise can occur only if kidney function is interfered with or if potassium is suddenly released or infused into the blood stream.

**Renal failure.**—Potassium is normally excreted by the kidney. In acute renal failure the rise in serum potassium may cause death from cardiac arrest.

**Adrenal failure.**—Insufficient secretion of aldosterone results in loss of sodium and retention of potassium. However, the rise of potassium in Addison's disease is rarely severe enough to cause symptoms.

**Intravenous infusions of potassium salts.**—When used without proper care for patients with potassium depletion, this form of treatment may cause fatal rises in the serum potassium level.

**Haemolysis.**—When a person drowns in fresh water, the uptake of water into the blood from the lungs (by osmosis) may result in haemolysis because of the reduction in blood tonicity. The release of the intracellular potassium may then cause cardiac arrest.

### Investigations

The concentration of potassium in blood is easily measured using a flame photometer. The normal value should be between about 3.5 and 5.5 mmol/l. The changes in the electrocardiogram (Fig. 126) are also useful in diagnosing hyperkalaemia.

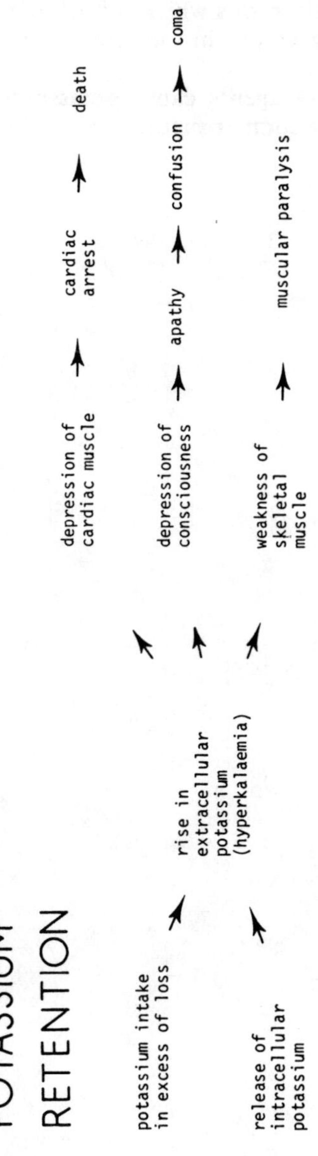


FIG. 128.—The consequences of potassium retention.

### Treatment

A variety of methods may be used to lower the potassium. They include:

*Potassium restriction in the diet.*—This is important in cases of acute renal failure when potassium excretion is depressed.

*Ion exchange resins.*—Certain resins which adsorb potassium can be given by mouth to adsorb the potassium ions in the several litres of intestinal fluids which are secreted daily.

*Glucose and insulin.*—These agents cause potassium to move from the extracellular to the intracellular compartment.

# 77. DISTURBANCE OF MAGNESIUM BALANCE

## MAGNESIUM ION DEPLETION

A FALL in the blood magnesium level can cause neuromuscular and mental disturbances. These include muscular tremor, twitching, convulsions, hallucinations and delusions. The physiological explanation of these effects is not understood but they are not unlike those seen in tetany caused by a low blood calcium level.

The way in which the blood magnesium level is maintained is little understood. Magnesium depletion occurs when magnesium loss exceeds intake. Conditions which cause potassium depletion are also likely to cause magnesium depletion. It may occur when the saline used to correct a fall in extracellular fluid does not include magnesium. It may also occur when intake is reduced in malabsorption syndromes. Where there is excessive secretion of aldosterone or prolonged use of diuretics, excessive magnesium may be lost in the urine.

If it is recognized it may be treated by giving magnesium salts by mouth or intravenously. Correction of the symptoms by magnesium is evidence of a correct diagnosis. When there is evidence of potassium depletion, one should consider the possibility of a concurrent magnesium depletion.

## MAGNESIUM ION RETENTION

Magnesium retention may cause clinical disturbances which are not peculiar to magnesium retention. They include drowsiness progressing into coma accompanied by muscular weakness and fall in arterial pressure. These effects are very similar to those caused by a rise in blood calcium.

Magnesium is normally taken in the diet in excess of the body requirements and the excess is excreted in the urine. Therefore in acute renal failure, magnesium retention, like potassium retention, is likely. In such cases the aim of treatment is to avoid giving magnesium in the diet until urinary output is re-established.

# 78. DISTURBANCE OF HYDROGEN ION BALANCE

## HYDROGEN ION DEPLETION (ALKALOSIS)

### DEFINITION

THE MEAN hydrogen ion concentration in arterial blood is about 40 nmol/1 (pH 7.4), with a range of normal extending from about 35 to 45 nmol/1 (pH 7.45 to 7.35). Hydrogen ion depletion (alkalosis) occurs when the concentration falls below the lower limit of normal. The concentration in venous blood varies with the metabolic activities of the tissues so that changes in venous blood $[H^+]$ do not indicate disturbances of acid-base balance in the body.

The hydrogen ion concentration in the extracellular fluid depends on the ratio of the concentrations of carbonic acid and bicarbonate ion in the fluid. This can be represented:

$$[H^+] \propto \frac{[H_2CO_3]}{[HCO_3^-]}$$

As the carbonic acid concentration in blood is proportional to the partial pressure of carbon dioxide in the blood, $P_{CO_2}$ may be substituted for $[H_2CO_3]$ in the equation. The carbonic acid level in the blood is normally held at about 1.3 mmol/l by the respiratory system. The bicarbonate level in the blood is normally maintained at about 25 mmol/l by the kidneys. By holding the ratio at its normal value of about 1:20, the respiratory system and the kidneys between them control the hydrogen ion concentration.

The hydrogen ion concentration in the body fluids may fall if (a) the concentration of carbonic acid is reduced (*respiratory alkalosis*) or (b) if the concentration of bicarbonate is raised (*metabolic alkalosis*). Both these changes lower the value of the ratio and therefore the hydrogen ion concentration (Fig. 129).

### EFFECTS

**General.**—As with many other electrolyte disturbances many of the symptoms associated with alkalosis are quite non-specific (Fig. 130). They include loss of appetite (anorexia, Gr. *a* = negative, *orexis* = appetite), feelings of sickness (nausea, Gr. *nausea* = sea sickness), vomiting, headache and dizziness.

**Tetany.**—This is the condition of increased excitability of muscle and nerve which occurs when the concentration of calcium ions in the blood falls. As the hydrogen ion concentration falls, the plasma proteins become stronger anions and their affinity for calcium ions increases. The resulting fall in free calcium ions in the plasma causes tetany.

**Ventilatory changes.**—In metabolic alkalosis respiration tends to be depressed. This *compensatory mechanism* tends to raise the carbonic acid level

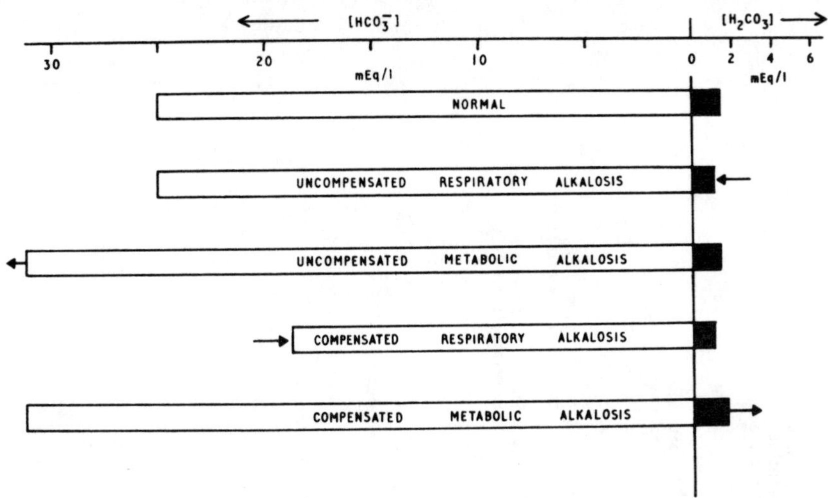


FIG. 129.—The changes in the concentrations of carbonic acid and bicarbonate that occur in uncompensated and compensated alkalosis.

(Fig. 129) in the blood so that the ratio of acid to base is restored towards its normal value. This brings the pH of the blood back to normal (*compensated metabolic alkalosis*).

**Urinary changes.**—In both respiratory and metabolic alkalosis, bicarbonate is excreted by the kidney. In the former case, the object is to reduce the plasma bicarbonate concentration to a value where the ratio it bears to the reduced carbonic acid concentration is near the normal value (*compensated respiratory alkalosis*). In the latter case, the object of the increased excretion is to lower the plasma bicarbonate to its normal value. The excretion of hydrogen ions and ammonium salts is reduced in both cases. Because hydrogen ions compete with potassium ions for secretion in the sodium-potassium exchange system in the distal tubules, potassium ion secretion is increased in alkalosis and may cause potassium ion depletion.

## CAUSES

**Metabolic alkalosis.**—Here the primary problem is a rise in the plasma bicarbonate level. There is usually a secondary and compensatory rise in plasma carbonic acid (Fig. 129). Metabolic alkalosis may follow the ingestion of large amounts of *sodium bicarbonate* in the treatment of indigestion or peptic ulceration. It also occurs in severe prolonged *vomiting* where the loss of the acid secretions of the stomach results in increased quantities of bicarbonate in the blood (Fig. 131). When the blood *potassium* level is low there is excessive secretion of hydrogen ion because potassium ion normally competes with hydrogen ion for excretion by the kidney. This also leads to an increase in the bicarbonate content of blood (Fig. 132).

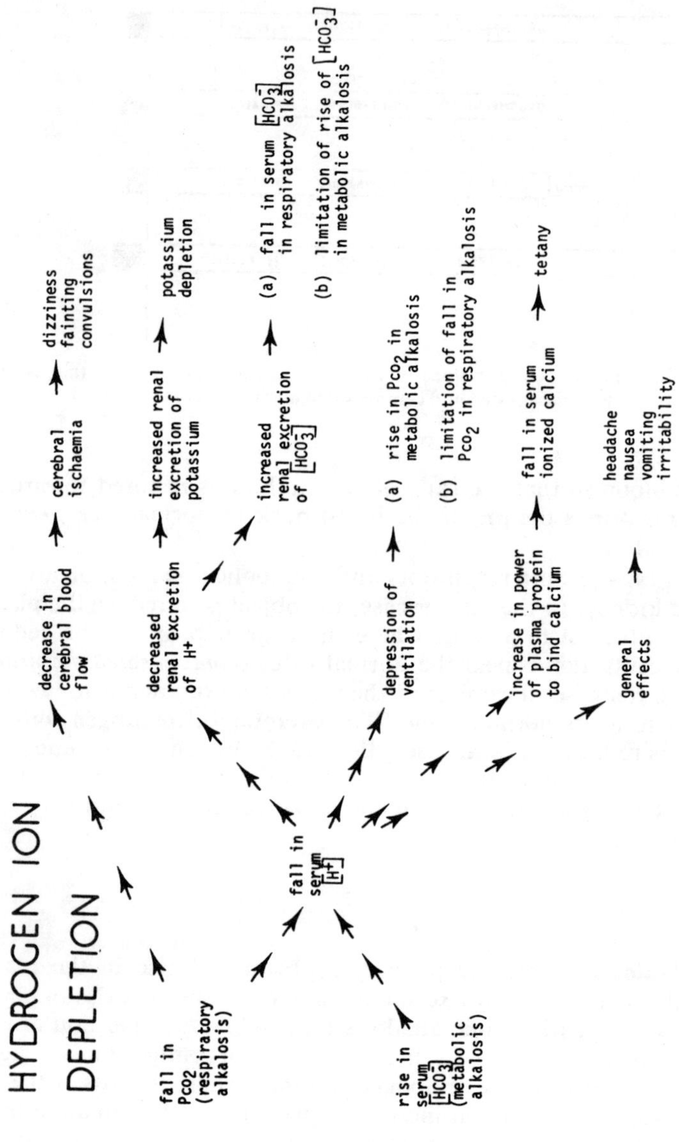


FIG. 130.—The consequences of respiratory and metabolic alkalosis.

**Respiratory alkalosis.**—Here the primary problem is a fall in the concentration of carbonic acid in the blood (Fig. 129). There is usually a secondary and compensatory fall in the plasma bicarbonate level. In all cases it is due to *hyperventilation.* This may occur in hysterical patients and in people working in hot humid environments or at altitude where the low atmospheric $P_{O_2}$ stimulates breathing. It may also result from over-enthusiastic *artificial ventilation.*

## Investigations

The hydrogen ion concentration in blood is proportional to the ratio of acid to base in any of its buffer systems. For the carbonic acid-bicarbonate system this can be written

$$[H^+] \propto \frac{[H_2CO_3]}{[HCO_3^-]}$$

since $H_2CO_3 \propto P_{CO_2}$ the equation can be written

$$[H^+] \propto \frac{[P_{CO_2}]}{[HCO_3^-]}$$

In describing the acid-base status of the blood, it is not enough to measure the hydrogen ion concentration. The cause of any disturbance can be understood better if the three variables ($[H^+]$, $[HCO_3^-]$ and $P_{CO_2}$) concerned in the acid-base equation are known. It is not necessary to measure all three because, if two are known, the third may be worked out by calculation. For example, it is possible to measure the $[HCO_3^-]$ and $[H^+]$ and calculate the $P_{CO_2}$ or measure the $[HCO_3^-]$ and $P_{CO_2}$ and calculate the $[H^+]$.

From the point of view of technique, it is probably most simple to find the $[H^+]$ and the $P_{CO_2}$ of a blood sample and from these values calculate the $[HCO_3^-]$. The steps for doing this are described below.

### Obtaining a Blood Sample

Measurement should be made on arterial blood. The acid-base status of venous blood is not regulated precisely and varies continually with the metabolic activity in the tissues from which it drains. Arterial blood may be obtained directly by puncturing a superficial artery, such as the brachial, and drawing the blood into a heparinized syringe. It is important not to expose the sample to air because this would allow the $P_{CO_2}$ to fall.

To avoid the slight difficulty involved in puncturing an artery, measurements are often made on venous or capillary blood which has been "arterialized". The idea here is to produce a local vasodilatation which so increases local blood flow that venous blood does not differ appreciably from that in the arteries. One way of doing this is to immerse a hand in water at 45°C for about 10 minutes. A blood sample can then be taken from a vein on the back of the hand. Another method is to warm an ear lobe and collect "arterialized" capillary blood from a small stab wound into a heparinized capillary tube (Fig. 133).

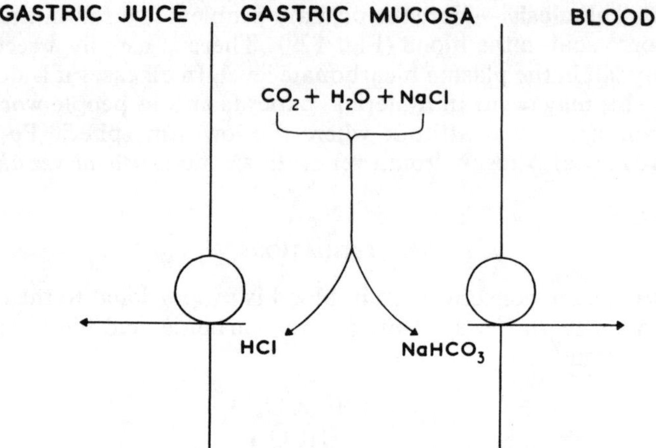


FIG. 131.—Hydrochloric acid secretion in the stomach. HCl and $NaHCO_3$ are synthesized from $CO_2$, $H_2O$ and NaCl in the oxyntic cells with the aid of carbonic anhydrase. HCl is actively pumped into the stomach and $NaHCO_3$ is taken up by the blood. In vomiting there is a net loss of HCl from the body and a net gain of $NaHCO_3$ by the blood. The rise in blood $HCO_3^-$ causes metabolic alkalosis.

Blood for acid-base studies should be analysed as soon as possible. Glycolysis in blood cells during storage increases blood acidity. If the measurements cannot be made immediately, the blood should be stored in iced water to reduce metabolic activity in the blood.

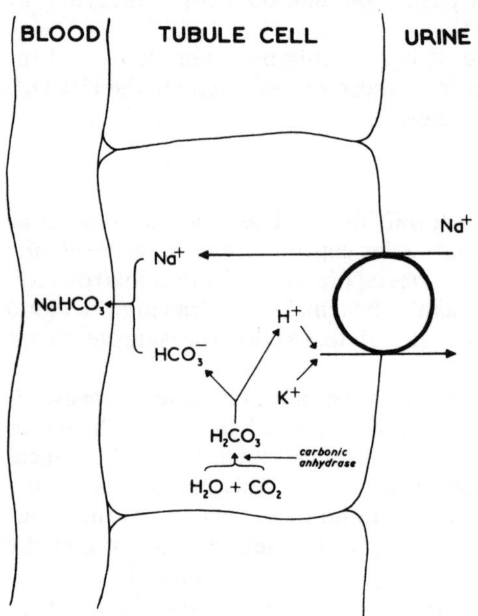


FIG. 132.—Bicarbonate formation in the distal convoluted tubules. $H^+$ and $HCO_3^-$ are formed from $CO_2$ and $H_2O$ under the influence of carbonic anhydrase. $H^+$ is actively pumped into the urine in exchange for $Na^+$. This $Na^+$, together with the $HCO_3^-$ is carried into the blood as $NaHCO_3$. $K^+$ competes with $H^+$ for exchange with $Na^+$ on the pump. If there is $K^+$ depletion, more $H^+$ is pumped because of lack of competition for places on the pump. More $HCO_3^-$ is then formed and passed into the blood as $NaHCO_3$. The rise in blood $HCO_3^-$ causes a metabolic alkalosis.

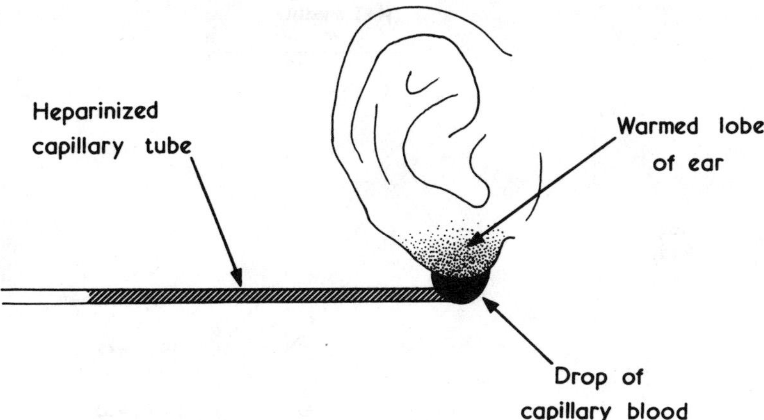


FIG. 133. —Collection of "arterialized" capillary blood from a drop gathering over a small stab wound in a previously warmed ear lobe. Blood flows by capillary attraction into the heparinized capillary tube without significant exposure to the atmosphere. When filled the capillary tube is sealed at each end with wax.

**Measurement of $[H^+]$**

The most convenient way of doing this is to use a glass electrode whose glass membrane is porous to hydrogen ions only. When placed in a solution containing hydrogen ions, the ions tend to move by diffusion into the electrode and set up a transmembrane potential (inside positive with respect to outside) whose size is proportional to $[H^+]$ in the solution.

Glass electrodes have been designed which allow small quantities of blood to be sucked into a small chamber where the blood makes contact with the glass membrane but is not exposed to the atmosphere. The temperature of the blood and electrode are maintained at body temperature by a thermostatic system.

**Measurement of $Pco_2$**

A variety of methods are available for this measurement.

(a) $Pco_2$ electrodes have been developed which generate a transmembrane potential which is proportional to the $Pco_2$ in the solution in which the electrode is immersed.

(b) Interpolation on a pH-log $Pco_2$ equilibration line. When pH is plotted against log $Pco_2$, straight lines are obtained for each particular value of $[HCO_3^-]$ (Fig. 134). To establish the $[HCO_3^-]$ line for the blood being analysed, two small fractions of the blood sample are equilibrated with gas mixtures of known but different $Pco_2$ values. The pH of these samples is measured. Plotting the pH against the $Pco_2$ for these samples gives two points on the $[HCO_3^-]$ line for that blood. Knowing the line and the pH of the test sample allows the $Pco_2$ of the test sample to be read off from the graph.

**Calculation of $[HCO_3^-]$**

If the pH and $Pco_2$ are known the $[HCO_3^-]$ may be read from a nomogram of the type shown in Fig. 135. For example, if the $Pco_2$ in the sample is 50 mm Hg

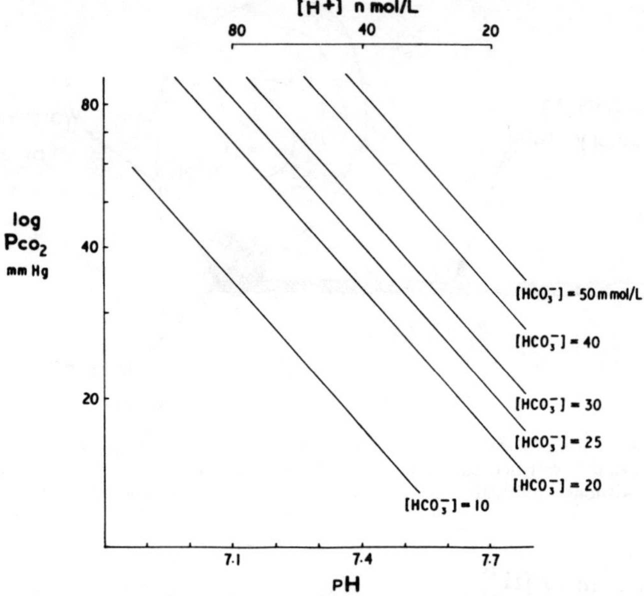


FIG. 134.—The relationship between log $P_{CO_2}$, pH [$H^+$] and [$HCO_3^-$]. Note that the $P_{CO_2}$ scale is logarithmic. For any value of [$HCO_3^-$] there is a straight line relationship between log $P_{CO_2}$ and pH.

and the pH is 7.3, a straight extrapolation of the line joining these two points cuts across the appropriate [$HCO_3^-$] value, i.e. 24 mmol/l. From these figures one can say that the patient had an uncompensated respiratory acidosis because the [$H^+$] is raised, the $P_{CO_2}$ is raised and the [$HCO_3^-$] is within normal limits.

Sometimes blood base is expressed as *standard bicarbonate.* This is the amount of bicarbonate contained in blood with a $P_{CO_2}$ of 40 mm Hg at 38°C. It is measured by equilibrating a blood sample to a known $P_{CO_2}$, i.e. 40 mm Hg, and measuring the pH at that $P_{CO_2}$. Figure 136 shows a [$HCO_3^-$]—pH diagram for

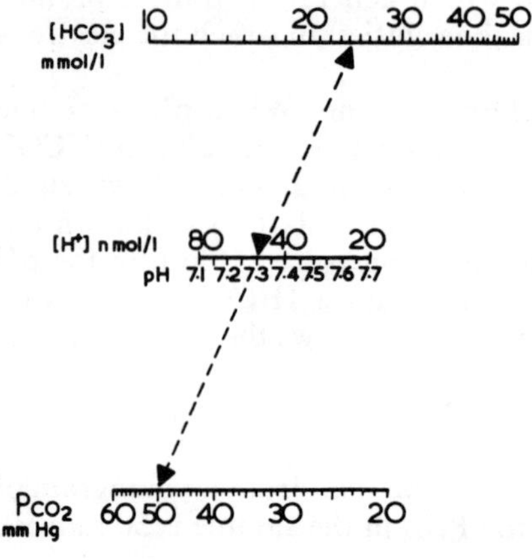


FIG. 135.—A nomogram for calculating pH [$H^+$], $P_{CO_2}$ and [$HCO_3^-$]. One of the quantities may be read off the nomogram if the values of the the other two are known.

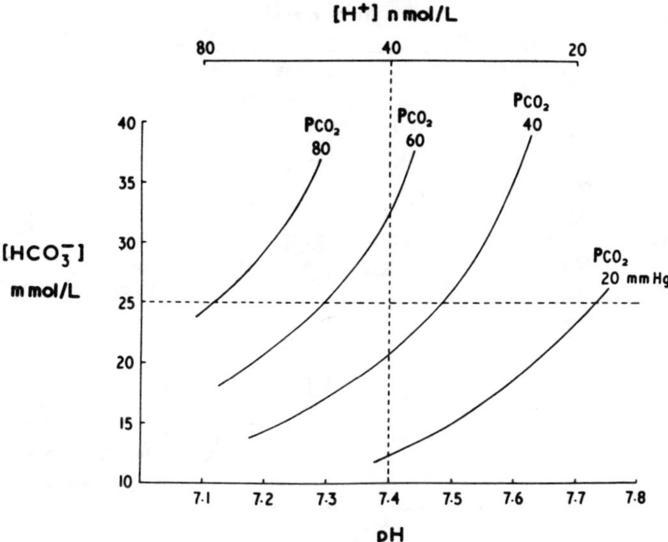


FIG. 136.—The relationship between [$H^+$] (pH), [$HCO_3^-$] and $Pco_2$ ([$H_2CO_3$]). If two of the quantities are measured, the third may be calculated.

different values of $Pco_2$. It can be seen that at a $Pco_2$ of 40 mm Hg, the [$HCO_3^-$] value (standard bicarbonate) can be read off for any value of pH. The normal range of standard bicarbonate is from about 22-26 mmol/l.

### Assessment of Acid-base Status

It may be helpful to plot the values obtained on a particular blood on a log $Pco_2$—pH diagram on which the normal ranges for $Pco_2$ pH and [$HCO_3^-$] are drawn (Fig. 137). This makes it easier to understand the true nature of any disturbance.

### Treatment

It is usual to treat the *cause* in these conditions. In particular potassium deficiency must be corrected. Ammonium chloride ($NH_4Cl$) may be used to antagonize alkalosis. In the body, ammonia ($NH_3$) is taken from the $NH_4Cl$ to form urea in the liver. This leaves hydrochloric acid as a residue. The bicarbonate level in the blood is reduced in buffering this acid and the pH falls. In practice this treatment is rarely required.

## HYDROGEN ION RETENTION [ACIDOSIS]

### Definition

This is the condition where the hydrogen ion concentration in arterial plasma rises above the upper limit of normal, i.e. 45 nmol/l (pH 7.35). If the prime fault is a rise in the carbonic acid concentration, it is known as respira-

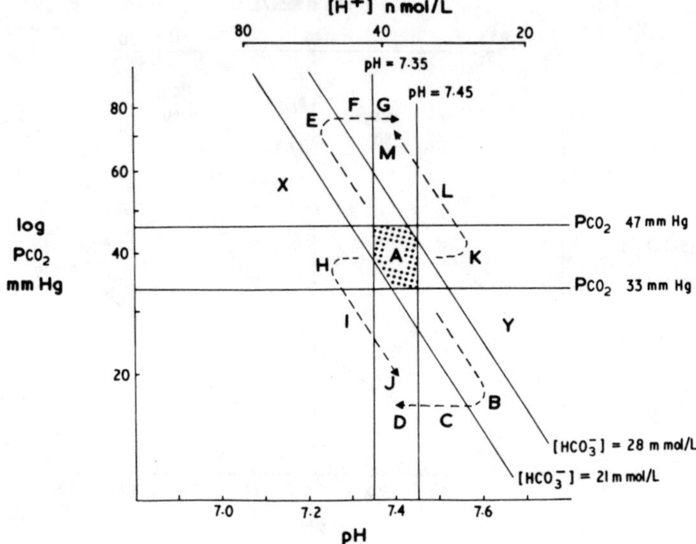


FIG. 137.—A log $P_{CO_2}$-pH diagram in which the range of normal values for $P_{CO_2}$, pH and $[HCO_3^-]$ have been marked. To use the diagram, the pH and $P_{CO_2}$ of a sample of the patient's arterial blood is measured and the result is plotted as a point on the diagram. If the point lies in the area labelled:
A, the acid-base status is normal.
B, C or D, there is uncompensated, partly compensated or compensated RESPIRATORY ALKALOSIS respectively.
E, F or G, there is uncompensated, partly compensated or compensated RESPIRATORY ACIDOSIS respectively.
H, I or J, there is uncompensated, partly compensated or compensated METABOLIC ACIDOSIS respectively.
K, L or M, there is uncompensated, partly compensated or compensated METABOLIC ALKALOSIS respectively.
X, there is a mixed respiratory and metabolic acidosis.
Y, there is a mixed respiratory and metabolic alkalosis.

tory acidosis. If the prime fault is a reduction in plasma bicarbonate, the condition is called metabolic acidosis (Fig. 138).

## Effects

The consequences of respiratory and metabolic alkalosis are summarized in Fig. 139.

**General effects.**—The increase in hydrogen ion concentration interferes with many cellular activities by moving the pH away from the optimal value for enzyme activity. The cells of the brain are the first to suffer and consciousness is depressed. The patients become drowsy and apathetic and eventually lapse into coma.

**Increase in renal excretion of hydrogen ion.**—The raised $[H^+]$ stimulates the renal excretion of hydrogen ion which is eliminated as ammonium and dihydrogen phosphate salts. This has two consequences. Firstly it limits the renal excretion of potassium (Fig. 132) which tends to be retained in the body and may depress the heart. Secondly, it is associated with increased production of bicarbonate by the renal tubules which is reabsorbed back into the blood. In respiratory acidosis this raises the plasma bicarbonate to a level where it may compensate for the raised $P_{CO_2}$. In metabolic acidosis, it may limit the fall in serum bicarbonate.

**Stimulation of ventilation.**—The rise in $[H^+]$ tends to stimulate the respiratory centre to increase ventilation. In metabolic acidosis, the hyperventilation (*Kussmaul breathing*) may lower the $P_{CO_2}$ to a value which may compensate for the reduced serum bicarbonate (Fig. 138). In respiratory acidosis, the stimulation of ventilation tends to limit the rise in $P_{CO_2}$.

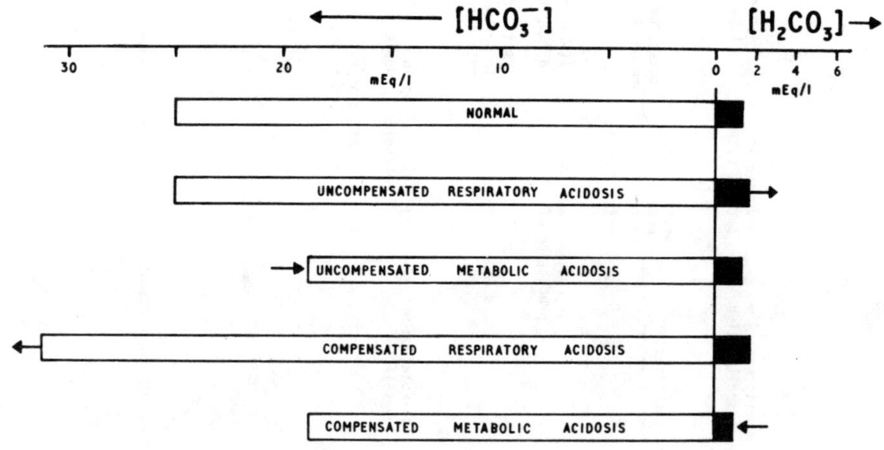


FIG. 138.—The changes that occur in carbonic acid and bicarbonate concentrations in uncompensated and compensated acidosis.

**Cardiovascular effects.**—These vary considerably with the cause and severity of the condition. The increase in $P_{CO_2}$ in respiratory acidosis is usually associated with dilatation of peripheral blood vessels, presumably because of a direct dilator action on the vessels. The low peripheral resistance is associated with a high cardiac output and warm peripheral tissues. If the condition is of long standing, overburdening of the heart may lead to a high output cardiac failure.

The rise in $[H^+]$ and the potassium retention depress the excitability and contractility of the heart and may aggravate the failure. In metabolic acidosis, low blood pressure, low cardiac output, disturbances of cardiac rhythm and cardiac arrest may occur.

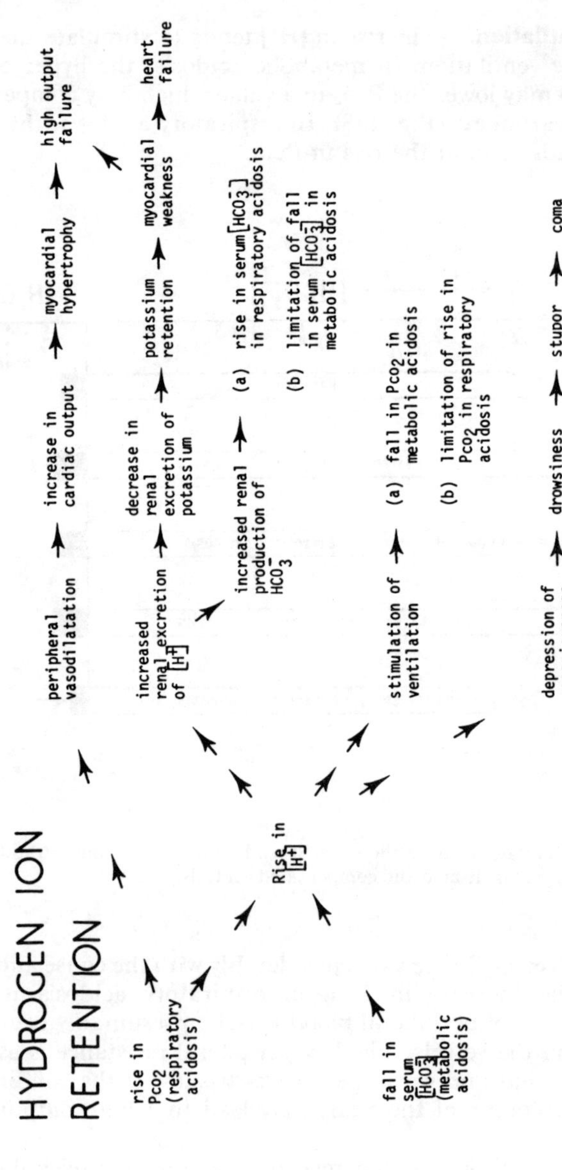


FIG. 139.—The consequences of respiratory and metabolic acidosis.

## Causes

### Metabolic Acidosis

(a) **Excessive production of acids.**—In *diabetes mellitus* and *starvation* the production of keto acids from excessive fat catabolism causes metabolic acidosis. Plasma bicarbonate is reduced in the buffering of these acids. In circulatory failure and hypoxia, anaerobic metabolism in muscle leads to the accumulation of lactic acid in the blood. *Lactic acidosis* is sometimes seen in the absence of circulatory failure and hypoxia, but the cause in these cases is not understood.

(b) **Ingestion of acid-producing substances.**—The metabolism of *ammonium chloride* and *methyl alcohol* (methylated spirits) in the body leads to the formation of hydrochloric and formic acid as shown below.

$$\underset{\text{ammonium chloride}}{NH_4Cl} \longrightarrow \underset{\text{urea}}{NH_3} + \underset{\text{Hydrochloric acid}}{HCl} \longrightarrow H^+ + Cl^-$$

$$\underset{\text{Methyl alcohol}}{CH_3OH} \xrightarrow{\text{oxidation}} \underset{\text{Formic acid}}{HCOOH} \longrightarrow H^+ + HCOO^-$$

(c) **Inadequate excretion of acids.**—Accumulation of the acid residues of protein digestion such as sulphates and phosphates may occur in *renal failure* and result in a reduction in plasma bicarbonate. *High plasma potassium* levels may reduce hydrogen ion secretion by the renal tubules because these ions compete with each other for secretion by the renal tubules. *Implantation of the ureters into the colon,* as may become necessary following the removal of the bladder, may be followed by reabsorption of the hydrogen ions in the urine to cause a metabolic acidosis.

Excretion of hydrogen ions is also reduced by *drugs* which inhibit carbonic anhydrase. They also decrease the reabsorption of bicarbonate causing a fall in plasma bicarbonate and a metabolic acidosis. In the condition of *renal tubular acidosis,* the tubules lose the ability to acidify the urine and a similar situation ensues.

In all cases where the $[HCO_3^-]$ is reduced, the anion deficit in the plasma is made good by an increase in the $[Cl^-]$. This variety of acidosis can be called a "hyperchloraemic" acidosis.

(d) **Loss of bicarbonate.**—This is seen in *intestinal obstruction* when the digestive juices rich in bicarbonate may be lost by vomiting. It can also occur when these juices are lost in severe diarrhoea.

### Respiratory Acidosis

(a) **Inadequate ventilation.**—This may result from depression of the respiratory centre by narcotic drugs, paralysis of the respiratory muscles, disease of the lungs and many other causes which are described in the section on the respiratory system.

(b) **Breathing gas mixtures with a high $CO_2$ content.**—This may occur during anaesthesia when air is rebreathed and carbon dioxide is not absorbed.

### Investigations

These are described in the section on hydrogen ion depletion.

### Treatment

The chief treatment must be that of the cause, e.g. relief of bronchial obstruction to correct a respiratory acidosis.

As most cases of metabolic acidosis are associated with disorders of salt and water balance, these may have to be corrected first. In some cases, the acidosis may be antagonized by oral administration of tablets or intravenous infusions of sodium bicarbonate or sodium lactate. The approximate amount to give may be estimated by multiplying the plasma deficit of bicarbonate in mmol/l by the volume of extracellular fluid.

# 79. DISTURBANCE OF CALCIUM BALANCE

## CALCIUM ION DEPLETION

### Definition

Of the kilogram or so of calcium contained in the adult body, less than 1 per cent is in solution in the body fluids; the rest is deposited as salts in bone. Normal plasma contains about 2.5 mmol/l (10 mg per cent). About half of this exists as free ions and most of the rest is bound to plasma proteins which act as anions at the pH of plasma. The plasma concentration of calcium can vary independently of the total amount of calcium in the body. For example, a reduction in total body calcium may occur with a reduced, normal or raised plasma concentration of calcium. It is thus necessary to consider two separate aspects of calcium depletion; firstly that which results from a fall in the calcium stores in the body, and secondly that which results from a fall in the concentration of ionic calcium in extracellular fluid.

### Effects

**Effects of Reduction in Total Body Calcium**

Because most of the calcium in the body is in bone, total body calcium depletion causes weakness in the skeleton. This manifests itself either by *spontaneous fracture,* where bone breaks in response to a trivial mechanical stress, or *deformity*, where the bones become bent out of their normal shape by the stresses to which they are normally subjected.

**Effects of Reduction in Calcium Ion Concentration**

Small reductions in calcium ion concentration cause an increase in the excitability of muscle and nerve which results in spontaneous contractions of muscles. The condition is known as *tetany*. It is thought that calcium ions in the extracellular fluid make the membranes of excitable cells less permeable to sodium ions. Thus, a fall in the concentration of calcium permits a greater influx of sodium. The depolarization so caused makes the membranes more excitable and, if sufficient, results in spontaneous excitation.

In a child, tetany is characterized by spasm of the small muscles of the hands and feet (*carpopedal spasm*), spasm of the muscles which pull the vocal cords together (*laryngismus stridulus*) and convulsions. In the adult, convulsions and vocal cord involvement are less common, but carpopedal spasm and painful muscle cramps throughout the body may occur.

Although calcium ions normally play an important role in blood coagulation and the contraction of muscle, falls in calcium ion concentration in the body are never severe enough to block coagulation processes or the ability of skeletal, smooth and cardiac muscle to contract. This is because the patient would die for other reasons before such levels were reached.

## Causes

### Causes of Reduction in Total Body Calcium

In all cases the calcium content of bone is reduced but there are two varieties of reduction. In one case there is a reduction in the amount of bony matrix in bone, the remaining matrix being adequately mineralized. This is known as *osteoporosis* and is really a form of bone wasting or atrophy. It is seen in old people (senile osteoporosis) or after prolonged bed rest where the stimulating effects of mechanical stress on bone formation are lost. It is also common in post-menopausal women. Impaired ability to manufacture body protein, as may occur in starvation, malabsorption syndromes and states of excessive corticosteroid secretion, also causes osteoporosis. In hyperparathyroidism, there is excessive osteoclastic activity resulting in breakdown of bone tissue.

The other variety of reduction in bone calcium is that where the bony matrix, though normal in amount, is inadequately mineralized. It is known as *osteomalacia* (Gr. *osteon* = bone, *malakos* = soft). It may result from inadequate absorption of calcium from the gut as in dietary calcium deficiency, vitamin D deficiency and in malabsorption syndromes. Frequent pregnancies may make the condition worse because of loss of calcium to the growing fetuses. Rickets is a juvenile form of osteomalacia.

### Causes of Reduction in Calcium Ion Concentration

This may occur if accidental *removal of the parathyroid glands* occurs at thyroidectomy. Parathormone normally maintains the blood concentration of calcium ion by its action on bone, kidneys and alimentary tract.

Conditions giving rise to osteomalacia (see above) may give rise to a low calcium ion concentration but sometimes the concentration is normal. Frank tetany is rarely seen in these conditions.

*Alkalosis* can cause tetany. It does this, not by reducing the concentration of calcium in the blood, but by increasing the affinity of plasma protein anions for calcium. This leads to a reduction in the concentration of *free* calcium ions in the plasma.

## Investigations

### X-ray Studies

X-ray pictures of the skeleton provide good evidence about the calcium stores in the body. Calcium salts are mainly responsible for the X-ray density of bone and osteoporosis and osteomalacia can be detected from such pictures.

### Serum Calcium Levels

The concentration of calcium in blood is easily measured in the laboratory. The range of normal extends from about 2.25 to 2.75 mmol/l (9-11 mg per cent). These figures refer to the sum of free ionic and protein bound calcium. Though one is interested primarily in the concentration of ionic calcium, this does not usually matter because the amount bound to plasma protein is a function of the relatively constant plasma protein concentration. However, if the plasma proteins are depleted, the total serum calcium level may be low whereas the ionic calcium concentration is normal. For this reason, the plasma protein concentration should be measured as well as that of calcium if there is any possibility of an abnormal protein value.

#### Serum Phosphate Level

As the product of the concentrations of calcium and phosphate ions in the blood is a constant (their solubility product), there is an inverse relationship between their concentrations. Measurements of serum phosphate levels therefore may help in diagnosis.

### Treatment

#### Treatment of Reduction of Total Body Calcium

(a) **Osteomalacia.**—Vitamin D and a calcium-rich diet are used to treat this condition. Vitamin D aids the absorption of calcium and the growth of developing bone. If the condition is due to malabsorption or excessive urinary loss, higher doses of vitamin D have to be used. In all cases, the blood calcium level needs to be checked periodically, because excessive absorption of calcium may lead to calcium deposition in the tissues. Rickets, which is a juvenile form of osteomalacia, is treated in the same way.

(b) **Osteoporosis.**—If there is an obvious cause, such as hyperadrenalism or hyperthyroidism, it should be dealt with. However, in most cases, such as senile and post-menopausal osteoporosis, the cause is not known. Anabolic steroids, sex hormones and dietary calcium have been used in an attempt to encourage new bone formation. Moderate exercise is also beneficial because the mechanical stresses it causes stimulate the formation of bone tissue.

#### Treatment of Reduction in Calcium Ion Concentration

A sudden fall in ionic calcium creates a medical emergency, especially in a child. To stop or prevent convulsions, calcium gluconate should be given intravenously. For long-term treatment large doses of vitamin D are usually sufficient to maintain the blood calcium level.

## CALCIUM ION RETENTION

### Definition

This condition usually implies a rise in the concentration of calcium ions in extracellular fluid. Though it may co-exist with a rise or no change in total body calcium, it often co-exists with a fall in total body calcium. For example, the serum calcium concentration is raised (*hypercalcaemia*) in hyperparathyroidism whereas the total body calcium is reduced.

## Effects

### Neuromuscular Disturbances

Hypercalcaemia is associated with muscular weakness and mental apathy, developing into drowsiness and coma. The physiological explanation of these effects may be a decreased excitability of nerve and muscle membranes due to decreased permeability of those membranes to sodium.

### Deposition of Calcium Salts in the Tissues

When the calcium or phosphate concentration rises in the blood and the solubility product of these ions is exceeded, calcium phosphate is precipitated out of solution. The effects of this depend on the tissues which are affected.

The kidney is most frequently affected giving rise to the condition of *nephrocalcinosis.* The deposition of calcium salts in the tubular cells results in renal failure. Calcium salts may also be deposited in the cornea, the walls of blood vessels and elsewhere.

### Formation of Renal Calculi

The increase in the urinary output of calcium salts in the urine in hypercalcaemia is associated with the formation of stones (calculi, L. *calculus* = pebble) in the kidney or ureter. These can damage kidney tissue or obstruct the outflow of urine. They often contribute to the renal failure caused by nephrocalcinosis.

### Alimentary Tract Disturbances

Loss of appetite, vomiting and abdominal pain are quite common in hypercalcaemia. There is no obvious explanation for this. It may result from damage caused by calcium salts deposited in the mucous membranes.

### Increase in Urinary Volume

The volume of urine increases and this may be associated with thirst. The increase is partly due to the increased osmotic load imposed by the raised calcium level and partly to a loss of ability in the kidney to concentrate urine.

## Causes

### Hyperparathyroidism

Parathormone acts on bone, kidney and alimentary tract to raise the serum calcium concentration. The output of parathormone may be raised by an increase in parathyroid secretory tissue. This sometimes occurs with tumours of the glands (primary hyperparathyroidism). It may also occur as an excessive response to conditions such as malabsorption or chronic renal failure which lower blood calcium (secondary hyperparathyroidism).

### Excessive Absorption of Calcium

This commonly results from *overdosage with vitamin D* which facilitates the absorption of calcium from the gut. In some conditions, the tissues of the gut seem to be over-sensitive to vitamin D. Excessive absorption of calcium can occur in these patients with a normal diet.

**Osteoporosis**

In the osteoporosis that occurs with prolonged immobilization, calcium is released from the bone and is excreted in the urine. Because the kidney can normally deal with all the calcium that is released, the blood concentration is little changed. Destruction of bone by cancer tissue may also release calcium and raise the blood and urine levels of calcium.

### INVESTIGATIONS

These are described in the section on calcium ion depletion.

### TREATMENT

Where hypercalcaemia is due to primary hyperparathyroidism, surgical removal of the affected glands may cure the condition. Where it is due to over-dosage of vitamin D, vitamin D therapy must be stopped. In certain types of hypercalcaemia, but not in cases due to hyperparathyroidism, glucocorticoids may reduce the blood calcium level by a mechanism which is not understood.

# 80. INCREASED INTERSTITIAL FLUID VOLUME (OEDEMA)

## Definition

Oedema (Gr. *oedema* = swelling) is the condition where the amount of interstitial fluid in the tissues is increased. It may occur locally or generally throughout the body.

## Effects

### Changes in the Skin

Oedema affecting superficial tissues becomes noticeable when the interstitial fluid increases by 10-15 per cent or more. The overlying skin becomes puffy and smooth as the wrinkles are stretched out. It has a doughy consistency and when sustained pressure is applied to the skin with a finger or thumb, an indentation is left when the pressure is removed. This is called "pitting on pressure" and is due to shifting of tissue fluid away from the site of pressure. When the pressure is removed, the fluid seeps back to fill out the indentation.

Puffiness of the skin develops in some areas more readily than others. Where the underlying tissues are lax and easily distended, as around the eyes, a large volume of tissue fluid can be accommodated to cause gross swelling. Where the skin is thick and the underlying tissue dense, as in the fingers and toes, not much swelling can occur. Nevertheless, fluid retention in the subcutaneous tissues of the fingers of a patient with generalized oedema may make it difficult for the patient to take off or replace rings and oedema can be demonstrated by pitting the skin overlying the front of the tibia.

Gravity is another factor which determines the distribution of oedema fluid. Oedema develops more readily in tissues where the capillary blood pressure is raised by dependency below the heart. Thus swelling may first be seen at the ankles of patients who are not confined to bed; when confined to bed, swelling is usually seen first over the sacral area.

Swollen oedematous skin is more readily damaged by injury than normal skin. When turgid with fluid, it lacks the elasticity and resilience that protects normal skin. Thus the tissues around the eyes of a boxer are more readily cut or damaged by further blows, once they have become swollen by oedema fluid.

### Increase in Body Weight

As extracellular fluid collects in the tissues, body weight rises appreciably, e.g. by 10 per cent or more. Daily measurements of body weight provide one of the best indices of the progress of a patient with generalized oedema.

### Impairment of Tissue Nutrition

Oedema jeopardizes tissue nutrition in a number of ways (Fig. 140). (a) The increase in interstitial fluid causes a rise in tissue pressure, especially if, as in

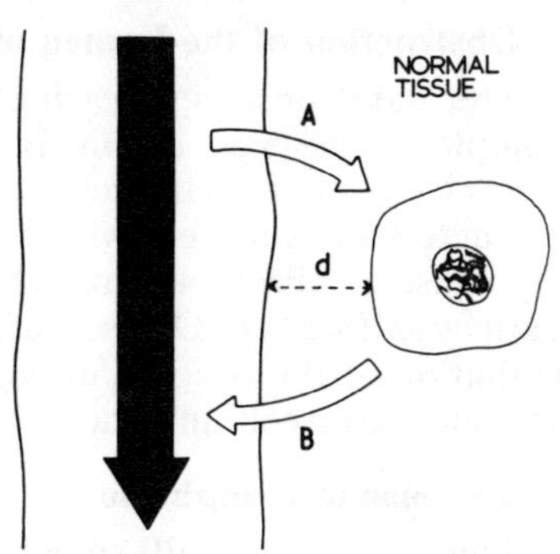


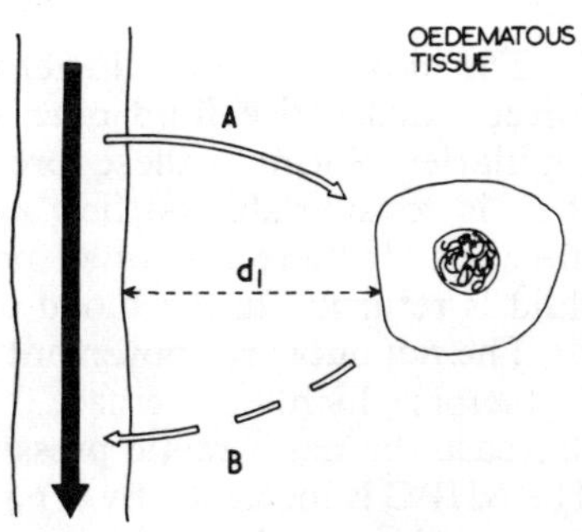


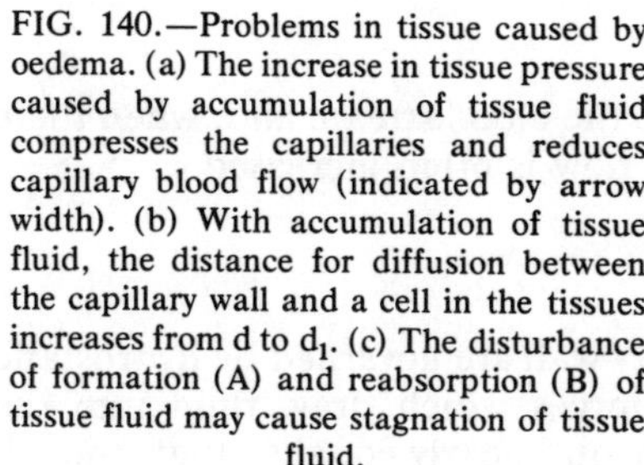
FIG. 140.—Problems in tissue caused by oedema. (a) The increase in tissue pressure caused by accumulation of tissue fluid compresses the capillaries and reduces capillary blood flow (indicated by arrow width). (b) With accumulation of tissue fluid, the distance for diffusion between the capillary wall and a cell in the tissues increases from d to $d_1$. (c) The disturbance of formation (A) and reabsorption (B) of tissue fluid may cause stagnation of tissue fluid.

brain, kidney, bone and limb muscles, the tissue is enclosed in a rigid or stiff capsule or sheath. The increase in tissue pressure tends to compress the local blood vessels and thus reduce the local blood flow. (b) As interstitial fluid accumulates, the distances separating cells from one another and from capillaries increases. This increases the distances for diffusion of nutrients and waste products and makes exchange between cells and blood more difficult. (c) The disturbance of fluid exchange across the capillary wall in oedema may cause stagnation of tissue fluid and thus impair the nutrition of the cells. Fluid is normally extruded from capillary blood vessels at their arterial end. After permeating through the tissue space it is returned to the blood at the venous end of the capillary by the osmotic pressure exerted by the plasma proteins. This movement of tissue fluid, which helps to nourish the cells, is dependent on the proper exchange of fluid across the capillary wall.

The effect of impairment of nutrition in oedematous tissues depends on the site. Oedematous skin is liable to develop ulcers and when the surface is cut, it is very slow to heal. Oedema of the brain can cause headache, confusion, loss of consciousness and, in severe cases, death of brain neurones. Infections of the kidney give rise to local oedema and the resulting impairment of cellular nutrition can lead to death of renal tissue. Bone marrow is also susceptible to this form of damage.

### Obstruction of the Lumen of Tubes and Passages

The nasal obstruction which occurs with the common cold is a familiar example of the way in which oedema of mucous membranes can obstruct internal passages. With colds, the pharyngotympanic tube may also become obstructed and interfere with hearing. Insect stings at the back of the mouth may cause localized oedema which completely obstructs the pharynx to cause death by suffocation. Oedema affecting the bronchial mucosa may interfere with ventilation of the lungs. Pulmonary oedema may interfere with gas exchange between the alveoli and the pulmonary capillaries.

### Increase in Lymph Flow

Lymphatics normally return tissue fluid to the blood stream and, when there is excessive formation of tissue fluid, lymph flow is often increased.

## CAUSES

The movements of fluid across the capillary wall are governed by hydrostatic forces which drive fluid out, and osmotic forces which draw fluid into the capillaries. Normally these forces are opposite and nearly equal so that, though there is considerable outflow and inflow, there is little *net* movement in either direction. If there is a small net outward movement, the excess of interstitial fluid is returned to the blood by the lymphatic system.

The net outward movement of fluid responsible for oedema may be caused by factors which, (a) increase the net hydrostatic pressure gradient (NHPG), (b) decrease the net osmotic pressure gradient (NOPG) or (c) do both (Fig. 141). The NHPG is increased by (1) a raised capillary hydrostatic pressure or by (2) a lowered tissue hydrostatic pressure. The NOPG is reduced by (3) a fall in blood colloid osmotic pressure or (4) a rise in tissue fluid colloid osmotic pressure.

### Increase in Capillary Hydrostatic Pressure

A rise in venous pressure causes pressure to build up in capillaries and is one of the most common causes of oedema. It may be due to local or generalized venous obstruction, vein valve incompetence or cardiac failure. In all cases it is most evident in dependent parts of the body where venous pressure is raised by gravitational forces. Arteriolar dilatation favours oedema because it allows more blood to enter the capillaries and increases the pressure within them. The high blood pressure in hypertension is confined to the arterial system; it does not result in oedema unless it causes cardiac failure.

When blood volume rises, as it does in congestive cardiac failure and in conditions where there is excessive secretion of adrenal cortical hormones, the mean hydrostatic pressure in the capillaries rises. This contributes to the oedema seen in these conditions.

In the lungs, capillary blood pressure is low compared to that in the systemic capillaries, resulting in a low NHPG. However, the osmotic pressure of the plasma proteins is normal, resulting in a normal NOPG. For this reason, oedema is less likely to occur in the lungs than in other tissues; relatively greater increases in capillary pressure are required for the NHPG to exceed the NOPG in the lungs.

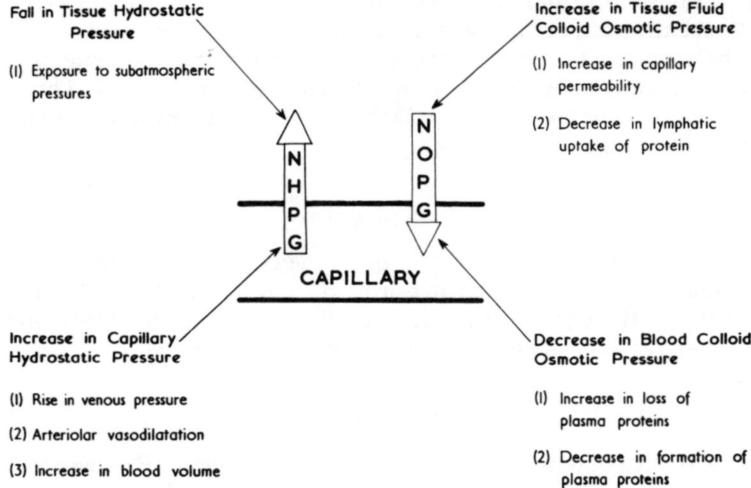


FIG. 141.—Causes of oedema. Oedema occurs when the net hydrostatic pressure gradient (NHPG) driving fluid out of the capillary greatly exceeds the net osmotic pressure gradient (NOPG) drawing fluid into the capillary. The NHPG rises if the capillary pressure rises or the tissue pressure falls. The NOPG falls if the blood colloid osmotic pressure falls or if the tissue colloid osmotic pressure rises.

### Fall in Tissue Hydrostatic Pressure

This rare type of oedema may be seen in tissues exposed to sub-atmospheric pressures. This causes a greater fall in tissue pressure than in capillary pressure. Applying suction to an area of skin can cause it to swell with oedema fluid.

### Decrease in Blood Colloid Osmotic Pressure

This may result from increased loss or decreased formation of plasma proteins. Plasma proteins are formed in the liver and the loss has to exceed the considerable ability of the liver to manufacture them, before their concentration in plasma falls. Prolonged loss of plasma proteins occurs in some forms of kidney disease where protein is lost in the urine. It also occurs in severe burns and ulcerative colitis where large amounts of protein may be lost with fluid seeping from damaged vessels on the burned skin or ulcer base.

Factors which impair the formation of plasma proteins, such as chronic liver disease and malnutrition, also cause oedema. From the point of view of colloid osmotic pressure, the albumin fraction is about four times as important as the globulin fraction. This is because there is twice as much albumin as globulin and its molecular weight is only half as great. Osmotic pressure depends on the number and not on the size of the particles.

### Increase in Tissue Fluid Osmotic Pressure

If the tissue fluid protein rises, the NOPG drawing fluid into the capillaries falls and oedema results. Such a rise occurs when capillary permeability is increased. This may occur with inflammation, or allergic reactions; histamine may be responsible in part for these phenomena. When protein escapes, the protein concentrations on the two sides of the capillary wall approximate. When the lymphatic system becomes obstructed by cancer cells or parasite infestation, the uptake of protein by the tissue lymph channels is reduced. Thus plasma proteins which escape from the capillaries become trapped in the tissue spaces and the NOPG falls.

### Investigations

Changes in body weight provide a good index of the progress of a patient with oedema. Analysis of the plasma proteins may also provide useful information. Sometimes the injection of a dye or a radio-opaque substance into a vein (lymph angiography) may show up a lymphatic obstruction.

### Treatment

The treatments of oedema are those of its many causes. However, oedema is reduced by general procedures which lower capillary blood pressure or raise the colloid osmotic pressure of the blood.

(a) *Reducing capillary pressure.*—Bed rest reduces the effect of gravity on capillaries in dependent tissues and may relieve oedema of the legs. Drugs which increase urine excretion by the kidneys reduce blood volume and thus capillary pressure. Blood volume may be reduced directly by withdrawing blood from a vein (venesection). Restriction of sodium intake may also reduce the volume of extracellular fluid and therefore blood volume.

(b) *Increasing plasma osmotic pressure.*—In severe cases of burns, the intravenous infusion of plasma or dextran solutions are used to increase colloid osmotic pressure. Diets high in protein may be used to foster the formation of plasma proteins when their level in the blood is low.

## RECENT LEADING ARTICLES

Disorders of magnesium metabolism in infancy. *Brit. med. J.*, 1973, **4**, 373.
Hydrochloric acid for metabolic alkalosis. *Lancet*, 1974, **1**, 720.
Lactic acidosis. *Brit. med. J.*, 1970, **4**, 258.
Inappropriate secretion of ADH. *Brit. med. J.*, 1972, **3**, 489.
Treatment of acute hypercalcaemia. *Lancet*, 1972, **2**, 314.

*Section XIII*

# DISORDERS OF THE URINARY SYSTEM

# 81. URINARY TRACT INFECTION

## Definition

In this condition, multiplication of micro-organisms leads to inflammatory changes in the walls of the bladder and urethra.

## Effects

When organisms multiply in the urinary tract, the infected walls become inflamed. In the bladder this is called *cystitis* and in the urethra, *urethritis.* A *scalding pain* is experienced while urine is being passed. This is due to hyperalgesia of the nerve endings in the urethra. The mere flow of urine becomes an adequate stimulus to excite them. The inflammation in the wall of the bladder seems to increase the sensitivity of the stretch receptors involved in the micturition reflex. Because of this, a wish to micturate is experienced much earlier during bladder filling than normal. This leads to frequent evacuations of the bladder (*frequency*), with only small volumes of urine being passed on each occasion. In severe cases a strong wish to micturate may be experienced even when the bladder is practically empty (*strangury,* Gr. *stranx* = drop, *ouron* = urine) so that each micturition may only produce a drop or so of urine.

The urine may be reddish because of blood, cloudy because of pus and have an unpleasant smell.

The infection may spread to involve the renal pelvis and kidney (*pyelonephritis*).

## Causes

The free flow of urine normally results in the elimination of micro-organisms before they can multiply excessively. However, any factor which limits the free flow of urine predisposes to urinary infection.

The organisms (usually *E. coli*) most commonly reach the bladder from below and the short urethra in the female makes her more prone to this than the male. However, the organisms may reach the bladder from above, if there is a chronic kidney infection; or from outside, if organisms spread from an inflamed colon to the bladder.

The introduction of a tube into the bladder (*catheterization*) allows organisms to track up the urethra to the bladder and a catheter left in for several days almost inevitably leads to infection.

## Investigations

A sample of urine taken in the mid-stream of micturition is collected and examined for micro-organisms and pus. The *mid-stream sample* is less likely to be contaminated with surface skin organisms than a sample collected at the beginning of micturition.

Urinary tract infection is very common in women of child-bearing age, so that it is not usually necessary to investigate for predisposing causes in such women. However, if the infection persists, it is necessary to carry out investigations to see if there is a predisposing cause such as a partial obstruction, a stone in the bladder or a bladder cancer. The inside of the bladder may be visualized through an instrument (*cystoscope*) inserted via the urethra. Investigations for predisposing causes are also required for urinary tract infections in patients other than women in the child-bearing age group; such infections suggest an abnormality in the tract.

### Treatment

Predisposing causes, if found, have to be dealt with. Otherwise the treatment involves the identification of the organism, the assessment of its sensitivity to antibiotics and the administration of the appropriate drug.

# 82. URINARY TRACT OBSTRUCTION

## Definition

In this condition, a mechanical obstruction in the lower urinary tract impedes the outflow of urine.

## Effects

### Dilatation and Hypertrophy

When the resistance to urinary outflow increases, the bladder responds in a similar fashion to the heart in hypertension by dilatation and hypertrophy. As it becomes more difficult for the bladder muscle to empty the bladder, the force and size of the urinary stream is decreased and urine tends to accumulate in the bladder. As with cardiac muscle, bladder muscle contracts with progressively more energy as it is stretched up to a certain limit (Fig. 142). Thus the dilatation of the bladder allows it to move further out along its normal bladder function curve, contract with greater vigour and generate a greater pressure.

If the bladder muscle is continuously given more work to do, it, like cardiac muscle, increases bulk by an increase in the size of the individual muscle cells (*hypertrophy*). When this happens, the bladder wall becomes noticeably thicker and thick bands of muscle (*trabeculae*) can be seen criss-crossing under the membrane lining the inner surface. With hypertrophy, the bladder function curve moves upward and to the left so that when it contracts, it can generate a greater pressure in the bladder at any given bladder volume (Fig. 142).

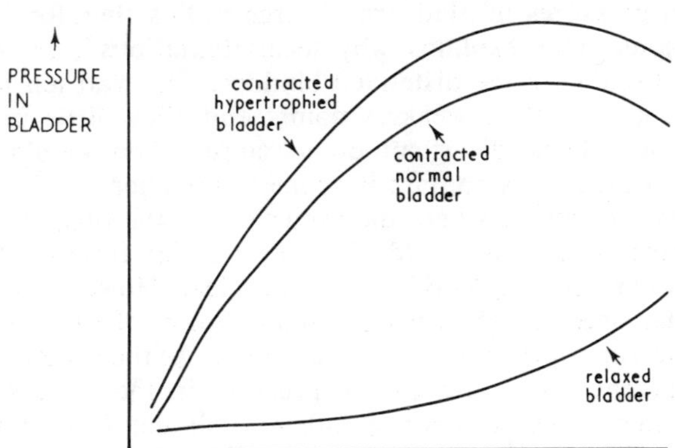


FIG. 142.—Changes in pressure in the bladder with changes in bladder volume. Note (1) The pressure in the relaxed bladder does not increase much with increasing volume until the elastic limits of the bladder are reached. (2) The contracted bladder can generate greater pressures the more it is stretched up to certain value. (3) With hypertrophy the contracted bladder can generate greater pressures than the normal bladder at any given bladder volume.

By the dilatation and hypertrophy mechanisms, the bladder is able to generate more power when it contracts in the attempt to overcome the increased resistance to outflow. As the bladder relies partly on stretch to generate the additional power, it has difficulty in emptying completely because the stretch advantage is lost at low bladder volumes. The stream of urine towards the end of micturition is very weak. It is for this reason that the residual volume in the bladder after micturition (*residual urine*) tends to increase as the obstruction becomes more severe.

With increasing dilatation these compensatory mechanisms become less effective. Increasing stretch brings progressively smaller returns in terms of increased contractile power. The law of Laplace has important consequences for the dilated bladder. The law may be expressed as:

$$T = \frac{Pr}{2}$$

where T is the tension in the wall, P the pressure in the bladder and r the radius of the bladder. Because of the law of Laplace, the muscle has to produce progressively more tension in the wall to generate the same pressure as the radius of the bladder increases. A time comes when the radius is so large that the muscle cannot generate enough pressure to overcome the obstruction. When this happens, it causes the surgical emergency of *acute obstruction.* The bladder keeps distending and gives rise to acute distress and pain. A tube (catheter) may have to be inserted into the bladder via the urethra to evacuate its contents. If this is not done, the bladder muscle may be severely damaged. Acute urinary tract obstruction is particularly likely to occur when the individual is in a situation where emptying of the bladder is delayed for social reasons until overdistension has occurred.

It can also be seen from the law of Laplace that the tension in the bladder wall increases not only as the internal pressure rises but also as the radius increases. High pressures in bladders of large radius therefore result in very high wall tensions. This explains why local dilatations (*diverticuli*) appear between the trabeculae in the distended bladder. The wall tension is so great that the tissues give at their weakest points and allow little protruding sacs (diverticuli) to form. These diverticuli do not empty when the bladder contracts and the stagnant urine they contain is prone to infection.

Normally, the structures where the ureter enters the bladder act as a valve which allows urine to enter the bladder but not to reflux from the bladder to the ureter. The way this valve works is by no means clear. However, when the outlet from the bladder becomes obstructed, accumulation of urine in the bladder leads to incompetence of the uretero-vesical valves. As the uretero-vesical valves become incompetent, there is a rise in pressure in the ureters. The smooth muscle in the walls of the ureters responds in a similar fashion to that in the bladder, i.e. by dilatation and hypertrophy. The dilated ureter with urinary obstruction is called a *hydro-ureter* and the condition when the pelvis of the ureter is dilated is called *hydronephrosis.*

**Infection**

As mentioned earlier, the free flow of urine does not permit micro-organisms in the urine enough time to multiply sufficiently to infect the urinary tract. With obstruction, the failure to empty the bladder completely allows residual urine to stagnate there. The stasis favours multiplication of micro-organisms and the development of urinary infections. The infections may affect the kidneys, ureters, bladder and urethra.

**Renal Failure**

With progressive urinary obstruction, the rising pressure in the bladder is transmitted back up the ureters to the kidneys. Eventually the pressure in Bowman's capsule rises to a value which interferes with glomerular filtration and renal failure results.

To understand the pressure changes produced by urinary obstruction, it is necessary to know something about the pressure-volume characteristics of the bladder. As urine accumulates in the bladder, the pressure in the relaxed bladder rises (Fig. 142). However, it should be noted that when the bladder is being filled, there is initially very little increase in pressure; it is only when the bladder is distended beyond its elastic limits that the pressure rises steeply with increasing volume. It used to be thought that the accommodation of large volumes with little change in pressure was due to a simultaneous relaxation of the smooth muscle. While this may be true in part, the phenomenon can be explained in physical terms by the law of Laplace. As $P = \frac{2T}{r}$, and wall tension (T) and radius (r) increase in about equal proportion in the early stages of bladder filling, the pressure (P) should not change much initially. However, when the bladder becomes stretched to the limits of its elasticity, small further increases in radius are associated with large increases in wall tension so that the bladder pressure rises rapidly.

Normally the bladder is emptied when it contains about 200-300 ml of urine and the pressure is something under 10 cm $H_2O$. However, it can accommodate about twice that volume with little further increase in pressure especially if the rate of filling is slow. When the volume exceeds about 600 ml, the pressure starts to rise more rapidly. For these reasons, back pressure on the kidneys is a late phenomenon when the outflow from the bladder is partially obstructed.

Urinary output starts to fall when the ureteric pressure rises above about 25 cm $H_2O$. The hydrostatic pressure in glomerular capillaries responsible for the formation of glomerular filtrate is normally opposed by the hydrostatic pressure of the plasma proteins and the intracapsular pressure (Fig. 143). This leaves a net filtration force of about 25 mm Hg. Thus a rise in ureteric pressure could stop glomerular filtration and hence renal function.

If the condition persists, the kidneys are permanently damaged. The condition is called *hydronephrosis*. The back pressure leads to dilatation of the pelvis of the ureters, enlargement of the calyces, and atrophy of the tubules. Eventually the kidney may become a thin-walled bag of fluid.

## CAUSES

The commonest cause of urinary obstruction is enlargement of the prostate gland which surrounds the urethra as it leaves the bladder. The enlargement often occurs in elderly men; its cause is not known. Following infections of the urethra, e.g. by gonorrhoea, scarring may cause a narrowing (stricture) in both the male and the female. Urinary outflow can also be blocked by stones and tumours in the ureters or bladder.

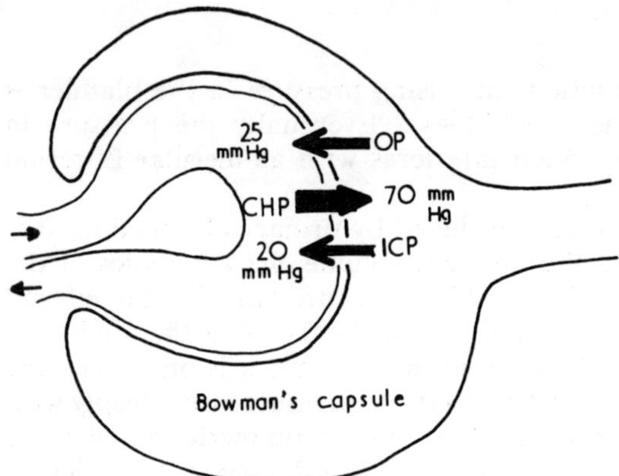


FIG. 143.—The forces involved in the formation of glomerular filtrate. The capillary hydrostatic pressure (CHP) of about 70 mm Hg, which tends to drive fluid into the capsular space, is opposed by the osmotic pressure of the plasma proteins (OP) of about 25 mm Hg and the intracapsular pressure (ICP) of about 20 mm Hg. This leaves a net filtration pressure of about 25 mm Hg.

## INVESTIGATIONS

Examination of the bladder through a device (cystoscope) inserted through the urethra may allow the cause of the obstruction to be visualized. Tests of renal function are also required to see if the back pressure has damaged the kidney.

Measurement of the pressure generated in the bladder during micturition provides objective evidence of the severity of the obstruction to outflow. The measurement is, however, rarely required because clinical evidence of urinary obstruction is usually conclusive. In a normal bladder, the peak pressure is usually in the range of 20-30 cm $H_2O$. With urethral strictures or prostatic enlargement, the peak pressure may be in the range of 60-120 cm $H_2O$.

## TREATMENT

The objects of treatment are to evacuate the bladder and remove the obstruction. When evacuating the bladder, great care is necessary to ensure sterility. In conditions where the urine is stagnating, organisms introduced with the catheter proliferate readily.

# 83. NEUROLOGICAL DISTURBANCES OF MICTURITION

## DEFINITION

**EMPTYING OF the bladder (micturition) is a reflex event, which is integrated in the sacral spinal cord and effected, in the main, by parasympathetic nerves. Activation of stretch receptors as the bladder fills eventually results in reflex contraction of the smooth muscle in the bladder wall and a relaxation of its sphincters. Normally the basic reflex can be inhibited or facilitated by higher centres exerting voluntary control. This chapter deals with the changes in micturition that occur when the nervous control of micturition is disturbed.**

## EFFECTS

**A simplified diagram of the neural control of the bladder is shown in Fig. 144. The spinal reflex arc consists of sensory fibres (S) whose stretch receptors in the wall of the bladder are stimulated by bladder distension and motor fibres ($M_1$ and $M_2$) which contract the emptying (detrusor) and relax the sphincter muscles respectively. The spinal reflex can be inhibited or facilitated by neurones from supraspinal centres in the cerebrum (C).**

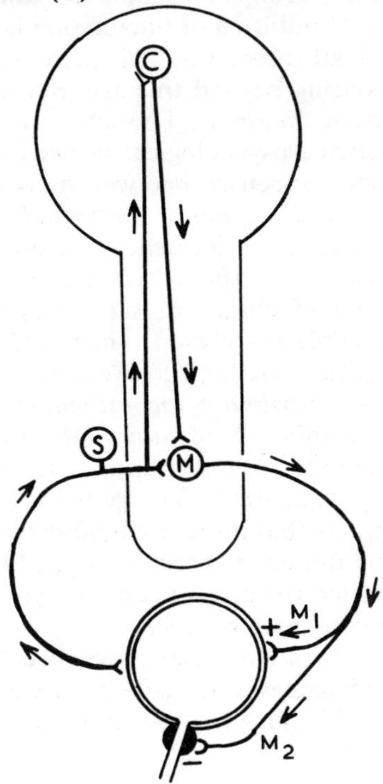


FIG. 144.—A simplified diagram of neural control of the bladder. Three main groups of neurones are involved. S represents the sensory fibres whose stretch receptors in the wall of the bladder signal distension of the bladder. M represents the autonomic motor fibres that excite (+) the emptying (detrusor) muscle in the bladder wall ($M_1$) and inhibit (—) the sphincters controlling exit from the bladder ($M_2$). C represents the supraspinal neurones in the cerebrum and elsewhere which can facilitate or inhibit the spinal reflex.

In a normal person, when the bladder has filled to contain about 300 ml of urine, the increased activity of the stretch receptors causes reflex contractions of the detrusor muscle and a conscious awareness of the need to micturate. If the higher centres accede to this, they facilitate the spinal reflex so that reflex contraction of the detrusor and relaxation of the sphincters results in micturition. If the higher centres decide that the time is not propitious, the spinal reflex is inhibited and the reflex contractions die away until further filling generates more urgent signals. When the bladder contains about 600 ml, the signals become almost irresistible.

When the nervous control of micturition is disturbed, a variety of effects can result depending on the nature of the disturbance. In some instances, the bladder may fail to empty so that *urinary retention* occurs. Retention may result in distension of the bladder which is appreciated as *pain*. In some cases, the emptying of the bladder cannot be controlled. This is known as *urinary incontinence*.

## CAUSES

### Disturbance of Supraspinal Neurone Function

In the very young, higher control is absent and the bladder empties automatically when filling is adequate to activate the spinal reflex. Higher control may also be lost in the very old as a result of the gradual deterioration of brain function with age and the degenerative changes in the brain blood vessels. Both these groups of people are unable to control bladder activity by will.

Inhibition of micturition is usually maintained during sleep after the age of about three, although there is a wide normal variation. The persistence of bed wetting beyond this age (*nocturnal enuresis,* Gr. *enourein* = to pass urine) is quite common. Though it is usually due to supraspinal factors, the cause is often a psychological rather than a physiological disorder. Most cases clear up spontaneously; bed wetting is very common in children and very rare in adults.

*Transection of the spinal cord* also cuts off the spinal reflex centres from supraspinal influence. In the stage of spinal shock, which lasts for several weeks, the spinal reflexes are inhibited, presumably because of loss of supraspinal facilitation. In this stage the muscles of the bladder and its sphincters are completely relaxed. The bladder fills passively and when the pressure within it is sufficiently high to overcome the outflow resistance, the excess urine dribbles away (*dribbling incontinence*). After some weeks, when reflex activity returns to the spinal cord, *automatic micturition* is seen; the bladder empties spontaneously when it is appropriately full without giving rise to conscious sensation.

Sometimes damage to the brain stem or spinal cord appears to knock out the tracts that normally inhibit the spinal reflex. These people are greatly troubled by frequency because slight filling of the bladder results in reflex micturition, which the patient cannot suppress. This condition is known as the *uninhibited neurogenic bladder*.

In some cases, excessive facilitation of the spinal reflex by supraspinal influences leads to frequency of micturition. This may be seen in anxious people and is not unknown in people waiting for interview or oral examination (Fig. 145).

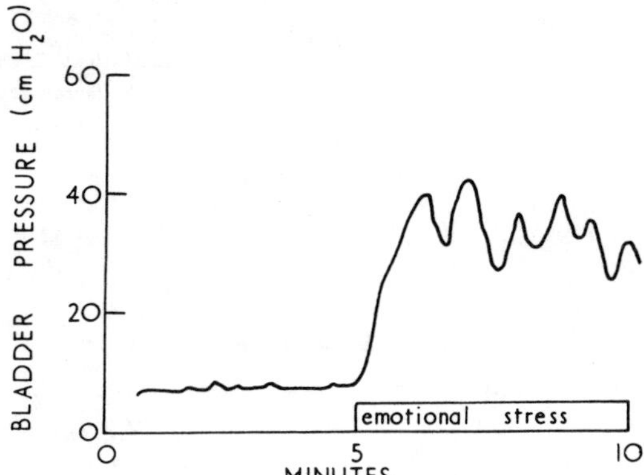


FIG. 145.—The effect of emotional stress on bladder tone.

*"Bashful bladder".*—Some people have considerable difficulty in initiating micturition in the presence of other people. This is due to an increase in the normal level of higher centre inhibition of micturition. It is not of any serious consequence but may make it difficult for the patient to pass urine readily during investigations of bladder function.

**Disturbance of the Spinal Reflex**

(i) *When the entire reflex arc is out of action* as occurs in the stage of spinal shock and when the lower spinal cord is crushed or otherwise destroyed, all the bladder muscles are flaccid and urine entering the bladder dribbles away in dribbling incontinence when its pressure is sufficient to overcome the outflow resistance. Because sensory impulses from the distended bladder do not reach consciousness no discomfort is experienced due to bladder distension.

(ii) *When the loss of function is on the sensory side,* the input of sensory impulses to provide conscious information of bladder distension and reflex emptying of the bladder is interrupted. In this condition, the bladder continues to fill but no reflex relaxation of the sphincters or contraction of the detrusor occurs. The bladder then becomes very distended, and, when the pressure within it is sufficient to overcome the sphincteric resistance, a *dribbling overflow incontinence* is seen. Because of the sensory damage, this type of distension of the bladder also does not cause pain. However, renal function may be impaired by the back pressure in the ureters.

(iii) *When the loss of function is on the motor side,* the picture is more confusing because the effects depend on which motor nerves are involved, and the results of the experimental work in animals are of little assistance because of variation between species in motor control of the bladder.

There are three sets of motor supply to the bladder and its sphincters (Fig. 146)—the parasympathetic fibres carried in the pelvic nerves, the sympathetic fibres carried in the hypogastric nerves and the somatic fibres carried in the pudendal nerves to the striated muscle comprising the external sphincter.

If the *parasympathetic* (pelvic nerves) are cut, the bladder loses its ability to empty effectively. It becomes distended and there is a dribbling overflow

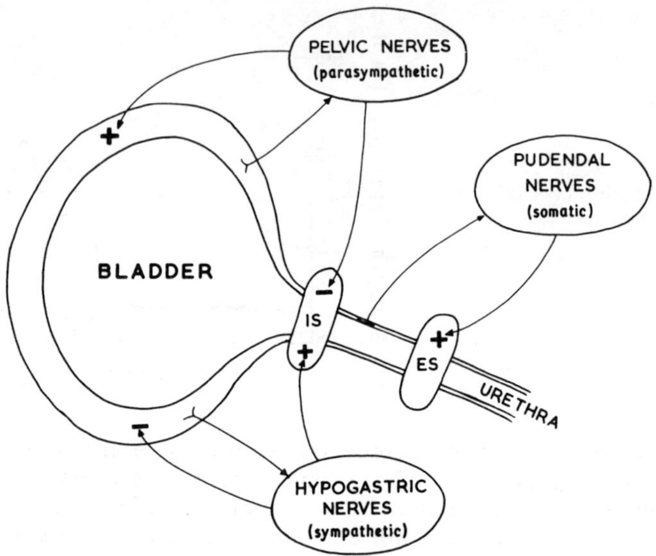


FIG. 146.—The nerve supply of the bladder and proximal urethra. The pelvic nerves carry (1) motor excitatory fibres to the smooth muscle in the bladder wall, (2) motor inhibitory fibres to the smooth muscle of the internal sphincter (IS) and (3) afferent fibres from the bladder wall. The hypogastric nerves carry (1) motor inhibitory fibres to the bladder wall, (2) motor excitatory fibres to the internal sphincter and (3) afferent fibres from the bladder wall. The pudendal nerves carry (1) motor excitatory fibres to the skeletal muscle of the external sphincter and (2) afferent fibres from the wall of the proximal urethra; impulses carried in these nerves result in reflex inhibition of tone in the external sphincter.

incontinence. Weak contractions of the detrusor muscle may return eventually, triggered perhaps by stimulation of local nerve networks in the wall. Proper micturition never returns and it is clear that the pelvic nerves are the main effector system in the micturition reflex. The pelvic nerves are also thought to carry the majority of the sensory fibres carrying information about bladder distension.

If the *sympathetic* (hypogastric) nerves are cut, micturition is surprisingly little affected. There have been some reports that the frequency of micturition is a little increased. Though these nerves do not seem too vital for micturition, they are important in the male for closing the internal sphincter and preventing the reflux of seminal fluid into the bladder during ejaculation. Cutting the sympathetic nerves or using drugs which block the action of sympathetic nerves may cause male infertility by allowing such reflux. The sensory fibres serving bladder pain seem to run mainly with the hypogastric nerves, because cutting these nerves relieves bladder pain.

Cutting the *somatic* (pudendal) nerves in man (or woman) does not produce dramatic changes in the micturition reflex. The ability to halt micturition in mid-stream is decreased, which suggests that voluntary contraction of the external sphincter is responsible for this ability. It should be stressed that voluntary micturition is not initiated by a voluntary relaxation of the skeletal muscle in the external sphincter. It is initiated by facilitation of the autonomic reflex, mediated through the pelvic nerves. This causes contraction of the detrusor and relaxation of the internal sphincter, which allows urine to enter the proximal urethra. Relaxation of the external sphincter is thought to be produced reflexly by afferent impulses from the proximal urethra which travel in the pudendal nerves whenever urine enters it.

In women, the external sphincter is not well developed and it is easily damaged or weakened, as are other perineal muscles and the internal sphincter during childbirth. Thus, after a woman has given birth to a number of babies, she may develop the condition of *stress incontinence.* She is unable to prevent the escape of small quantities of urine whenever there is a transient rise in her intra-abdominal pressure. This is a tedious condition, because the patient tends to wet herself slightly every time she laughs or coughs. The external sphincter seems to be useful for the voluntary suppression of urine escape.

Acute *retention of urine* sometimes occurs *after surgery,* especially with surgery involving pelvic structures. Many factors may contribute to this—irritation of the bladder sphincters causing local spasm, depression of reflex activity by the general anaesthetic and parasympathetic blockade by anti-cholinergic drugs given before the operation. Most cases clear spontaneously, but if the condition persists and is not treated, overdistension of the bladder wall may damage the bladder muscle.

## Investigations

**Record keeping.**—Micturition is such a familiar event that memory tends to be unreliable in recording the details of micturatory performances. Thus it is useful for the observer to keep accurate records of the times of micturition and the volumes passed.

**Cystography.**—X-ray opaque material introduced into the bladder permits the activity of the bladder to be visualized by X-ray screening (fluoroscopy) during micturition (*micturating cystogram*). Reflux of the material into the ureters during micturition indicates incompetence of the ureto-vesical valves and is a common consequence of urinary obstruction.

**Cystoscopy.**—A cystoscope is a device containing a light source and a lens system that can be inserted into the bladder through the urethra to visualize the inside of the bladder.

**Cystometry.**—Cystometry is an investigation to measure the pressure-volume characteristics of the bladder. It is not commonly used in clinical practice; its use is confined to special centres interested in the study of bladder function. For cystometry, a fine tube is inserted into the bladder, either through the urethra or the abdominal wall. Fluid is injected slowly into the bladder through the tube and the pressure in the bladder recorded at each bladder volume. The data is plotted as a *cystometrogram* (Fig. 147). This gives three indices of bladder function, the "tone" of the bladder, i.e. the rate of rise of pressure with increasing volume, the threshold value for evoking the micturition reflex and the maximum bladder pressure achieved in the micturition contraction.

Figure 147 shows how the cystometrogram may be affected by bladder disease. The normal bladder accommodates several hundred ml of urine with

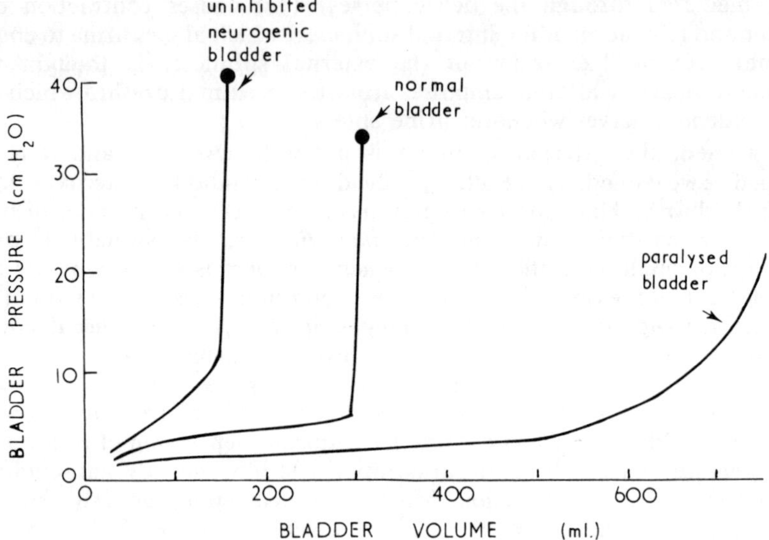


FIG. 147.—Cystometrograms from a normal bladder, an uninhibited neurogenic bladder and a paralysed bladder. These records give information on three points:

(a) The amount of bladder "tone" i.e. the rate of change of pressure with increasing volume: the greater the rate, the greater the "tone".

(b) The micturition threshold, i.e. the bladder volume at which the micturition contraction is elicited. At this point there is a sudden rise in pressure.

(c) The peak pressure achieved during the micturition contraction.

Note that in the uninhibited neurogenic bladder, tone is high and the micturition threshold low. In bladder paralysis, tone is low and a micturition threshold cannot be elicited.

negligible increase in pressure, has a micturition threshold of about 250 ml and has a peak micturition pressure of about 30 cm $H_2O$. If the bladder is infected or if inhibitory supraspinal pathways have been damaged by disease, bladder "tone" is increased and the micturition threshold is low. In bladder paralysis, "tone" is low and micturition contraction cannot be elicited.

## Treatment

If the micturition reflex is intact, but the supraspinal control is lost e.g. due to spinal cord transection, the patient may be trained to trigger the micturition reflex artificially at regular intervals. He may be able to elicit reflex contraction by sudden pressure on the lower abdomen, scratching the inner surface of the thigh or some other trick.

Where the micturition reflex is lost, collection of urine by an indwelling catheter in the bladder to a bag strapped to the patient may be the only permanent solution.

In patients with imperfect control of micturition, attempts have been made to control the activity of bladder sphincters by electrical means. Electrodes implanted in the perineum are triggered to excite the sphincters whenever involuntary leakage of urine occurs. Such devices may help some patients with persistent nocturnal enuresis. However, in the vast majority of cases of childhood enuresis, the condition clears up spontaneously as the child grows older and reassurance for the parents and the child is all that is necessary.

# 84. GRADUAL NEPHRON FAILURE (CHRONIC RENAL FAILURE)

## Definition

Chronic renal failure occurs when there is a slow and progressive reduction in the number of functioning nephrons. Because of the reserve of function, about 75 per cent of the nephrons must be put out of action before the kidneys are unable to deal with the day-to-day needs of the body. Thus if 50 per cent of nephrons are lost (one kidney removed) the glomerular filtration rate is maintained by increased filtration through the remaining nephrons.

## Effects

### Compensatory Hypertrophy

When parts of the kidneys are destroyed by disease, compensatory changes occur in the remaining parts. These changes also occur in the remaining kidney after removal of one kidney and in a solitary kidney transplanted into someone whose kidneys have been removed. About five weeks after removal of one kidney, the remaining kidney increases by about 50 per cent in weight. This is due mainly to an increase in the size and number of cells in the renal cortex. The growth is most marked in the proximal and distal convoluted tubules. By this process the functional efficiency of the kidney can increase to about 80 per cent of the function of both kidneys over the course of a year. The compensatory changes occur more rapidly in younger people on a high protein diet and are slowed by fasting and water restriction. The mechanism of the compensatory hypertrophy is not known.

### Retention of Waste Substances

Retention occurs when destruction of or damage to glomeruli causes a reduction in glomerular filtration rate (GFR). As the GFR falls, the amount of waste substances being filtered and excreted in the urine falls. For example, when the GFR is reduced by half, the concentration of urea in the blood about doubles (Fig. 148). It will be noticed that retention is much more dramatic at low values of GFR and with a high dietary protein content. If a reduction of GFR from 120 to 60 ml/min causes the blood urea to rise from 30 to 60 mg per cent, a fall in GFR from 30 to 15 ml/min causes the blood urea to rise from 120 to 240 mg per cent.

The rise in the concentration of waste products in the blood as the GFR falls ensures that there is a higher concentration of waste products in the diminished filtrate. This helps to maintain the *amount* of waste products being filtered when the GFR falls. In addition, these products, by retaining water osmotically in the renal tubules, tend to increase the volume of urine that is excreted. The effects of retention of the various waste products are considered below.

*Nitrogenous residues.*—These are mainly the end-products of protein metabolism such as urea and uric acid. The rate at which they accumulate depends partly on the amount of protein in the diet (Fig. 148). However, even if protein is excluded from the diet, urea is still formed from the breakdown of body proteins. This breakdown is accelerated if the patient has to metabolize his own protein to supply his energy requirements.

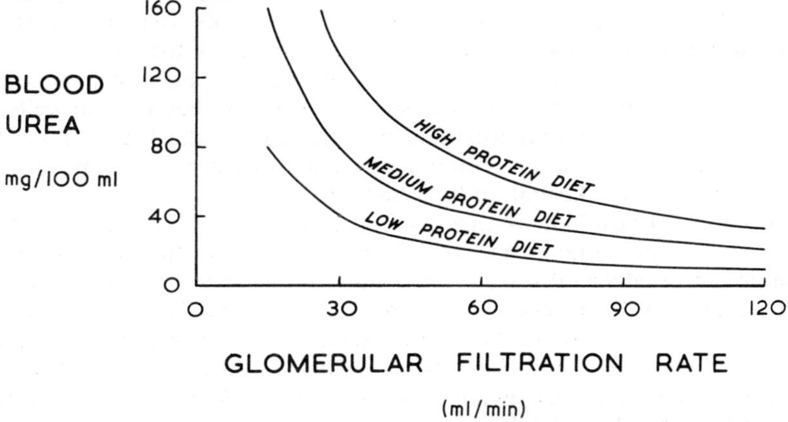


FIG. 148.—The effect of glomerular filtration rate (GFR) on the blood urea level on a high, medium and low protein diet. When the GFR is reduced by about half, the blood urea level about doubles.

Blood urea is easy to measure and is a useful indicator of the severity of renal failure but it is not particularly toxic. When urea is given to a normal person in doses sufficient to cause the blood levels seen in patients with renal failure, it does not produce the toxic effects of renal failure. Furthermore, the toxic effects of renal failure can be relieved in patients using an artificial kidney machine even if the urea concentration in the dialysing fluid is similar to that in the patient's blood.

An increase in blood urea increases urinary output by an osmotic effect in the renal tubules (*osmotic diuresis*) and gives rise to thirst, perhaps because of the rise in plasma osmolarity. Urea, like other osmotic diuretic agents, increases urinary output of both water and salt in the following way. When the concentration of urea in the filtrate rises, it limits the reabsorption of water from the tubules by its osmotic effect. Because the volume of fluid retained in the tubules is greater, fluid moves more rapidly down the nephrons than normal. As the amount of sodium reabsorbed depends in part on the *time* available for reabsorption, the more rapid transit of tubular fluid tends to limit the sodium reabsorption and the chloride and water reabsorption which are normally dependent on it. Thus when a large osmotic load of urea is presented to the kidneys, it results in an increased excretion of both water and salt.

It has been suggested that ammonia released from urea by bacteria containing urease in the skin and alimentary tract can irritate these membranes. This may account for the ammoniacal smell of the breath, the inflammation of the mucous membranes of the mouth, the vomiting and diarrhoea and the itching (pruritis, L. *Prurio* = I itch) of the skin seen in uraemic patients. In extreme cases, urea excreted in the sweat may crystallize on the skin to cause "urea frost".

*Potassium.*—Potassium is normally taken in food in excess of the body requirements and the surplus, plus that derived from the breakdown of cells, is excreted in the urine. Thus loss of renal function predisposes to potassium retention. However, elimination of potassium does not depend entirely on glomerular filtration; potassium can also be secreted by the distal tubules in exchange for sodium ions. In chronic renal failure, potassium retention is usually not severe until the terminal stages; it seems that tubular cells may continue to secrete potassium into the urine after glomerular filtration has been severely reduced. Potassium retention leads to apathy, depression of consciousness and weakness of skeletal and cardiac muscle.

*Phosphate.*—Phosphate is ingested in milk, vegetables, etc. and it is a residue of protein metabolism. In the renal tubules, the reabsorption of phosphate is normally controlled by parathormone so that a suitable blood level is maintained and the excess of intake over requirement is excreted. In renal failure, the phosphate level in the blood tends to rise. This may produce several important effects (Fig. 149). The product of the concentrations of calcium and phosphate in a solution is a constant, the *solubility product.* A rise in phosphate concentration therefore tends to cause a reduction in the ionic calcium

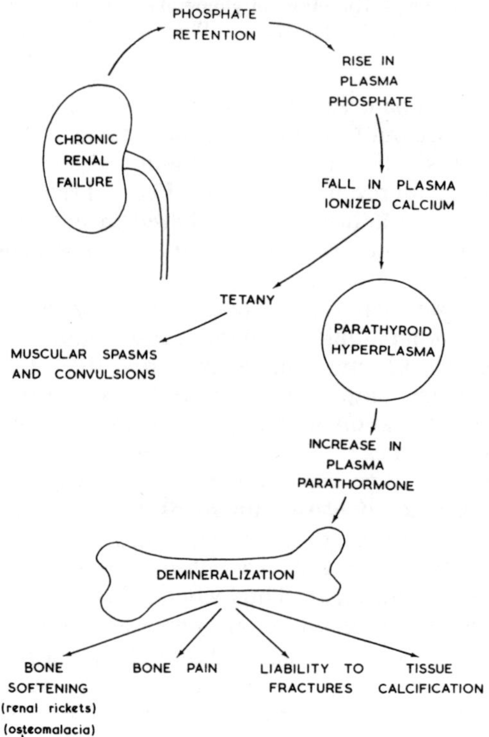


FIG. 149.—Possible consequences of phosphate retention in chronic renal failure. Tetany is not normally seen in chronic renal failure because the associated acidosis tends to compensate for the effect of phosphate retention on the calcium level. However, if the acidosis is corrected, tetany may occur.

concentration of ionized calcium in the blood. The resulting increase in the excitability of muscle and nerve (*tetany*) may cause muscular twitchings, spasms and convulsions, relieved by injections of calcium. In practice, tetany is rarely seen in chronic renal failure, because the fall in plasma pH usually prevents it; it does this by decreasing the ability of plasma proteins to bind calcium ions and thereby increasing the fraction of ionized calcium.

The reduction in the level of calcium stimulates the parathyroid glands to produce more parathormone and to increase in size (*secondary parathyroid hyperplasia*). The increased output of parathormone mobilizes calcium from the bones and leads to decalcification and softening of the vertebrae and other bones (osteomalacia, Gr. *malakos* = soft). Pain in the bones is common in these people and the weakened bones are liable to fracture, especially if the patient has a convulsion. In young people who are still growing, the bones do not grow well and become deformed (*renal rickets*). In the final stages, the decalcification of bone together with the high phosphate level in the blood favours the deposition of calcium salts in the tissues other than bones (metastatic calcification). After restoration of normal renal function by renal transplant, the parathyroids may remain overactive and cause hypercalcaemia ("tertiary hyperparathyroidism").

*Sulphate.*—The retention of sulphate and other acid radicles derived from protein metabolism tends to raise the hydrogen ion concentration in the body fluids (acidosis). Disturbance of acid-base balance in renal failure is considered in greater detail later in the chapter.

*Drugs.*—The kidneys normally play an important part in the elimination of drugs from the body. In chronic renal failure, drugs tend to be retained and their effects, for any given dose rate, may be greater than in normal people.

*Others.*—Many of the ill-defined effects of chronic renal failure have been attributed to the retention of equally ill-defined toxic end-products of metabolism. Phenols are toxic end-products of protein metabolism and have been associated with apathy and depression in patients. The sallow discoloration of the skin has been attributed to retention of pigments (urochrome) normally excreted in the urine. The depression of the bone marrow, with consequent anaemia, liability to infection, and tendency to bleed have also been attributed in part to the retention of toxic urinary constitutents. Many of the other effects of renal failure such as loss of appetite, vomiting, impaired intestinal absorption, muscular weakness, impaired function of the peripheral nerves, headache, dizziness, ulceration of the intestinal mucosa, inflammation of the pericardium and retardation of growth have been attributed to the same cause.

### Loss of Required Substances

All the crystalloid constituents of the blood are freely filtered in the glomerulus. Some protein, mainly albumin, is also filtered but all the rest and the formed elements in the blood are prevented by the glomerular epithelium from entering the capsular space. Most of the filtered molecules such as glucose, amino-acids, sodium chloride, protein, etc. which are required in the body, are reabsorbed back into the blood either actively or passively. In chronic renal failure, loss of tubular ability to reabsorb required substances is usually matched by such a reduction in glomerular filtrate that required substances are not lost. However, if the glomerular membranes are damaged in such a way that plasma protein can readily escape, the amount of protein being filtered may exceed the tubular capacity for reabsorption and protein may appear in the urine.

Loss of required substances is more likely to be seen when the rate of glomerular filtration is relatively normal at a time when tubular function is depressed. This may occur in the diuretic phase of acute renal failure.

### Failure of Plasma Bicarbonate Regulation

The hydrogen ion concentration in the body fluids depends on the ratio of carbonic acid to bicarbonate levels. This can be represented $[H^+] \propto \frac{[H_2CO_3]}{[HCO_3^-]}$

The respiratory system attempts to fix the carbonic acid level at about 1.3 mmol/l. The kidneys attempt to fix the bicarbonate level at about 25 mmol/l. By fixing both sides of the ratio, the two systems fix the hydrogen ion level in the blood. If the plasma bicarbonate level rises above 25 mmol/l, the excess is not reabsorbed but excreted in the urine. If the bicarbonate level falls below the set value, the kidneys actively reabsorb all the bicarbonate in the filtrate and manufacture bicarbonate in the distal convoluted tubule.

The acid residues of the average diet are normally excreted in the urine. In chronic renal failure, they accumulate and are used up in buffering reactions (Fig. 150). As the tubular mechanisms which regulate the blood bicarbonate are damaged by renal disease, replacement of the bicarbonate becomes progressively

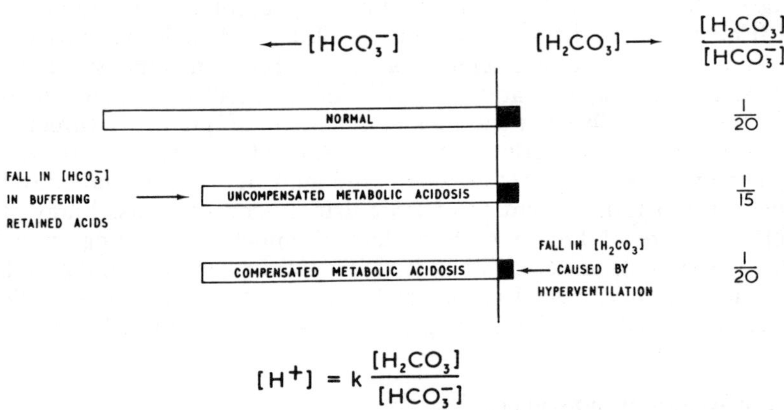


FIG. 150.—Acid-base disturbance in chronic renal failure. The hydrogen ion concentration $[H^+]$ depends on the ratio of carbonic acid $[H_2CO_3]$ to bicarbonate $[HCO_3^-]$ which is normally held at 1:20. The retention of acid residues of metabolism and the reduced ability of the kidney to manufacture $HCO_3^-$ in chronic renal failure cause a fall in plasma $[HCO_3^-]$. Therefore, the $[H^+]$ rises (uncompensated metabolic acidosis). The rise in $[H^+]$ stimulates the respiratory centre to cause hyperventilation. The resulting washout of carbon dioxide reduces the $[H_2CO_3]$. At this stage (compensated metabolic acidosis), the $[H^+]$ is about normal, even though the $[HCO_3^-]$ is reduced, because there is a corresponding reduction in $[H_2CO_3]$.

more difficult. This causes an "uncompensated" metabolic acidosis. The fall in pH stimulates the respiratory centre to cause hyperventilation. The resulting wash-out of carbon dioxide lowers the carbonic acid level in the blood so that the ratio of carbonic acid to bicarbonate returns to approximately its normal value. The metabolic acidosis is now "compensated". Though the bicarbonate level is reduced, the blood pH is about normal because there has been a corresponding reduction in carbonic acid.

In severe renal failure, compensation is incomplete and increasing acidosis develops. If the acidosis becomes severe, it may contribute to the coma and other signs of cerebral depression seen in the late stages of failure. With an average diet, acidosis is the disturbance of acid-base regulation usually seen in renal failure. However, damage to the kidneys limits the ability of the body to cope with either acid or alkaline loads.

### Failure to Regulate the Osmolarity of Body Fluids

When the osmolarity of the body fluids tends to rise or fall, the kidneys assist in homeostasis by excreting a concentrated or a dilute urine respectively. The fluid which is normally presented to the renal collecting ducts is very hypotonic because sodium has been pumped actively out of the distal convoluted tubule and water has been unable to follow. The interstitial fluid surrounding the collecting ducts becomes more and more hypertonic as the renal pelvis is approached because of the counter-current concentrating system in the medulla. In the absence of ADH, the walls of the collecting ducts are impermeable to water and the hypotonic fluid from the distal convoluted tubules passes out unaltered as a very dilute urine. In the presence of ADH, the walls of the collecting ducts become permeable to water. Fluid flowing through the ducts comes into equilibrium with the hypertonic fluid surrounding them as water permeates out into the interstitium. This results in the formation of a very concentrated urine. These functions depend on the ability of the tubular epithelium to transport sodium actively against a concentration gradient and on the renal collecting ducts being responsive to the action of antidiuretic hormone (ADH). Kidney damage frequently interferes with the first of these factors and hence the ability to excrete a urine whose osmolarity differs greatly from that of plasma. If the kidney cannot transport sodium so that it builds up an osmotic gradient in the renal medulla, it cannot form a concentrated or dilute urine (Fig. 151). The excretion of urine of a constant osmolarity similar to plasma irrespective of the body needs is known as *isosthenuria.* The specific gravity of protein-free plasma and of such urine is around 1.010.

As a consequence of the inability to form a concentrated or dilute urine the patient is less well able to withstand water deprivation or excessive fluid intake than a normal person. An excessive water intake may cause water intoxication and water deprivation may cause rapid dehydration.

The inability to concentrate urine results initially in large volumes of urine being excreted (*polyuria*). This helps to maintain the excretion of waste products but it may lead to excessive sodium loss and sodium depletion. Inability to concentrate urine during the night results in the need to micturate at times during the night (*nocturia*).

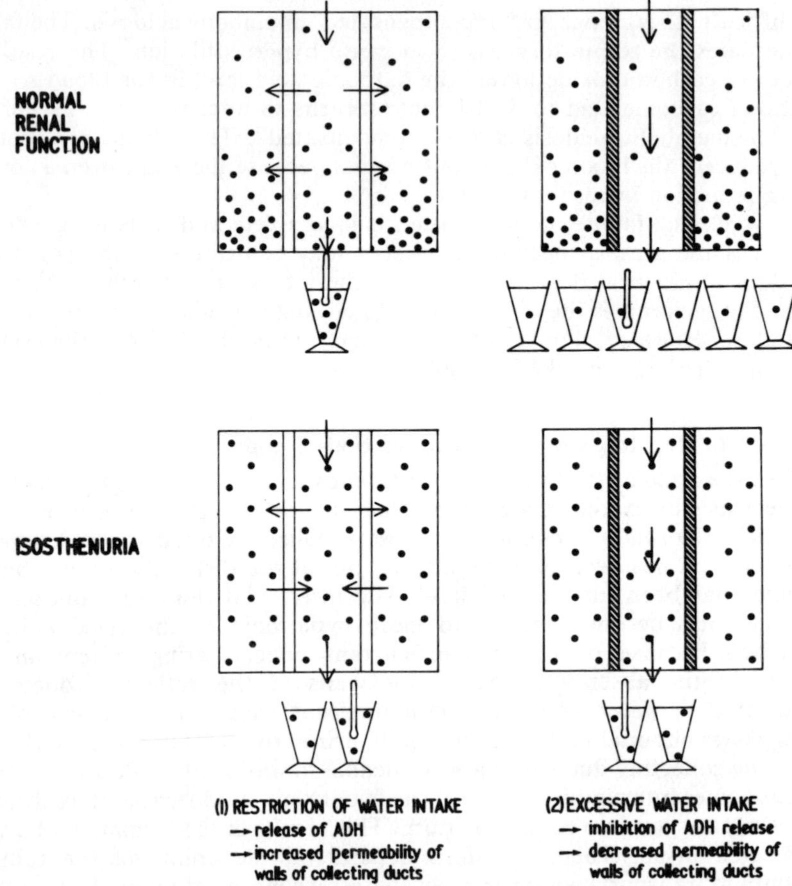


FIG. 151.—A possible explanation for the phenomenon of isosthenuria. It is postulated that there is loss of the function whereby sodium is pumped from inside the loop of Henle into the medullary interstitium. As a result, the fluid passing to the collecting duct is less dilute than usual and the fluid in the medullary interstitium is less concentrated than usual. Eventually the two are similar, the presence or absence of ADH is of no consequence, and the osmolarity of urine is that of protein-free plasma (glomerular filtrate).

## Failure to Regulate the Amount of Sodium in the Body

A rise in body sodium results in an expansion of extracellular fluid and vice versa. Changes in the volume of extracellular fluid are monitored by "volume receptors" in the heart and great veins and by the juxtaglomerular apparatus in the kidneys. The changes initiate appropriate corrections to adjust salt and water excretion by the kidneys and restore the extracellular fluid volume to normal.

Renal failure limits the ability to deal with sodium excess or deficiency. Excess intake may lead to an increased blood volume, hypertension and pulmonary oedema; sodium deficiency may lead to a fall in blood volume, concentration of the blood and peripneral circulatory failure. In practice, sodium retention is rarely seen in chronic renal failure. This is because the osmotic diuresis caused by the raised blood urea tends to maintain sodium excretion. Sodium depletion, on the other hand, can readily occur. Chronic renal failure may be suddenly made worse by vomiting or diarrhoea and result in rapid circulatory collapse. The circulatory failure accelerates the renal failure which in turn may lead to more vomiting. A vicious circle is thus set up and may be rapidly fatal.

**Failure of Blood Pressure Regulation**

When renal blood flow is reduced, the hormone renin is released from juxta-glomerular cells. The subsequent conversion of angiotensinogen to angiotensin in the blood stimulates the release of aldosterone from the adrenal cortex and results in a rise in blood pressure. Disease of the kidney may affect this complex mechanism and, indeed, hypertension often accompanies severe renal disease. However, hypertension in renal disease can not always be explained in such simple terms; more than one mechanism may be involved.

**Failure to Regulate Red Cell Production**

When the number of circulating red cells falls, the reduction in the oxygen-carrying power of the blood causes hypoxia in the renal tissues. The subsequent release of erythropoietin, by as yet unidentified cells in the kidney, stimulates the bone marrow to produce more erythrocytes. In severe kidney disease, the kidneys fail to respond to this stimulus and anaemia develops. The erythrocytes that are formed are normal in size and haemoglobin content. The anaemia in renal disease does not respond to iron or vitamin $B_{12}$ therapy. The retention of "toxic products" may account in part for the depressed function in bone marrow. Bleeding from the mucosa of the intestines may also contribute to the anaemia. The anaemia probably accounts in part for the tiredness and lack of energy that is common in these patients.

Occasionally a patient with a tumour in the kidney is found to have an excessive number of circulating erythrocytes (polycythaemia). It has been postulated that these tumour cells produce excessive quantities of erythropoietin. The increase in erythrocyte number may cause an increase in blood viscosity, hypertension and a liability to intravascular clot formation.

## Causes

Chronic renal failure occurs where there is a gradual destruction of kidney tissue. It may be caused by recurring bacterial infection in the kidney (chronic pyelonephritis) which leads to progressive scarring, affecting especially the distal parts of the nephrons. Damage to the nephrons also results from hypersensitive reactions (chronic glomerulonephritis, systemic lupus erythematosus), severe high blood pressure (nephrosclerosis), obstruction to urinary outflow (hydronephrosis), congenital defects (polycystic kidneys) and deposition of calcium salts in the renal tissues (nephrocalcinosis).

## Investigations

There are many tests for evaluating renal function which vary in their complexity and usefulness.

**Examination of the urine.**—The presence of abnormal constituents such as plasma albumin, haemoglobin, red cells, pus or casts of renal tubules in the urine provide evidence of disease of the kidneys or urinary tract.

**Measurement of blood urea.**—In excretory failure, waste products accumulate in the blood and the blood urea level is a good index of the degree of accumulation. It is important to realize that large reductions in glomerular filtration rate can occur without much increase in the level of urea. It is only when the GFR is very low that further reductions cause dramatic increases in blood urea.

In situations where the amount of urea being formed is altered, the blood urea level must be interpreted with caution. In liver failure, for example, urea formation is depressed and renal failure can exist with normal blood urea levels. In starvation, urea formation increases because of the breakdown of tissue protein for energy production; raised blood urea values during starvation need not indicate renal failure.

**Excretion rates.**—Renal function may be assessed by injecting intravenously a radio-opaque dye such as diodrast which is secreted by the renal tubules. In X-ray pictures, diodrast should appear as a shadow in the pelvis of the ureter if renal function is normal; the absence of a shadow indicates a failure of the kidneys to secrete diodrast. The function of the two kidneys can be compared in this test by comparing the density of the two shadows.

**Clearance rates.**—The clearance rate of a substance (C) is the volume of plasma which contains the amount of that substance excreted by the kidneys in a minute. It is calculated from the formula $C = \frac{UV}{P}$ where U = the concentration of the substance in the urine, V = the volume of urine (per min) and P = the concentration of the substance in plasma. In carrying out the test it is necessary to take samples of urine and blood to make the necessary analyses. If the clearance of a foreign substance is required, that substance must, of course, be injected into the blood before the samples are taken.

The clearance rate of *inulin* (a foreign polysaccharide) gives the glomerular filtration rate because inulin is freely filtered and neither reabsorbed nor secreted in the tubules. Normal adult kidneys usually have an inulin clearance in the range 100-150 ml/min. *Creatinine* derived from body metabolism is treated in a rather similar manner to inulin. Its clearance rate may be measured more easily than that of inulin and is a rough index of glomerular filtration rate in man. The clearance value of *para-amino hippuric acid* (*PAH*) gives the value for renal plasma flow because it is effectively eliminated from the blood in one circulation through the kidney by filtration and active secretion—provided the blood level is not unduly high. It should have a value in the adult of around 500-1000 ml/min.

**Urine concentration and dilution tests.**—Impaired renal function reduces the patient's ability to excrete a urine which differs in osmolarity from plasma. To test the ability of the kidneys to produce a concentrated urine, the patient may be subjected to water deprivation or given an injection of ADH. The osmolarity of protein-free plasma is about 300 mosmol/1 (sp. gr. = 1.010). The kidneys should be able to produce a concentrated urine with an osmolarity of above 900 mosmol/l (sp. gr. = 1.030). A urine: plasma osmolarity ratio of less than 3 in a urine concentration test suggests renal disease.

Theoretically, urine dilution tests may be carried out along the lines of the concentration tests. Normally they are little used because the range of change that can be produced by dilution is so limited compared with what can be achieved by concentration.

**Investigation of underlying disease.**—The appropriate investigations are determined by the type of disturbance of renal function and by the general clinical picture. They will include radiological studies and may in some cases assess the function of each kidney separately by collecting urine from catheters placed in the ureters. Renal tissue may be sampled through a needle (*renal biopsy*). For this procedure a needle is introduced into a kidney through a puncture site in the overlying skin. A device is introduced through the needle which takes a small "bite" of the kidney tissue. Histological examination of the tissue may reveal the cause and extent of the damage.

## Treatment

In mild cases the situation can be dealt with by routine *management,* i.e. dealing with the causes of failure and modifying the patient's life style to meet the limitations imposed by his kidneys. When destruction has progressed to a stage where routine management is inadequate to support life, a *dialysis* system which permits selective exchange between the patient's blood and an artifical saline solution becomes necessary to supplement renal function. In these patients *transplantation* of healthy kidneys from a donor can then be carried out when a suitable kidney becomes available.

### Routine Management

The main object is to control the patient's pattern of life so that he can live within the limitations imposed by his kidneys. Where possible the cause of the renal damage should be treated. This may involve the use of antibiotics to cure infections, immunosuppressive agents to reduce the severity of hypersensitive reactions, and surgery to remove stones in the kidney or obstructions to urinary outflow. Drugs to reduce blood pressure when hypertension is damaging the kidney may be given; however, in some cases, lowering the arterial pressure may cause a reduction in renal blood flow and a deterioration in renal function.

(a) *Food intake.*—The diet can be varied but proteins should be restricted in relation to the severity of the renal failure, e.g. intake may be halved from about 1 g/kg body weight to 0.5 g/kg, to allow for the diminished ability of the kidneys to excrete protein residues. Protein should not be excluded from the diet because this would result in increased utilization of the patient's tissue proteins. The energy value of the diet is made up by increasing the intake of fat and carbohydrate.

(b) *Water intake.*—The patient should drink rather more than the normal intake of fluid, i.e. 3 instead of 1.5 litres per day. This is necessary because the patient is less well able to concentrate urine and large volumes of urine are required to eliminate waste products; water restriction makes excretory failure worse. However, very large water intakes should also be avoided as the kidneys cannot eliminate them adequately. Water intake should be related to urinary output.

(c) *Electrolyte intake.*—As with water intake, the object is to give the amount of electrolyte that the kidneys can cope with. This may involve the measurement of sodium output in the urine, and giving the patient a diet which contains that amount. The physician takes over the normal role of the kidney in matching intake and output of electrolytes.

Potassium does not tend to rise much in chronic failure except in severe cases. It does not therefore cause as many problems as in acute renal failure. If dietary potassium is restricted, the serum potassium may even fall.

Calcium and vitamin D may be given in chronic renal failure in an attempt to raise blood calcium and aid calcium deposition in the decalcified bones. Unfortunately, this is not always successful because patients with chronic renal failure, for some unknown reason, may not respond adequately to vitamin D.

(d) *Correction of the acidosis.*—This may be attempted by giving sodium bicarbonate by mouth or intravenously. This provides relief from the symptoms of acidosis such as vomiting. However, care is necessary because it may elicit frank tetany by reducing the ionization of the depleted blood calcium.

(e) *Correction of the anaemia.*—As the anaemia is due to depression of the bone marrow, the use of iron or vitamin $B_{12}$ in treatment is of little value; the anaemia is not due to lack of these substances. Injections of the hormone erythropoietin would be a logical treatment but the hormone is not yet available for therapeutic purposes.

Blood transfusion is the method usually employed to raise the patient's haemoglobin level. However, transfusions should not be given lightly. They may cause antibodies to form, which would prejudice the success of a future transplant operation. The risk of hepatitis increases with the frequency of blood transfusion. In addition, raising the haemoglobin level to normal removes the stimulus to the bone marrow to form its own erythrocytes. For these reasons, blood transfusions should only be given when the patient's life is endangered by the anaemia and transplantation is not being contemplated. Even then, there is little to be gained by raising the haemoglobin level to more than about 60 per cent of normal.

## Dialysis

Dialysis is used as an artificial aid to kidney function when renal damage is such that routine management cannot support the required life style. The object of dialysis is to permit selective diffusion between the patient's blood and a suitable artificial fluid.

The two main approaches to this problem are (a) the *artificial kidney* where the patient's blood is circulated through a device outside the body which contains dialysis fluid and (b) *peritoneal dialysis* where dialysis fluid is circulated through the patient's peritoneal cavity and makes contact with peritoneal blood vessels.

*The artificial kidney.*—This is a system in which the patient's blood is pumped as a thin layer between cellophane membranes in contact with a stream of dialysis fluid. The composition of the dialysis fluid is such that retained substances in the blood diffuse across the semipermeable cellophane membrane into the dialysis fluid. The blood is usually taken from and returned to a suitable vein in the forearm into which an opening (fistula) had been made from an adjacent artery to maintain a high rate of flow through forearm veins. Anti-coagulants are used to prevent clotting and antibiotics to prevent infection. Such machines are remarkably good in permitting excretion. The urea clearance of such a machine may be four times or so that of normal kidneys. Raised serum potassiums are rapidly corrected. By raising the pressure of the blood perfusing the system, fluid which has accumulated in the patient can be driven off by ultrafiltration. Blood volume and hence blood pressure sometimes fall below normal. Most of the symptoms of chronic renal failure are relieved by dialysis though the nature of retained substances causing these symptoms is not known.

Treatment by dialysis is clearly indicated for tiding patients over acute renal failure until such time as function returns. It is also useful in preparing patients for the rigours involved in receiving a renal transplant. The value of repeated dialyses at intervals for patients with chronic renal failure when there is little hope of recovery to a useful life is still a matter for discussion in terms of ethics and economics. Patients usually need two sessions of dialysis a week, each lasting for several hours so that a considerable portion of the patient's time is given to the treatment. For this reason, there have been attempts to provide artificial kidney facilities at home so that the patient need not spend such a large proportion of his life in hospital.

*Peritoneal dialysis.*—In this treatment, dialysis fluid is driven through the peritoneal cavity. Though it presents fewer technical difficulties than the artificial kidney machines, it is uncomfortable, less efficient and carries a considerable risk of peritoneal infection.

**Renal Transplantation**

Though dialysis may preserve life in chronic renal failure, it does so at considerable cost to the patient and the state. The frequency and complexity of treatment, the restriction of dietary habits and social life and the danger of getting infected with a virus that causes liver damage (viral hepatitis) due to frequent blood transfusions make the prospect of getting fresh, healthy kidneys by transplantation very attractive.

The introduction of a transplanted kidney from an immunologically compatible donor is a very reasonable treatment for a young patient with chronic renal failure who is otherwise healthy. The surgery involved in this is relatively straightforward. However, the pre-operative selection of a suitable kidney and the post-operative prevention of infection and immunological rejection still pose formidable problems. Some of these are discussed in the chapter on the immunological system.

Both of the patient's kidneys are usually removed before transplantation because diseased kidneys may well cause hypertension which would damage the transplanted kidney.

# 85. SUDDEN NEPHRON FAILURE (ACUTE RENAL FAILURE)

## Definition

Though urinary output normally varies with fluid intake and other factors, a minimum volume of about half a litre per day is required to eliminate waste products adequately. Acute renal failure occurs when urine formation is suddenly reduced to below this level.

## Effects

In the extreme situation where renal function is suddenly and completely lost, the patient feels ill effects after a few days, gradually loses consciousness and dies within two weeks unless some renal function is restored. In many ways the changes are similar to those in chronic renal failure, but are more dramatic.

**Retention of waste products.**—Even in the absence of protein intake, nitrogenous residues of protein digestion accumulate rapidly. The blood urea rises by about 20-30 mg/100 ml per day. Phosphate retention results in a fall in the ionized calcium level and may cause tetany as in chronic renal failure.

Potassium retention is an especially dangerous problem. It may be made worse if the acute renal failure is associated with crush injuries or haemolysis where intracellular potassium is released from the disintegrating cells. Potassium retention may cause death by its action on the heart when the serum level rises from its normal value of about 4 to about 8-10 mmol/l. In the absence of any urinary output it takes about one week for such levels to be reached.

High potassium levels weaken the strength of heart muscle contraction and depress the initiation and spread of the cardiac impulse. They also produce characteristic changes in the electrocardiogram (Fig. 126). The T wave becomes larger and more peaked, indicating changes in ventricular repolarization.

**Metabolic acidosis.**—The inability to manufacture bicarbonate and the retention of acid residues of metabolism results in the rapid development of a severe metabolic acidosis. This stimulates an increase in ventilation (Kussmaul breathing) and causes depression of consciousness.

**Water and electrolyte disturbances.**—These vary depending on the diet and fluid intake of the patient and on other factors such as vomiting and diarrhoea. The patient can easily become under- or overhydrated because his kidneys are unable to make the necessary compensatory adjustments.

**Haemopoietic disturbances.**—In acute renal failure there is a rapid fall in the haemoglobin content of the blood. The anaemia is probably due to lack of erythropoietin but toxic depression of the bone marrow may also be involved. There is a greatly increased susceptibility to infection and a tendency to bleeding which would also suggest a general depression of bone marrow activity.

**Problems during the recovery (diuretic) phase.**—Where acute renal failure is due to damage to tubular cells, it is usually followed by a recovery or diuretic phase as the tubular cells regenerate. The urine volume rises rapidly but the regenerating cells have little ability to modify the composition of the filtrate. The urine is similar in composition to protein-free plasma. During this stage, so much water and salts may be lost that the patient becomes seriously depleted of sodium, potassium or water. Great care is therefore required in treatment to maintain water and salt balance.

## Causes

**Prerenal causes.**—Factors which reduce blood flow to the kidneys are a common cause of acute renal failure. When the mean arterial pressure falls below about 60 mm Hg, glomerular filtration virtually ceases. Thus acute heart failure, severe haemorrhage, burns, dehydration by vomiting and diarrhoea and excessive sweating may reduce urinary output to inadequate levels. The reduction in renal blood flow is usually due to both a reduction in the perfusion pressure and renal vasoconstriction mediated by sympathetic nerves in baroreceptor reflexes. Renal capillary pressure falls below the level which can overcome the opposing osmotic and hydrostatic pressures. It should be noted that damage to kidney tissue by lack of blood supply often contributes to the failure in "prerenal" renal failure.

**Renal causes.**—Here the reduction in urinary output is due directly to damage to renal tissue. Though this can follow an acute reduction in renal blood flow it can also be caused by certain poisons, drugs and hypersensitive reactions.

**Postrenal causes.**—A sudden obstruction in the lower urinary tract which raises the pressure in the ureters and hence renal tubules may stop glomerular filtration and cause acute renal failure.

## Investigations

In acute renal failure it is useful to make routine measurements of blood urea, potassium, sodium, pH, etc. to gauge the severity of the biochemical disturbances. It is also useful to monitor body weight and fluid intake and output so that approximate water and electrolyte balance can be maintained.

If the cause and severity of the condition is not obvious, it may be useful to perform a renal biopsy.

## Treatment

After infections, obstructions and other correctable defects have been dealt with, the problem in acute renal failure is to maintain some sort of biochemical normality in the body fluids until renal function is re-established. The problem may last for several weeks.

(a) *Food intake.*—The diet should exclude protein and provide enough energy from non-protein foods so that the patient does not require to break down his own proteins for energy production. This can be achieved by giving him a mixture of ground nut oil and glucose. The lack of appetite and nausea in acute renal failure makes such a diet unpalatable and it may have to be supplemented by giving glucose intravenously.

(b) *Fluid intake.*—To avoid overhydration, just enough water should be given by mouth, so that it, together with the water formed by the metabolism of foods (about 0.5 litre) just balances the fluid loss by respiration and transudation through the skin (about 1 litre). Fluid balance charts must be kept carefully. If fluid is lost by vomiting, this loss must be made up. It is important to realize that the kidneys in acute renal failure cannot be forced into activity by giving large amounts of fluid to "flush out" the damaged organs. Such treatment can lead to death from water intoxication.

(c) *Limitation of rise in serum potassium.*—To minimize the rise of extracellular potassium, all food rich in potassium such as fruit must be avoided in the diet. However, even if potassium is excluded from the diet, the serum level still rises due to release of the ion from disintegrating cells. If the level rises to over 6 mmol/l, a cation exchange resin which chelates potassium ions can be given by mouth. Glucose and insulin cause potassium to move from extracellular fluid into the cells; injections of insulin can therefore be used to lower the extracellular concentration of potassium. In very severe cases, e.g. those complicated with tissue injury or infection where cell destruction is a feature, dialysis may be the only way of preventing fatal increases in serum potassium.

(d) *Correction of the metabolic acidosis.*—Severe acidosis which is causing distress and mental confusion may be alleviated by infusing sodium bicarbonate intravenously but the inability to excrete sodium limits the usefulness of this procedure.

(e) *Prevention of salt depletion in the recovery (diuretic) phase.*—When the diuretic phase of acute failure comes, the inability to reabsorb sodium may lead to sodium depletion, contraction of extracellular fluid volume and circulatory collapse. This can be avoided by ensuring that sodium intake balances sodium output.

(f) *Dialysis.*—If the renal failure is so severe and prolonged that it cannot be dealt with by simpler methods, dialysis can be used to supplement renal function until the kidneys recover sufficiently to support life.

# 86. GLOMERULAR FAILURE (THE NEPHROTIC SYNDROME)

## Definition

IN THIS condition, damage to the glomerular membranes allows large quantities of plasma protein to escape with the glomerular filtrate. The other functions of the nephron are usually normal in the early stages but tend to be impaired with the progression of the underlying disease.

## Effects

The nephrotic syndrome is characterized by loss of protein in the urine, a consequent fall in the plasma protein level and hence oedema.

### Loss of Protein in Urine

Though protein cannot be detected in normal urine, albumin appears in normal glomerular filtrate in a concentration of about 20 mg per cent. Though this is small compared with the 4000 mg per cent found in plasma, the loss of filtered protein in the urine would eliminate an amount of albumin equivalent to one-third of the total plasma albumin every day. The filtered albumin is not, of course, excreted; it is reabsorbed by an active transport mechanism which can reabsorb up to a maximum of about 30 mg/min (not much above the normal load due to filtered protein).

When the glomerular membranes are damaged, protein may escape in the filtrate at a rate that exceeds the capacity of the tubular epithelium for reabsorption. When this happens, protein escapes in the urine. The main plasma protein which escapes is albumin; as its molecular weight is only half that of globulin, it can escape with the filtrate more readily.

Damage to the cells of the proximal convoluted tubules may accentuate loss of albumin in the urine by impairing their ability to reabsorb protein.

### Fall in the Level of Plasma Proteins

When protein is lost in the urine, the plasma protein level will fall when the urinary loss exceeds the capacity of the liver to replenish the plasma. Most of the plasma proteins are manufactured in the liver which can synthesize plasma proteins at the remarkable rate of about 100 g/day provided that it is supplied with the necessary substrates. In glomerular failure, albumin is rarely lost at rates greater than about 30 g/day. Nevertheless, this loss persists day after day and depletes the body stores of protein. Eventually the liver does not have the supply of substrates to manufacture albumin at a sufficient rate to replace the loss. Dietary protein deficiency will accentuate this body deficiency of protein.

When this happens, the plasma protein level can fall to levels as low as 1-2 g per cent. The loss is mainly in the albumin fraction but the globulin fraction also becomes depleted by impaired synthesis due to a deficiency of protein raw materials.

**Oedema**

As the plasma protein concentration falls, so does the colloid osmotic pressure that it exerts. Normally the osmotic pressure is such that it balances and counteracts the hydrostatic pressure gradient which tends to drive fluid out of the capillaries. As the colloid osmotic pressure falls, fluid moves out of the capillaries and accumulates in the tissue spaces to cause oedema. The resulting fall in the blood volume causes release of aldosterone; this leads to salt and water retention by the kidneys and an expansion of the extracellular fluid volume.

**Liability to Infection**

In severe cases, the liability to infection by micro-organisms is increased. The cause of this is not certain but it may involve (i) a decreased ability to manufacture and retain antibody, because of the protein depletion, and (ii) the poor nutrition of oedematous tissues.

**Increase in Blood Cholesterol**

In most cases the blood levels of lipids rise but the reason for this is not yet known.

**Chronic Renal Failure**

Tubular function is normally well maintained and the glomerular filtration rate remains high in the early stages of the nephrotic syndrome. However, in cases which do not clear up, these functions tend to diminish with time as glomerular and tubular damage become more widespread. Eventually the picture is that of chronic renal failure.

## Causes

The nephrotic syndrome can result from a number of causes such as poisoning by heavy metals, incompatible blood transfusions and hypersensitive reactions affecting the kidney. Structural changes have been seen in the podocytes and basement membrane in the glomeruli and the increase in permeability to protein has been attributed to these changes.

## Investigations

Investigations are carried out to differentiate the nephrotic syndrome from other types of renal damage. The blood urea is usually normal in the early stages whereas the blood cholesterol is high. The plasma protein concentration is usually very low, the reduction in the albumin fraction being relatively greater than that of the globulin fraction.

## Treatment

*Treatment of the cause.*—The problem here is that the cause of the condition is usually unknown. However, glucocorticoids often reduce the urinary loss of protein suggesting that hypersensitivity is responsible for some of the cases. When the condition is due to poisoning or drugs, withdrawal of the offending agents is required.

*Treatment of the oedema.*—There are two main approaches to the treatment of the oedema. Firstly, attempts may be made to raise the colloid osmotic pressure of plasma by increasing the dietary intake of protein and giving intravenous infusions of plasma. Secondly, attempts may be made to reduce the amount of sodium in the body by restricting sodium in the diet and giving diuretic drugs which increase the renal excretion of salt and water.

# 87. TUBULAR FAILURE

## Definition

The tubular cells, by a variety of transport mechanisms, can pump substances into and out of the blood in the peritubular capillaries. This section deals with the rare conditions where there is an inability of the tubules to transport one or more of these substances.

## Effects

The effects of tubular failure depend on which transport mechanism or mechanisms are defective.

**Renal glycosuria.**—In this condition, glucose appears in the urine and the disorder may be mistaken for diabetes mellitus. However, the fault lies in the tubular cells. Normally the maximum ability of the tubules to reabsorb glucose is such that all filtered glucose is reabsorbed when the blood glucose is less than about 180 mg per cent. In renal glycosuria, a fault in the glucose transport system results in a lowered tubular maximum (Tm) for glucose; the renal threshold for glucose is correspondingly lowered.

**Renal amino-aciduria.**—In the cells of the proximal convoluted tubules there are a variety of transport mechanisms to pump filtered amino-acids back into the blood. Defects in these transport mechanisms result in amino-acids such as xanthine or cystine appearing in the urine. The main effect is that stones of cystine or xanthine may form in, and obstruct, the urinary tract.

**Renal phosphaturia.**—This condition results in excessive excretion of phosphate in the urine. It is usually associated with an impaired ability to absorb calcium from the gut. Lack of calcium and phosphate leads to bone weakness.

**Renal tubular acidosis.**—In this condition the ability of the distal convoluted tubules to secrete hydrogen ions and manufacture bicarbonate ions is impaired. The patient is thus unable to acidify the urine and the low plasma bicarbonate gives rise to a metabolic acidosis.

**Nephrogenic diabetes insipidus.**—This is a very rare condition where the renal collecting ducts do not respond to antidiuretic hormone. Because of this the patient passes large quantities of very dilute urine and may become dehydrated.

## Causes

These conditions in some cases result from abnormalities in the genes responsible for the enzymes involved in tubular transport. The inherited characteristics may be dominant or recessive and may be attached to sex chromosomes or to autosomes. In other cases, the defects are secondary to conditions such as other metabolic abnormalities, vitamin deficiencies or circulating toxins.

### Investigations

Diagnosis of the above conditions depends on demonstrating excessive clearance of substances normally reabsorbed (e.g. glucose) or an inadequate clearance of substances normally secreted (e.g. $H^+$). Thus the coexistence of glucose excretion in the urine with a normal plasma level indicates renal glycosuria.

### Treatment

Treatment consists in removing underlying causes where possible. Otherwise the patients must be helped to compensate for the tubular defects. For example, renal tubular acidosis may be corrected by the oral administration of absorbable alkalis. An increased fluid intake helps to prevent stone formation in amino-aciduria.

## RECENT LEADING ARTICLES

Analgesics and the kidney. *Brit. med. J.,* 1973, **3,** 123.
Calcium in the kidneys. *Lancet,* 1972, **1,** 825.
Catheters and incontinence. *Brit. med. J.,* 1974, **3,** 703.
Dialysis and transplantation in Great Britain. *Lancet,* 1972, **2,** 582.
Electrical stimulation of the bladder. *Lancet,* 1972, **1,** 887.
Enuresis again. *Brit. med. J.,* 1973, **2,** 69.
Enuresis again. *Brit. med. J.,* 1974, **3,** 375.
Prognosis of acute renal failure. *Brit. med. J.,* 1973, **2,** 435.
Shortage of kidneys. *Brit. med. J.,* 1974, **3,** 703.
Sodium in chronic renal failure. *Lancet,* 1971, **1,** 1282.
Testing renal function. *Lancet,* 1974, **2,** 631.
Tomorrow's artificial kidney. *Lancet,* 1974, **2,** 446.

*Section XIV*

# DISORDERS OF THE ENDOCRINE SYSTEM

WHILE DISORDERS of the endocrine system may produce local effects by pressure from enlarged glands on surrounding structures, the general effects are usually the more important. These are produced by either an excess or a deficiency of hormone. Deficient hormone production leads to loss of the gland's normal effects. Excessive hormone production leads to an exaggeration of the normal effects—a caricature of the gland's function. It should be noted that abnormal cell growths (tumours) in endocrine glands may be associated with either excessive or deficient hormone production. The tumour may produce large amounts of hormone or it may interfere with normal glandular function and cause hormone deficiency.

# 88. THYROID GLAND

## DEFICIENT SECRETION

### Definition

THE THYROID gland produces the hormones thyroxine ($T_4$—it has 4 iodine atoms in its molecule) and triiodothyronine ($T_3$—3 iodine atoms) which are important in regulating the rate of metabolism in body tissues. Deficient secretion of these hormones (*hypothyroidism*) causes the condition of *myxoedema* when it develops in adults and the condition of *cretinism* when it develops in children.

### Effects

**In the Adult**

Thyroid hormones act on mitochondria to increase energy production in the form of high-energy phosphate bonds and heat during the process of oxidative phosphorylation. Thus these hormones increase the energy available for cellular function and at the same time increase heat production by the body. In hypothyroidism, most body organs carry out their normal functions, but at a suboptimal level. The basal metabolic rate and hence heat production decline in proportion to the severity of the disease. Deficient heat production means that in an environment where the average individual is comfortably warm, the hypothyroid patient has cold hands and feet due to the vasoconstriction required to conserve heat. Sweating is uncommon and the patient often complains of difficulty in keeping warm. In severe hypothyroidism, central body temperature falls (hypothermia) and the patient lapses into coma and, unless appropriate treatment is urgently started, death results.

The condition of hypothyroidism is sometimes referred to as *myxoedema* (Gr. *myxa* = mucus) because the patient has a puffy oedematous appearance which is due, not to oedema fluid, but to the deposition of mucoid material in the interstitial tissues; the reason for this is obscure. The myxoedema tissue can interfere with normal function, e.g. in the vocal cords it can cause hoarseness and in the inner ear it can cause deafness and disturbance of balance.

Though the physiological link between thyroid hormones and autonomic nervous activity is unknown, many of the effects of hypothyroidism resemble loss of normal autonomic activity (Fig. 152).

Alertness diminishes and the individual tends to fall asleep easily. Impairment of brain function is made worse in severe cases by a low blood glucose level and by a low body temperature. Depression of consciousness may result in abnormal behaviour ("myxoedematous madness") and progress to unconsciousness.

The drowsy appearance is reinforced by a tendency for the upper eyelid to droop due to relaxation of the sympathetically innervated smooth muscle fibres which help to elevate it.

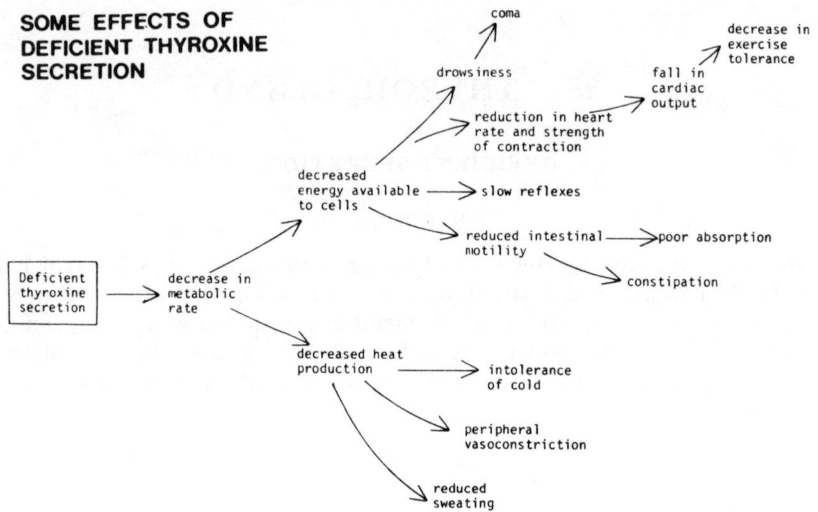


FIG. 152.—A classification of some effects of hypothyroidism. Note that the effects of hyperthyroidism include reversal of the above effects.

Sympathetic stimulation speeds reflex responses in skeletal muscle. In hypothyroidism, tendon reflexes, especially the ankle jerk, show a delayed onset, and muscle contraction and relaxation are prolonged.

The rate and force of the heart beat decrease; the electrocardiogram tends to be of low voltage.

Blood sugar tends to fall in severe hypothyroidism. Intestinal motility decreases and there is a consequent tendency to impaired absorption. This would contribute to the low blood glucose level. The frequency of bowel evacuation tends to decline.

### In the Child

The foregoing description applies to adults who have developed myxoedema. The situation is somewhat different in the case of children. If a child is born and grows up with a severe lack of thyroid hormones, mental and physical development are permanently stunted, the end result being a "cretin"—a mentally deficient dwarf. Skeletal development is delayed, e.g. X-ray signs of a calcified epiphysis may be late in appearing.

### Local Effects

In some cases the gland is absent from birth. In others it is enlarged, due either to the disease which is damaging the gland or to a compensatory hypertrophy. This occurs when the formation of hormones is interfered with by deficiency of iodine or by abnormality of thyroid enzymes. Because the gland lies in the neck, enlargement (*goitre*) can be readily seen and felt. Occasionally, thyroid tissue lies behind the sternum and the enlarged gland (retrosternal goitre) may press on and obstruct the trachea.

### Causes

A deficiency of thyroid hormones may be due to *primary* hypothyroidism, caused by congenital absence of the gland, interference with thyroid function by disease, or lack of iodine in the diet. Dietary iodine deficiency tends to occur in inland regions where little fish (a rich source of iodine) is consumed. Addition of iodine to table salt helps to prevent it. In this condition, the gland may increase markedly in size. Deficiencies of the enzymes required for thyroxine synthesis may also cause hypothyroidism.

Hypothyroidism may be *secondary* to inadequate release of thyroid stimulating hormone by the pituitary. This is referred to in the section on deficient anterior pituitary secretion. In this condition, adrenal corticotrophic hormone is usually deficient also, so that features of adrenal insufficiency modify the picture found in primary hypothyroidism.

### Investigations

*Blood levels of thyroxine and triiodothyronine* provide a direct indication of thyroid activity. These hormones are carried mainly on circulating globulins (e.g. thyroxine-binding globulin). The degree of saturation of the carriers may be measured; saturation is low in hypothyroidism and high in hyperthyroidism.

The fact that the thyroid hormones are rich in iodine makes iodine a convenient indicator of thyroid activity and of the blood level of thyroid hormones. Circulating thyroid hormones are largely bound to plasma proteins and measurement of the *protein bound iodine* level (PBI) is a good indicator of the level of circulating thyroid hormones, being lowered in hypothyroidism and raised in hyperthyroidism.

Another method of assessing thyroid function is to administer orally a small dose of a radio-active isotope of iodine. The *rate of uptake of iodine* by the gland (which is proportional to its hormone-producing activity) may be assessed by measuring radio-activity over the gland, or by collecting the patient's urine and measuring its radio-activity. In both tests, the level of radio-activity is below normal in hypothyroidism and above normal in hyperthyroidism.

Measurement of the *basal metabolic rate* provides an indication of thyroid activity but, on grounds of convenience and accuracy, has largely been superceded by the chemical tests described above.

Measurement of the *level of thyroid stimulating hormone* (TSH) reveals a high level in primary hypothyroidism, in which there is increased pituitary drive on the inadequate thyroid gland. The reverse is, of course, the case in hypothyroidism secondary to pituitary insufficiency.

The *blood cholesterol* level tends to rise above normal in hypothyroidism for reasons which are obscure.

### Treatment

When hypothyroidism is due to iodine deficiency, the remedy is to supply adequate dietary iodine. Otherwise, treatment consists of replacement therapy—the administration of hormones to replace the naturally produced hormones which are lacking. Triiodothyronine ($T_3$) or thyroxine ($T_4$) may be used. $T_3$ is more rapidly acting and may be used where treatment is urgent, but $T_4$ is commonly used for long-term treatment.

Thyroid replacement therapy is generally embarked upon cautiously, the initial dose being low and subsequent doses increasing gradually. The reason is that the increased metabolism stimulated by a large dose of thyroxine demands greatly increased cardiac effort. If the heart muscle has been weakened by disease, heart failure may result. If the coronary arteries are narrowed, the increased myocardial metabolism may lead to oxygen requirements which the coronary vessels cannot meet and myocardial damage, possibly fatal, may be precipitated. For this reason, the dose of thyroid hormone given may have to be less than that required to raise body metabolism to normal.

In the case of infants suffering from hypothyroidism, treatment must be started in the early months of life if serious and permanent physical and mental retardation are to be avoided.

## EXCESSIVE SECRETION

### Definition

**Hyperthyroidism,** or *thyrotoxicosis* as it is sometimes called, is due to excessive secretion of the thyroid hormones. It appears that in most cases excessive thyroxine ($T_4$) is mainly responsible for the condition, but excessive triiodothyronine ($T_3$) may be the main culprit in some cases.

### Effects

#### In the Adult

Many of the effects of hyperthyroidism are the reverse of the effects of hypothyroidism (Fig. 152) and can be related to excessive production of energy for cellular activities and of heat. In hyperthyroidism there is a tendency towards "uncoupling of oxidative phosphorylation" i.e. oxidative processes tend to result in more heat production and less high-energy phosphate bond production than normal.

Metabolic rate is increased in proportion to the severity of the condition. Due to the high rate of metabolism, appetite and food intake are often markedly increased, but despite this, weight typically falls, as fat reserves are utilized.

The cardiac output rises proportionately to serve the increased metabolic needs of the tissues. The increased work load on the heart may result in heart failure of the high output variety in severe cases. If there is pre-existing heart disease, the disease may be unmasked by the increased metabolic needs.

The excessive heat production with the increased rate of basal metabolism leads to peripheral vasodilatation and an increased tendency to sweat so that the peripheries are warm and often moist. The combination of vasodilatation and a high cardiac output leads to a forceful pulse (hyperdynamic circulation). A hot environment leads to severe discomfort and cold environments are sought.

Some of the effects mimic overactivity of the autonomic nervous system. These effects include some of the most troublesome and exhausting effects of hyperthyroidism. Exaggeration of the normal alerting effect of sympathetic activity on brain function leads not to any useful degree of high alertness, but to an unpleasant irritable, tense, anxious state which militates against normal activities. In extreme cases, the patient may be in a state of excited confusion (delirium).

Increased activity of the smooth muscle in the upper eyelids leads to wide opening of the eyes; the margins of the iris become visible and the eyes appear to stare.

Accentuation of reflex responses of skeletal muscle exaggerates physiological tremor so that the outstretched hands exhibit a fine, rapid tremor. Intravenous infusions of adrenaline produce a similar tremor.

Increased sympathetic stimulation of the heart causes a marked increase in the rate and force of cardiac contraction and a consequent unpleasant awareness of the heart's activity (palpitations). The ECG voltage may increase, but this is not a recognized feature of the condition, probably because of the large scatter of normal voltages. In the presence of mild heart disease, the increased cardiac effort may precipitate failure.

Motility in the alimentary tract is increased. The rate of glucose absorption rises and tends to raise the blood glucose level. The frequency of defaecation tends to increase.

Other effects of hyperthyroidism include muscle pain and weakness and loss of bone density. The mechanism is obscure; possibly the high rate of metabolism is associated with excessive use of protein as an energy source.

One feature of thyrotoxicosis which cannot adequately be explained by the action of thyroid hormones is *exophthalmos*—the eyes are pushed forward by excess tissue at the back of the orbit. This accentuates the staring appearance. In severe cases, movement of the eyes may be restricted and vision impaired by damage to the cornea, no longer adequately moistened by blinking movements of the eyelids, whose movements are restricted by the over-prominent eyes.

**In the Child**

Hyperthyroidism in childhood is rare, but when it occurs it may lead, in addition to the features described in the adult, to an increased rate of growth, especially in height. Bone maturation (e.g. appearance of calcification in epiphyses) is in advance of chronological age.

## CAUSES

Although in theory hyperthyroidism could be caused by excessive secretion of hypothalamic thyrotropin releasing hormone (TRF) or pituitary thyroid stimulating hormone (TSH), most cases are due to overactivity of the thyroid gland independent of these factors. Some cases seem to be due to the development of antibodies to some component of the thyroid gland (an example of autoimmune disease, in which the body produces antibodies against its own tissues). A long-acting thyroid stimulator (LATS) has been found in the plasma of some patients with hyperthyroidism; it is associated with the globulin fraction of the plasma proteins. Such a substance may cause the release of excess thyroid hormones from inside the follicles and may also be responsible for the production of exophthalmos, which often persists after the blood level of thyroid hormones has been reduced to normal.

### Investigations

Investigations are, in general, the same as for hypothyroidism, the results deviating from normal in the opposite direction.

### Treatment

Three methods are available for suppressing the excess release of hormones from the thyroid gland. These are *surgical removal* of the greater part of the gland, the *use of antithyroid drugs* to interfere with the formation of hormones by the gland, and the *use of oral radio-active iodine* (in a much greater dose than is used in the diagnostic test) which is temporarily trapped in the gland and destroys a considerable proportion of it. The radio-active iodine treatment is extremely simple but carries to a small extent the hazards of radiation exposure, e.g. interference with bone marrow function. As hyperthyroidism is probably caused in many cases by an immune process which is destructive to the gland, it is perhaps not surprising that some patients eventually develop hypothyroidism and require therapy with thyroxine.

While the above procedures are being prepared for or are taking effect, some patients may be seriously affected by severe features of hyperthyroidism. The rapid heart rate may be particularly distressing and heart failure may develop in a patient whose cardiac reserve is impaired by an inadequate blood supply to heart muscles. In these circumstances, the use of a *beta adrenoceptor blocking agent* may greatly improve the situation by removing the sympathetic drive from the heart (this drive acts on beta adrenoceptors in the sinu-atrial node and ventricular muscle).

## CALCITONIN SECRETION

The "C" cells, which adjoin the thyroid follicles, produce the hormone (thyro)calcitonin which lowers the blood calcium level by reducing the mobilization of calcium salts from bone. Just as glucagon is overshadowed by insulin, so thyrocalcitonin seems to be of much less functional importance than parathormone and diseases due to its deficiency or excess have not emerged as distinct entities. Calcitonin may be of value in certain diseases where a high blood calcium level or loss of bone calcium cannot be otherwise treated.

# 89. PARATHYROID GLAND

## DEFICIENT SECRETION

### Definition

**Hypoparathyroidism** is characterized by a deficient secretion of the parathyroid hormone (parathormone), a polypeptide with a molecular weight of around 8,500. This hormone raises the level of ionized calcium in the blood by promoting release of calcium from bones, by increasing excretion of phosphate by the kidneys and by increasing intestinal absorption of calcium.

### Effects

The effects of hypoparathyroidism are due to a reduction in the level of ionized calcium in the blood. The condition may be aggravated by alkalosis, e.g. due to hyperventilation, because alkalosis moves blood pH further from the isoelectric point of the plasma proteins, increases their net negative charge and hence increases their ability to bind calcium. A low level of ionized calcium leads to the condition referred to as *tetany*, in which there is increased excitability of muscles and nerves. Spontaneous activity in motor nerves may lead to muscle spasms and spontaneous activity in sensory nerves may lead to sensations of tingling or "pins and needles" (paraesthesiae).

In adults the condition is suggested by tingling or muscle cramps in the limbs. The hands may go into a typical position of spasm with fingers and thumb straight and metacarpophalangeal joints flexed. The diagnosis of tetany may be confirmed if tapping on the facial nerve leads to spasm of facial muscles and if rendering nerves ischaemic by occluding the circulation to the arm sends the hand into spasm in a few minutes.

In children the spasm may affect also the larynx, interfering with breathing, and convulsions may occur.

There is no interference with the clotting mechanism. Severe muscle spasms would prove fatal before the blood calcium had fallen to a level which interferes with clotting.

### Causes

Parathyroid failure may be *primary*, or *secondary* to surgery of the thyroid gland in which either the glands (of variable number and situation) or their blood supply are interfered with. If some parathyroid tissue remains, it may subsequently increase its secretory activity so that the insufficiency is only temporary.

### Investigations

When features suggestive of tetany are present, the finding of a low blood calcium level confirms the diagnosis. The blood phosphate level is raised.

Measurement of the blood calcium level includes both ionized and protein-bound calcium. Normally about half the blood calcium is ionized; most of the rest is bound to plasma protein. Abnormalities in protein level should be taken into account before drawing conclusions about the ionized calcium level and parathyroid activity. Thus if plasma protein levels are low, a reduced total calcium level may coexist with a normal ionized calcium level.

### Treatment

The immediate treatment for hypocalcaemia is the slow intravenous injection of a calcium salt. If marked alkalosis is present, ammonium chloride may be given by mouth. On metabolism, this gives rise to free hydrogen ions.

If the condition is permanent, it is not necessary to give repeated injections of calcium; calcium and large doses of vitamin D (calciferol) by mouth may be given instead. The vitamin D causes increased calcium absorption from the intestine and the blood calcium level rises. Because marked hypercalcaemia may develop and cause serious tissue calcification, the blood calcium level must be checked at intervals to ensure that it is neither too high nor too low.

## EXCESSIVE SECRETION

### Definition

The condition of **hyperparathyroidism** is due to the effects of excess secretion of parathormone. The condition may be primary (the excessive secretion is due to abnormality of the parathyroids) or secondary to renal disease or malabsorption of calcium from the gut, which leads in some cases to increased parathyroid secretion.

### Effects

Parathyroid hormone (parathormone) raises the blood calcium level by acting directly on bone to cause release of calcium salts, by promoting calcium absorption in the alimentary tract and by increasing excretion of phosphates by the kidneys. Lowering of the plasma phosphate level due to increased excretion favours solution of calcium phosphate from bone by moving the reaction below in the direction shown:

$$\text{Ca phos.} \rightleftharpoons Ca^{2+} + \text{phos.}^{2-}$$

The equation is a simplified one because bone salts are complex in nature and phosphates in plasma exist in several forms. The level of ionized calcium is raised when parathormone activity is excessive.

Effects of hyperparathyroidism are related to a high blood calcium level (hypercalcaemia), loss of bone calcium and deposition of calcium in the urinary tract.

**High blood calcium level.**—This tends to cause clouding of consciousness and nausea. It may also account for the increased occurrence of gastro-intestinal disturbances, including peptic ulcers, found in hyperparathyroidism, although evidence on the precise mechanism of these conditions is lacking.

**Loss of bone calcium.**—The raised blood calcium ion level is achieved by depleting calcium stores in bone. Fractures and pain may be experienced in areas of severe depletion.

**Deposition of calcium in the urinary tract.**—When the blood calcium level rises, renal excretion of calcium increases. Increased concentrations of calcium and phosphates in the urine increase the risk of precipitation of calcium phosphate in the form of renal stones in the renal pelvis. These stones may pass into the ureter thereby obstructing it and causing exceedingly painful contractions of its muscle (renal colic). Calcium may also be deposited in the renal tubules and renal function impaired.

Hyperparathyroidism has been described as "a disease of bones, stones and abdominal groans." However, bone disease and stone disease do not often coexist in the same patient.

## Causes

*Primary* hyperparathyroidism is usually due to a functioning tumour in one of the glands.

Hyperparathyroidism may be *secondary* to *malabsorption* states or *renal failure.* Calcium malabsorption leads to a low blood calcium and hence hypersecretion by the parathyroid glands in an effort to restore a normal calcium level. Hyperparathyroidism secondary to renal failure is not fully understood. It is thought that renal failure impairs calcium absorption by the gut. In addition there may be impaired phosphate excretion. This would raise the blood phosphate level and lower the calcium level as the equation

$$\text{Ca phos.} \rightleftharpoons Ca^{2+} + \text{phos.}^{2-}$$

moved to the left (the product of calcium and phosphate in concentrations must remain constant).

## Investigations

The typical findings in primary hyperparathyroidism are a raised blood calcium level, a lowered blood phosphate level and excessive calcium in the urine (the calcium intake must be considered in relation to calcium excretion).

X-ray of various bones often reveals patterns of loss of density which are characteristic of hyperparathyroidism. Renal stones or diffuse renal calcification may also be seen.

In secondary hyperparathyroidism the blood phosphate tends to be raised and calcium may be reduced. Bone changes are similar to those in primary hyperparathyroidism.

## Treatment

Whether or not the disease is due to malabsorption or renal failure, treatment consists of surgical removal of parathyroid tissue. Sometimes parathyroid glands in unusual sites, such as behind the sternum, may be difficult to find. After the operation, *tetany* may occur and require urgent treatment.

# 90. ADRENAL CORTEX

## DEFICIENT SECRETION

### Definition

Deficient adrenal cortical secretion can involve the *glucocorticoid, mineralocorticoid* and *androgenic* groups of hormones. When the disease is *primary* (Addison's disease), i.e. due to local adrenal disease, all these functions are lost. When the disease is *secondary* to pituitary failure to produce ACTH, mineralocorticoid activity is little affected because secretion of these hormones is largely independent of ACTH.

### Effects

**Glucocorticoid deficiency.**—The effects of an inadequate secretion of glucocorticoids are largely due to an inability to deal with physical stress. When the deficiency is relatively mild so that secretion of hormones is more or less adequate for ordinary conditions, the lack of reserve may become apparent during a fairly severe stress, such as an injury causing a broken bone, or an attack of pneumonia. In these circumstances the patient fails to respond normally to the stress and may die unless glucocorticoid therapy is given.

When hormone secretion falls markedly, slight stress may lead to collapse, typically with severe weakness, vomiting, low blood pressure and low blood sugar level. In severe cases, death results rapidly unless replacement therapy is given.

The mechanism whereby glucocorticoids protect the body from physical stress is unknown. Glucocorticoids tend to divert the body's metabolism away from protein synthesis to glucose formation and metabolism. It may be that this diversion of resources is essential in enabling the body to combat stress.

**Mineralocorticoid deficiency.**—Mineralocorticoid deficiency is due not only to reduced secretion of the main mineralocorticoid, aldosterone, but to loss of the mineralocorticoid effects of the glucocorticoids, e.g. cortisol.

Aldosterone by its stimulation of sodium reabsorption in the renal tubules is essential for the maintenance of a normal body sodium level and hence a normal extracellular fluid and blood volume. When it is deficient, excessive sodium is lost in the urine, extracellular fluid volume falls, and with it the blood volume. This in turn leads to a low blood pressure and peripheral circulatory failure.

**Androgen deficiency.**—Adrenal androgens are largely responsible in both sexes for the growth of axillary and pubic hair which appears at puberty. In adrenal failure, adrenal androgen production falls in association with the fall in glucocorticoid production (both are controlled by ACTH) and axillary and pubic hair become sparse.

**Pigmentation.**—A pointer to the diagnosis in some cases is the presence of excessive skin pigmentation. This occurs when the disease is due to primary adrenal failure rather than adrenal failure secondary to inadequate ACTH. In primary adrenal failure there is increased release of adrenocorticotrophic hormone (ACTH). ACTH and melanocyte stimulation hormone (MSH) are thought to be produced as part of the same molecule in man. This would explain the raised level of MSH responsible for the increased pigmentation. In adrenal failure secondary to ACTH deficiency there is no increase in pigmentation.

## Causes

### Primary Adrenal Failure

In primary adrenal failure, both glands must be affected because, as with the adjoining kidney, one healthy organ can take over the function of two. A variety of diseases including tuberculosis may destroy the adrenals; they may be removed surgically in certain conditions.

### Secondary Adrenal Failure

This is due to failure of the pituitary to produce adequate ACTH.

### Iatrogenic Adrenal Failure

"Iatrogenic" disease is disease due to the doctor's treatment (Gr. *iatros* = physician, *gennan* = to produce). When prolonged glucocorticoid therapy is given for any reason, the release of ACTH is suppressed and the inner zones of the adrenal cortex, having no work to do, suffer "disuse atrophy" (Gr. *atrophia* = lack of nourishment). When the steroid therapy is withdrawn, the gland may fail to recover, or may recover partially, being unable to increase secretions adequately under stress. In order to avoid this, steroid therapy is sometimes replaced by ACTH therapy. However, in this case the pituitary cells which produce ACTH in response to hypothalamic corticotrophin release factor (CRF) may atrophy so that the end result is again a risk of adrenal failure.

## Investigations

Measurement of the blood cortisol level and of urinary breakdown products usually indicates whether adrenal insufficiency is present. As cortisol secretion undergoes marked diurnal variation, the time at which the blood sample for the cortisol determination is taken must be noted, so that the result can be related to the normal range of values *at that time of day.*

Administration of ACTH helps to distinguish between primary failure and failure due to inadequate ACTH. In the latter case, administration of ACTH will raise glucocorticoid secretion, there will be a prompt fall in the blood eosinophil count, and excretion of steroids will increase.

An unexplained finding is that patients lacking in glucocorticoids cannot excrete a water load, e.g. 1 litre given by mouth, as rapidly as normal people. Thus a water excretion test may help to confirm or refute the diagnosis of adrenal insufficiency.

### Treatment

Treatment generally consists of regular administration of a glucocorticoid, e.g. cortisone, with or without supplementary dietary sodium chloride or a mineralocorticoid as required. As well as this basal treatment, patients require an increase in dose if they develop an illness or are injured or have a surgical operation. It is essential that in these circumstances the patient informs his doctor of the adrenal failure, otherwise death may result. In addition, the patient should carry a statement of his condition and treatment so that if he is found unconscious, the essential glucocorticoid treatment may be given without delay.

## EXCESSIVE SECRETION

### Definition

Excessive secretion by the adrenal cortex can produce a variety of conditions due to the variety of hormones produced by the gland, glucocorticoids, mineralocorticoids and sex corticoids. Three main effects are possible, due to these three types of hormones. Three main patterns can be distinguished:

**Excessive glucocorticoid secretion.** (*Cushing's syndrome*—Gr. *syndrome* = occurring together, i.e. a number of symptoms and signs which regularly occur together).—Because glucocorticoids have some mineralocorticoid properties, excess glucocorticoids may produce similar effects to excess mineralocorticoids. In some cases, excess sex hormones are also produced, so that Cushing's syndrome may include excessive effects of all three types of adrenal cortical hormone. *Cortisol* is the main naturally occurring glucocorticoid.

**Excessive mineralocorticoid secretion.** (*Conn's syndrome*).—Effects here are almost entirely mineralocorticoid effects, because mineralocorticoids have only slight glucocorticoid effects. *Aldosterone* is the main naturally occurring mineralocorticoid.

**Excessive sex corticoid secretion.** (*Adrenogenital syndrome*).—As the name suggests, these hormones affect the secondary sex characteristics. Faults in the complex chemistry of adrenal hormone synthesis may lead to an excess of either male or female hormones.

### Effects

#### Glucocorticoid Effects

These effects are not fully understood but seem to derive mainly from an action whereby amino-acids in the body are converted into glucose, the process being termed *gluconeogenesis*. When this effect is excessive, the body tends to suffer from protein shortage and glucose excess; some of the excess glucose is converted into fat. These effects are outlined in Fig. 153.

The excess glucose appears partly as a raised blood sugar level (*hyperglycaemia*) which may exceed the renal threshold so that *glycosuria* results, and partly as excess body fat, confined largely to the face (giving a rounded appearance—"moon face") and trunk.

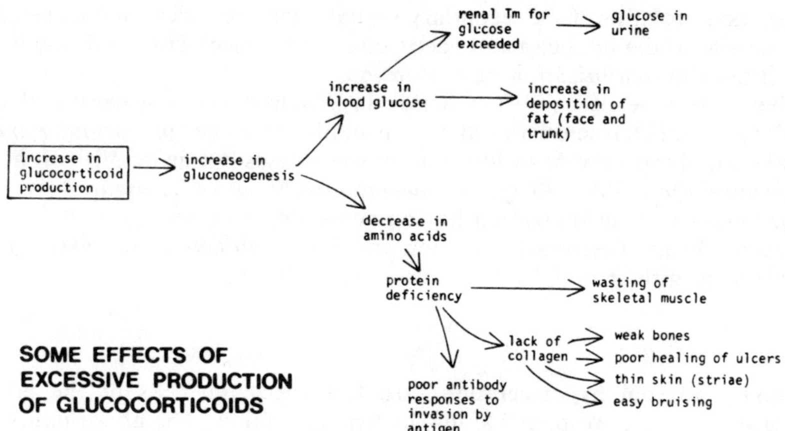


FIG. 153.—Outline of the effects of excess glucocorticoids, based on the diversion of amino-acids away from protein synthesis and towards glucose formation.

The increased tendency to mobilize amino-acid resources for glucose formation interferes with protein formation in many tissues and produces widespread effects. Muscle bulk is reduced and weakness results. Collagen formation is depressed. This interferes particularly with scar formation so that repair of injuries, including alimentary tract ulceration, is slowed down. The skin becomes thin, and irregular linear streaks (striae—L. = streaks) appear on the overstretched skin, especially on the abdomen. Weakness of blood vessel walls results in an excessive tendency to bruising. Formation of lymphocytes, and especially eosinophil granulocytes, is depressed, as is the production of immunoglobulins, and resistance to infection is reduced. For reasons which are not clear, formation of neutrophils and of red blood cells is increased. In bone, loss of the essential collagen matrix causes weakness, and fractures may result from little or no apparent injury.

### Mineralocorticoid Effects

Excess secretion of mineralocorticoid hormones, of which aldosterone is the most important, stimulates the renal tubular reabsorption of sodium, including the mechanism whereby sodium is reabsorbed from the tubular fluid in exchange for potassium and hydrogen ions. Excessive activity of this mechanism leads to sodium chloride and water retention, a rise in body sodium and fall in potassium and hydrogen ion levels. The extracellular fluid and blood volumes rise and hypertension and, in some cases, oedema result. The fall in body potassium leads to weakness of skeletal muscle and is reflected in a reduced plasma potassium level (hypokalaemia). The excessive loss of hydrogen ions leads to an alkalosis.

In summary, the major effects of excess mineralocorticoid activity are weakness, hypertension and a hypokalaemic alkalosis.

### Sex Hormone Effects

Androgenic effects are much commoner than oestrogenic effects. Women tend to develop a male pattern of body hair (hirsutism—L. *hirsutus* = hairy) and may acquire other male secondary sexual characteristics such as recession of the hairline at the forehead and deepening of the voice. The condition is then referred to as *masculinization* or *virilization.*

When this type of condition arises in children, boys generally develop secondary sexual characteristics at an unusually early age (*precocious puberty*) and girls usually suffer virilization, and in some cases it may be difficult at first to determine the child's sex (*pseudohermaphroditism*—Gr. *pseudes* = false, *hermaphroditos* = an individual having gonads of both sexes).

Excess adrenal oestrogen secretion causing *feminization* of males occurs very rarely; in girls it may lead to precocious puberty.

### Causes

**Primary adrenal hypersecretion.**—In this case, the cells of the adrenal cortex are themselves responsible for the hypersecretion. The abnormality may range from an excess of normal activity by apparently normal cells which are still partly influenced by ACTH to grossly excessive and abnormal activity by cancer cells which are completely independent of ACTH.

**Secondary adrenal hypersecretion.**—This is due to excessive secretion of ACTH by the pituitary. The main hormonal excess consists of glucocorticoids.

**Iatrogenic glucocorticoid excess.**—This is due to the administration of synthetic glucocorticoids for various conditions including severe bronchial asthma and rheumatoid arthritis. The features of excess glucocorticoid action are then seen as side-effects of the drug and their severity is related to the dose used.

### Investigations

The presence of excess adrenal glucocorticoid secretion may be confirmed by finding a raised blood level of cortisol and excess steroid breakdown products in the urine. A low eosinophil count and radiological signs of reduced bone density help to confirm the diagnosis.

The *dexamethasone suppression test* may help to distinguish between normality and abnormality in doubtful cases and may throw light on the nature of the adrenal abnormality if present. Dexamethasone is a potent synthetic glucocorticoid which is capable by a negative feedback action of suppressing release of adrenocorticotrophic hormone (ACTH) from the anterior pituitary at a dose level which has little effect on total blood and urinary steroid levels. Therefore, if given to a normal person (for about 2 days), it will cause a marked fall in blood and urinary steroid levels (Fig. 154). When the adrenal overactivity is little influenced by central steroid feedback and ACTH levels, dexamethasone causes much less suppression and in the case of a malignant tumour there may be no effect.

*When ACTH is administered,* an overactive but otherwise normal adrenal tends to give an exaggerated response, whereas a tumour is, again, unresponsive.

In a patient suffering from hypertension or muscle weakness, the finding of a reduced plasma potassium level, normal or raised sodium level, raised bicarbonate and raised pH, strongly suggest excessive aldosterone activity.

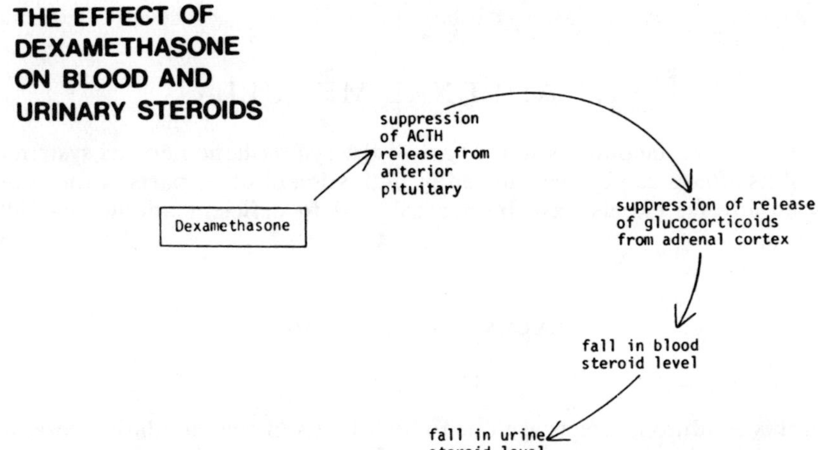


FIG. 154.—Normal response to the administration of dexamethasone, a potent synthetic glucocorticoid. In the case of a secreting adrenal tumour which is independent of ACTH control, glucocorticoid levels are not suppressed by dexamethasone.

## Treatment

When an adrenal tumour is present it is, if possible, removed surgically. If there is marked overactivity of both adrenals, they must both be removed. When there is only moderate overactivity, this may be suppressed by administration of an appropriate dose of glucocorticoid. This acts by reducing the pituitary production of ACTH.

If pituitary overactivity is responsible for the disease, partial destruction of the pituitary by irradiation may reduce the excessive release of ACTH.

In cases where both adrenals have been removed, replacement therapy is necessary. In many cases, only glucocorticoid therapy need be given. The mineralocorticoid "side-effects" of the glucocorticoid may maintain normal body levels of sodium, potassium, water, etc. Otherwise a mineralocorticoid must be added to the treatment. There is no need to consider adrenal medullary replacement therapy. The adrenal medulla is simply part of the sympathetic nervous system and its functions can be taken over by the rest of the system.

# 91. ADRENAL MEDULLA

THE ADRENAL medulla is but a part of the sympathetic nervous system and most of its effects can be reproduced by activation of other parts of the system. Thus no adverse effects have been attributed to deficient adrenal medullary secretion.

## EXCESSIVE SECRETION

### DEFINITION

In this condition, effects are due to high levels of the circulating *catecholamines,* adrenaline and/or noradrenaline. Excessive secretion may be constant or intermittent.

### EFFECTS

Effects depend on which catecholamine predominates and on whether secretion is intermittent or fairly constant.

Noradrenaline stimulates mainly alpha receptors, leading to a rise in peripheral resistance, hypertension and often headache. If this is severe, there may be associated vomiting and sweating. Excessive adrenaline secretion stimulates alpha and beta adrenoceptors and leads to a fall in peripheral resistance and increased strength of cardiac contraction. This causes a high pulse pressure and the patient becomes aware of his heart beat (palpitations). In addition, there may be a raised level of blood glucose, glucose in the urine and a rise in the basal metabolic rate, effects which are similar to those of hyperthyroidism.

Intermittent secretion typically leads to dramatic episodic symptoms. Constant secretion of noradrenaline is one—very rare—cause of sustained hypertension. Constant secretion of adrenaline may mimic hyperthyroidism.

### CAUSES

This condition is caused by a tumour, called after its histological appearance, a phaeochromocytoma (Gr. *phaios* = dark, *chroma* = colour, *kytos* = hollow vessel, i.e. cell, *onkoma* = swelling, i.e. tumour). The tumour is usually situated in one of the adrenal medullae but is occasionally found elsewhere in the abdomen.

### INVESTIGATIONS

Diagnosis is confirmed by finding a raised blood level of catecholamines (the collective term for adrenaline and noradrenaline which are amine derivatives of catechol) or a raised urinary level of their breakdown products. If secretion is intermittent, the results may be abnormal only during an attack. If hypertension is present, prompt lowering of blood pressure in response to an alpha blocking agent is typical.

### Treatment

The tumour should be removed. In some cases it is very difficult to find.

Removal is potentially hazardous, because handling of the tumour may release enormous quantities of catecholamines. To avoid excessive cardiac stimulation or hypertension, alpha and beta adrenoceptor blocking drugs can be given. It may subsequently be necessary to administer noradrenaline, should the blood pressure fall because of the loss of sympathetic activity due to the blocking drugs.

# 92. PITUITARY GLAND

DISORDERS of pituitary function are rare, despite the complex functions of this tiny gland. Abnormalities are often due to a tumour and, in view of the position of the pituitary, within the skull, adjoining the brain and near the optic chiasma, local effects are important. A tumour may cause deficient secretion by interfering with normal glandular function, or it may cause excessive secretion by itself producing the hormone.

Theoretically, excess or deficient production of any one of the pituitary hormones could occur. In fact, most variations from the normal level of single hormones may be described as physiological—a rise in the trophic hormone level when thyroid, glucocorticoid or gonadal function is deficient, and a fall in the level when the corresponding hormone is present in excessive amounts. These trophic hormones (TSH, ACTH, FSH, LH) do not of themselves produce any appreciable general effects in the body when present in excess amounts.

Deficient pituitary activity tends to take the form either of anterior lobe deficiency or posterior lobe deficiency, when lack of antidiuretic hormone causes diabetes insipidus. Both lobes of the pituitary may, of course, be affected together.

Excess pituitary activity usually consists of either excess adrenocorticotrophic hormone (ACTH) production or excess growth hormone production. Excess ACTH leads to excess glucocorticoid secretion, the effects of which have already been described. Excess antidiuretic hormone (ADH) is occasionally produced by the posterior pituitary ("inappropriate ADH secretion"). It leads to excessive water retention.

### Local Effects of Pituitary Tumours

The optic chiasma lies close to the pituitary and a pituitary tumour not infrequently presses on it and interferes with its function. The crossing fibres are affected initially and this restricts lateral vision by damaging the fibres serving the medial halves of the retinae. The condition is referred to as *bitemporal hemianopia* (Gr. *opsis* = vision)—Fig. 155.

If the tumour is large it may produce signs of increased intracranial pressure —forward protrusion of the optic discs, headaches, vomiting. A typical local effect is enlargement of the pituitary fossa which can be seen in a lateral X-ray of the skull.

## DEFICIENT ANTERIOR LOBE SECRETION

### DEFINITION

The condition to be described results from impaired secretion of all anterior pituitary hormones due to, for example, complete destruction of the anterior lobe of the pituitary. The secretions involved are:

adrenocorticotrophic hormone (ACTH),
the gonadotrophic hormones (follicle-stimulating hormone, FSH, luteinizing hormone, LH),
growth hormone (GH),
prolactin,
melanocyte-stimulating hormone (MSH), and
thyroid-stimulating hormone (TSH).

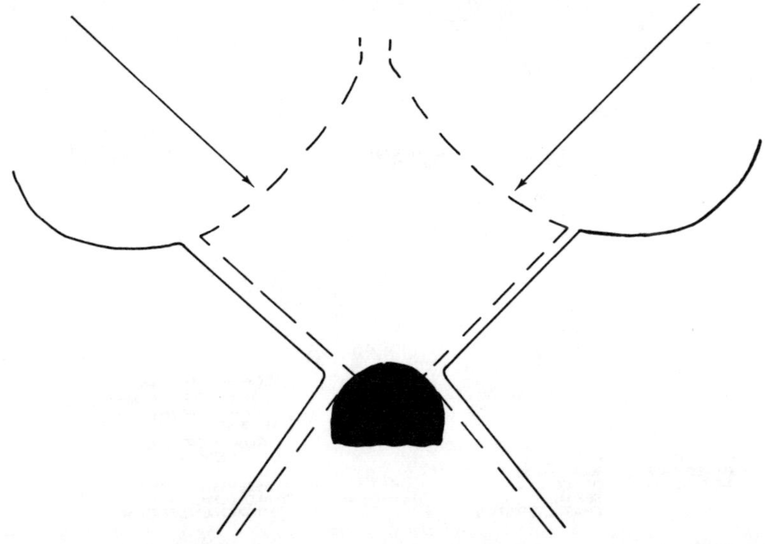

FIG. 155.—Bitemporal hemianopia produced by a pituitary tumour (black). The tumour damages the fibres which cross the midline. These fibres carry impulses from the nasal side of the retina which detects light (arrows) from the temporal (lateral) parts of the field of vision.

## EFFECTS

These differ in the adult and in the child. In the adult, the condition is referred to as *Simmond's disease* and the main effects are due to deficiency of thyroid and glucocorticoid hormones and to depression of gonadal function. In the child, stunting of growth is the main effect (*pituitary dwarfism*).

The combined effects of thyroid and glucocorticoid hormonal deficiencies—reduced tissue metabolism with a low blood sugar level and a low blood pressure—tend to make the patient easily fatigued. If the pituitary failure is severe, the patient may pass into coma in which the depression of brain function is related to the fall in body temperature, blood glucose and blood pressure. As might be expected, even slight physical stress may lead to collapse, coma and death. Lack of MSH leads to marked pallor, and lack of adrenal androgens contributes, with impaired gonadal hormones, to loss of axillary and pubic hair in both sexes.

Deficiency of gonadotrophic hormones leads to infertility, loss of sexual drive and regression of secondary sexual characteristics in both sexes. In the female, loss of cyclical ovarian activity leads to cessation of menstruation. If, as may happen, the condition arises after childbirth, impaired mammary gland activity makes lactation impossible. Prolactin deficiency presumably contributes to this failure of lactation.

Figure 156 shows how various features of anterior pituitary deficiency may be related to deficiency of thyroid, adrenal, gonadal and pituitary hormones.

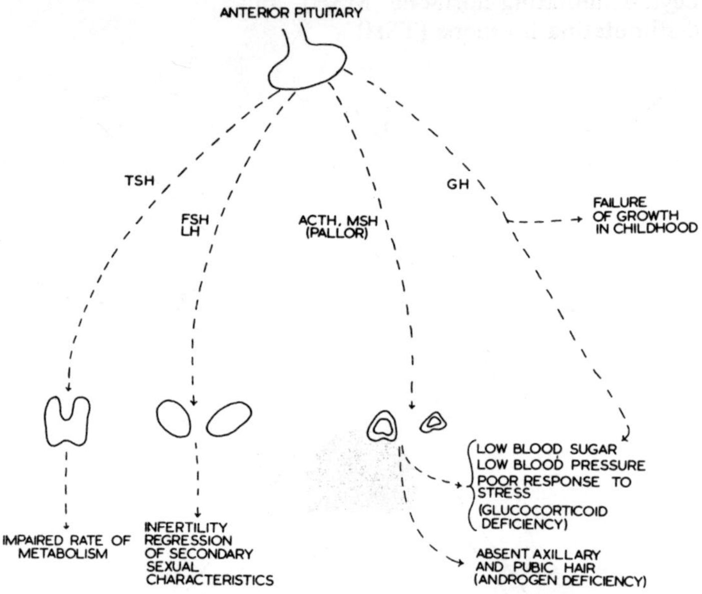


FIG. 156.—Mechanisms of causation of the main groups of consequences of anterior pituitary failure. Dotted lines indicate deficient hormone activity.

When pituitary activity is impaired in *childhood,* growth is retarded due to lack of growth hormone. Such individuals are normally proportioned although subnormal in size. In addition to the dwarfing, there may also be evidence of thyroid and adrenal deficiency. Secretion of gonadotrophic hormones may also be impaired, so that secondary sexual development fails to occur.

## CAUSES

A variety of diseases may damage the pituitary. A not uncommon cause of anterior pituitary failure is postpartum necrosis—that is, death of the gland in a woman who has just given birth (L. *partus* = childbirth; Gr. *nekrosis* = death). There has often been some complication of the birth such as excessive haemorrhage, but why the anterior pituitary should be selectively destroyed in these circumstances is not known. The gland may also be damaged by a head injury or removed surgically for a variety of reasons. Sometimes a non-functioning tumour compresses and damages the normal functioning tissue. As with other tissues, there is a big reserve of function and usually more than 90 per cent of the gland must be destroyed before signs of deficiency appear. In some cases it may take many years for signs of deficiency to appear.

### Investigations

A low protein-bound iodine or thyroxine level in the blood indicates thyroid underactivity. As both adrenal and gonadal steroid production are interfered with, urinary steroid secretion is markedly reduced. The blood glucose level may be reduced. In a child who is not growing adequately, a low growth hormone level may confirm that pituitary deficiency is the cause of the dwarfing.

### Treatment

The most urgent requirement is glucocorticoid replacement. This is necessary to avoid collapse, which may occur without warning or in relation to other disease or surgery. Thyroid replacement therapy is also required, but it is important to start this gradually and only after glucocorticoid therapy has been given, otherwise the increased metabolism with its increased glucocorticoid requirements may lead to sudden collapse. Except in the elderly, sex hormone replacement therapy is usually given to maintain normal secondary sexual development. This will not, of course, restore fertility. Fertility can be restored by the use of gonadotrophic hormones. Occasionally an overdose of these hormones causes multiple births.

In the child, human growth hormone (HGH) may be given to accelerate growth. Animal growth hormones are ineffective due to having different molecular structures from HGH.

## DEFICIENT POSTERIOR LOBE SECRETION

### Definition

Although the posterior pituitary produces the two hormones vasopressin (antidiuretic hormone) and oxytocin, lack of oxytocin has not emerged as a clinical entity. The effects to be described are due to lack of antidiuretic hormone (ADH), a condition referred to as *diabetes insipidus,* (*diabetes,* Gr. = profuse urine, *insipidus,* L. = tasteless).

### Effects

In the absence of ADH, the renal collecting ducts are impermeable to water. Hence, the dilute urine passing through them is lost to the body instead of being concentrated by the osmotic passage of water from the tubules into the hyper-osmotic interstitial fluid of the renal medulla (Fig. 157). The urine passed tends to have a constant specific gravity of less than 1.005. its volume may be as much as 20 litres/24 hours when ADH deficiency is severe.

Loss of such vast quantities of urine (*polyuria*) leads to water depletion. There is a fall in extracellular fluid (ECF) and intracellular fluid (ICF) volume and, because the urine is dilute, the osmotic pressure of these fluids rises. Blood volume and tonicity change in parallel with the body fluid changes and a low blood volume and high plasma osmotic pressure are both powerful stimulators of thirst. In order to satisfy this drive the patient must drink a sufficient

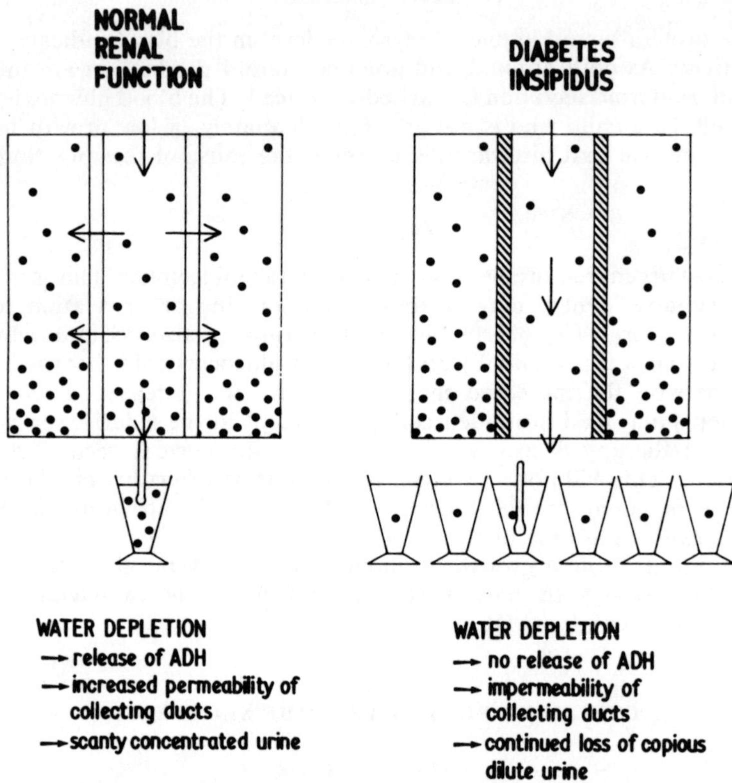


FIG. 157.—Comparison of events at renal collecting duct level—normal (left) and diabetes insipidus (right). ADH makes the duct wall permeable to water which enters the interstitial space (open arrows). In the absence of ADH this cannot occur, and urine specific gravity is low (indicated by shading and by hydrometer) and volume copious.

quantity of fluid to replace the excess urinary loss. Patients may be obliged to carry bottles of water with them when they are away from a reliable source of water for as little as an hour or two.

## Causes

Diabetes insipidus is due to deficient release of ADH (vasopressin) by the posterior pituitary nerve terminals of hypothalamic neurones. The deficiency may be due to damage to these neurones in the hypothalamus or in the pituitary. It appears that some neurones may release ADH from nerve endings in the pituitary stalk, because loss of the entire stalk plus posterior pituitary leads to diabetes insipidus, whereas if the stalk is preserved, deficiency is unlikely.

### Investigations

Confirmation of ADH deficiency requires tests which are potentially dangerous and require careful supervision in hospital. The *water deprivation test* involves measuring the specific gravity of urine while the patient is deprived of water. When deprived of water, the normal person soon produces urine of increasing concentration. The patient with diabetes insipidus cannot. If the test is prolonged, the patient may die from water depletion; hence careful supervision is necessary. Such patients also fail to concentrate their urine when given intravenous infusions of hypertonic saline, but, in contrast, injection of ADH causes marked urinary concentration, provided the kidneys can respond to ADH. (Rarely, in nephrogenic diabetes insipidus the fault lies with the kidneys).

### Treatment

Vasopressin (ADH) is given regularly by injection, or by inhalation of a nasal spray or snuff. The dose is adjusted to prevent excessive fluid loss—or excessive retention.

Treatment of diabetes insipidus is also possible in some cases by drugs which potentiate the action on the kidney of otherwise insufficient amounts of ADH.

## EXCESSIVE GROWTH HORMONE SECRETION

### Definition

Excessive secretion of growth hormone may commence in childhood or adulthood. In childhood, there is excessive growth, including growth in height—*gigantism.* In adult life, further growth in height does not occur, but body organs increase in size, the enlarged jaws, hands and feet being particularly prominent—*acromegaly.*

### Effects

Growth hormone promotes growth of body tissues other than the brain.

When the condition arises in *childhood,* the most striking feature is a rapid rate of growth. If the condition is not treated, the individual may reach a height of 7-8 feet (2-2.5 m) or more (*gigantism*).

When the condition arises in an *adult,* growth in height and in limb length is prevented by the fact that the epiphyses have closed. Growth takes place laterally with thickening of bones throughout the body. The changes are most obvious in the extremities—the facial bones grow, and skin thickening adds to the coarsening of the features. The hands and feet enlarge, the fingers becoming greatly thickened and the palms enlarged with greatly thickened skin and subcutaneous tissue. This enlargement of the extremities gives rise to the term *acromegaly* (Gr. *akron* = extremity, *megale* = large).

The viscera—heart, lungs, liver, kidneys, intestine, etc.—increase in weight to twice normal or more. Body weight is increased by the general tissue growth, rather than by fat deposition. All these changes take place so gradually that it may be years before the patient is aware of what is happening.

Although endocrine organs such as the thyroid, adrenals and gonads, share in the general enlargement, their rate of secretion is not excessive and may be deficient if the growth-hormone-secreting tumour interferes with other pituitary functions.

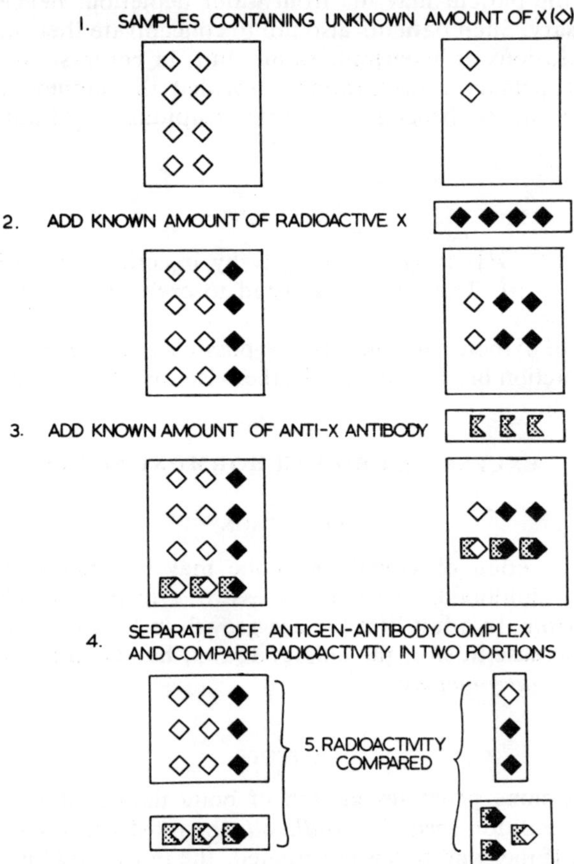


FIG. 158.—A diagrammatic representation of the principles of radioimmunoassay. Two samples are shown, their content of hormone being proportional to the number of open diamonds in the rectangle. X and radio-active X have similar affinities for anti-X. The right hand sample yields more radio-activity in the complexed fraction, indicating a lesser original content of X.

## CAUSES

The condition is due to excessive activity of the eosinophilic pituitary cells which normally produce growth hormone. This almost always leads to enlargement of the pituitary.

### Investigations

The condition may be confirmed by the finding of a raised growth hormone level in the blood. Such a finding helps to distinguish children or adolescents of unusual height due to growth hormone excess from those with other reasons for large stature including impaired sex hormone secretion and constitutional tallness. Measurement of growth hormone is made by means of radioimmunoassay. The method, which is rather a complicated one, is shown in principle in Fig. 158.

X-ray of the skull may show enlargement of the pituitary fossa due to a tumour.

### Treatment

The only effective treatment is partial or complete destruction of the eosinophilic pituitary cells by radiation or by surgical removal. As with other pituitary tumours, surgical intervention is required urgently if damage to the optic chiasma is interfering with vision.

# 93. PANCREATIC ISLET CELLS

## DEFICIENT BETA CELL SECRETION

### DEFINITION

DEFICIENT beta cell secretion leads to an inadequate blood level of insulin, which leads to *diabetes mellitus,* by far the commonest of the endocrine diseases.

However, the term diabetes mellitus probably covers a variety of different conditions; the mechanisms operating are only partly understood. These conditions share the hallmark of hyperglycaemia and deficient insulin activity. However, the beta cells may not be primarily at fault but may fail to compensate for some other disturbance.

### EFFECTS

Most of the immediate effects of insulin deficiency can be explained by the failure of glucose to enter certain cells, notably skeletal muscle cells, in sufficient amounts. This has two initial consequences: (a) glucose accumulates outside the cells, (b) glucose is not utilized inside the cells. The events which follow these two abnormalities are summarized in Fig. 159 and will now be considered in turn. It should be noted that the severity of the condition varies widely and in mild cases there may be no symptoms and relatively little disturbance of body functions.

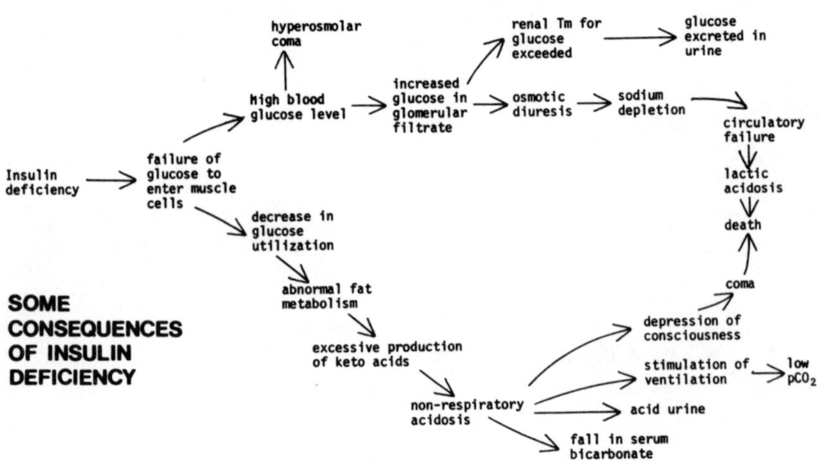


FIG. 159.—Diagram of the main sequences of events which result from failure of glucose to enter certain cells due to failure of insulin action.

**Hyperglycaemia (excess glucose in the blood).**—Once the level of glucose in the blood exceeds a certain value (around 180 mg/100 ml—the highest level at which the proximal convoluted tubules can reabsorb all the glucose presented to them by the normal rate of glomerular filtration), glucose passes through the entire nephron and is lost in the urine, giving rise to the term diabetes mellitus—profuse sweet urine (*diabetes,* Gr. = profuse urine; *mellitus,* L. = sweet). When prolonged, this drain on the body's energy reserves leads to loss of weight and the appetite tends to increase. The presence of glucose in the fluid in the collecting ducts leads to an osmotic diuresis. This is because the reabsorption of water in the collecting ducts under the influence of antidiuretic hormone depends on an osmotic gradient between the fluid in the ducts and the concentrated saline in the interstitial fluid around them. When there is glucose in the duct fluid, its osmotic pressure diminishes the osmotic gradient and reduces the amount of water which can be reabsorbed. Loss of water accounts for the characteristic excessive thirst experienced by diabetics. If the diabetic cannot take sufficient fluid, dehydration develops, evidenced by a rising heart rate and falling blood pressure.

**Cellular lack of glucose.**—Because glucose cannot enter cells adequately, other sources of energy—fat and protein—are metabolised to a greater extent than normally. Metabolism of protein leads to wasting and weakness of skeletal muscle. However, body proteins tend to be spared, and the main food metabolized is fat. The metabolized fat is not all broken down to $CO_2$ and $H_2O$, but other end products accumulate—acetone (which may be smelt in the breath and detected chemically in the urine) and organic acids (keto-acidosis). The excess hydrogen ions which these acids provide are particularly harmful.

Buffering limits the accumulation of hydrogen ions by combining them with ionized acidic groups in protein and haemoglobin:$-COO^- + H^+ \rightarrow -COOH$; and with bicarbonate ions: $HCO_3^- + H^+ \rightarrow H_2CO_3 \rightarrow H_2O + CO_2$ (excreted in lungs). This process tends to lower the "alkali reserve" of bicarbonate ions. Buffering takes place inside cells, in interstitial fluid, and in the blood. The fall in the plasma bicarbonate concentration causes an uncompensated metabolic acidosis. Compensation is effected by the increase in ventilation stimulated by the acidosis which reduces the $Pco_2$ so that the ratio of carbonic acid to bicarbonate ion returns towards its normal value. The kidneys also help by excreting hydrogen ions and manufacturing bicarbonate ions in an attempt to restore the plasma bicarbonate level to normal.

If the hydrogen ion concentration continues to rise despite these compensatory mechanisms, the acidosis impairs cellular function, particularly in the brain. As with other disturbances in the internal environment of brain cells, thought processes are interfered with, then drowsiness increases, followed by unconsciousness and death.

**Hyperosmolar coma and lactic acidosis.**—These are two ways in which diabetes mellitus may occasionally affect patients.

In *hyperosmolar coma,* the blood osmolarity is markedly raised due to a very high glucose level, e.g. 1500-2500 mg/100 ml. The high extracellular osmotic pressure draws water out of body cells and the resulting cellular dehydration can cause coma and death. The situation tends to arise in maturity onset diabetes when large quantities of glucose are consumed by someone with sufficient insulin to allow some glucose to enter cells and so prevent ketosis but not sufficient to maintain a normal blood glucose level. The treatment is to restore extracellular tonicity to normal by administering hypotonic solutions intravenously.

**Lactic acidosis,** as the name implies, is a condition where excessive accumulation of lactic acid causes acidosis. It tends to arise in severely ill diabetics when anaerobic metabolism occurs on a large scale due to poor tissue perfusion. It is an indication of severe derangement of body metabolism and tends to be rapidly fatal.

**Late complications of diabetes mellitus.**—Patients suffering from diabetes mellitus have an increased risk of developing a variety of degenerative diseases. Such diseases include degeneration and blockage of cerebral, coronary and peripheral arteries, and damage to the retina, the kidneys and peripheral nerves. There is also an increased risk of infections, especially skin infections. Many of these problems are probably related to damage to the small blood vessels in the region (*microangiopathy;* Gr. *angeion* = vessel, *pathos* = disease). Sometimes such complications draw the doctor's attention to previously unrecognized diabetes mellitus and it is important to suspect diabetes when such conditions are present.

**Juvenile onset and maturity onset.**—The situation in individual patients who develop diabetes mellitus varies widely. Two main modes of onset can be recognized. In childhood and early adult life, the disease generally comes on relatively rapidly and is more severe when it begins (*juvenile onset*). A child, apparently previously well, may become ill over the course of a few days and pass into keto-acidotic coma before appropriate treatment is commenced. At the other extreme, an elderly person may feel perfectly well and mild signs of the disease are found only during routine tests (*maturity onset*). Keto-acidosis is rare in this type of diabetes.

## Causes

Although it is common knowledge that diabetes mellitus is related to lack of insulin, the precise mechanism of the disease remains unknown. Several mechanisms may operate. While in certain patients the cause of the disease may be primarily failure of the pancreas to secrete insulin, in other patients factors which oppose the action of insulin seem to be important. These factors may be hormones which favour hyperglycaemia, e.g. growth hormone, glucocorticoids, glucagon, or they may be other circulating substances which antagonize insulin's action. Certainly some patients require much more insulin than would be necessary simply to replace the normal amount secreted by the pancreas in health. Thus, although diabetes mellitus is due to deficient insulin *action*, it is probably not always due to deficient *secretion.* In most cases in which the pancreas is examined after death, some abnormality of the beta cells is found. This could be the cause of the disease, or a result of increased demand for insulin due to the presence of antagonists. The basis of the disease appears to be genetic in many cases, because there is a higher than average incidence of diabetes in the relatives of diabetics, particularly young diabetics.

Just as the features of diabetes in the elderly differ from those in young people, so the causes appear to differ. In the elderly, the disease is typically associated with overweight, which appears to increase insulin requirements and throw excessive strain on the beta cells. When calorie intake is decreased the disease becomes milder or may disappear completely.

## INVESTIGATIONS

These vary depending on circumstances and will be considered under the headings (a) *initial screening,* (b) *confirmation of the diagnosis,* and (c) *the severe case.*

**Initial screening.**—In the great majority of cases, the first definite evidence of diabetes mellitus is provided by the finding of sugar in the urine. If the test for sugar or, specifically, for glucose is strongly positive, then diabetes mellitus is almost certainly present, particularly if acetone is also present. If the test is only weakly positive, the position is less clear. Such a finding is not uncommonly made at a routine medical examination for insurance purposes. If the patient is anxious about the examination, increased sympathetic activity, particularly release of adrenaline from the adrenal medulla, may raise the blood glucose above the renal threshold. Another cause for glycosuria not due to diabetes is a low renal Tm for glucose. For some reason the proximal convoluted tubule cells are less efficient than usual at reabsorbing glucose so that some passes into the urine when the blood glucose level is in the upper part of the normal range. In such cases the test may be repeated and further tests used to confirm or refute the diagnosis.

**Confirmation of the diagnosis.**—The mainstay of the definite diagnosis of diabetes mellitus in doubtful cases is the glucose tolerance test. This measures the response of the patient to a standard glucose load. After fasting overnight, the patient provides an initial sample of urine and an initial blood sample is taken. The patient then receives the glucose, e.g. 50 g in water by mouth. Samples of blood and of urine are then obtained at intervals, e.g. half-hourly for two hours. The blood and urine are then analysed for glucose and the results are plotted as a graph. In diabetes, the fasting blood sugar is typically raised, the maximal level after glucose ingestion is increased, and it takes longer than usual for the blood sugar level to return to the fasting value. Glucose is present in the urine during part or all of the test. When there is a low renal Tm for glucose, the blood sugar measurements fall within the normal range, although glucose is found in the urine.

**The severe case.**—This is the case in which the patient is severely ill, possibly unconscious, due mainly to a combination of acidosis (sometimes referred to as keto-acidosis or ketosis) and dehydration. The patient may be already known to be a diabetic. The diagnosis is confirmed by finding a high blood sugar level and acetone in the urine. The severity of the acid-base disturbance may be estimated by measuring the pH and the serum bicarbonate level. Plasma electrolyte levels are usually measured; they give some indication of body electrolyte levels but are not entirely reliable because, e.g. a low intracellular $K^+$ level may be masked by a normal or high plasma level. Plasma specific gravity gives some indication of water loss because this results in concentration of the plasma and a rise in the specific gravtiy.

These tests are repeated during treatment and indicate whether the patient is responding satisfactorily or whether modification of the treatment is required.

## Treatment

Treatment of diabetes includes dealing with a great variety of problems. The present consideration will include three distinct situations: (a) *treatment of keto-acidosis,* (b) *routine treatment with insulin,* (c) *routine treatment of the elderly obese diabetic.* Finally, *shortcomings in the treatment of diabetes* will be referred to.

### Treatment of Keto-acidosis

This is the treatment of the patient whose diabetes is seriously out of control, leading to severe illness with dehydration and possibly coma. Two main aspects of treatment are of overwhelming importance—*replacement of insulin* and *replacement of body fluids.*

Insulin should be given as soon as the diagnosis of keto-acidosis has been made. *Insulin is the main requirement for correction of the acidosis.* Once cellular metabolism has returned to normal, excessive production of hydrogen ions ceases and the body systems can correct the acid-base imbalance, provided that dehydration is also corrected so that the kidneys can function effectively. In some cases an alkali is given to speed correction of the acidosis, but this is of secondary importance.

Fluids are required urgently and must be given intravenously as the patient cannot take and absorb sufficient via the alimentary tract (the patient may be unconscious or fluids given by mouth may be vomited, an effect of acidosis on the alimentary tract). Sodium chloride solution (isotonic saline—0.9 per cent), is the usual fluid given. Insulin aids the entry of $K^+$ into cells. As the intracellular $K^+$ deficit is made up, the plasma level may fall and appropriate amounts are then added to the infused fluid. Frequent assessment of the state of the blood and the urine is needed to ensure adequate treatment and avoid excessive insulin treatment which could convert hyperglycaemia into hypoglycaemia.

### Routine Treatment with Insulin

This is the form of long-term treatment necessary for the younger patient with fairly severe diabetes. The aim of treatment is to provide a calorie intake and a dose of insulin (usually given once or twice daily) appropriate for the individual's degree of physical activity. The insulin used is generally obtained from cattle or pig pancreas. The structure of these insulins varies very slightly from that of human insulin—a few amino-acids in the sequence are different. As these are foreign proteins they may lead to antibody formation. However, the difference in structure from human insulin is so slight that antibody formation rarely presents a serious problem. Changing to insulin prepared from a different species usually helps in such cases.

Insulin treatment requires the sterilization of syringe and needle and the subcutaneous injection of insulin, often by the patient. To save the patient trouble, injections are not usually prescribed more than once or twice per day.

The insulin is in a form which is gradually released from the injection site throughout the day. This steady release of insulin cannot precisely match the body's requirements. To avoid the risk of serious hypoglycaemia, the dose of insulin used is such that the blood sugar rises at times above normal and glycosuria occurs. The patient tests his urine regularly for sugar and acetone. If there is never any sugar in the urine, the dose of insulin may well be slightly too high, if acetone and much sugar are present, the dose is too low. Blood glucose estimations are carried out at intervals whose length depends on how well the diabetes is controlled. In general, the intelligent patient whose diet and activities are fairly constant, is best controlled and need attend for blood glucose estimations least often.

### Routine Treatment of the Elderly Obese Diabetic

In these patients, relatively mild disease is caused by partial failure of insulin secretion to meet the body's requirements. As the disease is produced or aggravated by obesity, a low-calorie diet with consequent weight reduction is the first requirement. If this by itself does not relieve the condition, *oral hypoglycaemic drugs,* inadequate in the severe diabetic, usually control the hyperclycaemia. It is not clear how these drugs work, but some seem to act by stimulating release of insulin from the pancreas.

### Shortcomings in the Treatment of Diabetes

The treatment of diabetes described above clearly does not entirely restore blood insulin levels to normal. Insulin levels are determined by the rate of release of injected hormone rather than by the fluctuating demands of the body in relation to meals and activity. The insulin is absorbed into the peripheral circulation rather than being secreted by the pancreas. In cases where insulin antagonists may be important, the antagonists are not removed but are overwhelmed by large doses of insulin. These deviations from ideal treatment may account for the fact that even when their blood glucose levels are well controlled, diabetics are rather more prone than the general population to the complications previously described.

## EXCESSIVE BETA CELL SECRETION

### Definition

The effects of excessive beta cell secretion are due to an abnormally high blood insulin level. These effects can be reproduced by the injection of excessive insulin and, in fact, by far the commonest cause of excessive insulin in the blood is the treatment of diabetes mellitus with insulin. Most diabetics treated with insulin suffer from *insulin reactions* when the level of circulating insulin is temporarily too high for the patient's needs. The subsequent description of effects, etc. refers to both spontaneous and medically-induced excessive insulin levels.

## Effects

The actions of insulin on the body are not fully understood but most of its effects can be explained by its facilitation of the entry of small nutrient molecules—glucose, amino-acids, fatty acids—into skeletal muscle, fat and some other cells. Excess insulin activity depletes the bloodstream of nutrients. Insulin does not facilitate entry of nutrients into brain cells, which depend for their metabolism on an adequate blood level of glucose in particular. The symptoms of a low blood glucose level (*hypoglycaemia*) fall into two groups—(a) those directly due to the low blood sugar, (b) those due to the increased sympathetic nervous system activity which is stimulated by the hypoglycaemia and which tends to counteract it.

**Hypoglycaemia.**—As brain cells are more sensitive to glucose lack than any other cells in the body, the main direct effects of hypoglycaemia are a general depression of cerebral function, particularly the higher critical functions. The cerebral depression is similar to that produced by alcohol, hypoxia, etc. and manifests itself by inappropriate mood and behaviour which vary from patient to patient. Excessive hunger may be due to stimulation by the low glucose level of hypothalamic appetite-increasing areas. As the blood sugar level falls, the condition may progress to drowsiness, unconsciousness and, as the vital centres in the brainstem are depressed, death. Repeated attacks of coma may lead to permanent brain damage.

**Sympathetic nervous system overactivity.**—The direct compensatory effects are a rise in blood sugar due to increased glycogen breakdown and increased alertness due to the alerting action of the sympathetic nervous system. Excessive alerting activity leads to a feeling of nervousness. In addition, the action of the sympathetic nervous system on voluntary muscle leads to tremor. Its action on the cardiovascular system leads to an increased heart rate, an increased pulse pressure and skin vasoconstriction leading to pallor. Excessive sweating is typical.

## Causes

By far the commonest cause of an excessively high blood insulin level is the treatment of diabetic patients with insulin. Most patients who receive regular treatment with insulin experience at least occasional attacks of hypoglycaemia. When a patient has received a certain dose of insulin, the blood sugar may fall abnormally if (a) energy intake is inadequate, e.g. due to a gastro-intestinal upset or (b) energy utilization is excessive, e.g. due to increased muscular activity. Illness of any kind is liable to change insulin requirements because both the metabolic rate and food intake tend to change. Blood glucose can be maintained in the normal range only if insulin dose and diet are appropriately matched at all times.

Oral hypoglycaemic agents can also, on occasion, produce a severe fall in the blood sugar level with effects similar to those of excess insulin.

Excessive activity of the beta cells of the pancreas is a rare cause of a raised blood insulin level. Typically this is due to an *insulin-secreting tumour* of the beta cells.

### Investigations

Measurement of the blood glucose level readily confirms or excludes the presence of hypoglycaemia. If an insulin-secreting tumour is present, a prolonged period of fasting, e.g. for 1-2 days, may be necessary to produce an abnormally low blood glucose level. A rapid response of symptoms to administration of glucose confirms the diagnosis of hypoglycaemia.

### Treatment

Hypoglycaemia is relieved by glucose. If the patient is conscious, this should be given by mouth. Diabetic patients who suffer from hypoglycaemic reactions learn to recognize early symptoms and take something sweet to eat or drink. It is advisable for them to carry around some sugar lumps or sweets to be taken at the first suggestion of an attack.

If the patient is unconscious, the glucose is given intravenously and produces a dramatic recovery. In situations where intravenous glucose is not available, an injection of adrenaline may raise the blood sugar sufficiently to restore consciousness so that food may be taken by mouth. Treatment of hypoglycaemia is urgent if brain damage is to be prevented. If the diagnosis is to be confirmed by a blood glucose measurement, the treatment must be given before the result is available. Glucose is unlikely to harm someone whose symptoms are not in fact due to hypoglycaemia.

The treatment of a beta cell tumour is surgical removal.

## ALPHA CELL SECRETION

The pancreatic alpha cells and also certain cells in the intestinal wall produce *glucagon,* which tends to raise the blood glucose level by promoting breakdown of glycogen. Glucagon in high doses has other effects, including increasing the strength of contraction of heart muscle. However, its normal role in the body has not been elucidated and as yet no clear disease entities due to its deficiency or excess have emerged.

# 94. GONADS

## DEFICIENT TESTICULAR SECRETION

### DEFINITION

THE TESTIS exerts its hormonal effects through the secretion of *testosterone* by its interstitial cells. Deficient testicular secretion exerts its effects through the reduced level of circulating male hormones (*androgens*). It should be noted that, although the adrenal cortex produces androgenic hormones, the quantity produced is too small to prevent signs of hormone deficiency when testicular function is impaired.

### EFFECTS

The effects of androgen deficiency depend on whether the deficiency arises before or after puberty.

In *immature males,* absence of testicular secretion prevents the development of secondary sexual characteristics. The increased level of sex hormones at puberty leads to cessation of growth in height by promoting closure of the epiphyses of long bones, so when these hormones are absent, limb length and height tend to be greater than average. In addition, body build tends toward the female type. The voice remains high-pitched. Body hair follows the female distribution. The penis and testes are small. Sexual desire is slight or absent, and infertility the rule. Other typical male drives, aggressiveness and ambition, are also reduced. The term "*eunuch*" is applied to the individual with the characteristics described in this paragraph.

In *mature males,* loss of testicular hormones causes only slight or moderate regression of the secondary sexual characteristics. The growth of facial and body hair may be slowed but is not usually abolished. The *desire* for sexual intercourse (*libido*) and the *ability* to maintain an erection of the penis (*potency*) tend to persist at a somewhat reduced level. The effects of androgens thus tend to be permanent—this applies also in females treated with androgens for any reason.

It is possible that a falling level of testosterone secretion with age may account for the gradual regression of secondary sex characteristics seen in some ageing men.

### CAUSES

Both testes must be destroyed or severely damaged before loss of hormonal function or infertility are apparent. Injury and various diseases may cause this. In addition, failure of the testes to develop normally may be due to genetic abnormality, most commonly when there is an extra X chromosome in the male, leading to the chromosome constitution 44 plus XXY (Klinefelter's syndrome). Finally, inadequate testicular hormone secretion may be due to inadequate production of gonadotrophic hormones by the anterior pituitary gland.

### Investigations

Measurement of the blood testosterone level will indicate whether there is a deficiency of this hormone.

When testicular hormone output is reduced due to local disease of the testes, a high level of gonadotrophic hormones may be found in the urine due to lack of negative feed-back on the hypothalamus and pituitary. When the cause is pituitary failure, the level of gonadotrophic hormones is below normal.

Determination of the sex chromosome constitution will reveal whether the condition has a genetic origin.

### Treatment

Treatment consists of replacement therapy using androgens. This treatment will restore the secondary sexual characteristics but will not, of course, restore fertility if the seminiferous tubules have been destroyed or are abnormal for genetic reasons.

## DEFICIENT OVARIAN SECRETION

### Definition

The ovary exerts its endocrine effects by means of the oestrogen and progestogen groups of hormones. Deficiency of both types of hormone usually occur together.

### Effects

These parallel in many respects, the effects of deficient testosterone secretion in that deficiency in *childhood* leads to failure of development of secondary sexual characteristics, whereas in the *adult* these characteristics do not disappear but tend to regress, and there is decreased libido. The onset of menstruation (the *menarche*—Gr. *arche* = beginning) may be delayed. Fertility may be impaired. Menstrual loss tends to be slight or absent.

Particularly when the loss of ovarian hormones is sudden, the patient may suffer from instability of circulation and sweating in the skin, producing what are described as *hot flushes* and cold sweats. There may be considerable emotional upset with irritability or depression. Why the withdrawal of ovarian hormones should produce these effects is unknown.

### Causes

Cessation of normal ovarian function is the cause of the normal ending of menstruation in women (the *menopause*), generally at around 40-50 years of age. This is a primary ovarian failure—the level of pituitary gonadotrophic hormones rises markedly due to withdrawal of the sex steroids which act on the hypothalamus and pituitary in a negative feedback fashion.

Non-physiological causes of deficient ovarian secretion include destruction of both ovaries by disease, surgical removal, genetic absence of the ovaries (Turner's syndrome—only one X chromosome is present) and failure of the pituitary to release adequate amounts of gonadotrophins.

### Investigations

Plasma oestrogen and progestogen levels are typically low.

Disorders of menstruation are a sensitive index of ovarian function. Various tests indicate whether normal cyclical ovarian hormone secretion is occurring. These include the observation of cyclical changes in basal body temperature and in the lining of the uterus and vagina.

A characteristic feature of primary ovarian failure is a very high level of urinary gonadotrophins, indicating active pituitary "efforts" to increase ovarian activity when the absent hormones no longer exert a negative feedback effect on the hypothalamus.

### Treatment

Synthetic oestrogens usually relieve the general symptoms of ovarian hormone deficiency, e.g. after the menopause. The dosage used is the lowest which will relieve symptoms. If it is thought desirable for psychological reasons to restore "menstruation" in a younger woman, the replacement therapy is given cyclically—three weeks of treatment, followed by a week without treatment. The withdrawal bleeding is not exactly the same as true menstruation, because the endometrium does not undergo the same cyclical changes.

If ovulation is not occurring, gonadotrophic hormones may restore fertility.

Where the ovaries are absent, hormone therapy can produce secondary sexual characteristics, but cannot, of course, cure the infertility.

## EXCESSIVE TESTICULAR SECRETION

### Definition

The effects of excessive testicular secretion are due to a raised level of circulating testosterone.

### Effects

As with deficient testicular secretion, effects depend on whether the condition arises before or after puberty.

In *boys,* excess secretion of androgens, mainly testosterone, leads to premature development of secondary sex characteristics including enlargement of the genitalia, growth of body and facial hair and deepening of the voice. However, this *sexual precocity* is not true puberty, because androgens cannot produce spermatogenesis in the immature testis—pituitary gonadotrophic hormones are required for this. There is also exceptional development of skeletal muscles because of the anabolic effect of androgens. Such a child may look like an "*infant Hercules*".

In the *adult male,* excess testicular secretion does not produce any noticeable effects apart from inducing a degree of breast development known as *gynaecomastia* (Gr. *gynaikos* = woman, *mastos* = breast). The testis, like the adrenal cortex, produces a variety of steroids with overlapping effects. Testosterone in high doses can cause gynaecomastia, as may other testicular steroids with oestrogenic properties. Testosterone itself can be metabolized in the body to yield products with oestrogenic properties.

### Causes

This rare condition is caused by a tumour of the interstitial cells of the testis. In most cases the only noticeable feature is enlargement of the testis due to the tumour.

Because of their anabolic effects, a variety of synthetic steroids ("*anabolic steroids*") with somewhat similar properties to testosterone are sometimes administered in an effort to stimulate protein formation. They may thus be used in elderly patients where protein formation is thought to be deficient and, in younger age groups, are not infrequently taken by weight-lifters and others who hope thereby to increase muscle bulk and strength. They have some androgenic properties and may lead to a degree of masculinization in the female. As in some other situations where hormones are administered in the absence of obvious hormone deficiency, their value is doubtful.

### Investigations

The blood testosterone level is elevated, as is the urinary steroid excretion.

### Treatment

Treatment consists of surgical removal of the tumour and an effort to deal with spread of the tumour it if is malignant.

## EXCESSIVE OVARIAN SECRETION

### Definition

Excessive ovarian secretory activity leads to raised levels of the oestrogen and progestogen groups of steroid hormones.

In some cases excessive male (androgenic) hormones may be produced by the ovary.

### Effects

As with testicular hormones, excessive production of ovarian hormones has the most dramatic effects when it occurs in *childhood;* precocious development of secondary sexual characteristics may be produced in girls.

In *women of child-bearing* age, cyclical ovarian activity is responsible for the cyclical changes in the uterus. Menstruation is a consequence of the fall in female sex hormone levels in the second half of the cycle. A constant high level of circulating ovarian hormones abolishes cyclical ovarian activity and hence causes cessation of menstruation (*amenorrhoea*). The ovarian hormones suppress hypothalamic secretion of luteinizing hormone and follicle stimulating hormone-releasing hormone (LH/FSH-RH), thus abolishing the rise of the gonadotrophic hormones which cause ovulation; infertility thus results.

Ovarian tumours which produce androgenic hormones cause male secondary sexual characteristics (*virilization*) in girls or women. The clitoris enlarges, there is a male distribution of hair and deepening of the voice. Cyclical ovarian activity and hence menstruation cease, because of the negative feedback effect of the abnormal steroids.

### Causes

Ovarian hormones are produced in greatly increased amounts in normal pregnancy, in which, of course, they suppress ovulation. However, the commonest non-physiological cause in many countries nowadays of excess ovarian hormone levels in the body is the use of *oral contraceptive therapy.* This consists of a synthetic oestrogenic steroid, sometimes combined with a progestogenic steroid, the object being to suppress ovulation by inhibiting hypothalamic secretion of the hormone which stimulates release of pituitary gonadotrophic hormones. The treatment is generally stopped for one week in four. This allows withdrawal bleeding to occur, while the otherwise high hormone levels suppress ovulation. The reason for allowing withdrawal bleeding to occur is that many women would regard cessation of menstruation as abnormal.

Apart from pregnancy and oral contraceptive therapy, excess ovarian hormones are generally produced by ovarian tumours.

### Investigations

Plasma progestogen and oestrogen levels are typically raised, as are urinary steroid levels.

The presence or absence of cyclical ovarian activity may be assessed as described under "deficient ovarian secretions".

### Treatment

Treatment consists of removal of a tumour, if present.

# 95. MISCELLANEOUS

## HYPOTHALAMUS

IN THE present state of knowledge it is difficult to distinguish with certainty between a primary pituitary abnormality and abnormality secondary to hypothalamic disease. In the case of posterior pituitary hormones, damage to cell bodies in the hypothalamus can produce similar effects to damage of their axons in the pituitary. Further, the hypothalamus and pituitary are so close together that disease, e.g. a tumour, may affect both regions simultaneously. In the rare cases in which hypothalamic functions, such as appetite regulation and temperature regulation accompany pituitary disease, it may be suspected that the endocrine function of the hypothalamus is also affected.

## KIDNEY

There is evidence that *erythropoietin* is produced in the kidney, although the cells which produce it have not been identified. *Excessive activity* probably accounts for the polycythaemia which occasionally accompanies renal tumours. *Deficient activity* is similarly thought to cause the bone marrow depression and anaemia which accompany severe bilateral kidney disease. A successful kidney transplant cures the anaemia as well as the renal failure. Otherwise, until erythropoietin replacement therapy is available, the anaemia of renal failure must be treated by blood transfusion.

Some cases of renal hypertension may be caused by disturbance of renal endocrine activity, such as excessive *renin* secretion.

Some cases of rickets (renal rickets) and *osteomalacia* may be related to failure of the diseased kidney to play its normal role in the conversion of vitamin D to the active form which facilitates calcium absorption from the intestine and calcium deposition in bone.

## ENDOCRINE SECRETION BY TUMOURS

Occasionally a tumour, benign or malignant, produces large amounts of hormone, even though the tissue from which the tumour arises does not normally produce the hormone in detectable amounts. For example, a lung cancer may produce ACTH, or an ACTH-like substance, which leads to adrenal hyperactivity. Certain pancreatic islet tumours may secrete a gastrin-like substance which causes excess acid secretion and peptic ulcers (Zollinger-Ellison syndrome). Certain intestinal or lung tumours (carcinoid tumours) secrete large amounts of 5-hydroxytryptamine which may cause bronchospasm, diarrhoea and episodes of pallor and flushing of the skin. Other tumours may produce antidiuretic hormone and cause excessive water retention.

In fact, it appears that cancer cells are capable of producing polypeptides which mimic virtually all known endocrine polypeptide hormones.

## RECENT LEADING ARTICLES


Apathetic thyrotoxicosis. *Lancet,* 1970, **2,** 809.
Calcium infusion in the diagnosis of primary hyperparathyroidism. *Lancet,* 1972, **1,** 578.
Cushing's syndrome in childhood. *Lancet,* 1972, **2,** 267.
Diabetes mellitus, disease or syndrome. *Lancet,* 1971, **1,** 583.
Diagnosis of thyrotoxicosis. *Brit. med. J.,* 1973, **2,** 131.
Endocrine exophthalmos. *Brit. med. J.,* 1972, **3,** 68.
Glucagon. *Lancet,* 1972, **2,** 637.
Glucagon and diabetes. *Brit. med. J.,* 1973, **3,** 310.
Glucagon and growth hormone. *Brit. med. J.,* 1973, **1,** 188.
H.G.H. and growth. *Lancet,* 1972, **1,** 187.
Human growth hormone. *Brit. med. J.,* 1971, **2,** 236.
Human prolactin. *Brit. med. J.,* 1971, **3,** 201.
Hyperosmolar coma. *Lancet,* 1972, **2,** 1071.
Hypothalamic releasing hormones. *Brit. med. J.,* 1972, **1,** 65.
Incomplete diabetes insipidus. *Lancet,* 1971, **1,** 53.
Infection and diabetes. *Brit. med. J.,* 1974, **3,** 76.
Intermittent hormones. *Brit. med. J.,* 1971, **3,** 493.
Interpretation of serum protein-bound iodine. *Lancet,* 1971, **1,** 1341.
Microaneurysms in diabetic retinopathy. *Brit. med. J.,* 1971, **3,** 548.
Obesity and diabetes mellitus. *Lancet,* 1971, **1,** 381.
Parathyroid hormone and vitamin D. *Lancet,* 1972, **1,** 1000.
Pathogenesis of diabetes mellitus. *Brit. med. J.,* 1971, **3,** 594.
Plasma chloride levels in hyperparathyroidism. *Brit. med. J.,* 1971, **3,** 444.
The latest on LATS. *Lancet,* 1974, **2,** 443.
Thymic hormones. *Brit. med. J.,* 1974, **3,** 75.
Thyroid function tests. *Lancet,* 1970, **2,** 806.
Thyrotrophin immunoassay. *Brit. med. J.,* 1971, **4,** 761.
Thyrotrophin-releasing hormone. *Brit. med. J.,* 1973, **1,** 62.
Thyrotrophin-releasing hormone and the pituitary. *Lancet,* 1972, **1,** 782.
Treatment of Cushing's syndrome. *Brit. med. J.,* 1971, **2,** 233.
Treatment of diabetic retinopathy. *Brit. med. J.,* 1973, **3,** 421.
Treatment with calcitonin. *Brit. med. J.,* 1973, **1,** 371.
Treatment with growth hormone. *Brit. med. J.,* 1971, **3,** 547.
Triiodothyronine. *Lancet,* 1971, **1,** 898.

*Section XV*

# DISORDERS OF THE REPRODUCTIVE SYSTEM

SUCCESSFUL reproduction requires the successful completion of a series of complex physiological events which include fertilization of the ovum, adaptation by the mother to pregnancy, growth of the fetus, labour and adaptation by the baby to life outside the uterus. Failure at any stage may jeopardize the success of the entire project. The problems that may arise at the various stages are discussed in the following chapters.

# 96. PROBLEMS OF FERTILITY

## INFERTILITY

### Definition

Statistically, it is unlikely that a single act of coitus will result in fertilization of an ovum. Many couplings are usually required to achieve conception. For this reason a couple should not be considered infertile until they have practised regular coitus for about two years without conception, in the absence of any form of birth control.

### Effects

About 15 per cent of all married couples do not have children and infertility is the underlying cause in the majority of cases. The inability to have children or the fear of inability can give rise to great personal, marital and social stresses and unhappiness. These stresses vary depending on the social background and inclinations of the couple concerned, but couples are usually prepared to go to great lengths to overcome infertility.

### Causes

Failure to conceive may result from infertility in one or other of the partners, male infertility being about as common as female infertility. It may also result from difficulties in performing the sexual act (coitus) even though both partners are fertile.

**Female Infertility**

For conception to occur it is necessary for the ovary to produce ova and the female genital tract to permit free passage to the spermatozoa and ova. For further development to occur, it is necessary that the uterine endometrium is in a suitable state for the embedding of the fertilized ovum. Failure of any of these conditions results in female infertility (Fig. 160). Failure to produce ova may result from failure of the pituitary gland to release gonadotrophic hormones or failure of the ovary to respond to them. This accounts for only about 15 per cent of all cases of female infertility.

Mechanical obstruction to the passage of germ cells in the reproductive tract is the commonest cause of female infertility. Free passage of the germ cells along the tract may be impeded by obstruction of the uterine tubes by the inflammatory changes following infection (e.g. tuberculosis or gonorrhoea). Infections of the tissues at the cervix of the uterus may alter the composition of the cervical mucus so that sperms cannot penetrate it easily.

Pathological changes in the endometrium may occur following infections of the genital tract or when a foreign body is present in the uterine cavity. These may prevent normal implantation and cause infertility. Foreign bodies placed in the uterus may be used as contraceptive devices.

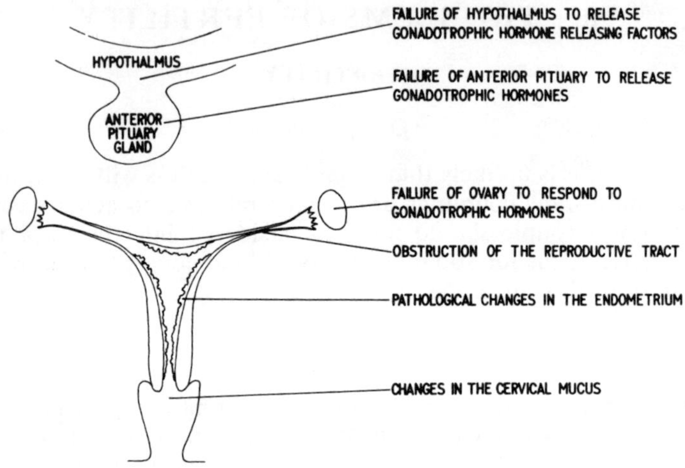


FIG. 160.

## Male Infertility

Male fertility depends on adequate production of gonadotrophic hormones, responsive testicles, adequate production of seminal fluid by accessory glands and a patent delivery system (Fig. 161). As in the female, damage to the hypothalamus or anterior pituitary may reduce the output of gonadotrophic hormones. When this happens, the spermatogenic function of the testis fails.

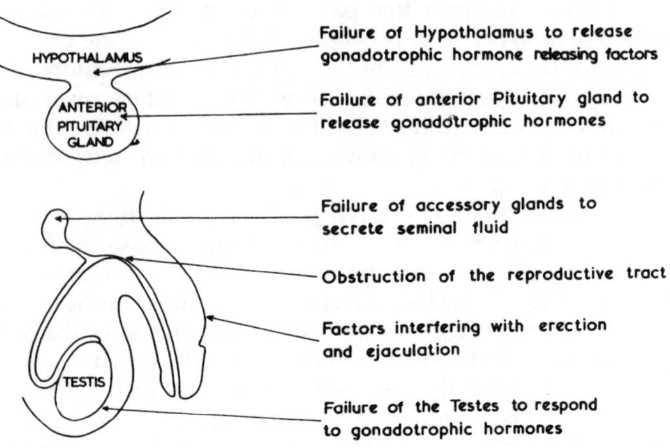


FIG. 161.

Failure of spermatogenesis also occurs when the testes fail to descend into the scrotum. The higher temperatures they encounter in the abdomen inhibit the formation of sperms. Damage to the seminiferous tubules by inflammation may also cause male infertility. Such inflammation may occur as a complication of mumps. Most of the volume of seminal fluid is derived from glands in the prostate and seminal vesicles. Damage to these glands may reduce the volume or alter the composition of seminal fluid in such a way that the survival and motility of sperms is threatened. Inflammatory changes in the ducts such as the vas deferens, which convey seminal fluid, may result in their obstruction and hence male infertility. Ligature of these vessels is commonly used to sterilize the male. Finally, any psychological or neurological disturbance which interferes with the reflexes responsible for erection of the penis and ejaculation of the seminal fluid will reduce male fertility.

### Difficulties with Coitus

The reproductive object of coitus is the deposition of seminal fluid near to the mouth of the uterus. Erection of the penis is necessary for its introduction into the vagina and inability to erect the penis (*impotence*) makes coitus impossible. The causes of impotence are more commonly psychological than physiological. Another difficulty which interferes with coitus is premature ejaculation. Ejaculation may occur before the penis enters the vagina. On the female side, a tight and narrow hymen, pain on intercourse (*dyspareunia*) and dislike of the sexual act (*frigidity*) may make coitus difficult. It should be noted, however, that female orgasm is not essential for, and indeed may not help, fertilization.

## INVESTIGATIONS

### In the Female

In cases of female infertility it may be possible to establish (a) whether ovulation is occurring, (b) whether the reproductive tract is patent and (c) whether the endometrium is healthy.

Evidence as to whether *ovulation* is occurring may be gained from several sources. Assay of the blood or urine levels of gonadotrophic and ovarian hormones may indicate a lack of the normal fluctuations associated with a normal menstrual cycle. The body temperature rises normally about a degree or so at the time of ovulation. The absence of such a rise is evidence of failure to ovulate. The squamous cells shed by the vagina show characteristic staining reactions throughout the menstrual cycle, and the normality of the cycle can be gauged by examining these cells. Changes in the composition of the cervical mucus also occur about the time of ovulation. The absence of these changes is also suggestive of failure to ovulate.

The *patency* of the reproductive tract may be tested by monitoring the passage of a radio-opaque fluid through the reproductive tract using X-ray examination. Failure of the fluid to traverse the uterine tubes is suggestive of obstruction in the tubes.

The state of the *endometrium* may be investigated by taking a sample of the tissue and subjecting it to histological examination. This can be done as a simple surgical procedure where the cervix is dilated and the endometrium scraped away (dilatation and curettage). The normal endometrium should show a proliferative phase in the first part of the cycle and a secretory phase in the second.

### In the Male

If there are no obvious anatomical or other abnormalities, analysis of the seminal fluid is a useful first investigation. Semen may be obtained when ejaculation (plus orgasm) is produced by the patient stimulating his glans penis (*masturbation*) or when the sex act is interrupted just prior to ejaculation (*coitus interruptus*). It should be collected in a plastic container (rubber condoms are inimical to sperms). A normal ejaculate should consist of about 4 ml of fluid with a sperm count of more than $20 \times 10^6$ sperm/cc. The majority of the sperm should remain motile for more than an hour and not more than 20 per cent of the sperms should be abnormal or deformed. Low sperm counts or a high incidence of abnormal sperms are nearly always associated with infertility.

If the male is not producing a satisfactory seminal fluid, further investigations may be tried. The fructose content of semen is thought to reflect the function of the glands in the seminal vesicles. The hyaluronidase content of semen is thought to reflect the function of the prostatic glands. Normal semen, when left to stand, forms a clot which later liquefies again. Hyaluronidase is responsible for liquefaction and the failure of semen to liquefy suggests abnormality of prostatic secretion. Testicular biopsy may indicate pathological changes in the germinal epithelium. Assays of gonadotrophic hormone and testosterone levels in the blood and urine may indicate an endocrine basis for the infertility. However, hormone assays, especially those for gonadotrophic hormones are still expensive and troublesome to carry out.

## TREATMENT

### Treatment of the Couple

The couple can be advised on the optimum time in the menstrual cycle for attaining conception. This is about the time of ovulation. A guide to the time of ovulation can be obtained from daily body temperature records kept by the female. Near ovulation the body temperature rises and coitus can be concentrated around this time.

Guidance may also be necessary on coital problems and surgical correction may be needed for anatomical features such as a tight hymen which makes coitus difficult. The stress associated with the stigma of childlessness may become a serious difficulty for the infertile couple. Reassurance should therefore be given even though the treatment of infertility is not always successful. After adoption of a child, an infertile couple frequently conceive a baby of their own. This may be due to the relief of stress and a consequential increase in the effectiveness of their coital activity.

### Treatment of the Female

If the problem is failure of ovulation, an attempt may be made to induce ovulation with hormones. Various combinations of gonadotrophins have been tried for this with varying success. The drug chlomiphene (Clonid) may also induce ovulation; it is thought to do this by an action which results in the release of gonadotrophins from the pituitary gland. A problem in this type of treatment is dosage; the gonadotrophins may cause the release of a large number of ova. Fertilization can then result in a large number of fetuses. In such multiple pregnancies, the chances of any fetus surviving are small.

If the problem is an obstruction in the reproductive tracts, a number of surgical manoeuvres such as excision of an obstructed segment can be tried to relieve the obstruction but they are not usually successful. In addition, infections of the tract which might affect its patency or the condition of the endometrium should be attended to.

### Treatment of the Male

The correction of obvious abnormalities may help. As with obstructive lesions in the female genital tract, those in the male genital tract are difficult to correct by surgery.

Where the production of sperms is inadequate, a number of agents have been tried to stimulate spermatogenesis with very limited success. Gonadotrophic hormones are necessary for normal spermatogenesis and may be tried where hormone deficiency is suspected. The fact that some testosterone is also necessary for spermatogenesis has led to the trial of this hormone for male infertility.

Where the volume of ejaculate is small, several ejaculates can be collected and introduced into the female genital tract by *artificial insemination.* Artificial insemination of the female with semen collected from the husband is a possible way of circumventing infertility where the husband is unable to copulate but can obtain semen by masturbation or where the wife is unable or unwilling to have sexual intercourse. This is referred to as artificial insemination by husband (AIH) to distinguish it from insemination by semen obtained from a donor male (AID). Under certain conditions, semen can be frozen and stored in semen banks for long periods before being used for artificial insemination.

## HYPERFERTILITY (OVER-POPULATION)

### Definition

Until recently human populations have been effectively curbed by famine and pestilence. The improvements in sanitation, food production and the prevention and treatment of disease have greatly disturbed the earlier balance. Though birth rates have not changed much, the decrease in death rates has led to an explosive increase in the world population (Fig. 162). Because measures to increase the death rate are not acceptable, measures to limit the birth rate have become an urgent world problem. Thus birth control must be practised in most countries to limit fertility so that family size is not greater than that which the parents and society can manage. Looked at from this point of view, hyperfertility and its treatment is a much greater problem for mankind than infertility.

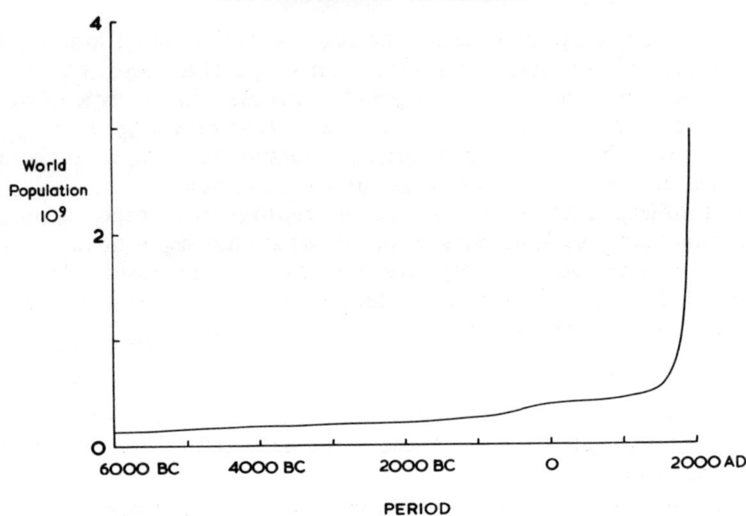


FIG. 162.—Estimated population growth since 6000 B.C.

## EFFECTS

Unchecked hyperfertility results in a surfeit of children. Early marriage, good health, prolongation of the reproductive life-span and good obstetric care could make it possible for a single couple to produce thirty or more children. When a mother gives birth to a large number of children, there are consequences not only for the mother but also for the children, the family unit and society.

**For the mother,** frequent pregnancy usually results in overwork from looking after her large family. Associated poverty may lead to malnutrition and anaemia. Wear and tear on her reproductive tract and pelvic floor may result in gynaecological disorders. The uterus may *prolapse* down through the pelvic floor and the associated deformation of the urinary sphincters may result in inability to control bladder function (*stress incontinence*). The incidence of *cancer of the cervix* is greater in women who have had numerous pregnancies. During subsequent pregnancies such women frequently experience complications because of their flabby uterine and abdominal muscles. The baby may lie in a position which does not allow easy transit through the pelvis. This, together with the tendency for such babies to be big, often causes *difficult labour.*

**For the child,** a large family can provide a wealth of companionship. However, the good features are usually outweighed by the associated poverty, inadequate nutrition and dilution of parental care and attention. Children in large families tend to have lower intelligence quotients than those in small families. This may be because parents have less time to devote to each individual in a large family rather than because of some intrinsic reason. The lower I.Q. together with the dilution of the family resources for providing education means that children from large families are handicapped in achieving their full potential.

**For the family unit,** a large number of children usually results in a low standard of living, although in some countries, welfare benefits mitigate this to some extent.

**For society** as a whole, the present explosion in world population poses very serious problems. The problem of increasing food production in the geometric fashion necessary for the expanding numbers is well known. The provision of adequate housing, water, sanitation and other facilities becomes more and more difficult. The psychological effects of urban overcrowding are being studied extensively now and the evidence suggests that personality disorders are more common in overcrowded conditions. In general, it seems that the rapid increase in world population is likely to result in a gradual reduction in the quality of life available to each individual.

## Causes

Though there has been a gradual improvement in the standard of health and nutrition during this century in most parts of the world, there is not much evidence that fertility has increased as judged by the birth rates. However, birth rates may not be a good index of fertility because they may be affected by birth control and social factors. Nevertheless, it is mainly the fall in death rates in recent times which had made the present level of fertility excessive. Between 1953 and 1959 the number of deaths from malaria in India fell from 800,000 to 10,000. The introduction of antibiotics and other medical advances have led to a great increase in life expectancy in most parts of the world.

## Treatment

The object of treatment is to prevent or limit conception. Though there are a great variety of methods of varying effectiveness available for this, they fall into one of three basic categories. They may act (a) by preventing release of germ cells from the gonads, (b) by preventing the sperms from fertilizing the ova or (c) by preventing implantation and growth of the fertilized ova.

### Prevention of Germ Cell Release

Oestrogens and progestogens (the "pill") have been found to provide very effective contraception when taken orally by women. They are thought to act by suppressing the output of gonadotrophic hormones from the anterior pituitary gland. This results from the normal negative feed-back mechanism whereby a raised level of sex hormones suppresses release of hypothalamic releasing hormones and hence suppresses pituitary release of gonadotrophins. This removes the necessary stimulus for ovulation. It is possible that ovarian hormones aid contraception by other actions as well.

Theoretically, an analogous pill could be used by men. Androgens given by mouth could inhibit the output of gonadotrophic hormones in men and thus inhibit spermatogenesis. However, an androgen suitable as an oral contraceptive for men is not yet available. Those that are available tend to cause liver damage.

Permanent prevention of germ cell release may be obtained by surgical removal of the gonads (*castration*). However, this results in loss of the endocrine function of the ovaries and testes as well as germ cell production. This causes structural changes in the body due to recession of the secondary sexual characteristics. It also produces, in severe form, the symptoms that cause distress at the end of the reproductive period (the *climacteric*). These include vasomotor disturbances such as "hot flushes", nervous irritability, indigestion and constipation. Castration causes more marked symptoms than the normal climacteric because the withdrawal of gonadal hormones is much more abrupt.

### Prevention of Fertilization

In this category of contraception, the object is to prevent the sperm reaching the ovum. Much ingenuity has been shown in the pursuit of this object and a great variety of methods are employed. This in itself would suggest that none is perfect. The better known methods include:

*Coitus interruptus.*—This is a rather hazardous method of birth control which involves withdrawing the penis from the vagina just before ejaculation occurs. It is often unsuccessful.

*Use of the safe period.*—This method of contraception involves limiting coitus to those days when conception cannot occur. Conception is only possible when coitus occurs within a period of a few days around the time of ovulation. The success of this method of contraception depends on the ability to recognize this period. It usually occurs about the middle of the menstrual cycle and this can be checked by monitoring body temperature and noting the rise which occurs at ovulation. The method is not very effective because it is difficult to predict the safe period with precision.

*Use of mechanical barriers.*—The ascent of sperm in the female genital tract may be prevented by covering the penis with a sheath (*condom*) or the cervix of the uterus with a *diaphragm*. These methods may be augmented by the use of a chemical spermicide placed high in the vagina. Though these types of contraception are quite convenient they are not always effective.

Some of the contraceptive action of oral progesterone is thought to be due to a change in the composition of the mucus secreted by the glands at the mouth of the cervix which makes it more difficult for the sperms to penetrate.

A permanent barrier may be made by dividing and ligaturing both vasa deferentia in the male (vasectomy) and both uterine tubes in the female (tubal ligation). Though highly effective, this type of contraception is only for people who want no further children because a return of fertility by further surgery cannot be guaranteed.

### Destruction of the Fertilized Ovum

Procedures which destroy the fertilized ovum are not strictly methods of "contraception" in that they do not prevent conception (fertilization) but act by destroying the developing organism.

An object such as a plastic coil, placed in the uterine cavity, is a method of birth control almost as effective as oral contraceptive drugs. It is not certain how such an *intra-uterine device* (IUD) works but it is thought to produce changes in the endometrium which make it difficult for the fertilized ovum to implant successfully. The ovum is then discharged.

There are many other possible approaches in this category of birth control. Drugs may be given which hasten the transport of the fertilized ovum to the uterus so that it arrives before the uterine endometrium has developed to a stage suitable for implantation. Drugs such as prostaglandins may be used to cause uterine contractions strong enough to expel the products of conception. *Surgical abortion* also falls into this category of birth control; the uterine endometrium containing the developing organism can be sucked away using negative pressure or scraped away using a curette.

The importance of limiting population growth in the world today has meant that research into better methods of contraception is being actively pursued. It is a rapidly developing subject and improvements in existing techniques may be expected. With an ideal contraceptive system, the partners would take some form of "pill" whenever they *wanted* to conceive; at all other times their fertility would be suppressed by some agent, e.g. a depot of hormone implanted under the skin. Active steps would then be required to have a child unlike the present system where active steps are taken to prevent conception. There would be fewer children but they would all be wanted.

# 97. PROBLEMS IN MATERNAL ADAPTATION TO PREGNANCY

NEARLY EVERY system of the body is involved as the mother makes the complex adaptation to house the growing baby and act as its lungs, kidneys, alimentary tract, etc. Usually the adaptation occurs without mishap, but occasionally pregnancy gives rise to certain problems. These will now be described briefly.

## UNMASKING OF MATERNAL DISEASE

Pregnancy gives a considerable amount of additional work to the maternal organs. The cardiovascular, respiratory and urinary systems have to support not only the growing fetus, but also the additional maternal tissue required to house the fetus.

In this situation, pre-existing diseases which have reduced the functional reserve of organs but have not given rise to symptoms and signs, may be unmasked. Thus a valvular defect in the heart, which formerly only reduced exercise tolerance, may cause heart failure during pregnancy. Pregnancy may convert a mild hypertension to a severe hypertension, a mild anaemia into a severe anaemia and so on.

## HYPERTENSION, OEDEMA AND PROTEINURIA (PRE-ECLAMPSIA)

This condition which may develop in the second half of pregnancy is characterized by weight gain due to retention of fluid in the body, a rise in arterial blood pressure and the appearance of protein in the urine. In severe cases it develops into eclampsia which is characterized by convulsions. Though there are many theories, the cause of this condition which affects about 5 per cent of all pregnancies is not understood. It is dangerous to the fetus in that it can damage the placenta. The gradual loss of placental function may retard fetal growth or result in fetal death. A potential danger to the mother is the possibility of convulsions (eclampsia).

Good treatment greatly reduces the dangers of pre-eclampsia. With the adoption of methods to lower blood pressure and induction of premature labour for severe cases, little harm should occur to mother or child. Pre-eclampsia seems to cause no permanent damage in the mother.

## ACCUMULATION OF AMNIOTIC FLUID (POLYHYDRAMNIOS)

In this condition, the volume of fluid in the amniotic sac is excessive, e.g. more than 2 litres. It may result in painful distension of the abdomen and the pressure exerted by the enlarged uterus on pelvic vessels may cause the leg veins to be congested with blood. The excessive distension of the uterus tends to lead to early (premature) labour. The freedom of the fetus to move in the sac may allow it to adopt an abnormal position which makes labour more difficult. Like a dilated heart, a dilated uterus works at a mechanical disadvantage and this may add to difficulties in labour.

The cause of the fluid accumulation is quite unknown though the condition is often associated with diabetes, multiple pregnancy and severe fetal abnormalities.

There is no effective treatment which makes the excess fluid disappear. Bed rest makes the effects of pressure less troublesome. Drainage of some of the fluid through a needle does not help much because it may precipitate labour and if not, the fluid rapidly accumulates again. If the pregnancy has progressed to a stage where the fetus is sufficiently mature to survive outside the uterus, labour may be induced.

## DEFICIENCY OF HAEMOGLOBIN (ANAEMIA)

Normally the uptake of iron by the alimentary tract is sufficient to balance the loss of iron in desquamated cellular debris. During pregnancy, increased intake of *iron* is needed for the additional maternal haemoglobin, the requirements of the fetus and placenta, and in anticipation of the blood loss during labour and the iron loss via lactation. If this iron is not forthcoming, iron deficiency anaemia develops during pregnancy. It may be prevented by ensuring a suitable diet and giving iron supplements during pregnancy and lactation. *Folic acid* is required by multiplying cells. If the increased demand in pregnancy is not met, red cell multiplication is impaired.

## INFECTION OF THE URINARY TRACT

A number of factors contribute to this. Firstly, the relaxation of smooth muscle by the high progesterone levels in the blood may reduce peristalsis and favour dilatation of the ureters. Secondly, pressure by the expanding uterus may obstruct the ureters as they pass into the pelvis. This allows urine to accumulate in the relaxed ureters and cause further distention. The resulting urinary stasis provides favourable conditions for the multiplication of bacteria.

## VOMITING

Vomiting occurs in about 80 per cent of pregnancies but is usually mild and confined to the first three months. Such vomiting typically occurs on rising from bed (*morning sickness*). In the rare cases where vomiting is severe and persistent, it is called *hyperemesis gravidarum* (excessive vomiting of the pregnant). When severe, it gives rise to the usual complications of vomiting such as dehydration and ketosis.

The cause of vomiting in pregnancy is unknown. It may be related in part to a relaxing effect of progesterone on the smooth muscle of the alimentary tract.

This would tend to favour gastric distension, thereby initiating the vomiting reflex. Reduced muscle tone in the alimentary tract may also be a cause of the reflux of acid into the oesophagus which causes chest discomfort ("heartburn"), and constipation, both of these conditions being associated with pregnancy.

## PLACENTAL HAEMORRHAGE

Bleeding sometimes occurs from the genital tract in late pregnancy and may constitute a serious threat to the life of both the mother and the child. It is usually caused by separation of part of the placenta from the uterine wall. This is more likely to occur when the placenta is situated near the mouth of the uterus (*placenta praevia*) and in mothers who have developed pre-eclampsia.

## INTRAVASCULAR BLOOD CLOTTING (VENOUS THROMBOSIS)

During normal pregnancy there is an increase in the concentration of many of the coagulation factors such as fibrinogen and prothrombin and in the platelet count in the blood. These changes favour the clotting process and help to limit blood loss in labour. Occasionally intravascular clotting may occur spontaneously in pregnancy. It may begin in the veins of the calf. When it does, it causes pain and swelling in the affected leg. The main danger of the condition is that part of the clot may break free and travel as an embolus in the blood to lodge in a branch of the pulmonary artery. If the embolus is very large, it may cause death.

## ABORTION

### Definition

Abortion denotes termination of pregnancy before the fetus is mature enough to survive outside the uterus. After that time the fetus has a chance of survival and termination is referred to as premature labour. Conventionally, the dividing line between abortion and premature labour is the 28th week of pregnancy. However, given suitable care, the occasional resilient abortus can survive to become an infant!

About 15 per cent of all pregnancies end in abortion and this usually occurs between the 8th and the 13th week.

### Effects

Abortion usually results in the fetus and its surrounding membranes being expelled from the uterus. It is usually heralded by bleeding from the vagina and painful contractions of the uterus. If it occurs before the 12th week of pregnancy, it may cause little more disturbance than a heavy menstrual period. If the fetus is more than 12 weeks mature, stronger uterine contractions are required to expel it and the abortion may resemble a miniature labour; the membranes rupture and the fetus is expelled separately from the placenta.

Abortion accounts for about one fifth of all maternal deaths in pregnancy. This is partly because it may cause serious haemorrhage if the pregnancy is well advanced. However, the high mortality also reflects the results of clumsy and criminal efforts to induce abortion. Unskilled use of drugs and operative interference may cause poisoning, infection and severe tissue trauma. In some cases, air enters the severed maternal vessels to cause fatal air embolism.

### Causes

Abortion usually occurs when the fetus is malformed for genetic or other reasons, a compensating mechanism by which pregnancy is not unduly prolonged when there is no prospect of a healthy child. In other cases, injury to or anatomical abnormality of the uterus may halt pregnancy prematurely. Rupture of the membranes which allows the fluid surrounding the fetus to escape nearly always results in abortion. Drugs which stimulate the uterus to contract vigorously have a similar effect. Serious illness in the mother tends to cause abortion, again this can be regarded as a compensatory mechanism to conserve maternal resources, provided the abortion does not itself give rise to serious problems.

In many cases of abortion which occur towards the end of the third month of pregnancy, the cause is not obvious. Often it tends to recur in pregnancy after pregnancy. "Hormonal imbalance" has been suggested as a possible cause but the evidence for this is not certain. Nevertheless, the third month of pregnancy is a critical time for hormonal balance. In the early part of pregnancy, the primitive placenta produces large quantities of chorionic gonadotrophic hormone (HCG) which stimulates the corpus luteum to secrete oestrogens and progestogens. These hormones have actions on the endometrium and the uterine muscles which are essential for the continuation of pregnancy. In the third month, the placenta takes over as the main production site for progesterone. One could imagine a failure in the change-over from the corpus luteum to the placenta causing a breakdown in progesterone secretion and thus abortion. However, this is only speculation.

### Investigations

Manual examination of the cervix may be very helpful. If it is not dilated and if the vaginal bleeding is not accompanied by much pain, abortion is not inevitable. In these cases treatment is directed at saving the pregnancy. If, on the other hand, the cervix is dilated and the vaginal bleeding is accompanied by considerable pain, the abortion is *inevitable* and treatment is directed at evacuating the uterus.

### Treatment

If abortion is threatened or appears likely because of previous unsuccessful pregnancies, bed rest, sedation and reassurance is the most useful treatment.

Ovarian hormones have been given in an attempt to prevent abortion, but it is not certain that they have any influence on the course of events.

If it is decided that abortion is inevitable, the uterus may be evacuated by scraping out the contents (*curettage*) when the pregnancy is less than 12 weeks in duration. If it is more than 12 weeks, it is better to allow the uterus to evacuate itself spontaneously.

# 98. PROBLEMS IN FETAL GROWTH AND DEVELOPMENT

DISORDERS affecting either the mother or the fetus may prejudice normal fetal development and lead to retarded growth, malformation or death. These effects are considered in more detail in the chapter on disease before birth. If the disorder causes gross malformation, the fetus usually dies and abortion follows. The mere fact that pregnancy continues increases the likelihood that there is no serious malformation. However, maternal or fetal disorders which limit the nutrition of the child usually result in the fetus being relatively small for its age.

## CAUSES

Disorders affecting primarily the fetus include congenital abnormality, e.g. mongolism; infection, e.g. german measles; adverse drug effects, e.g. deformities due to thalidomide; and lack of oxygen, e.g. due to kinking of the umbilical cord. Disorders affecting primarily the mother include malnutrition, cardiovascular, respiratory and blood diseases, which limit the delivery of oxygen to the placenta and diseases such as pre-eclampsia, which result in damage to the placenta.

## INVESTIGATIONS

As the fetus is relatively inaccessible for clinical inspection, indirect methods are required to assess his growth. X-ray studies present radiation hazards for the fetus and should be made only where essential.

The size of the fetus may be gauged by scanning the maternal abdomen with ultrasonic waves. The fetus reflects the waves and a picture of his shape and size is obtained.

Examination of the amniotic fluid may reveal biochemical abnormalities and genetic abnormalities (in cast-off fetal cells) which indicate fetal disease.

## TREATMENT

There are a number of general approaches to the problem of preventing congenital malformation and disease. From the eugenic point of view, avoidance of inbreeding limits the possibility of undesirable recessive genetic characters coming together to produce abnormality. Avoidance of breeding by people who are known carriers of genetic disorder has the same effect. *Genetic counselling* is the name of the advisory service given to people who are in danger of transmitting genetic disorders.

The prevention of maternal infection in the early days of pregnancy when fetal organization is occurring is another approach. This may be achieved by some degree of isolation during the critical period or by previous immunization against the more dangerous organisms.

Another approach is to avoid rigorously the unnecessary use of drugs, X-rays and other agents in early pregnancy which might conceivably injure the fetus. Though this sounds easy, the widespread use of drugs and the fact that the mother may not know that she has conceived in the early days of pregnancy, make it difficult for the mother to avoid these hazards. A final approach is to induce abortion if it is known for certain that the child has a severe malformation or deformity.

In later pregnancy, the main dangers arise from an inability to supply adequate nutrition for the logarithmic growth of the fetus. The main approach here is to bring on labour prematurely so that the fetal needs may be catered for artificially outside the mother. The main clinical problem in this case is making the decision about when to induce labour. The earlier the induction, the less danger of the fetus being harmed by the maternal inadequacy but the greater the danger of the infant dying from prematurity.

# 99. PROBLEMS IN LABOUR

THOUGH THERE are many theories, there is no convincing explanation yet for the initiation of labour after about forty weeks of pregnancy. The uterus becomes progressively more excitable towards this time and this may be a consequence of hormonal changes. If labour comes on early, it causes difficulty because of the immaturity of the premature infant. The infant is also endangered if the onset of labour is delayed to cause postmaturity. If labour is unduly prolonged, there are dangers for both the mother and the child, irrespective of the time of onset. These three conditions, *early labour, late labour* and *prolonged labour* are dealt with in this chapter.

## EARLY LABOUR (PREMATURITY)

### DEFINITION

A labour which occurs three or more weeks before the expected date of delivery may be considered premature. As it is often difficult to be certain about the date of commencement of pregnancy, labour is considered premature regardless of dates if the baby weighs less than 2.5 kg (5.5 lb). About one in ten labours is premature. If labour occurs more than 12 weeks before term, it is referred to as an abortion because the fetus has little chance of survival.

### EFFECTS

#### Maternal

Premature labour has few consequences for the mother. The mechanism of labour appears quite normal; it does not seem to require 40 weeks to mature. The delivery of the baby is relatively easier and quicker because of its small size.

#### Fetal

A premature infant has a smaller chance of survival than a mature one because it is less well equipped to cope with the rigours of labour and adaptation to extra-uterine life. The reasons for this include:

*General weakness.*—A premature infant tends to be more feeble than a mature infant. He is thus more likely to be damaged during labour. The thinness of the skull may allow brain damage or tearing of cerebral blood vessels. When born, he has less vigour to give to breathing, feeding etc.

*Respiratory problems.*—Lung surfactant is necessary to lower the surface tension of the fluid lining the alveoli to a level which permits the alveoli to be inflated. It normally appears in the fetal lungs towards the end of pregnancy. In a premature infant it may not be present in sufficient quantity and the work required to inflate the lungs is greatly increased. This, together with the feebleness of the muscles in the premature, accounts for the respiratory problems so frequently seen in these infants. Collapse of the lungs predisposes to infection. There is also difficulty in clearing secretions from the respiratory tract because of feeble coughing.

If excess oxygen is given to the premature infant to treat respiratory failure, there is a danger that fibrous tissue may be laid down behind the lens in the eyes (retrolental fibroplasia) and cause blindness.

*Fluid balance problems.*—The immature kidney is less well able than a mature kidney to form a concentrated or a dilute urine in accordance with the body needs. Therefore, the more premature the infant, the less well can he regulate the volume and composition of his body fluids. Vomiting or diarrhoea, which a normal infant might tolerate, may be lethal to the premature.

*Body temperature problems.*—Thermoregulatory reflexes are poorly developed in the normal newborn and are even worse in the premature. The lack of subcutaneous fat means that the skin has poor insulating qualities and the muscular feebleness means that the muscles do not generate much heat. These infants require a suitably warm environment if their body temperature is not to fall.

*Feeding problems.*—Because of their rapid growth rate, premature infants require relatively more food than a mature infant. Because of the small size of the stomach, the premature infant cannot manage large feeds; he needs frequent small feeds. His feebleness makes it more difficult for him to obtain adequate food by suckling and this may be made worse if the swallowing reflex is poorly developed. The premature infant is likely to become anaemic because the main fetal reserves of iron are normally transferred to the fetus via the placenta in the last few weeks of pregnancy. The premature infant therefore requires iron in the diet.

*Liver problems.*—The immaturity of the liver in the premature infant has several consequences. The inability to excrete bilirubin efficiently may lead to jaundice. The poor ability to form glucose may result in a low blood glucose level. The poor ability to synthesise prothrombin can result in a tendency to bleed spontaneously.

*Infection problems.*—The premature infant has a lower resistance to infection than a mature infant. This is partly due to the immaturity of his defence mechanisms and because he may not have received adequate maternal immunoglobulins via the placenta. These immunoglobulins normally provide passive immunization for the fetus and are transferred mainly towards the end of pregnancy.

Despite these problems, good nursing and medical care for premature children may allow a large percentage of those which weigh more than 1.5 kg to survive.

## Causes

As the cause of the initiation of normal labour is unknown, it is not surprising that in the majority of cases of premature labour, the early onset cannot be explained. In a number of diseases, of course, early labour is induced by the doctor because a maternal disorder such as pre-eclampsia endangers the life of the fetus. Certain conditions predispose to spontaneous early pregnancy, though they do not necessarily cause it. Unusual enlargement of the uterus as may occur with twins or multiple fetuses or excessive accumulation of amniotic fluid (polyhydramnios) often lead to early labour. On average, twins are born about three weeks earlier than single children. This suggests that stretch of the uterine wall has some part to play in the initiation of labour.

However, rupture of the amniotic sac which allows the amniotic fluid to escape also encourages early labour; this may follow trauma to the abdomen, but may occur spontaneously if an abnormality of the cervix results in a lack of support for the amniotic sac (cervical incompetence). Loss of amniotic fluid reduces uterine volume but the mechanism by which it initiates labour is unknown.

### Treatment

As the mechanism of the initiation of labour is not known, it is not possible to work on this mechanism for therapeutic purposes. However, there are two main aspects of treatment; firstly the discouragement of premature labour in patients where it is likely to occur or is threatened and secondly the care of the premature child.

(a) *Discouragement of premature labour.*—Bed rest and sedation are thought to be effective here. Drugs have also been tried which depress the excitability of uterine muscle. Drugs which stimulate beta adrenotropic receptors and inhibit the uterus may do this.

(b) *Care of the premature child.*—The main problems are to ensure an adequate body temperature, an adequate food and fluid intake and prevent infection. The *body temperature* problem is overcome by keeping the baby in an incubator at a temperature a little below body temperature. *Nutrition* has to be given in small frequent feeds. Glucose solutions may have to be used if the baby does not tolerate dilute milk. If the baby is unable to suck or swallow adequately, the feeds may be given via a tube to the stomach. No more fluid should be given than the immature kidneys can handle and care must be taken that dehydration does not occur. *Prevention of infection* is best achieved by keeping the premature infant isolated from the general life of the hospital and making sure that masks are worn and sterile precautions observed by the attendants caring for the child.

## LATE LABOUR (POSTMATURITY)

### Definition

Labour may be considered delayed if the pregnancy extends beyond 42 weeks. Because of difficulty in being sure of the exact date of commencement of pregnancy, a birth weight of more than 4 kg (8.75 lb) is suggestive of postmaturity. It is a fairly common condition, occurring, like prematurity, in about one in ten of all pregnancies.

The length of time by which pregnancy can be extended is of some medicolegal importance. In a case where a court was considering the legitimacy of a child born to a woman 47 weeks after her husband's death, the court gave the benefit of the doubt to the child (or the mother!). However, 43 weeks is usually taken as about the upper limit for duration of pregnancy.

## EFFECTS

### Maternal

Mothers do not like delayed labour because they tend to find the discomfort of late pregnancy increasingly irksome and become quite impatient to have the pregnancy end. Because of the larger, more solid baby, labour tends to be more difficult and prolonged, especially for those mothers whose pelvis is relatively small. In many cases of postmaturity, labour has to be induced by drugs or surgical interference and this exposes the mother to additional risk.

### Fetal

The death rate of postmature babies is more than twice that of babies born at the normal time. The reasons for this are not altogether clear. It may be that the higher incidence of prolonged and difficult labour results in more fetal distress and damage. It has also been suggested that the placenta becomes progressively less able to supply the growing fetal needs and that "senile" changes in the placenta limit oxygen delivery to the fetus. The evidence about this is conflicting and many people feel that the placenta normally has a large functional reserve that should easily cover the needs of postmaturity.

## CAUSES

The causes of delayed labour are not known. It tends to occur if the fetal head does not make its normal descent into the pelvis towards the end of pregnancy but remains above the pelvic brim. A mother who has a delayed labour in one pregnancy is likely to repeat the delay in subsequent pregnancies.

## INVESTIGATIONS

As postmaturity carries dangers and requires treatment, its diagnosis is of some importance. This is not always easy because of uncertainty about the date of conception.

It is difficult to assess maturity by clinical estimation of the size of the uterus in late pregnancy because such estimation is not sufficiently precise and the normal variation in size is so great. For this reason, decisions have to be made on the history of the pregnancy. X-ray examination of the fetus may help; maturity may be assessed by the presence or absence of ossification centres in certain bones, the length of the fetus and the size of its skull. Ultrasonic scanning of the maternal abdomen is also used to measure fetal size.

## TREATMENT

Because of the dangers to both the mother and child, attempts are made to start labour artificially if pregnancy extends beyond 41 or 42 weeks. This may be attempted by drugs or surgical measures.

(a) *Drug induction.*—Drugs which stimulate uterine smooth muscle to contract may start labour. These include synthetic oxytocin and prostaglandins.

(b) *Surgical induction.*—The most common method used is to rupture the amniotic sac via the cervix and allow the amniotic fluid to escape. For an unknown reason it initiates labour in the majority of cases within 24 hours. However, it increases the risk of infection because organisms may spread into the ruptured membranes. In some cases, merely dilating the cervix and separating the membranes in that area from the uterine wall by manual manipulation may be effective. Again the rationale is obscure.

If these measures fail, the baby may have to be delivered through an opening in the wall of the uterus (Caesarian section).

## PROLONGED LABOUR

### DEFINITION

Labour consists of three stages. In the *first stage,* uterine contractions cause the membranes to rupture and the cervix and lower part of the uterus to dilate, not necessarily in that order. The first stage is very variable in length and it is sometimes hard to say when it begins but it usually lasts from 6-12 hours. In the *second stage,* the baby is propelled by strong expulsive forces through the birth canal making a variety of rotational changes in position as it goes. This stage lasts 1-2 hours but may be much shorter if the baby is small or if the mother has had many previous pregnancies. In the *third stage,* the placenta is expelled and the uterine muscle contracts down hard to limit bleeding. It rarely lasts more than one hour. Prolongation of labour can occur in any or all of the three stages.

Though there is such great variation in the normal duration of labour in different individuals and in the same individual from time to time, labour may be considered prolonged if it lasts for more than 24 hours. First labours tend to be longer than subsequent ones.

### EFFECTS

#### Maternal

*Exhaustion.*—Labour is hard work and the mother may become very weak and exhausted if it continues for too long without effect. The mother may not be able to eat or drink during labour because she may require general anaesthesia for surgical intervention; this may accentuate the loss of strength. When her carbohydrate stores become used up, the abnormal metabolism of fat results in the formation of keto acids to cause a metabolic acidosis (ketosis).

*Dehydration.*—The continued loss of fluid via the respiratory tract, skin and kidneys may lead to severe dehydration in prolonged labour in the absence of water intake. This is made worse if vomiting occurs.

*Infection.*—The longer the labour continues, especially after the membranes rupture, the more likely is infection to spread from the lower reproductive tract to the uterus.

*Uterine damage.*—If the contractions are unable to expel the baby, the increase in their strength may lead to uterine damage. When the baby is finally expelled, the damaged uterine muscle may be unable to contract down vigorously to close off the blood vessels on the placental site. This probably accounts for the high incidence of severe bleeding from the uterus after delivery (*postpartum haemorrhage*) when labour has been prolonged.

*Surgical interference.*—Prolonged labour usually has to be treated by some form of surgical intervention. For example, forceps may have to be applied to the baby's head so that the attendant may apply traction to the baby. In some cases, an abdominal operation may be required to remove the baby from the uterus. This operation is called *Caesarian section,* not, as is commonly believed because Julius Caesar was born by this method. It is so named because of an ancient Roman law (Lex Caesarea) which said that a woman dying at or about the time of normal labour was to be delivered of her baby by an incision through the abdominal wall and uterus. The surgical interference in prolonged labour carries surgical and anaesthetic risks for the mother.

### Fetal

*Fetal distress.*—In prolonged labour, the placenta has to work under increasingly difficult conditions because of the uterine contractions. Because of this, the fetus may develop "respiratory failure" characterized by a rise in $Pco_2$ and a fall in $Po_2$ in fetal blood. It may be aggravated if there is a partial separation of the placenta or if the cord becomes compressed by the head. In severe cases it may cause permanent brain damage or death.

Fetal distress manifests itself during labour by phenomena whose mechanisms are little understood. There is increased activity of the fetal alimentary tract so that its green fluid contents (***meconium,*** Gr. ***mekonion*** = poppy juice) are discharged into the amniotic sac. The green staining of amniotic fluid is a sign of fetal distress. Another sign is an alteration in fetal heart rate. The normal fetal heart rate is about 120-160 beats/min. Serious fetal distress usually causes slowing of the heart rate.

*Infection.*—Prolonged labour carries the risk of infection to the child as organisms have more chance to spread up the genital tract.

*Surgical hazards.*—The operative procedures which are often necessary to extricate the child may damage it, e.g. by compression of the head by forceps. The anaesthetic and pain-relieving drugs required to treat the mother may so depress the fetal brain that the infant may fail to breathe after delivery.

## Causes

Prolonged labour may result from a *disproportion* between the diameter of the maternal birth canal and the diameter of the fetal part being presented to it or from a failure of uterine muscle to produce strong and properly co-ordinated contractions.

Thus a prolonged labour is likely if the fetus is very big or if the baby is presented to the pelvis in an unusual position (*malpresentation*). A similar prolongation may occur if the maternal bony pelvis is narrowed or deformed or if the pelvic cavity is obstructed by an abnormal tissue mass or by a full bladder or rectum.

The uterine muscle may fail to achieve expulsion of the fetus because of weakness of the contractions or because of a lack of co-ordination between muscle action in the upper and lower parts of the uterus. The physiological basis of these muscle failures is not well understood.

### INVESTIGATIONS

Prolonged labour may endanger fetal life. When it does so, measures need to be taken rapidly to deliver the child. For this reason, investigations which can provide rapid and clear information on the presence or absence of fetal distress are desirable. A number are available which include:

**Fetal heart monitoring.**—This provides a continuous record of fetal heart rate. The counter may be triggered by a microphone device which is activated by fetal heart sounds or by an electrode which picks up the fetal ECG.

**Intra-uterine pressure monitoring.**—A record of intra-uterine pressure may be obtained from a small pressure transducer inserted into the amniotic fluid via the vagina. It provides a useful indication of the strength and frequency of uterine contractions.

**Fetal blood monitoring.**—Samples of blood taken from the fetal scalp as it advances down the birth canal can be analysed for blood gas tensions and pH.

Though these and other specialized techniques may give early warning of fetal distress, they require expensive apparatus and, what is even more important, staff adequately skilled in their use.

### TREATMENT

The treatment of prolonged labour obviously depends on the cause but there are several general points.

The mother's general condition must be closely watched for signs of dehydration and ketosis. If they become severe, intravenous glucose solutions may correct them. Rest and relief of pain become important if exhaustion is to be avoided. The strong sedative and analgesic drug pethidine, is useful for this. If the membranes have ruptured, antibiotics should be given to prevent infection. If it is not too late, an attempt may be made to correct a malpresentation if one is present. If the prolongation is due to the weakness of uterine contractions, drugs such as synthetic oxytocin may be used to stimulate the uterine muscle.

If labour continues without success despite these measures and signs of fetal or maternal distress occur, delivery of the baby should be assisted either by applying traction to the baby's head or by delivering the baby by Caesarian section.

# 100. PROBLEMS IN ADAPTATION TO EXTRA-UTERINE LIFE

BIRTH AND the subsequent adaptation to the outside world pose formidable problems for the newborn child, especially if he is born prematurely. Thus the death rate is much greater about the time of birth than it is throughout pregnancy or throughout the first few decades of life. This chapter deals with some of the disorders which prevent the child from adapting smoothly from intra- to extra-uterine life.

## FAILURE TO BREATHE

The failure of breathing to commence or be maintained adequately is the commonest cause of death in the newborn. It is called *asphyxia neonatorum.* It accounts for about one-third of all deaths in this period. Though asphyxia means literally no pulse (Gr. *a* = negative, *sphyxiz* = pulse) it is used to describe the whole series of symptoms and signs which occur when a person stops breathing. Infants who fail to breathe or stop breathing shortly after starting, become pale and flabby as muscle tone disappears and circulatory failure develops.

Many factors occurring during labour may contribute to this type of death. The brain may be injured by compression and the brain tissue damaged by lack of oxygen. The respiratory centre may be depressed by drugs given to the mother to relieve pain. If the child is born prematurely, its neuromuscular systems are weak and poorly developed. The respiratory tract may be blocked by amniotic fluid and meconium. Lack of pulmonary surfactant may make the lungs difficult to inflate.

Treatment should be aimed mainly at prevention. If the newborn fails to breathe, its respiratory tract should be cleared of fluid either by drainage in the head down position or by suction through a tube inserted through the mouth. The respiratory centre may be stimulated by applying sensory stimuli to the skin or injecting a respiratory stimulant drug into the umbilical vein. Once the child starts to breathe it may be given oxygen to breathe via a face mask.

## CONGENITAL MALFORMATION

The term "congenital" (L. *congenitus* = born together) describes a condition with which an individual is born. Though a fetus may live with a severe malformation in the protected environment of the womb, it may not be able to survive when it has to face the outside world. Congenital malformations account for about one-fifth of deaths in the period about birth (perinatal period).

Congenital abnormalities can take many forms. The ureters may not be patent so that the baby cannot eliminate urine. The gut may be blocked by a septum so that there is intestinal obstruction. The anterior abdominal wall may be absent so that the peritoneum is open to the exterior. The skull vault and brain may have failed to develop (*anencephaly*) to produce an "anencephalic monster". The lower part of the spinal cord may open out on to the skin of the back (*spina bifida*). Obstruction to the flow of cerebrospinal fuid may cause hydrocephaly. In the case of twins, they may be joined together (*Siamese twins*) in such a way that they have a common circulatory system and cannot be separated. In many of these cases the malformation does not seriously interfere with intra-uterine life but is not compatible with extra-uterine life and the infants die shortly after birth.

These malformations are the result of genetic abnormality or the effects of certain drugs and infections in early pregnancy.

In severe cases, there may be little useful treatment. The maintenance of life by medical skill in a severely handicapped child is a problem in ethics. Less severe problems, e.g. some of the congenital disorders affecting the heart, are amenable to surgery.

## BIRTH INJURIES

Mechanical damage to the infant during delivery accounts for about a tenth of all deaths in the perinatal period and a considerable amount of residual incapacity.

As with congenital malformation it may take many forms. Compression of the head with forceps or other forces may tear a blood vessel in the brain and cause a fatal cerebral haemorrhage. Damage to the pyramidal motor system may cause a bilateral spastic paralysis and the child, if he survives, becomes a *"spastic"*. The abdominal contents may also sustain damage and a fatal haemorrhage can occur from a torn liver into the peritoneal cavity.

Many injuries are not fatal but cause troublesome disability. Traction on the arm may tear the brachial plexus to give a paralysed arm. The forceps blade may damage the facial nerve to cause facial paralysis. Limb bones may be fractured.

As with congenital abnormalities, the best treatment for birth injuries is prevention if at all possible; there is no effective way to repair the central nervous system.

## INFECTION

The newborn child, especially the premature newborn child, is susceptible to infection. There is poor active immunity and conditions in hospitals are such that micro-organisms are easily spread. The infection may enter by a variety of routes; the umbilicus, the lungs, the urinary tract, the alimentary tract, the eyes and the skin may be involved. Usually there is little fever and not much increase in the white cell count. The infant becomes apathetic, dehydrated and progressively more feeble. Vomiting and convulsions may be seen.

Most infections can now be prevented by good hygiene and dealt with by appropriate antibiotics if they occur. However, poor homeostasis in the newborn makes them less able to cope with infections and the associated bodily upsets than older children. As a result, infections account for about 5 per cent of deaths in the perinatal period.

## HAEMOLYTIC DISEASE

The destruction of fetal red cells by maternal immunoglobulins may result in the fetus being born dead, anaemic or jaundiced. If the child is born live but severely anaemic, death may occur later unless the child's blood is modified by an exchange transfusion.

## HAEMORRHAGE

*Haemorrhagic disease of the newborn* is the name given to the bleeding tendency which may occur in the first few days of life.

Bleeding, which may be severe, can occur from the umbilicus, blood may be coughed up from the lungs, vomited up from the stomach, passed in the urine from the kidney or passed in the faeces from the lower alimentary tract.

The tendency appears to be due to a prothrombin deficiency. This may be due to an inability of the immature liver to produce prothrombin in adequate amounts.

The tendency to bleed can be prevented if the mother is given an injection of synthetic vitamin K shortly before birth or if the baby is given a similar injection at birth. This is usually done routinely when labour is premature or if instruments such as forceps are used to assist delivery, so that bleeding of the newborn is no longer an important problem.

# 101. PROBLEMS IN BREAST FEEDING

WITH THE development of bottle feeding and hygiene, breast feeding has become a less important ingredient for successful reproduction. However, two types of problem can arise in this context. Firstly, the mother may wish to breast feed her child but finds that she cannot provide it with adequate milk (*failure of breast feeding*). Secondly, she may wish to feed the baby by bottle and require *suppression of lactation.*

## FAILURE OF BREAST FEEDING

### DEFINITION

Breast feeding may be considered a failure if the mother and child are not contented and the child fails to thrive.

### EFFECTS

When the baby is getting insufficient milk by breast feeding, it tends to fret and cry before and after every feed. There is increased restlessness and sleeplessness and the baby does not appear to be contented. The body weight falls or does not rise adequately and very small quantities of urine are passed as the child attempts to conserve water. The stools consist of slimy mucus, stained with bile.

### CAUSES

**Causes in the infant.**—If the child is weak or premature it may not have sufficient strength to obtain milk by suckling the breast. If the infant has a cleft palate or other congenital abnormality affecting the mouth and palate, it may not be able to generate the negative pressures in the mouth that are necessary for suckling. Some babies in which no obvious cause can be demonstrated are said to suckle badly because of laziness!

**Maternal causes.**—Many factors may influence the ability of the mother to provide her baby with adequate milk. If her nipples become sore and cracked, they may become too painful to permit adequate suckling. Milk production is best in a restful and pleasant environment and may fail if the mother finds the environment unpleasant and hostile. The explanation for this is not certain. It may be that emotional stress interferes with the reflex release of the hormone oxytocin which is responsible for milk ejection or with the release of prolactin which is necessary for milk formation. Another possibility is that sympathetic nervous activity and adrenaline causes vasoconstriction in mammary tissue so that circulating oxytocin cannot reach the myoepithelial cells. Poor breast development and general malnutrition are other possible causes of failure. Very rarely, failure of lactation may be due to pituitary damage which reduces the output of prolactin, the hormone necessary for milk formation.

### Investigations

To find out how much milk is actually being delivered, the baby can be weighed before and after feeds, the difference equalling the milk intake.

### Treatment

If the child is unable to suckle it must be fed by other means such as an intragastric tube. If the mother is producing insufficient milk, this may well be remedied. Encouragement and provision of a warm, pleasant and stable environment are useful first steps. She should be encouraged to drink fluids. Hormone therapy has been tried but does not seem to help much and is probably not justified.

If these measures still fail to provide adequate nutrition, breast feeding may be supplemented or replaced with bottle feeding.

## SUPPRESSION OF LACTATION

If the mother decides not to breast feed, lactation rarely presents a serious problem because the main stimulus to lactation comes from suckling of the nipples. Thus if the nipples are not suckled, the breasts usually do not long remain congested and painful. If they do, an attempt to suppress milk formation may be made by giving the mother a synthetic ovarian hormone which depresses the pituitary production of prolactin. The efficacy of such treatment is doubtful, and simple pain relief may be just as good. Bandaging and strapping of the breasts may also help by limiting nipple stimulation and milk production.

## RECENT LEADING ARTICLES


Abdominal decompression in pregnancy. *Brit. med. J.*, 1974, **2**, 238.
A birth control plan. *Lancet*, 1972, **1**, 675.
Abortion. *Lancet*, 1971, **2**, 646.
Active management of labour. *Brit. med. J.*, 1972, **4**, 126.
A human right. *Brit. med. J.*, 1974, **3**, 699.
A.I.D. *Lancet*, 1973, **1**, 755.
A time to be born. *Lancet*, 1974, **2**, 1183.
Blood pressure and the pill. *Brit. med. J.*, 1973, **1**, 693.
Choice of contraceptives. *Brit. med. J.*, 1974, **3**, 642.
Coagulation defects in the newborn. *Lancet*, 1971, **2**, 85.
Evaluating the I.U.D. *Lancet*, 1974, **1**, 394.
Fetal/maternal incompatibility. *Lancet*, 1972, **2**, 958.
Gonorrhoea of the pharynx. *Brit. med. J.*, 1974, **2**, 239.
Haemorrhage in the newborn. *Brit. med. J.*, 1971, **4**, 1.
Heartburn of pregnancy. *Brit. med. J.*, 1973, **2**, 378.
Management of neonatal jaundice. *Brit. med. J.*, 1974, **1**, 469.
Maternal deaths. *Lancet*, 1972, **2**, 370.
Measuring placental function. *Brit. med. J.*, 1972, **1**, 193.

Oral contraceptives containing only progestogens. *Brit. med. J.*, 1972, **3**, 190.
Overpopulation. *Lancet*, 1972, **1**, 781.
Post-coital contraception. *Lancet*, 1972, **2**, 314.
Pre-eclampsia and the kidney. *Brit. med. J.*, 1974, **1**, 468.
Premenstrual symptoms. *Brit. med. J.*, 1973, **1**, 689.
Progestogen only contraception. *Lancet*, 1971, **1**, 25.
Progress in population control. *Lancet*, 1972, **1**, 829.
Progress in pre-eclampsia. *Lancet*, 1973, **1**, 754.
Prostaglandins in abortion. *Lancet*, 1971, **2**, 536.
Stopping the pill. *Brit. med. J.*, 1974, **2**, 517.
The human predicament. *Lancet*, 1972, **2**, 1015.
The initiation of labour. *Lancet*, 1974, **1**, 124.
Towards a population policy. *Lancet*, 1971, **1**, 1109.
Treatment of infertility. *Lancet*, 1970, **2**, 508.
Troubles with I.U.C.D.s. *Brit. med. J.*, 1973, **2**, 2.
Vasectomy. *Lancet*, 1972, **2**, 1074.

# INDEX

# INDEX

**Abortion,** criminal, 557
inevitable, 557
spontaneous, 556
surgical, 553
**Absorption,** tests, 201
**Acceleration,** adverse effects, 71
**Accident,** cerebrovascular, 257
**Acetone,** excess due to cellular lack of glucose, 529
**Acholuric** jaundice, 365
**Acid-base** status, graphical representation, 450
measurement, 445
**Acidity,** gastric, measurement, 230
reduction, 231
**Acidosis,** 449
compensation, 450-451
due to acute renal failure, 492
due to chronic renal failure, 484
due to diabetes mellitus, 529
due to diarrhoea, 218
due to intestinal obstruction, 226
hyperchloraemic, 453
lactic, 114, 530
metabolic and respiratory, 449-453
renal tubular, 498
**Acromegaly,** 525
**ACTH** stimulation test in adrenal failure, 513
in adrenal hyperactivity, 516
**Acupuncture,** 284
**Adaptation** to heat, 41, 48
**ADH** secretion, deficiency, 523
"inappropriate" (excessive), 520
**Adrenal** cortex, deficient secretion, 512
excessive secretion, 514
**Adrenal** failure, iatrogenic, 513
**Adrenal** medulla, excessive secretion, 518
**Adrenaline,** effects of excess, 518
in treatment of hypoglycaemia, 535
**Adrenogenital** syndrome, 514
**Aerophagy,** 216
**Age** and disease patterns, 13
**Ageing,** 25
**Agglutination,** 394
**Agglutinins,** 394
cold, 407
**Agnosia,** 289
**Agranulocytosis,** 375
due to hypersensitivity, 404
**AID,** AIH, 549
**Air** embolism during decompression, 61
**Akinesia,** 263
**Aldosterone,** 514
**Alkalosis,** 442
compensation, 442
due to potassium ion depletion, 435
due to pyloric obstruction, 226
due to vomiting, 212
metabolic and respiratory, 442-445
graphical representation, 450
**Allergy,** 400
**Allogenic** homograft, 411
**Alpha** adrenoceptor blockade in diagnosis and treatment of phaeochromocytoma, 519
**Amaurosis,** 321
**Amblyopia,** 321
ex anopsia, 331
**Amelia,** 17
**Amenorrhoea** due to excess ovarian hormones, 540
in starvation, 198
**Amino-aciduria,** 498
**Amniotic** fluid examination, 19
**Anabolic** phase, trauma, 82
**Anaemia,** 357
classification, 361
due to bone marrow depression, 367
haemolytic, 365
due to hypersensitivity, 404
in chronic renal failure, 487
in pregnancy, 555
macrocytic, 363
microcytic, 360
normocytic, 365
pernicious, 363, 407
**Anaesthesia,** dissociated, 273
**Anaphylactic** reaction, 400
shock, 401
**Androgen,** adrenal, deficiency, 512
excess, 515
**Androgen,** testicular, deficiency, 536
excess, 538
**Anencephaly,** 569
**Angina** pectoris, 134

**Ankle** jerk, slowed in hypothyroidism, 504
**Ankylosing** spondylitis, 344
**Ankylosis,** 343
**Anorexia** nervosa, 199
**Anosmia,** 337
**Anoxia,** 163
**Antibody,** 393
**Anticoagulant** therapy, 388
**Antigen,** Australia, 414
**Antigen-antibody** reaction, 394
**Antigens,** 393
lymphocyte, 412
transplantation, 411
**Antitoxin,** 394
**Anxiety** state, 90
**Aphasia,** 289
**Apraxia,** 289
**Arrest,** cardiac, 131
**Arthritis,** 341
septic, 345
**Arthrodesis,** 345
**Arthus** phenomenon, 405
**Artificial** insemination, 549
**Artificial** ventilation, 191
causing respiratory alkalosis, 445
**Asphyxia** neonatorum, 568
**Asteriognosis,** 274
**Asthma,** bronchial, 162, 172
cardiac, 123
due to hypersensitivity, 400
**Astigmatism,** 330
**Ataxia,** cerebellar, 268
sensory, 273
**Atelectasis,** 177
**Atheroma,** in hypertension, 103
**Athetosis,** 264
**Audiometry,** 313
**Australia** antigen, 414
**Autograft,** 411
**Autoimmunity,** 407
**Automatic** micturition, 474
**Automatism,** 290
**Axillary** hair, loss in adrenal deficiency, 512
**Axis** deviation, 147
**Axis,** electrical, of heart, 146

**B-cells,** 395
failure, 397
**$B_{12}$** absorption, 201

**Babinski** response, 256
**Barrier** nursing, 376
**Basal** metabolic rate in thyroid disease, 505
in excessive adrenaline secretion, 518
**Bed** sore, 137
**Bence-Jones** protein, 409
**Beta** adrenoceptor blockade during removal of phaeochromocytoma, 519
in treatment of hyperthyroidism, 508
making asthma worse, 187
stimulation, 187
**Beta** cell (pancreas), deficient secretion, 528
excessive secretion, 533
**Bicarbonate,** regulation, failure, 484
standard, 448
**Binocular** vision, 334
**Birth** injuries, 569
**Bladder,** bashful, 475
uninhibited neurogenic, 474
**Bleeding** time, 382
**Blind** loop, malabsorption, 200
**Blindness,** 321
colour, 335
night, 323
**Block,** alveolar-capillary, 178
heart, 127, 146
**Blood** colloid osmotic pressure decrease, 463
donation, 414
pressure, high, 97
transfusion, 414
**Blue** bloaters, 166
**Body** temperature, and ovulation, 538, 552
measurement, 45
**Bone** function disturbance, 347
**Bradycardia,** sinus, 134
**Breast** feeding, failure, 571
**Bronchiectasis,** 174
**Bronchitis,** chronic, 173
**Bruising,** due to glucocorticoidid excess, 515
**Bruit,** in anaemia, 359

**Caesarean** section, 566
**Calcitonin,** 508
in Paget's disease of bone, 349
**Calcium** distribution and ageing, 28

**Calcium** (*cont.*)
ion depletion, 455
level in blood, effect of plasma protein, 456
retention, 457
**Calculus,** renal, 348, 458
cystine, xanthine, 498
**Caloric** test, 318
**Cancer,** due to radiation, 87
**Cancer,** radiotherapy, 86
**Capillary,** raised, hydrostatic pressure. 462
resistance, 382
**Carbon** dioxide pressure measurement, 447
retention (hypercapnia), 165
**Carbon** monoxide, poisoning, 370
transfer by lung, 181
**Carbonic** anhydrase inhibitors, causing acidosis, 453
in glaucoma, 326
**Carboxyhaemoglobin,** 370
**Carcinoid** tumours, 541
**Cardiac arrest,** 131
due to metabolic acidosis, 451
due to potassium ion retention, 438
**Cardiac** massage, 132
**Cardiac** resuscitation and point of death, 35
**Cardiac** tamponade, 134
**Cardiac** vector, 146
**Carpopedal** spasm, 455
**Castration,** 552
**Catabolic** phase of trauma, 81
**Cataract,** 322
**Catecholamines,** 518
**Catheterization,** cardiac, 127
central venous, 118
urinary tract, 467
**Central** venous pressure measurement, 118
**Cerebellar** system disturbance, 267
**Cerebrospinal** fluid disorders, 300
**Cerebrovascular** accident, 257
**Cholesterol,** raised in hypothyroidism, 505
raised in nephrotic syndrome, 496
**Chorea,** Huntington's, 265
Sydenham's, 265
**Choreiform** movements, 264
**Chronic** bronchitis, 173
**Circulatory** failure, 110
central, 121
**Circulatory** failure (*cont.*)
local, 133
peripheral, 110
due to mineralocorticoid deficiency, 512
in chronic renal failure, 486-7
in sodium depletion, 423
**Claudication,** intermittent, 136
**Clearance** rates, renal, 488
**Climacteric,** 552
**Clonus,** 255
**Clotting** time, 382
**Coagulation,** disseminated intravascular, 386
time, 382
**Coeliac** disease, 200
**Coil,** contraceptive, 552
**Coitus,** difficulties, 547
interruptus, for contraception, 552
for seminal analysis, 548
**Cold,** agglutinins, 407
common, 157
stress, 48
**Colic,** renal, 511
**Colitis,** ulcerative, 408
**Collagen** deficiency due to excess glucocorticoids, 515
**Collapse,** pulmonary, 177
**Colour** blindness, 335
**Coma,** 290
hyperosmolar, 529
hypoglycaemic, 534
hypopituitary, 521
hypothyroid, 503
keto-acidotic, 529
**Compensation,** 6-10
inappropriate, 10
**Complement,** 395
**Compliance** of lungs, 182
**Concentration-dilution** tests, 488
**Condom,** 552
**Congenital** abnormalities, 15, 568
**Congestive** heart failure, 121, 124
**Coning** of medulla, 305
**Conn's** syndrome, 514-5
**Consciousness,** depression of, 289
in hypothyroidism, 503
**Consolidation,** pulmonary, 177
**Constipation,** 221
**Contact** dermatitis, 406
**Contact** lens, 330
**Contraception,** 551
**Contraceptive** therapy, oral, 540, 551

**Contracture,** in lower motor neurone disease, 248
ischaemic, 137
**Convulsions,** 295
**Coombs'** test, 368
**Corneal** grafting, 325
**Cor** pulmonale, 164
**Cortisol,** 514
**Coryza,** 157
**Counter-irritation,** 284
**Cover** test, 333
**Creatinine** clearance, 488
**Cretinism,** 503-4
**Critical** closing pressure, 114
**Cromoglycate** for hypersensitivity, 402
**Curettage,** diagnostic, 548
for inevitable abortion, 558
**Cushing's** syndrome, 514-5
**Cyanosis,** 163
**Cystine,** in urine, 498
**Cystitis,** 467
**Cystogram,** micturating, 477
**Cystometrogram,** 477, 478
**Cystometry** 477,
**Cystoscope,** 468
**Cytolytic** reaction, 403

**Dead** space, 170
**Deafness,** 311
conductive, 311
due to myxoedema, 503
sensorineural, 312
traumatic, 312
**Death,** 34
diagnosis of, 35
legal, 36
medical, 35
physiological, 34
**Decompensation,** 7
**Decompression** sickness, 61
**Degenerative** diseases, 15
**Dehydration,** in chronic renal failure, 485
in prolonged labour, 565
in sodium ion depletion, 423
in water deprivation, 75
**Delirium,** 290
**Demyelinating** diseases, 258
**Denervation** hypersensitivity, arterioles, 139
skeletal muscle, 249
**Depression** of the nervous system, 288
**Dermatitis,** contact, 406
**Detoxication,** failure of, 232
**Deuteranopia,** 335
**Dexamethasone** suppression test, 516
**Diabetes** insipidus, 523
nephrogenic, 498
**Diabetes** mellitus, 528
in obese patients, 533
juvenile and maturity onset, 530
late complications, 530
**Dialysis,** for renal failure, 490
**Diaphragm,** contraceptive, 552
**Diarrhoea,** 218
osmotic, 198
spurious, 221
**Diodrast** excretion, indicating renal function, 488
**Diplopia,** 332
**Disaccharidase** deficiency, 200
**Disc** lesions, intervertebral, 352
**Discrimination,** two-point, 274
**Dislocation,** 343
**Disproportion,** causing prolonged labour, 566
**Diuresis,** osmotic, in chronic renal failure, 481
in diabetes mellitus, 529
**Diuretic** phase, acute renal failure, 492
**Diurnal** variation in cortisol secretion, 513
**Diverticuli,** bladder, 470
**Dopamine,** deficiency and excess, 265
**Double** vision, 332
**Drowning,** fresh water, causing haemolysis, 438
**Drugs** and the elderly, 32
in chronic renal failure, 483
in liver failure, 233
**Dumping** syndrome, 231
**Dwarfing,** due to cretinism, 504
due to pituitary disease, 521-2
**Dynamic** lung volumes, 180
**Dysdiadochokinesia,** 268
**Dysmetria,** 268
**Dyspareunia,** 547
**Dyspnoea,** 162

**Eclampsia,** 554
**Eclamptic** fits, 297
**Eclipse** burn, 323
**Effusion,** pleural, 177
**Elastic** tissue and ageing, 28
**Elderly,** disease in, 30

**Electrocardiography,** 141
low voltage in hypothyroidism, 504
potassium disturbances, 433
standard leads, 143
unipolar leads, 149
**Electroencephalography,** 292
**Electrolyte,** composition of body fluids, 420
disturbances, 419, 421
units, 419
**Electromyography,** 251
**Embolism,** 137, 384
air, 137, 557
fat, 137
pulmonary, 384
systemic, 385
**Emotional** stress, 90
**Emphysema,** 174
**Encephalopathy,** hypertensive, 103
**Endometrium,** sampling, 547
**Enema,** 223
**Enuresis,** 474
**Environment,** cooling power, 46
**Epilepsy,** 295
functional (idiopathic), 296
grand mal, 298
petit mal, 298
secondary, 296
**Epiphysis,** delayed calcification in cretinism, 504
early calcification in childhood hyperthyroidism, 507
**Eunuch,** 536
**Exophthalmos,** 507, 508
**Extramedullary** haemopoiesis, 367
**Extrapyramidal** system disturbances, 262
**Extrasystoles,** 145

**Failure,** cardiac, 121
due to hypertension, 99
due to potassium ion depletion, 433
high cardiac output, 126
liver, 232
of physiological function, 6-10
peripheral circulatory, 110
renal, acute, 492
chronic, 480
respiratory, 161
ventilatory, 170
ventricular, 99
**Fainting** due to upright posture, 70
**Fasciculation,** 248
**Fasting,** in diagnosis of insulin-secreting tumour, 535
**Feminization,** due to adrenal hormones, 516
due to liver failure, 233
**Fetal** distress, 566
heart monitoring, 567
**Fibrillation,** atrial, 146
skeletal muscle, 248
ventricular, 146
**Fibrinogen,** deficiency, 381
radio-active, 387
**Fibrocystic** disease, lung effects, 174
malabsorption, 200
**Fibrosis,** pulmonary, 178
**Fibrositis,** 352
**Figlu** test, 364
**Fits,** 295
**Flaccidity,** muscle, 248
**Flatulence,** 216
**Flocculation,** barium, 201
**Flushes,** due to carcinoid tumours, 541
hot, 537, 552
**Focal** motor seizures, 295
**Folic** acid deficiency, causing macrocytic anaemia, 363
in pregnancy, 555
**Forced-expiratory** volume (FEV), 180
**Frequency,** urinary, 467
**Frigidity,** 547
**Frost-bite,** 51
**Frozen** shoulder, 344

**G.A.B.A.** deficiency, 265
**Gallop** rhythm, pre-systolic, 123
**Gangrene,** 136
**Geiger** counter, 87
**Genetic** counselling, 559
**Gigantism,** 525
**Glaucoma,** 323, 326
**Glomerular** failure (nephrotic syndrome), 495
**Glomerulonephritis,** causing chronic renal failure, 487
**Glucagon,** 535
**Glucocorticoid** deficiency, 512
excess, 514
iatrogenic, 516
primary and secondary, 516
**Gluconeogenesis,** 514

**Glucose** tolerance test, in diabetes mellitus, 531
in malabsorption, 201
**Glycosuria,** at routine examination, 531
due to diabetes mellitus, 529, 531
due to glucocorticoid excess, 514
renal, 498
**Goitre,** 504
**Gonadotrophins,** increased level, 538
**Gout,** 344
**Granulocyte** deficiency, 375
**Gravity,** decreased, 72
disorders caused by, 68
increased, 71
**Gynaecomastia,** 539

**Habit** spasms, 353
**Haemoglobin,** deficiency, 358
inactivation, 370
increased, 372
**Haemolysis,** due to fresh water drowning, 438
**Haemolytic** disease of the newborn, 365
**Haemophilia,** 381
**Haemorrhage,** from alimentary tract, 229
placental, 556
postpartum, 566
**Haemorrhagic** disorders, 379
disease of the newborn, 381, 570
**Haptens,** 393
**Hashimoto's** disease, 407
**Hay** fever, 401
**Headache,** 282
ocular, 328
**Hearing** disorders, 311
**Heartburn,** in pregnancy, 556
**Heat** cramps, 42
disorders caused by, 41
exhaustion, 42
stress, 41
stroke, 43
syncope, 42
**Hemianopia,** bitemporal, 520
**Hemiparesis,** 254
**Hemiplegia,** 254
**Hepato-lenticular** degeneration, 265
**Hercules,** infant, 538
**Hiccup,** 215
**Hirschsprung's** disease, 222
**Hirsutism,** due to adrenal androgens, 515
**Hoarseness,** due to laryngitis, 159
due to myxoedema, 503
**Homeostasis** and physical stress, 40
**Homeostatic** efficiency, 15
**Homograft,** 411
allogenic, 411
isogenic, 411
**Human** growth hormone, assay, 526-527
in treatment of pituitary dwarfing, 523
**Humidity** and heat stress, 44
**Humidity,** relative, 46
**Hunger,** 206
**Huntington's** chorea, 265
**Hydramnios,** 554
**Hydrocephalus,** 302
**Hydrogen** ion concentration, control, 442
measurement, 445
normal, 442
depletion, 442
retention, 449
**Hydronephrosis,** 470, 471
**Hydro-ureter,** 470
**5-Hydroxytryptamine,** secreted by carcinoid tumours, 541
**Hygrometer,** 46
**Hyperactivity** in the nervous system, 295
**Hyperacusis,** 316
**Hyperalgesia,** 278
**Hyperbaric** oxygen, 371
**Hypercalcaemia,** 457
treated by glucocorticoids, 459
**Hypercapnia,** 165
**Hyperdynamic** circulation, in anaemia, 359
in hyperthyroidism, 506
**Hyperemesis** gravidarum, 555
**Hyperfertility,** 549
**Hyperglycaemia,** due to diabetes mellitus, 529
due to glucocorticoid excess, 514
**Hyperkalaemia,** 438
**Hypermetropia,** 327
**Hyperosmolar** coma, 529
**Hyperparathyroidism,** 510
secondary to renal disease, 483
"tertiary", 483
**Hypersensitivity,** 400
immediate, 400
type 1 (anaphylactic reaction), 400
type 2 (cytolytic reaction), 403
type 3 (serum sickness reaction), 404
type 4 (sensitized lymphocyte reaction), 405
**Hypertension,** 97
portal, 234

**Hypertension** (*cont.*)
pulmonary, 164
renal, 487
**Hyperthyroidism,** 506
beta adrenoceptor blockade in treatment, 508
**Hypertonicity,** intracellular, 428
**Hypertrophy,** bladder, 469
compensatory renal, 480
left ventricular, 97
right ventricular, 164
ventricular, ECG signs, 150
**Hyperventilation,** causing respiratory alkalosis, 445
**Hypocapnia,** 170
**Hypogammaglobulinaemia,** 397
**Hypoglycaemia,** 534
**Hypoglycaemic** drugs, oral, 533
**Hypokalaemia,** 433
**Hypoparathyroidism,** 509
**Hypothalamus,** possible endocrine disturbances, 541
**Hypothermia,** 52
in hypothyroidism, 503
**Hypothyroidism,** 503
primary and secondary, 505
**Hypoxia,** 163
tissue, in anaemia, 357
**Hysteria,** 92

**Iatrogenic** adrenal failure, 513
**IgA,** IgG, IgM, 394
**Immaturity,** 22
**Immersion** foot, 49
**Immune** response, normal, 393
excessive, to body tissues, 407
excessive, to foreign antigens, 400
**Immunity,** 393
cell-mediated, 395
impaired, 398
deficient, 397
humoral, 393
impaired, 397
**Immunoglobulins,** 393
abnormal, 409
**Immunological** surveillance, 395
tolerance, 397, 413
**Impotence,** 547
**Incontinence,** dribbling, 474, 475
urinary, 474
**Infarction,** myocardial, 133
ECG changes, 148
pulmonary, 385
**Infection,** upper respiratory tract, 157
urinary tract, 467
**Infertility,** 545
female, 545
male, 546
**Inflammation,** 80
**Insemination,** artificial, 549
**Insomnia,** 79
**Insulin,** excessive, 533
plus glucose causing a fall in extracellular potassium, 437, 440, 532
reactions, 533
test for vagotomy, 230
therapy, 532
**Insulin-secreting** tumour, 534
**Intra-uterine** device, contraceptive, 552
**Inulin** clearance, 488
**Iodine** deficiency, causing hypothyroidism, 505
**Iodine,** radioactive, diagnostic use, 505
therapeutic use, 508
**Iodine** uptake in thyroid disease, 505
**Ion** exchange resins, 440
**Ionizing** radiation, 85
**Iron** deficiency, causing microcytic anaemia, 360
in pregnancy, 555
**Ischaemia,** 133
gut, 136
**Isogenic** homograft, 411
**Isosthenuria,** 485
due to potassium ion depletion, 433
**IUD,** 552

**Jaundice,** 241
acholuric, 365
**Joint** function disturbance, 341

**Katathermometer,** 46
**Kernig's** sign, 306
**Keto-acidosis** in diabetes mellitus, 529
treatment, 532
in hyperemesis gravidarum, 555
in prolonged labour, 565
in starvation, 197
**Kidney,** artificial, 491
transplantation, 491
**Klinefelter's** syndrome, 536
**Knee-jerk,** pendular, 269
**Kussmaul** breathing, 451
**Kyphosis,** 348

**Labour,** early, 561
induction, 565
late, 563
prolonged, 565
stages, 565
**Lactation,** suppression, 572
**Lactic** acidosis, 114, 530
**Laplace,** law of, bladder function, 470, 471
cardiac function, 99
**Laryngismus** stridulus, 455
**Laryngitis,** 159
**LATS,** 507
**Leads,** E.C.G., 143, 149
**Lens** extraction, 325
**Levodopa,** 265
**Libido,** 536
**Liver** function, impaired, 232
**Long-acting** thyroid stimulator, 507
**Loudness** recruitment, 316
**Lower** motor neurone disturbance, 247
**Lumbago,** 352
**Lupus** erythematosus, systemic, 408
**Lymphocytes,** antigens, 412
B-cells, 395
sensitized, 395
reaction, 405
T-cells, 395
**Lysins,** 394
**Lysis,** 394

**Macroglobulin** (IgM), 394
**Magnesium** ion depletion, 441
retention, 441
**Malabsorption,** 198
causing secondary hyperpara-thyroidism, 511
**Malingering,** 92
**Malpresentation,** causing prolonged labour, 566
**Masculinization,** due to adrenal hormones, 516
due to ovarian hormones, 540
**Mast** cells, degranulation, 400
**Masturbation,** 548
**Mean** cell volume, 362
corpuscular haemoglobin concentration, 362
**Meconium,** and fetal distress, 566
**Megacolon,** 222
**Megaloblastic** bone marrow, 364
**Melaena,** 219
**Menarche** delayed, 537
**Ménière's** disease, 320
**Menopause,** 537
**Methaemoglobin,** 370
**Methyl** alcohol causing metabolic acidosis, 453
**Microangiopathy,** 530
**Micturition,** automatic, 474
neurological basis, 473, 476
neurological disturbances, 473
**Migraine,** 282
**Milk-alkali** syndrome, 231
**Mineralocorticoid** deficiency, 512
excess, 515
**Moon** face, 514
**Morning** sickness, 555
**Mountain** sickness, 65
**Mucoviscidosis,** lung effects, 174
malabsorption, 200
**Multiple** myeloma, 409
**Murmurs,** 151
in anaemia, 359
**Muscae** volitantes, 336
**Muscle,** strain, 351
tone decreased, 248, 269
tone increased, clasp-knife, 255
tone increased, cog wheel, lead pipe, 263
weakness, 248, 352
potassium ion depletion, 433
potassium ion retention, 438
**Muscular** dystrophy, 353
**Myasthenia** gravis, 352
**Myelogram,** 287
**Myeloma,** multiple, 409
**Myoclonic** jerks, 295
**Myopia,** 327
**Myxoedema,** 503

**Nausea,** 211
**Nephrocalcinosis,** 458
**Nephron** failure, gradual, 480
sudden, 492
**Nephrotic** syndrome, 495
**Nerve,** conduction velocity, 251
degeneration, 249
regeneration, 249
**Neuritis,** brachial, 352
cervical, 352
peripheral, 250, 274
**Neurogenic** bladder, uninhibited, 474
**Neuromuscular** disorders, 351

**Neutropenia,** 375
**Newborn,** disease in, 21
**Night** starts, 295
**Nitrogen** narcosis, 59
washout test, 181
**Nocturia,** 485
**Noradrenaline,** effects of excess, 518
**Normal** values, 4
**Nutrition,** excessive, 204
inadequate, 73, 197
inadequate in childhood, 24
**Nystagmus,** cerebellar, 269
vestibular, 318

**Obesity,** 204
**Obstruction,** alimentary tract, 224
urinary tract, 469
acute, 470
**Obstructive** airways disease, 171
**Occlusion** (for squint), 334
**Occult** bleeding, 362
**Oedema,** 460
due to protein deficiency, 496
in congestive heart failure, 122, 124
pulmonary, 123
**Oestrogen** deficiency, 537
excess, 539
**Opsonins,** 394
**Orgasm,** female, not essential for fertilization, 547
**Orthopnoea,** 163
**Orthoptic** training, 334
**Osmolarity,** failure of regulation in chronic renal failure, 485
**Osmotic** fragility, 365
**Osteitis** deformans, 349
**Osteoarthritis,** 344
**Osteoarthrosis,** 344
**Osteolytic** change, 343
**Osteomalacia,** 349
in chronic renal failure, 483, 541
treatment, 457
**Osteophytes,** 344
**Osteoporosis,** 349
treatment, 457
**Osteosclerotic** change, 343
**Otosclerosis,** 312
**Ovarian** secretion, deficient, 537
excessive, 539
**Over-population,** 549
**Ovulation,** evidence for, 547
induction of, 548
**Oxygen** therapy, controlled, 191
toxicity, 59
**Oxytocin,** for induction of labour, 565
in prolonged labour, 567

**P** wave of ECG, 141
**Paget's** disease of bone, 349
**Pain,** 275
cutaneous, 277
deep, 280
from alimentary tract, 229
from muscle strain, 351
from peripheral nerve damage, 272
joint, 342
visceral, 280
**Palpitations,** due to adrenaline, 518
in anaemia, 359
in hyperthyroidism, 507
**Pancreatitis,** 238
**Papilloedema,** 303
**PAH** (Para-amino hippuric acid) clearance, 488
**Paraesthesia,** 272
in tetany, 509
**Paralysis** agitans, 265
familial periodic, 437
**Paralytic** ileus, 226
due to potassium ion depletion, 433
**Parathyroid** gland, deficient secretion, 509
excessive secretion, 510
**Parkinsonism,** 262
**Parosmia,** 337
**PBI,** 505
**$Pco_2$** measurement, 447
**Peak** flow rate (PFR), 180
**Perforation,** of alimentary tract, 229
**Perimetry,** 325
**Peritoneal** dialysis, 491
**Petechiae,** 379
**pH,** measurement, 445
normal, 442
**Phaeochromocytoma,** 518
**Phantom** limb pain, 279
**Phlebography,** 387
**Phocomelia,** 17
**Phosphate** retention, effects, 482
**Phosphaturia,** renal, 498
**Pigmentation,** due to adrenal deficiency, 512
**Pink** puffers, 166

**Pituitary,** deficient anterior lobe secretion, 520
deficient posterior lobe secretion, 523
excessive growth hormone secretion, 525
tumours, effects, 520
**Placenta** praevia, 556
**Plasma** cells, 393
**Plasma** proteins, deficiency in nephrotic syndrome, 495
effect on calcium ion level, 456
**Platelet** deficiency, 380
**Pneumothorax,** 177
tension, 178
**Polycythaemia,** 372
secondary, 164
**Polyhydramnios,** 554
**Polyuria,** due to hypercalcaemia, 458
due to potassium ion depletion, 433
in diabetes insipidus, 523
in early chronic renal failure, 485
**Postmaturity,** 563
**Postural** drainage, 188
**Postural** hypotension, 68, 117
**Potassium** ion depletion, 433
ECG changes, 433
elimination in renal failure, 482
retention, 438
ECG changes, 438
in acute renal failure, 492
**Potency,** 536
**Precipitins,** 395
**Precocious** puberty, female, 539
male, 516, 538
**Pre-eclampsia,** 107, 554
**Prematurity,** 22, 561
**Presbycusis,** 313
**Presbyopia,** 328
**Pressure,** high barometric, disorders caused by, 58
low barometric, disorders caused by, 64
**Proctalgia** fugax, 220
**Progestogen** deficiency, 537
excess, 539
**Prolapse** of uterus, 550
**Prostaglandins,** for induction of labour, 565
**Protanopia,** 335
**Protein-bound** iodine, 505
**Prothrombin** deficiency, 381
time, 382
**Pseudohermaphroditism,** 516
**Psychological** stress, 90
**Psychosomatic** disease, 91
**Puberty,** precocious, female, 539
male, 516, 538
**Pubic** hair, loss in adrenal deficiency, 512
**Pulmonary** oedema, 123
on mountains, 66
**Pulmonary** wedge pressure, 127
**Purpura,** 379
senile, 380
vascular, 380
**Pyelonephritis,** 467
**Pyramidal** system disturbance, 253
**Pyrexia,** 42
**Pyrogens,** 45

**Q** waves in myocardial infarction, 149
**QRS** complex, 141
various forms, 142

**Retrolental** fibroplasia, 322
**Rhesus** factor, haemolytic disease due to, 365
**Rheumatoid** arthritis, 344
autoantibodies, 408
**Rhinitis,** 157
**Rickets,** 348
renal, secondary hyperparathyroidism, 483
abnormal vitamin D metabolism, 541
**Rigidity,** muscle, clasp-knife, 255
cog wheel, 263
lead pipe, 263
**Rigors,** 375
**Rinne** test, 313
**Romberg's** sign, 273
**Runt** disease, 397

**Safe** period, 552
**Satiety,** 206
**Schilling** test ($B_{12}$ absorption), 201, 364
**Sciatica,** 352
**Seminal** fluid analysis, 548
**Sensitized** lymphocyte reaction, 405
**Sensory** system disturbance, 271
**Septicaemia,** 375
**Serum,** antilymphocyte, 413
sickness, 404

**Shock,** 110
anaphylactic, 401
irreversible, 115
spinal, 286
**Shunting,** 169
**Siamese** twins, 569
**Simmond's** disease, 521
**Sinus** arrhythmia, 145
rhythm, 144
tachycardia, 146
**Sinusitis,** 158
**Skin** tests for hypersensitivity, 401
**Sleep,** lack of, 78
**Slit** lamp microscopy, 325
**Smell** disturbance, 337
**Snuff,** vasopressin, 525
**Sodium,** bicarbonate ingestion causing alkalosis, 443
ion depletion, 423
regulation, failure in chronic renal failure, 486
retention, 426
**Solubility** product, calcium and phosphate, 482
**Spasms,** flexion, 257
habit, 353
**Spastic** children, 258, 569
**Spasticity,** 255
**Sperm** count, 548
**Spina** bifida, 569
**Spinal** cord transection, 286
**Squinting,** 331
**St. Vitus** dance, 265
**Standard** bicarbonate, 448
**Starling** curve, cardiac dilatation and failure, 99
cardiac hypertrophy, 97
circulatory failure, 110, 113
and digitalis, 129
**Starvation,** 197
**Static** lung volumes, 180
**Steatorrhoea,** 198
**Steroids,** anabolic, 539
**Strabismus,** 331
concomitant, 332
paralytic, 333
**Strangury,** 467
**Stress,** disorders caused by, 39
emotional, 90
incontinence, 477, 550
physical, 39
and glucocorticoids, 512-514
and pituitary deficiency, 521
**Striae,** due to glucocorticoid excess, 515
**Stroke,** 257
**Stupor,** 290
**Suffocation,** 163
**Sulphaemoglobin,** 370
**Surfactant** deficiency in newborn, 22, 561
**Surgery** in the elderly, 33
**Sydenham's** chorea, 265
**Sympathectomy,** 139
**Sympathetic** ophthalmia, 324
**Sympathetic** overactivity due to hypoglycaemia, 534
**Sympathetic** release test, 138
**Syndrome,** definition, 514
**Systemic** lupus erythematosus, 408

**T-cells,** 395
failure, 398
**T** wave of ECG, 141
**Tachycardia,** sinus, 134, 146
**Taste** disturbances, 338
**Telangiectasia,** 380
**Temperature** of body core, 45
**Tendon** reflexes, exaggerated, 255
slowed in hypothyroidism, 504
**Tenesmus,** 220
**Tenosynovitis,** 352
**Teratogenicity,** 19
**Testicular** secretion, deficient, 536
excessive, 538
**Tests,** respiratory function, 179
**Tetany,** 455
due to alkalosis, 442
due to hypoparathyroidism, 509
rare in chronic renal failure, 483
**Thalamic** syndrome, 273
**Thrombocytopenia,** 380
due to hypersensitivity, 404
**Thrombosis,** 384
deep venous, 387
**Thyrocalcitonin,** 508
**Thyroid** gland, deficient secretion, 503
excessive secretion, 506
**Thyroid** stimulating hormone, 505
**Thyrotoxicosis,** 506
**Thyroxine,** 503, 505
**Tic** douloureux, 285
**Tics,** 353
**Tinnitus,** 315
**Tone,** *see* Muscle
**Tonometry,** 325
**Tonsillitis,** 159

**Trabeculae,** bladder, 469
**Tracheitis,** 159
**Trachoma,** 322
**Transfusion** reactions, 416
**Transplantation,** renal, 491
tissue, 411
**Trauma,** general effects, 80
**Tremor,** cerebellar, 268
in hyperthyroidism, 507
intention, 268
Parkinsonism, 264
**Trench** foot, 50
**Triiodothyronine,** 503, 506
**Tritanopia,** 335
**TSH** in thyroid disease, 505
**Tubal** ligation, 552
**Tuberculin** reaction, 406
**Tubular** failure, renal, 498
**Tumours,** endocrine secretion by, 541
**Tunnel** vision, 323
**Turner's** syndrome, 538
**Tympanoplasty,** 315

**Ulcer,** in alimentary tract, 228
**Ulcerative** colitis, 408
**Underperfusion** of tissues, 114
**Unmasking** effect of birth, 21
pregnancy, 554
**Upper** respiratory tract disorder, 157
**Uraemia,** due to renal failure, 480, 492
post-renal, 493
pre-renal, 493
in sodium depletion, 425
**Urea,** blood level, interpretation, 480-1, 488
retention, effects, 481-2
**Urethritis,** 467
**Urinary** incontinence, 474
**Urinary** retention, acute obstruction, 470
neurological, 474
post-operative, 477
**Urinary** tract infection, 467
obstruction, 469
**Urticaria,** 401
**Uveitis,** 324

**Vagotomy,** 231
causing diarrhoea, 219
insulin test, 230
**Vasectomy,** 552
**Vasovagal** syndrome, 70
**Vector,** cardiac, 146
**Venous** admixture, 169
**Ventilation,** artificial, 191, 445
**Ventilation**/perfusion relationships, 169
**Ventricular** dilatation, 99
failure, 99
failure and digitalis, 129
fibrillation, 145-6
function curve, circulatory failure, 110, 113
hypertrophy, 97
**Vertebro-basilar** insufficiency, 317
**Vertigo,** 319
**Vestibular** disorders, 317
**Virilization,** due to adrenal hormones, 516
due to ovarian hormones, 540
**Vision** disturbances, 321
**Visual** acuity, 324
**Vitamin** A and D toxicity, 210
**Vitamin** D overdosage, 458
**Vomiting,** 211
causing metabolic alkalosis, 443
causing potassium ion depletion, 435-6
in pregnancy, 555

**Water** depletion, 75, 428
deprivation test, 525
intoxication, 430
in chronic renal failure, 485
load, delayed excretion, 513
retention, 430
**Water-brash,** 211
**Weber** test, 313
**Weight,** body, normal values, 208
excessive, causes, 208
due to oedema, 460
**Wilson's** disease, 265

**Xanthine** in urine, 498
**Xenograft,** 411
**Xerophthalmia,** 322
**Xylose** absorption, 201

**Zollinger-Ellison** syndrome, 541